Climate Change Adaptation, Risk Management and Sustainable Practices in the Himalaya

Sanjeev Sharma • Jagdish Chandra Kuniyal
Pritam Chand • Pardeep Singh
Editors

Climate Change Adaptation, Risk Management and Sustainable Practices in the Himalaya

Springer

Editors
Sanjeev Sharma
Centre for the Study of Regional Development, School of Social Sciences
Jawaharlal Nehru University
New Delhi, India

Jagdish Chandra Kuniyal
G.B. Pant National Institute of Himalayan Environment
Almora, Uttarakhand, India

Pritam Chand
Department of Geography, School of Environment and Earth Sciences
Central University of Punjab
Bathinda, Punjab, India

Pardeep Singh
Department of Environmental Studies
PGDAV College, University of Delhi
New Delhi, Delhi, India

ISBN 978-3-031-24658-6 ISBN 978-3-031-24659-3 (eBook)
https://doi.org/10.1007/978-3-031-24659-3

This Springer imprint is published by the registered company Springer Nature Switzerland AG
The registered company address is: Gewerbestrasse 11, 6330 Cham, Switzerland

Contents

1 **Paraglacial Response to Recent Climate Change in the Upper Ganga Catchment** 1
Maria Asim, Subhendu Pradhan, and Shubhra Sharma

2 **Trend of Climatic Components in Sub-Himalayan West Bengal: Evidence from Jalpaiguri District During the Last Century (1901–2000)** 21
Manoranjan Ghosh and Rakhohori Bag

3 **Climate Change Impact on Major River Basins in the Indian Himalayan Region: Risk Assessment and Sustainable Management** 45
K. Amrutha, Rasmi Patnaik, A. S. Sandeep, and Jitendra Kumar Pattanaik

4 **Forestry Policies and Practices to Promote Climate Change Adaptation in the Indian Western Himalayan States** 65
Manoj Kumar

5 **Climate-Induced and Geophysical Hazards and Risk Reduction Financing in Mountain Regions** 89
Kamleshan Pillay, Hari Ballabh, and Srinivasan Pillay

6 **Assessment of Existing Himalayan Glacier Inventories for Glacier Studies: A Case Study from the Ravi Basin of North-Western Himalaya (India)** 109
Ishtiaq Ahmed, Vikram Sharma, Rinku Kumar, Devi Lal, Rajan Bhandari, and Pritam Chand

7 **Landslide Susceptibility Mapping of Tehri Reservoir Region Using Geospatial Approach** 135
Gaurav Tripathi, Achala Shakya, Ritambhara Kumari Upadhyay, Suraj Kumar Singh, Shruti Kanga, and Sandeep Kumar Pandey

8 **Socioeconomic, Livelihood, and Ecological Transformation in the Sikkim Himalayan Region** 157
Lak Tsheden Theengh, Priyadarshinee Shrestha, Passang D. Lepcha, Rajarshi Chakraborty, and Udai Gurung

9 **Socioeconomic Aspect of Disaster Risk in Kashmir: Contextualizing Village Vulnerability in Sindh Basin** 199
Mohd Sharjeel Sofi, Sami Ullah Bhat, Irfan Rashid, and Jagdish Chandra Kuniyal

10 **Indigenous Knowledge System and Livelihood Option of Natives of Lahaul and Spiti District, Himachal Pradesh** 213
Sumati Rathore and Sarla Shashni

11 **Impacts of Climate Change on Plants with Special Reference to the Himalayan Region** 237
Zishan Ahmad Wani, Taskeen Fatima, Omkar Verma, Virbala Sharma, and Shreekar Pant

12 **Promotion, Utilization, and Commercial Cultivation of Local Spices with Special Reference to Eryngo (*Eryngium foetidum* L.) as a Measure for Livelihood Improvement Towards Achieving the Goal of Sustainable Development in the Indo-Burma Biodiversity Hotspot: A Case Study From Manipur, North-East India** 253
Pangeijam Bijaya Devi, Panna Deb, and Huidrom Birkumar Singh

13 **Seasonal Variation of Ecosystem Fluxes of a Himalayan Banj-Oak-Dominated Vegetation** 267
Priyanka Lohani, Sandipan Mukherjee, and K. Chandra Sekar

14 **Human-Wildlife Conflict in the Western Himalaya: A Systematic Review of Research and Conservation Interventions Implemented Over Three Decades** 281
Christi Sylvia and Rishi Kumar Sharma

15 **Groundwater Potential Assessment Using an Integrated AHP-Driven Geospatial Techniques in the High-Altitude Springs of Northwestern Himalaya, India** 337
Dharam Chand, Renu Lata, Rajat Dhiman, and Kireet Kumar

16 Climate-Induced and Geophysical Disasters and Risk Reduction Management in Mountains Regions 361
Ambikapathi Ramya, Ramesh Poornima, Ganesan Karthikeyan, Sengottiyan Priyatharshini, Kalyanasundaram Geetha Thanuja, and Periyasamy Dhevagi

17 Modelling of Reference Crop Evapotranspiration in Humid-Wet Tropical Region of India 407
Ujjal Deka Baruah, Anup Saikia, and Nitashree Mili

Index 419

Contributors

Ishtiaq Ahmed Department of Geography, School of Environment and Earth Sciences, Central University of Punjab, Bathinda, Punjab, India

K. Amrutha Department of Geology, School of Environment and Earth Sciences, Central University of Punjab, Bathinda, Punjab, India

Maria Asim Department of Geography, Institute of Science, Banaras Hindu University, Varanasi, Uttar Pradesh, India

Rakhohori Bag Doctoral Fellow (UGC-SRF) at Centre for the Study of Regional Development, School of Social Sciences, Jawaharlal Nehru University, New Delhi, India

Hari Ballabh District Disaster Management Authority, Haridwar, Uttarkhand, India

Ujjal Deka Baruah Department of Geography, Cotton University, Guwahati, Assam, India

Rajan Bhandari Department of Geography, DAV College, Panjab University, Chandigarh, India

Sami Ullah Bhat Department of Environmental Science, University of Kashmir, Srinagar, India

Rajarshi Chakraborty UNDP, SECURE Himalaya, New Delhi, India

Dharam Chand G. B. Pant National Institute of Himalayan Environment, Himachal Regional Centre, Mohal-Kullu, Himachal Pradesh, India

Pritam Chand Department of Geography, School of Environment and Earth Sciences, Central University of Punjab, Bathinda, Punjab, India

Panna Deb Centre for Biodiversity and Natural Resource Conservation, Assam University, Silchar, Assam, India

Department of Ecology & Environmental Science, Assam University, Silchar, Assam, India

Pangeijam Bijaya Devi Krishi Vigyan Kendra, Utlou, Manipur, India

Periyasamy Dhevagi Department of Environmental Sciences, Tamil Nadu Agricultural University, Coimbatore, Tamil Nadu, India

Rajat Dhiman G. B. Pant National Institute of Himalayan Environment, Himachal Regional Centre, Mohal-Kullu, Himachal Pradesh, India

Taskeen Fatima Conservation Ecology Lab, Department of Botany, Baba Ghulam Shah Badshah University, Rajouri, Jammu and Kashmir, India

Manoranjan Ghosh Fellow (ISCF) at National Institute of Urban Affairs, New Delhi, India

Udai Gurung Forest Department, Government of Sikkim, Sikkim, India

Shruti Kanga Department of Geography, School of Environment and Earth Sciences, Central University of Punjab, Bathinda, India

Ganesan Karthikeyan Department of Environmental Sciences, Tamil Nadu Agricultural University, Coimbatore, Tamil Nadu, India

Kireet Kumar G. B. Pant National Institute of Himalayan Environment, Almora, Uttarakhand, India

Manoj Kumar GIS Centre, Forest Research Institute (FRI), Dehradun, India

Rinku Kumar Department of Geography, School of Environment and Earth Sciences, Central University of Punjab, Bathinda, Punjab, India

Jagdish Chandra Kuniyal G.B. Pant National Institute of Himalayan Environment (NIHE), Almora, Uttarakhand, India

Devi Lal Department of Geography, School of Environment and Earth Sciences, Central University of Punjab, Bathinda, Punjab, India

Renu Lata G. B. Pant National Institute of Himalayan Environment, Himachal Regional Centre, Mohal-Kullu, Himachal Pradesh, India

Passang D. Lepcha Darjeeling, West Bengal, India

Priyanka Lohani G.B. Pant National Institute of Himalayan Environment, Almora, Uttarakhand, India

Nitashree Mili Department of Geography, Cotton University, Guwahati, Assam, India

Sandipan Mukherjee Ladakh Regional Centre, G. B. Pant National Institute of Himalayan Environment, Leh, India

Sandeep Kumar Pandey Gujarat Institute of Disaster Management, Gandhinagar, Gujarat, India

Shreekar Pant Coordinator, Centre for Biodiversity Studies, Baba Ghulam Shah Badshah University, Rajouri, Jammu and Kashmir, India

Rasmi Patnaik Independent Consultant, Bathinda, India

Jitendra Kumar Pattanaik Department of Geology Central, University of Punjab, Lahore, Punjab, Pakistan

Kamleshan Pillay Global Change Institute, University of Witwatersrand, Johannesburg, South Africa

Srinivasan Pillay Department of Environmental Sciences, University of KwaZulu-Natal, Durban, South Africa

Ramesh Poornima Department of Environmental Sciences, Tamil Nadu Agricultural University, Coimbatore, Tamil Nadu, India

Subhendu Pradhan Department of Geography, Institute of Science, Banaras Hindu University, Varanasi, Uttar Pradesh, India

Sengottiyan Priyatharshini Vanavarayar Institute of Agriculture, Coimbatore, Tamil Nadu, India

Ambikapathi Ramya Department of Environmental Sciences, Tamil Nadu Agricultural University, Coimbatore, Tamil Nadu, India

Irfan Rashid Department of Botany, University of Kashmir, Srinagar, India

Sumati Rathore G. B. Pant National Institute of Himalayan Environment, Himachal Regional Center, Mohal-Kullu, Himachal Pradesh, India

Anup Saikia Department of Geography, Gauhati University, Guwahati, Assam, India

A. S. Sandeep Department of Geography, School of Environment and Earth Sciences, Central University of Punjab, Bathinda, Punjab, India

K. Chandra Sekar Garhwal Regional Centre, G. B. Pant National Institute of Himalayan Environment, Srinagar-Garhwal, India

Achala Shakya School of Computer Science, University of Petroleum and Energy Studies, Dehradun, India

Rishi Kumar Sharma Wildlife and Habitats Program WWF India, New Delhi, India

Shubhra Sharma Department of Geography, Institute of Science, Banaras Hindu University, Varanasi, Uttar Pradesh, India
Geosciences Division, Physical Research Laboratory, Ahmedabad, India

Vikram Sharma Department of Geography, Institute of Science, Banaras Hindu University, Varanasi, India

Virbala Sharma Central University of Himachal Pradesh, Kangra, Himachal Pradesh, India

Sarla Shashni G. B. Pant National Institute of Himalayan Environment, Himachal Regional Center, Mohal-Kullu, Himachal Pradesh, India

Priyadarshinee Shrestha WWF-India, Khangchendzonga Landscape Program, Sikkim Office, Gangtok, India

Huidrom Birkumar Singh CSIR-North East Institute of Science & Technology, Branch Laboratory, Imphal, Manipur, India

Suraj Kumar Singh Centre for Climate Change and Water Research, Suresh Gyan Vihar University, Jaipur, India

Mohd Sharjeel Sofi Department of Environmental Science, University of Kashmir, Srinagar, India

Christi Sylvia Wildlife and Habitats Program WWF India, New Delhi, India

Kalyanasundaram Geetha Thanuja Department of Agricultural Microbiology, Tamil Nadu Agricultural University, Coimbatore, Tamil Nadu, India

Lak Tsheden Theengh WWF-India, Khangchendzonga Landscape Program, Sikkim Office, Gangtok, India

Gaurav Tripathi Department of Geoinformatics, Central University of Jharkhand, Ranchi, India

Ritambhara Kumari Upadhyay Centre of Advanced Studies in Geology, Panjab University, Chandigarh, India

Omkar Verma Geology Discipline Group, School of Sciences, Indira Gandhi National Open University (IGNOU), New Delhi, India

Zishan Ahmad Wani Conservation Ecology Lab, Department of Botany, Baba Ghulam Shah Badshah University, Rajouri, Jammu and Kashmir, India

Chapter 1
Paraglacial Response to Recent Climate Change in the Upper Ganga Catchment

Maria Asim, Subhendu Pradhan, and Shubhra Sharma

Abstract Paraglacial zones are indirectly conditioned by glaciers and glacial processes and are ecologically and geologically unstable. Multiple phases of glacier advances/recessions have left an enormous quantity of unconsolidated sediments in the form of moraines, talus/scree fans, and outwash gravel terraces (collectively known as paraglacial sediment). Thus, the paraglacial zones are transport-limited and not sediment-limited. Small triggers such as an extreme rainfall event or rain on snow may lead to significant sediment mobilization in form of debris avalanches/ debris flows inflicting severe damage to life and infrastructures. Therefore, it is pertinent to have a real-time assessment of paraglacial sediments locked in valleys vacated by glaciers in the recent geological past. So that in case of their mobilization during an extreme hydrometeorological condition, a downstream threat perception can be assessed, and the vital infrastructures are protected. With this objective, the present study attempts to estimate the amount of paraglacial sediments stored in the two key tributary river valleys—Dhauli Ganga and Mandakini, in the upper Ganga catchment. Also, locations that can act as natural damming sites for force amplification in case of sediment mobilization during extreme weather events are demarcated. Using the published literature supported by remote sensing techniques, it has been observed that the Dhauli Ganga and Mandakini valleys have stored $\sim 1467 \times 10^6$ m^3 paraglacial sediment, making these valleys extremely vulnerable to extreme weather events. The study cautions that with global warming as a reality, where frequencies and magnitude of extreme weather events are predicted to increase, extreme caution is required while planning and executing the infrastructure developmental projects, particularly the hydropower projects in these valleys.

M. Asim · S. Pradhan
Department of Geography, Institute of Science, Banaras Hindu University, Varanasi, Uttar Pradesh, India

S. Sharma (✉)
Department of Geography, Institute of Science, Banaras Hindu University, Varanasi, Uttar Pradesh, India

Geosciences Division, Physical Research Laboratory, Ahmedabad, India

S. Sharma et al. (eds.), *Climate Change Adaptation, Risk Management and Sustainable Practices in the Himalaya*,
https://doi.org/10.1007/978-3-031-24659-3_1

Keywords Paraglacial zone · Upper Ganga catchment · Extreme events · Climate change

1 Introduction

The paraglacial environment is ecologically fragile and geologically unstable because of frequent re-sedimentation and transfer of glacial sediments within and beyond the high-mountain landscapes (Barnard et al., 2004). The non-glacial processes that dominate the paraglacial zone are directly conditioned by glaciation (Church & Ryder, 1972). The mass movement and fluvial erosion help in re-sedimentation/mobilization of the paraglacial sediments downstream resulting in the net denudation of mountain ranges (Ballantyne, 2002; Slaymaker, 2009). In addition to the inherent fragility of paraglacial zones, the biodiversity changes constantly because of climatically induced changing glacier boundary conditions (Slaymaker & Kelly, 2007). Studies indicate that with the anticipated global warming, the stability of paraglacial sediment zones would be adversely affected by a reduction in the cohesiveness of rocks (Fischer et al., 2012), changing glacier mass balance (O'Connor & Costa, 1993), and thus enhance the mobilization of sediment in densely populated lower valleys (Rana et al., 2021a).

The Himalayan terrain has witnessed multiple phases of glacier advances/recessions and hence deposited an enormous quantity of unconsolidated paraglacial sediments comprising moraines, talus/scree fans, and outwash gravel terraces (e.g., Barnard et al., 2004; Juyal et al., 2010; Ali et al., 2013; Rana et al., 2019). The glacially scoured valleys provide accommodation space for sediment sequestration, which was estimated as ~44% (Blöthe & Korup, 2013). Often these sediments in mountain environments are mobilized as debris flows, particularly with the onset of summer as glacial ablation starts. Such mobilization of a large volume of sediment poses a serious threat to the free flow of water in the Himalayan rivers as observed during the June 2013 Kedarnath disaster (Sundriyal et al., 2015). The constricted flow path along a river course provides obstruction for the sediment, which results in the creation of temporary dams (Jakob & Hungr, 2005; Passmore et al., 2008; Sundriyal et al., 2015; Rana et al., 2021a). The breach of such temporary obstruction creates force amplification, alluvial bulking, and lateral valley slope erosion (Rickenmann & Zimmermann, 1993). Although clear evidence of global warming and the associated increase in the frequency/magnitude of debris flows generated from paraglacial zones are yet to be demonstrated conclusively (Huggel et al., 2012). However, the mobilization of paraglacial sediment during extreme rainfall events is indeed showing an increasing trend (Trenberth et al., 2007; Armstrong, 2010; Allen et al., 2019; Masson-Delmotte et al., 2022; Sabin et al., 2020).

Historically, there is evidence to suggest that indeed paraglacial zone responded intermittently to extreme geological and/or weather events, for example, the 1894 Birahi Ganga flood, the 1970 Alaknanda flood, and the 1978 Kanodia Gad flood. More recently, catastrophic mobilization of the paraglacial sediment inflicting severe damage to life and infrastructures was observed during the unusual weather event in

June 2013 in the Mandakini valley (Sundriyal et al., 2015 and reference therein) and in February and April 2021 in the Dhauli Ganga (Rana et al., 2021a; Sharma et al., 2021). In this chapter, we have tried to provide a broad assessment of the paraglacial zone in the Alaknanda and Mandakini valleys (upper Ganga catchment) in context with historical and recent events. The objectives of the present study are (i) to estimate the volume of paraglacial sediment stored in two key tributary river valleys, Dhauli Ganga and Mandakini; (ii) to estimate the sediment mobilized during the June 2013 Mandakini and February 7, 2021 Rishi Ganga flash floods; and (iii) to demarcate constricted valley segments that have potential for landslide-induced damming.

2 Study Area

2.1 Geology and Geomorphology

The Mandakini River originates from the Chorabari glacier (3840 m asl) and is joined by the Alaknanda River at Rudraprayag (600 m asl) (Fig. 1.1). The Mandakini basin has an area of ~1638 km^2. The Main Central Thrust (MCT) is the major tectonic boundary that separates the low-grade metamorphic rocks in the south from the Higher Himalayan Crystalline (HHC) rocks in the north (Bist & Sinha, 1980) (Fig. 1.2a). Geomorphologically, the Mandakini valley can be divided into three broad zones. These are from northwest to southeast the upper glaciated zone (>3500 m asl, above Kedarnath), middle paraglacial zone (<3500–2000 m asl, between Kedarnath and Gaurikund), and lower fluvial zone (<2000 m asl, below Sitapur) (Fig. 1.3) (Sundriyal et al., 2015). Mehta et al. (2012) identified four major glacial events in the valley. The oldest glacier advances extended down to an elevation of around 2800 m asl at Rambara. Following this, the subsequent advances were confined within 3000–3500 m asl indicating that the upper catchment of Mandakini valley is filled with a significant amount of paraglacial sediment.

The Dhauli Ganga, in the upper catchment, drains the Tethyan Sedimentary Sequence (TSS) comprising dominantly of limestone, sandstone, and shale. It crosses South Tibetan Detachment (STD), draining the HHC dominated by quartzite, granite, gneisses, and marbles (Fig. 1.2b). The MCT (Vaikrata Thrust) lies in the southwest, which forms the boundary between HHC and the Lesser Himalayan meta-sedimentary rocks (Valdiya, 1980). In the Rishi Ganga catchment, the lithology is dominated by gneisses and schist, followed by kyanite, biotite, and muscovite schist intruded by leucogranite (Mukherjee et al., 2019).

Four stages of glaciations with decreasing magnitude were identified in the upper Dhauli Ganga catchment, which is preserved in the form of morainic ridge and associated paraglacial landforms such as outwash plain, alluvial fan, and scree deposits (Fig. 1.4) (Bisht et al., 2015, 2017). Apart from this, a minor advancement was also observed during the cool and moist Little Ice Age, which is marked by unconsolidated and poorly defined moraine mounds proximal to glacier snout (Sati

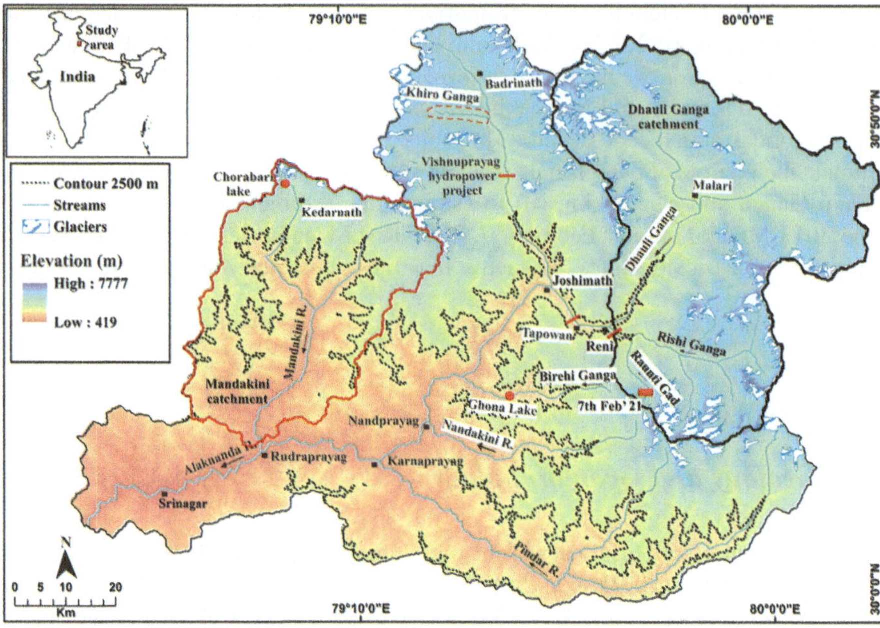

Fig. 1.1 The catchment map of the Alaknanda river with boundaries of Mandakini (red boundary) and Dhauli Ganga (black boundary) is shown. The 2500 m contour (dotted black line) marks the approximate extent of the paraglacial zone. Modern glaciers and rivers, which generated flash floods in July 1970 (Gohna lake), June 2013 (Khiro Ganga and Chorabari lake), and February 2021 events are also marked. The locations of February 7, 2021 in Raunthi Gad valley rockfall (red box) and the affected hydropower (red solid lines) of Reni and Tapovan are marked

et al., 2014; Bisht et al., 2017). The valley has experienced early Holocene deglaciation associated with the insolation-driven Indian Summer Monsoon (ISM), which is represented by deposition of outwash gravel (Juyal et al., 2004, 2009) and valley-wide fluvial aggradation (Juyal et al., 2010).

2.2 Climate

The study area receives ~80% rainfall by the ISM, while the mid-latitude westerlies contribute the remaining 20%, which largely occurs in form of snow in the paraglacial zone (above 2000 m asl) (Kumar et al., 2021; Taloor et al., 2021). The rainfall trend reconstructed using the past 120 years (1901–2019) data from the Climate Research Unit Time-series (CRU-TS) tiles from north of MCT shows that during the summer (June to September), precipitation varies from 61.95 (2007) to 182 mm/month (1917) from the mean average (Fig. 1.5a). The average annual summer precipitation is ~125 mm/month. The average precipitation of ~27 mm/month during winter (November to February) is mainly received by mid-latitude

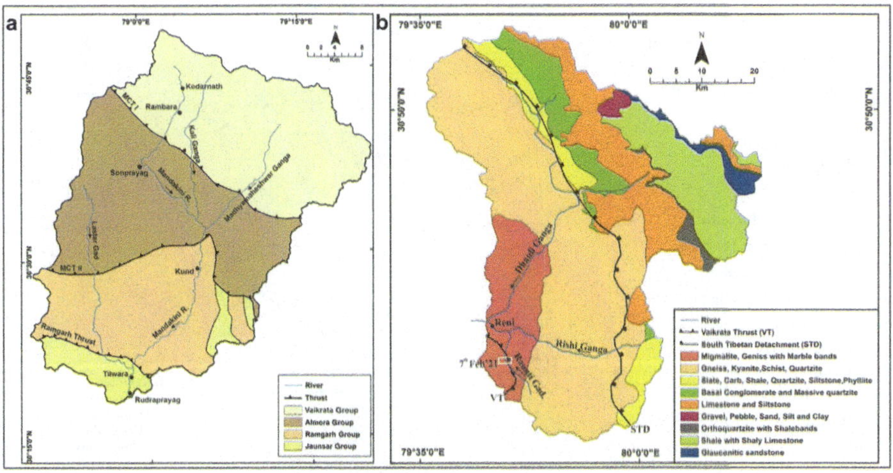

Fig. 1.2 The geological maps of (**a**) Mandakini catchment, where Main Central Thrust-I (MCT-I) in the north separates Vaikrata and Almora group, MCT-II in the middle separates Almora and Ramgarh group, and Ramgarh thrust to the south separates Ramgarh and Jaunsar group. Multiple tributary streams join the Mandakini River-Madhyamaheshwar Ganga, Kali Ganga, and Mandakini River and (**b**) Dhauli Ganga catchment with two major tectonic structures—The South Tibetan Detachment (STD) in the north-south direction and Vaikrata Thrust/Main Central Thrust in the south-western part. The location of February 7, 2021 event is marked (with box) in Raunthi Gad valley. (Modified after Valdiya, 1980, 1981)

westerlies (Fig. 1.5b). The average annual spring precipitation (March to May) has an increasing trend from 2013 onwards (Fig. 1.5c). Similarly, a gradual rise in the winter temperature is observed from the past 120 years.

3 Methodology

In this study, we used freely available satellite data for geomorphic mapping and digital elevation model (DEM) for sediment volume estimation.

3.1 Geomorphological Mapping

Various geomorphic features such as modern glaciers, ablation zone, hanging glaciers, scree/talus cones, fluvial terraces, and glacial lakes were mapped using Landsat 8 OLI (30 m) along with Google Earth imagery. The images used here are acquired with <30% cloud cover.

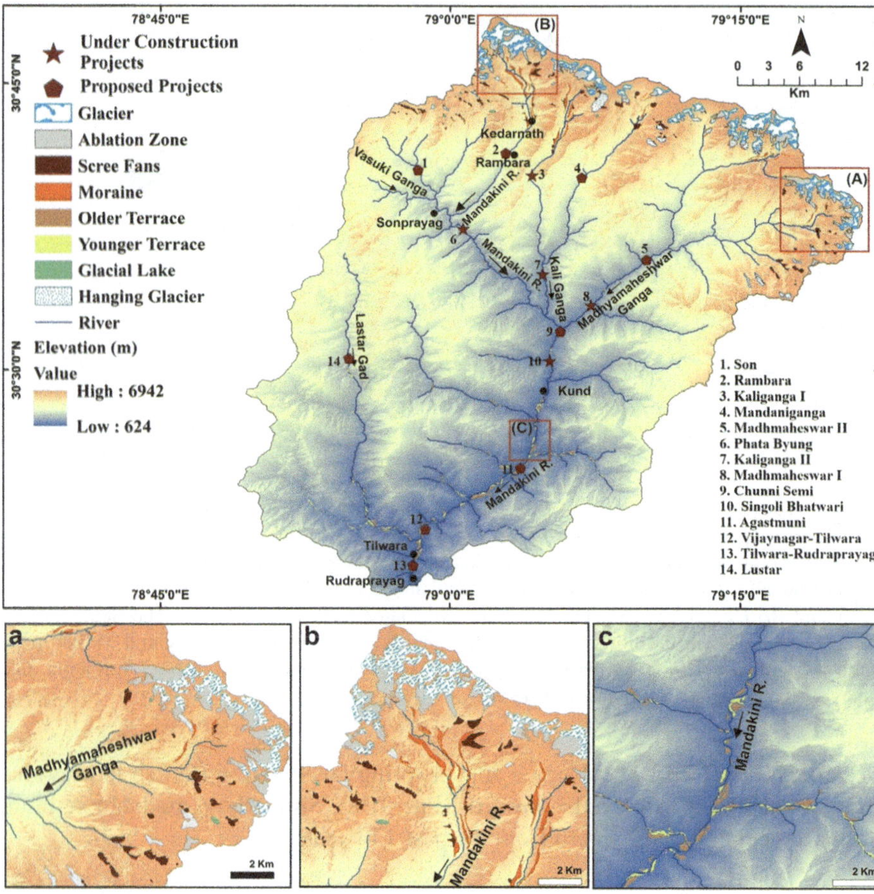

Fig. 1.3 The geomorphological features of Mandakini valley were extracted based on Landsat 8 and Google Earth imagery. The paraglacial sediments are stored in several landforms: scree cones, hanging glaciers, valley beds, lateral moraines, etc. Subsets (**a**), (**b**), and (**c**) show closeup of various landforms. The locations of under-construction and proposed hydropower projects are labeled from 1 to 14. (Source: South Asia Network on Dams, Rivers, and People (SANDRP))

3.2 Geospatial Analyses

The 12.5 m ALOS PALSAR DEM was used for sediment volume calculation in Arc GIS 10.8. The volume of paraglacial sediment stored in the valley was estimated by multiplying the surface area with the thickness of various digitized paraglacial features on Google Earth. Thickness is estimated based on the elevation profile of moraines, which is measured at multiple locations from the valley floor to the moraine crest. The remote sensing-based estimate is validated by comparing the moraine thickness measured in the field (e.g., Sati et al., 2014). Similarly, the top

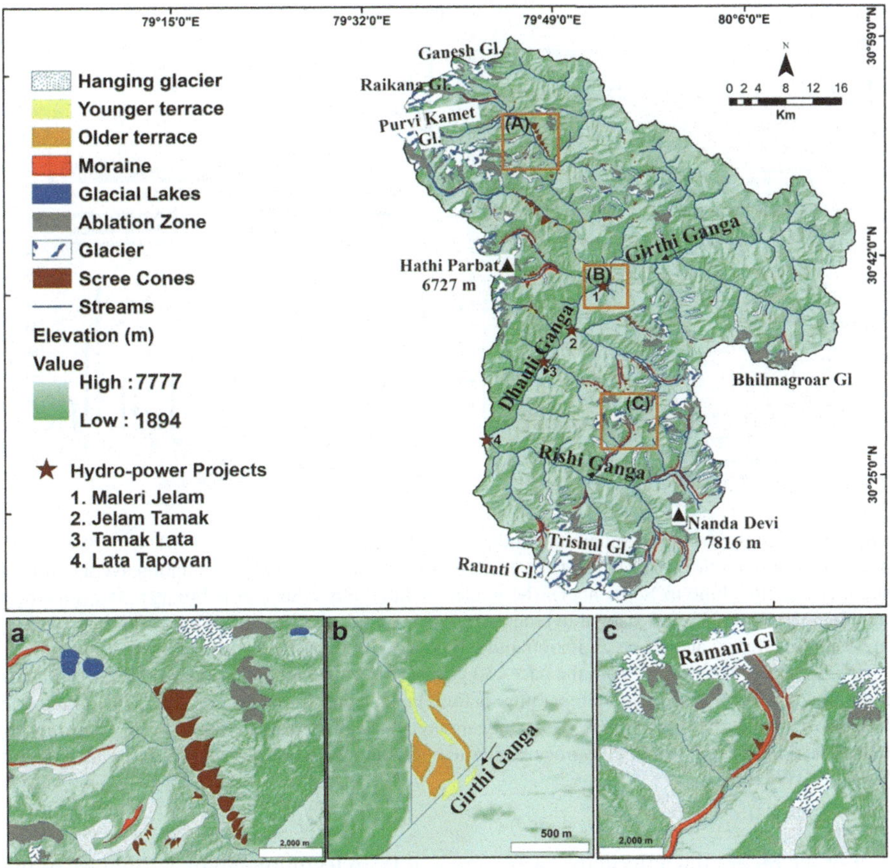

Fig. 1.4 The geomorphological features of Dhauli Ganga valley extracted based on Landsat 8 and Google Earth imagery. The valley sequestered a large quantity of paraglacial sediment in the form of moraines, hanging glaciers, river terraces, scree cones, etc. Subsets (**a**), (**b**), and (**c**) show closeup of geomorphic landforms. The locations of proposed hydropower projects are labeled from 1 to 4. (Hydropower project Source: South Asia Network on Dams, Rivers, and People (SANDRP))

surfaces of the older and younger terraces were digitized and then multiplied by the average thickness calculated by drawing elevation profiles. For the thickness of other geomorphological features, field observations from the published literature were used. The thickness of scree cones varies from 10 to 20 m in the Dhauli Ganga valley (Bisht et al., 2015), where the mean value (15 m) is used for calculating the volume of debris stored in the scree cones. The average thickness for the valley bed and hanging glaciers, which store thick layer debris, is taken as 1.8 m (Mehta et al., 2012). The errors are calculated at one standard deviation for various landforms, and error is propagated using standard formula in the final volume estimation.

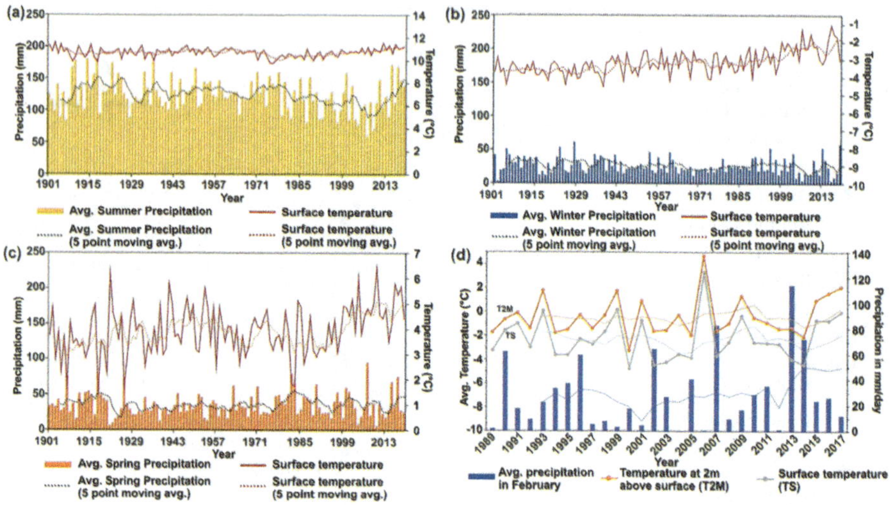

Fig. 1.5 Monthly average temperature and precipitation data from the year 1901 to 2019 of upper Mandakini region (North of MCT). The vertical bar shows the seasonal precipitation average for (**a**) summer months (June to September), (**b**) winter months (November to February), and (**c**) spring months (March to May). Seasonal average temperatures are marked by the red solid line. (**d**) The trend of the average annual precipitation and temperature both surface and air (above 2 m from the surface) for February month from the past ~30 years (1989–2017) has also been analyzed, which is found fluctuating in nature, and temperature is found increasing from 2004 onward. The dotted line represents 5-point moving average. The study area receives precipitation mainly during the summer months and increasing seasonal average winter temperature can be observed. (Source: CRU TS and NASA POWER)

For the recent flash flood that was triggered from Raunthi valley, a tributary of the Rishi Ganga, the sediment removed has also been calculated using pre- (February 1, 2021) and post-disaster (February 10, 2021) Google Earth imagery and cross-checked with the field observations (Rana et al., 2021b).

3.3 *Climate Trend Analysis*

The meteorological datasets of the past ~120 years (1901–2019) of precipitation and temperature for summer, winter, and spring seasons are generated from CRU TS (www.uea.ac.uk) and NASA POWER (https://power.larc.nasa.gov/). The NASA datasets acquired are modelled with ~85% accuracy (power.larc.nasa.gov), and CRU datasets are modelled with 89% accuracy (Mutti et al., 2020) respectively with respect to the ground station data. Annual average temperature and precipitation for February month have also been generated from NASA POWER data in order to understand the climate variation specifically for February in the Raunthi valley.

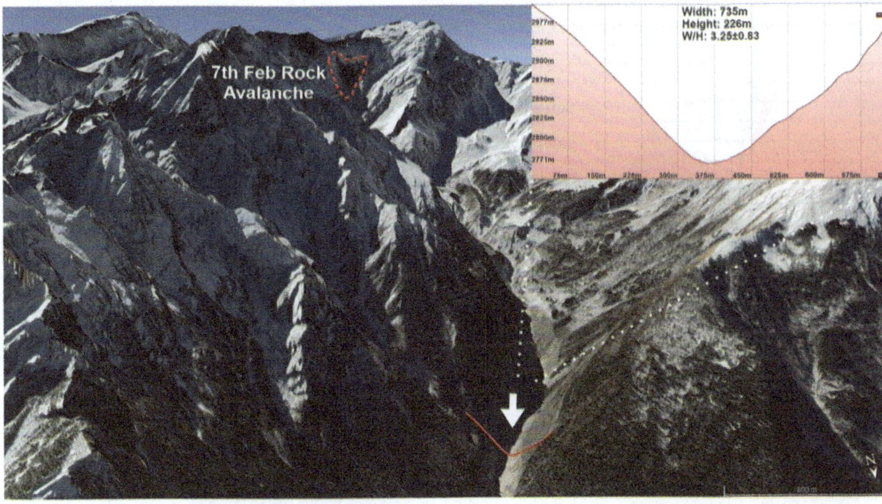

Fig. 1.6 Figure shows the gorge location (solid red line) and associated valley profile of the corresponding gorge, on Google Earth imagery for the calculation of valley width ratio (W/H) near the confluence of Raunthi Gad and Rishi Ganga. The distance between both valley flanks at first knick points is taken as width, and the difference between first knick point elevation and valley floor elevation is taken as height. The location of the February 7, 2021 rock avalanche has also been marked by a red-dotted polygon. The valley width ratio of Raunthi is taken as a standard mean and then 1σ error is considered

3.4 Constricted Valley Segment Demarcation

Valleys that have the potential for obstructing the down valley mobilization of paraglacial sediments and landslides that could lead to the formation of temporary lakes are demarcated considering the following geomorphic factors: (i) slopes with higher landslide susceptibility (Chahal et al., 2017; Maheshwari, 2019), (ii) gorge sections having valley slopes >60° (Liu et al., 2004), and (iii) valley profiles have valley-width ratio similar to Raunthi gad (3.25 ± 0.83) (Fig. 1.6).

4 Result

4.1 Sediment Volume Estimation

Mandakini Valley The total area covered by scree cones in the catchment is found to be ~$5.9 \pm 0.03 \times 10^6$ m^2 (Fig. 1.7a and Table 1.1), and the maximum volume of sediment sequestered is estimated to be ~$88 \pm 0.5 \times 10^6$ m^3. The total area covered by lateral moraines is ~$3.7 \pm 0.08 \times 10^6$ m^2, and the volume of sediment stored is ~$181 \pm 4.2 \times 10^6$ m^3 (Table 1.1). Similarly, the area enclosed by hanging glaciers

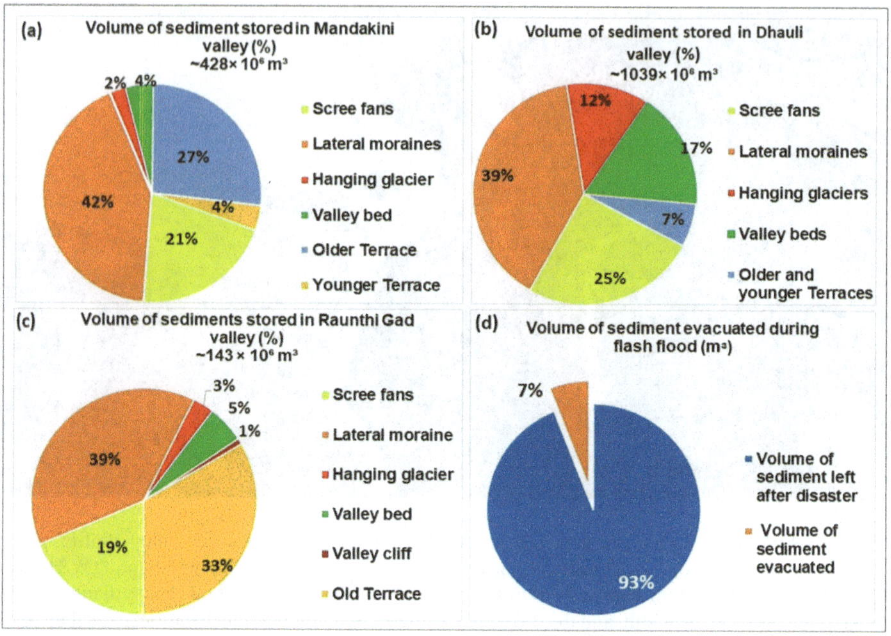

Fig. 1.7 The percentage of sediments stored in different landforms of (**a**) Mandakini valley, (**b**) Dhauli Ganga valley, and (**c**) Raunthi Gad valley. The pie chart (**d**) shows the percentage of volume mobilized during the event of February 7, 2021

Table 1.1 This shows the total area covered and volume calculated for the paraglacial sediment stored in the Mandakini catchment

S. no.	Landform	Area ($\times 10^6$ m^2)	Thickness (m)	Volume ($\times 10^6$ m^3)
1	Scree cone	5.9 ± 0.03	15	88 ± 0.46
2	Moraines	3.7 ± 0.08	37.4 (3–110)	181 ± 4.16
3	Hanging glacier	6 ± 0.26	1.8	10.4 ± 0.47
4	Valley bed	10 ± 0.76	1.8	18.3 ± 1.4
5	Older terrace	3.7 ± 0.04	21 (4–65)	114 ± 1.38
6	Younger terrace	2.1 ± 0.01	5.6 (1–30)	16 ± 0.1
	Total	**31 ± 0.8**		**428 ± 4.6**

and valley beds is estimated to be ~$6 \pm 0.3 \times 10^6$ and ~$10 \pm 0.8 \times 10^6$ m^2, respectively, with the stored volume of sediment of ~$10.4 \pm 0.5 \times 10^6$ and ~$18.3 \pm 1.4 \times 10^6$ m^3 (Table 1.1). The area occupied by older and younger terraces (Fig. 1.3) is ~$3.7 \pm 0.04 \times 10^6$ and ~$2.1 \pm 0.01 \times 10^6$ m^2, respectively, which store sediment volume of ~$114 \pm 1.4 \times 10^6$ and ~$16 \pm 0.1 \times 10^6$ m^3. Summing up, the total volume present in various geomorphic landforms (paraglacial zone) of the Mandakini valley is estimated to be ~$428 \pm 4.6 \times 10^6$ m^3 (Table 1.1).

Table 1.2 This shows the total area covered and the volume calculated for the paraglacial sediment stored in the Dhauli Ganga catchment

S. no.	Landform	Area ($\times 10^6$ m^2)	Thickness (m)	Volume ($\times 10^6$ m^3)
1	Scree cone	17.5 ± 0.07	15	263 ± 1
2	Moraines	11.5 ± 0.1	15 and 60	409 ± 4.7
3	Hanging glacier	70 ± 0.9	1.8	125 ± 1.7
4	Valley bed	98 ± 4.9	1.8	176 ± 8.9
5	Older terrace	2 ± 0.05	26 (4–80)	65 ± 1.5
6	Younger terrace	0.15 ± 0.005	8 (2–32)	1.2 ± 0.04
	Total	**198 ± 5**		**1039 ± 10.3**

Dhauli Ganga Valley The total area covered by scree cones in the upper catchment of Dhauli Ganga is ~17.5 ± 0.07 × 10^6 m^2 (Fig. 1.7b and Table 1.2), which translates into ~263 ± 1 × 10^6 m^3 of sediment. The total area covered by the lateral moraines is estimated to be ~11.5 ± 0.1 × 10^6 m^2, and the volume of sediment stored in it is ~409 ± 4.7 × 10^6 m^3 (Table 1.2). Similarly, the area occupied by the hanging glaciers and valley bed (ground moraines) covered with the drape of sediments is estimated as ~70 ± 0.9 × 10^6 and ~98 ± 4.9 × 10^6 m^2, respectively, with the stored volume of sediment being ~125 ± 1.7 × 10^6 and ~176 ± 8.9 × 10^6 m^3. The area occupied by the older terrace (Fig. 1.4) is ~2 ± 0.05 × 10^6 m^2, which stores ~65 ± 1.5 × 10^6 m^3 sediment volume. The area occupied by the younger terrace (Fig. 1.4) is ~0.15 ± 0.01 × 10^6 m^2, which stores ~1.2 ± 0.04 × 10^6 m^3 sediment volume. Thus, adding the sediment volume of all the landforms in the paraglacial zone of Dhauli Ganga is estimated to be ~1039 ± 10.3 × 10^6 m^3 (Table 1.2).

4.2 Post-disaster (February 7, 2021) Removal of Sediments

The total area of the paraglacial sediments sequestered before February 7, 2021 flash flood in the Raunthi valley is estimated as ~11 ± 0.2 × 10^6 m^2, which stores ~143 ± 11.4 × 10^6 m^3 volume of the sediment (Fig. 1.7c and Table 1.3). The Google Earth imagery of February 10, 2021 is used to demarcate the area affected by the flash flood of February 7 in the Raunthi Gad valley. The landforms affected in the valley, which experienced erosion, were mapped critically to calculate the amount of total volume mobilized by comparing the pre (February 1) and post (February) 2021 imagery. The total amount of sediment mobilized due to flash floods is estimated to be ~11.45 ± 1.31 × 10^6 m^3 (Fig. 1.7d and Table 1.4).

Table 1.3 This shows the area covered and volume of paraglacial sediment stored before the February 7, 2021 flash flood in the Raunthi Gad valley

S. no.	Landform	Area ($\times 10^6$ m^2)	Thickness (m)	Volume ($\times 10^6$ m^3)
1	Scree cone	1.8 ± 0.03	15	26.7 ± 0.5
2	Moraines	0.92 ± 0.17	15 and 60	55.2 ± 10.3
3	Hanging glacier	2.4 ± 0.01	1.8	4.3 ± 0.25
4	Valley bed	4.29	1.8	7.7
5	Valley cliff	1.2 ± 0.12	1	1.24 ± 0.12
6	Old terrace	0.5 ± 0.05	150 and 50	48 ± 4.8
	Total	**11 ± 0.2**		**143 ± 11.4**

4.3 Potential Zones for Landslide-Dammed Lake

Based on published landslide inventory (Chalal et al., 2017; Maheshwari, 2019), classified slope map, and gorge profiles, a preliminary attempt has been made to outline constricted valley segments in the Mandakini (Fig. 1.8) and Dhauli Ganga valley (Fig. 1.9). These sites can serve as future river blockade sites as in the case of the events like June 2013 or February 2021 in the region, thus amplifying the effect of an event (Figs. 1.3 and 1.4).

5 Discussion

The paraglacial valleys in the higher Himalayas are highly vulnerable to extreme weather events as receded glaciers in the geological past have left behind an enormous quantity of unconsolidated paraglacial sediment (Sati et al., 2014; Bisht et al., 2017). These sediments are mobilized by creeping due to permafrost thawing and freezing, debris flows, and hyperconcentrated flash floods. The hydrometeorological disasters like cloud bursts, flash floods, landslides, extreme precipitation events, etc. are frequently increasing in the Higher Himalayas. From 2009 to 2018, around 30 extreme events have been observed in the Uttarakhand region alone (Pandey & Mishra, 2018). Considering that extreme weather events are going to increase both in frequency and magnitude (Allen et al., 2019), it is extremely important that we must treat these regions with the utmost sensitivity, particularly for developmental activities (river valley projects and other infrastructure development).

The Raunthi Gad valley alone stores 14% of the total sediment stored in the Dhauli Ganga catchment in the form of old terraces (33%), lateral moraines (39%), hanging glaciers, scree fans, etc. (Fig. 1.7c and Table 1.3). The flash flood event on February 7, 2021 removed ~7% of the total sediment stored in Raunthi valley (Fig. 1.7d and Table 1.4). According to Pandey et al. (2021), the volume of rock

Table 1.4 This shows the total volume of paraglacial sediment evacuated during the flash flood of 07th Feb 2021 from the Raunthi Gad valley

		Spot height (asl)	Coordinates	Coordinates	Pre-disaster area	Post-disaster area	Difference	Thickness	Volume
S. no.	Feature	(m)	X	Y	(m^2)	(m^2)	(m^2)	(m)	(m^3)
1	Landform 1	4334	30.39	79.74	89,009.61	80,389.31	8620.3	20	172,406
2	Landform 2	4244	30.39	79.74	102,089.86	90,571.03	11,518.83	20	230,376.6
3	Landform 3	3641	30.41	79.72	213,247.7	146,930.38	66,317.32	150	9,947,598
4	Landform 4	3645	30.41	79.72	319,652.04	304,672.78	14,979.26	50	748,963
5	Landform 5	3678	30.41	79.73	136,729.08	132,729.38	3999.7	40	159,988
6	Landform 6	3489	30.42	79.73	215,806.84	193,548.56	22,258.28	2	44,516.56
7	Landform 7	3348	30.42	79.72	112,228.72	70,038.93	42,189.79	2	84,379.58
8	Landform 8	3410	30.42	79.72	48,180.75	41,280.44	6900.31	9	62,102.79
	Total				**1,236,944.6**	**1,060,160.81**	**176,783.79 (~0.18 ± 0.02 × 10^6)**		**11,450,330.53 (~11.45 ± 1.31 × 10^6)**

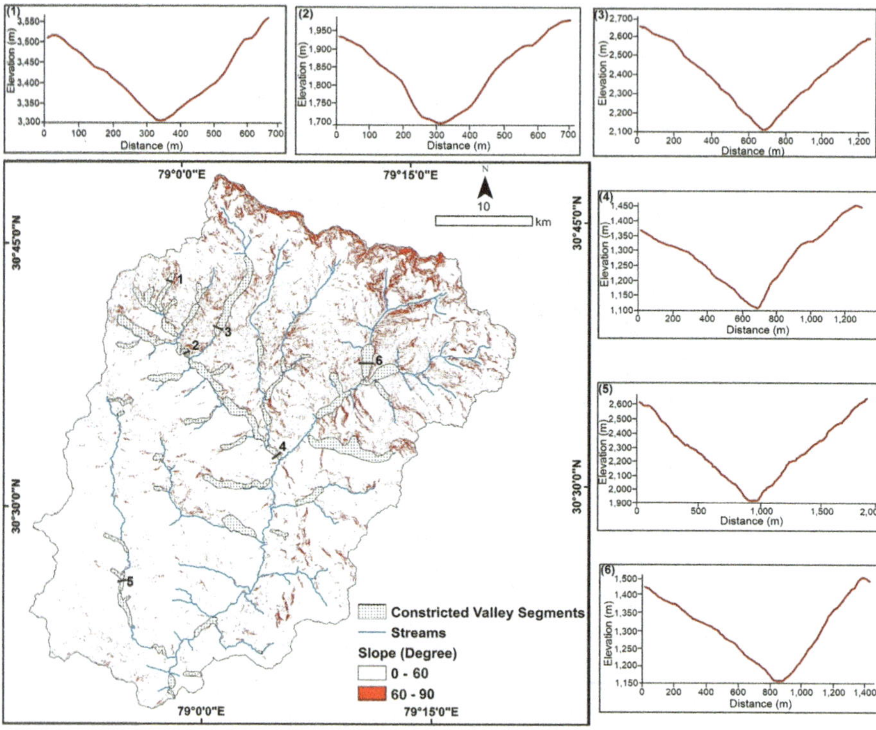

Fig. 1.8 The map shows the constricted valley segments in the Mandakini valley representing the zones of potential landslide-induced damming. Six gorge sections having a similar valley width ratio as that of Raunthi Gad are labeled as 1–6, and their corresponding profiles are shown

and ice mass is ~0.23×10^8 m^3, which was released during the event in a 10 km-long stretch of the valley. The total volume of paraglacial debris evacuated was calculated as ~0.1×10^8 m^3; thus, it shows around 44% of the total debris generated was added by paraglacial sediment alone in the valley. According to Rana et al. (2021b), the mid-channel segment of Rishi Ganga and Dhauli Ganga witnessed the deposition of ~0.12×10^8 m^3 volume sediment with a significant proportion of sediment being coarse (~5 m large boulders; e.g., Rana et al., 2021b), suggesting short-distance transport and deposition in transport-limited settings (Raunthi valley is not sediment-limited due to the presence of moraines, e.g., Rana et al., 2021a). A similar observation was made during the June 2013 Mandakini valley flash flood, in which according to Sundriyal et al. (2015), more than 0.35 million cubic meters of paraglacial sediments were transported into the lower valleys.

The event of February 7 was unique in the sense that there was no perceptible external trigger such as extreme rainfall or seismicity. However, climate change over mountains affects the distribution, rate, and processes of weathering and denudation in (in)direct ways (Knight & Harrison, 2014). Studies show that glaciers are

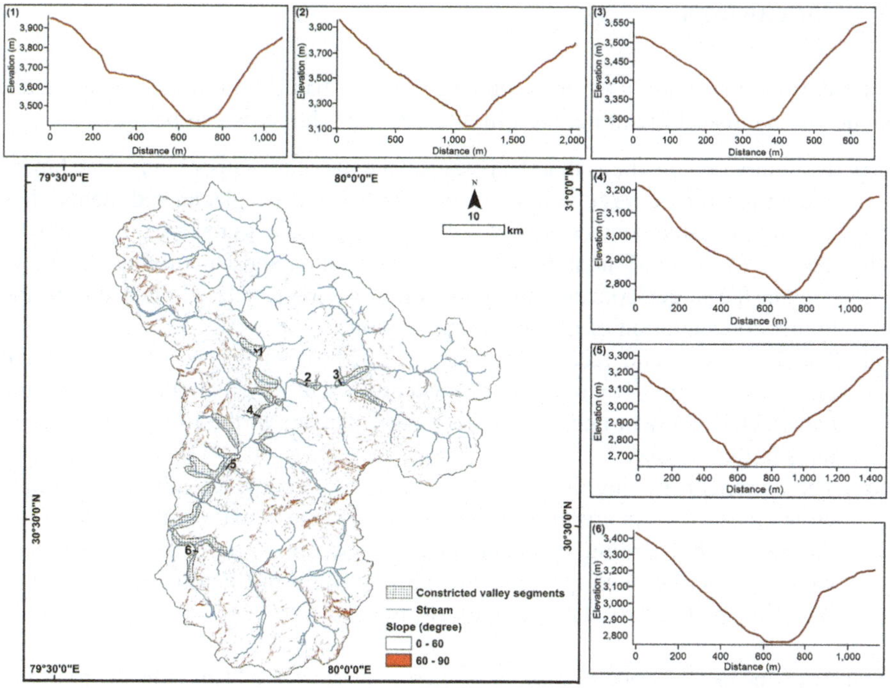

Fig. 1.9 The map shows the delineated constricted valley segments of Dhauli Ganga valley representing the zones of potential landslide-induced damming. Six gorge sections having a similar valley width ratio as that of the Raunthi Gad are labeled as 1–6, and their corresponding profiles are shown

retreating at a faster rate than before from past decades with increasing summer temperature and will further speed up if climate warming continues (Ren et al., 2004). The rate of warming is further suggested to be higher for areas above 4000 m asl (Sabin et al., 2020). The long-term trend (1901–2019) of precipitation and temperature for the region suggests increasing summer precipitation along with the rise in winter temperature. The spring season, particularly the month of February, also shows an increase in the temperature over the last 30 years (Fig. 1.5d). There are several reported incidences in the spring of 2021 of high rainfall triggering debris flows in the higher Himalayas, for example, in Chamba on July 28, 2021 (Himachal Pradesh), an avalanche on April 23, 2021 (Girthi Ganga), Siachen avalanche on April 27, 2021, etc. With the rising temperature and increase in summer rainfall, it seems the occurrence of these events is more than a mere coincidence especially when a definite relationship is shown between warmer snow surface temperature (0 °C) and the occurrence of avalanches (Ballesteros-Cánovas et al., 2018).

6 Conclusions

Based on the preliminary estimate obtained largely based on remote sensing technique and published literature, the following broad inferences can be drawn.

(i) The study demonstrates that the paraglacial zone of the upper Ganga catchment is dominated by a large volume of unconsolidated paraglacial sediments, thus making the catchment extremely vulnerable toward extreme weather events.
(ii) Compared to the Mandakini valley, the Alaknanda and Dhauli Ganga valleys contain the larger concentration (volume) of paraglacial sediment due to the higher concentration of glaciers in this region.
(iii) The June 2013 and recent (February 7, 2021) flash floods are the classical examples of how extreme hydrometeorological events can mobilize the paraglacial sediment and thus inflict severe damage to the life and infrastructures in the region.
(iv) Therefore, considering that global warming has become a reality and multiple projections are converging toward the fact that extreme events are going to increase with warming global temperature, we must be extremely cautious while interfering with paraglacial valleys for infrastructure developmental activities, particularly the hydropower projects.

Acknowledgments The work is part of the MSc dissertation, 2021 of Maria Asim, supervised by Shubhra Sharma. The authors are grateful to Banaras Hindu University for its infrastructural facilities.

References

Ali, S. N., Biswas, R. H., Shukla, A. D., & Juyal, N. (2013). Chronology and climatic implications of Late Quaternary glaciations in the Goriganga valley, Central Himalaya, India. *Quaternary Science Reviews, 73*, 59–76.

Allen, M., Antwi-Agyei, P., Aragon-Durand, F., Babiker, M., Bertoldi, P., et al. (2019). *Technical summary: Global warming of 1.5 C. An IPCC special report on the impacts of global warming of 1.5 °C above pre-industrial levels and related global greenhouse gas emission pathways. (In the context of strengthening the global response to the threat of climate change, sustainable development, and efforts to eradicate poverty).*

Armstrong, R. L. (2010). *The glaciers of the Hindu Kush-Himalayan region: A summary of the science regarding glacier melt/retreat in the Himalayan, Hindu Kush, Karakoram, Pamir, and Tien Shan mountain ranges* (p. 16). International Centre for Integrated Mountain Development (ICIMOD). ISBN: 9789291151738.

Ballantyne, C. K. (2002). A general model of paraglacial landscape response. *The Holocene, 12*(3), 371–376.

Ballesteros-Cánovas, J. A., Trappmann, D., Madrigal-González, J., Eckert, N., & Stoffel, M. (2018). Climate warming enhances snow avalanche risk in the Western Himalayas. *Proceedings of the National Academy of Sciences, 115*(13), 3410–3415.

Barnard, P. L., Owen, L. A., & Finkel, R. C. (2004). Style and timing of glacial and paraglacial sedimentation in a monsoon-influenced high Himalayan environment, the upper Bhagirathi Valley, Garhwal Himalaya. *Sedimentary Geology, 165*(3–4), 199–221.

Bisht, P., Ali, S. N., Shukla, A. D., Negi, S., Sundriyal, Y. P., Yadava, M. G., & Juyal, N. (2015). Chronology of late Quaternary glaciation and landform evolution in the upper Dhauliganga valley, (Trans Himalaya), Uttarakhand, India. *Quaternary Science Reviews, 129*, 147–162.

Bisht, P., Ali, S. N., Rana, N., Singh, S., Poonam, Sundriyal, Y. P., Bagri, D. S., & Juyal, N. (2017). Pattern of Holocene glaciation in the monsoon-dominated Kosa Valley, central Himalaya, Uttarakhand, India. *Geomorphology, 284*, 130–141.

Bist, K. S., & Sinha, A. K. (1980). Some observations on the geological and structural setup of Okhimath area in Garhwal Himalaya. *Himalayan Geology, 10*, 467–475.

Blöthe, J. H., & Korup, O. (2013). Millennial lag times in the Himalayan sediment routing system. *Earth and Planetary Science Letters, 382*, 38–46.

Chalal, P., Rana, N., Bisht, P., Bagri, D. S., Wasson, R. J., & Sundriyal, Y. (2017). Identification of landslide-prone zones in the geomorphically and climatically sensitive Mandakini valley, (central Himalaya), for disaster governance using the Weights of Evidence method. *Geomorphology, 284*, 41–52.

Church, M., & Ryder, J. M. (1972). Paraglacial sedimentation: A consideration of fluvial processes conditioned by glaciation. *Geological Society of America Bulletin, 83*(10), 3059–3072.

Fischer, L., Purves, R. S., Huggel, C., Noetzli, J., & Haeberli, W. (2012). On the influence of topographic, geological and cryospheric factors on rock avalanches and rockfalls in high-mountain areas. *Natural Hazards and Earth System Sciences, 12*(1), 241–254.

Huggel, C., Clague, J. J., & Korup, O. (2012). Is climate change responsible for changing landslide activity in high mountains? *Earth Surface Processes and Landforms, 37*(1), 77–91.

Jakob, M., & Hungr, O. (2005). *Debris-flow hazards and related phenomena* (p. 413). Springer. ISBN 3-540-20726-0.

Juyal, N., Pant, R. K., Basavaiah, N., Yadava, M. G., Saini, N. K., & Singhvi, A. K. (2004). Climate and seismicity in the higher Central Himalaya during 20–10 ka: Evidence from the Garbayang basin, Uttaranchal, India. *Palaeogeography, Palaeoclimatology, Palaeoecology, 213*(3–4), 315–330.

Juyal, N., Pant, R. K., Basavaiah, N., Bhushan, R., Jain, M., Saini, N. K., Yadava, M. G., & Singhvi, A. K. (2009). Reconstruction of Last Glacial to early Holocene monsoon variability from relict lake sediments of the Higher Central Himalaya, Uttrakhand, India. *Journal of Asian Earth Sciences, 34*(3), 437–449.

Juyal, N., Sundriyal, Y., Rana, N., Chaudhary, S., & Singhvi, A. K. (2010). Late Quaternary fluvial aggradation and incision in the monsoon-dominated Alaknanda valley, Central Himalaya, Uttrakhand, India. *Journal of Quaternary Science, 25*(8), 1293–1304.

Knight, J., & Harrison, S. (2014). Mountain glacial and paraglacial environments under global climate change: Lessons from the past, future directions and policy implications. *Geografiska Annaler: Series A, Physical Geography, 96*(3), 245–264.

Kumar, D., Singh, A. K., Taloor, A. K., & Singh, D. S. (2021). Recessional pattern of Thelu and Swetvarn glaciers between 1968 and 2019, Bhagirathi basin, Garhwal Himalaya, India. *Quaternary International, 575*, 227–235.

Liu, J. G., Mason, P. J., Clerici, N., Chen, S., Davis, A., Miao, F., Deng, H., & Liang, L. (2004). Landslide hazard assessment in the Three Gorges area of the Yangtze river using ASTER imagery: Zigui–Badong. *Geomorphology, 61*(1–2), 171–187.

Maheshwari, B. K. (2019). Earthquake-induced landslide hazard assessment of Chamoli district, Uttarakhand using relative frequency ratio method. *Indian Geotechnical Journal, 49*(1), 108–123.

Masson-Delmotte, V., Zhai, P., Pörtner, H. O., Roberts, D., Skea, J., et al. (2022). *Global warming of 1.5 °C. An IPCC Special Report on the impacts of global warming of 1.5 °C above pre-industrial levels and related global greenhouse gas emission pathways. (In the context of strengthening the global response to the threat of climate change, sustainable development, and efforts to eradicate poverty)*, 1(5).

Mehta, M., Majeed, Z., Dobhal, D. P., & Srivastava, P. (2012). Geomorphological evidences of post-LGM glacial advancements in the Himalaya: A study from Chorabari Glacier, Garhwal Himalaya, India. *Journal of Earth System Science, 121*(1), 149–163.

Mukherjee, P. K., Jain, A. K., Singhal, S., Singha, N. B., Singh, S., Kumud, K., Seth, P., & Patel, R. C. (2019). U-Pb zircon ages and Sm-Nd isotopic characteristics of the Lesser and Great Himalayan sequences, Uttarakhand Himalaya, and their regional tectonic implications. *Gondwana Research, 75*, 282–297.

Mutti, P. R., Dubreuil, V., Bezerra, B. G., Arvor, D., de Oliveira, C. P., & Santos e Silva, C.M. (2020). Assessment of Gridded CRU TS data for long-term climatic water balance monitoring over the Sao Francisco Watershed, Brazil. *Atmosphere, 11*(11), 1207.

O'Connor, J. E., & Costa, J. E. (1993). Geologic and hydrologic hazards in glacierized basins in North America resulting from 19th and 20th century global warming. *Natural Hazards, 8*(2), 121–140.

Pandey, V. K., & Mishra, A. (2018). Trends of hydro-meteorological disaster in Uttarakhand, India. *International Journal of Current Research in Science and Technology, 4*(12), 1–7.

Pandey, P., Chauhan, P., Bhatt, C. M., Thakur, P. K., Kannaujia, S., Dhote, P. R., Roy, A., Kumar, S., Chopra, S., Bhardwaj, A., & Aggrawal, S. P. (2021). Cause and process mechanism of rockslide triggered flood event in Rishiganga and Dhauliganga River Valleys, Chamoli, Uttarakhand, India using satellite remote sensing and in situ observations. *Journal of the Indian Society of Remote Sensing, 49*(5), 1011–1024.

Passmore, D. G., Harrison, S., Winchester, V., Rae, A., Severskiy, I., & Pimankina, N. V. (2008). Late Holocene debris flows and valley floor development in the northern Zailiiskiy Alatau, Tien Shan mountains, Kazakhstan. *Arctic, Antarctic, and Alpine Research, 40*(3), 548–560.

Rana, N., Sharma, S., Ali, S. N., Singh, S., & Shukla, A. D. (2019). Investigating the sensitivity of glaciers to climate variability since the MIS-2 in the upper Ganga catchment (Saraswati valley), Central Himalaya. *Geomorphology, 346*, 106854.

Rana, N., Sharma, S., Sundriyal, Y., Kaushik, S., Pradhan, S., Tiwari, G., Khan, F., Sati, S. P., & Juyal, N. (2021a). A preliminary assessment of the 7th February 2021 flashflood in lower Dhauli Ganga valley, Central Himalaya, India. *Journal of Earth System Science, 130*(2), 1–10.

Rana, N., Sundriyal, Y., Sharma, S., Khan, F., Kaushik, S., Chand, P., Bagri, D. S., Sati, S. P., & Juyal, N. (2021b). Hydrological characteristics of 7th February 2021 Rishi Ganga flood: Implication towards understanding flood hazards in higher Himalaya. *Journal of the Geological Society of India, 97*(8), 827–835.

Ren, J., Qin, D., Kang, S., Hou, S., Pu, J., & Jing, Z. (2004). Glacier variations and climate warming and drying in the central Himalayas. *Chinese Science Bulletin, 49*(1), 65–69.

Rickenmann, D., & Zimmermann, M. (1993). The 1987 debris flows in Switzerland: Documentation and analysis. *Geomorphology, 8*(2–3), 175–189.

Sabin, T. P., Krishnan, R., Vellore, R., Priya, P., Borgaonkar, H. P., Singh, B. B., & Sagar, A. (2020). Climate change over the Himalayas. In R. Krishnan et al. (Eds.), *Assessment of climate change over the Indian region* (pp. 207–222). Springer. ISBN: 978-981-15-4326-5.

Sati, S. P., Ali, S. N., Rana, N., Bhattacharya, F., Bhushan, R., Shukla, A. D., Sundriyal, Y., & Juyal, N. (2014). Timing and extent of Holocene glaciations in the monsoon dominated Dunagiri valley (Bangni glacier), Central Himalaya, India. *Journal of Asian Earth Sciences, 91*, 125–136.

Sharma, S., Sati, S. P., Sundriyal, Y. P., & Dobhal, H. (2021). The 23rd April'21 snow avalanche, Girthi Ganga post the 7th February'21 Rishi Ganga flash flood: Are these events linked to climate warming in the Western Himalaya? *Journal of the Geological Society of India, 97*(9), 975–979.

Slaymaker, O. (2009). Proglacial, periglacial or paraglacial? *Geological Society, London, Special Publications, 320*(1), 71–84.

Slaymaker, O., & Kelly, R. (2007). *The cryosphere and global environmental change* (p. 25). Wiley-Blackwell. ISBN: 978-1-405-12976-3.

Sundriyal, Y. P., Shukla, A. D., Rana, N., Jayangondaperumal, R., Srivastava, P., Chamyal, L. S., Sati, S. P., & Juyal, N. (2015). Terrain response to the extreme rainfall event of June 2013: Evidence from the Alaknanda and Mandakini River Valleys, Garhwal Himalaya, India. *Episodes Journal of International Geoscience, 38*(3), 179–188.

Taloor, A. K., Joshi, L. M., Kotlia, B. S., Alam, A., Kothyari, G. C., Kandregula, R. S., Singh, A. K., & Dumka, R. K. (2021). Tectonic imprints of landscape evolution in the Bhilangana and Mandakini basin, Garhwal Himalaya, India: A geospatial approach. *Quaternary International, 575*, 21–36.

Trenberth, K.E., Jones, P.D., Ambenje, P., Bojariu, R., Easterling, D., Tank, A.K., Parker, D., Rahimzadeh, F., Renwick, J.A., Rusticucci, M., & Soden, B. (2007). Observations: Surface and atmospheric climate change. In *Climate change 2007: The physical science basis. Contribution of working group 1 to the 4th assessment report of the intergovernmental panel on climate change*. Cambridge University Press.

Valdiya, K. S. (1980). *Geology of Kumaun lesser Himalaya*. Wadia Institute of Himalayan Geology.

Valdiya, K. S. (1981). Tectonics of the central sector of the Himalaya. *Zagros Hindu Kush Himalaya Geodynamic Evolution, 3*, 87–110.

Chapter 2
Trend of Climatic Components in Sub-Himalayan West Bengal: Evidence from Jalpaiguri District During the Last Century (1901–2000)

Manoranjan Ghosh and Rakhohori Bag

Abstract The monsoonal rainfall in North-East India, including the study region (Jalpaiguri district, located in the eastern Himalayan foothills), is decreasing, and heavy precipitation event days have been increasing. In this study, different climatic components, such as average monthly minimum temperature, maximum temperature, precipitation, and wet days' frequency of Jalpaiguri district over a century (1901–2000), have been investigated in order to figure out the changing pattern of climate. For robust interpretation, the Mann-Kendall trend and Sen's slope statistical methods were applied for trend analysis of each climatic component. At the same time, a hundred-year temperature anomalies (average, minimum, maximum) have also been calculated. The linear trend of the annual average temperature anomaly in Jalpaiguri has indicated that the temperature anomaly was increased by about 0.006 °C/year. It was highest in 1999 (+1.2 °C); simultaneously, the temperature has risen significantly by 0.01 °C in February and November in the last century. The Mann-Kendall test reveals a significant drop in the annual rainfall by 0.93 mm in June, followed by 0.18 mm in August during the last century (1901–2000). In 1955, July received a record-breaking rainfall (1105.93 mm), the highest monthly rainfall in the last century.

Keywords Trend analysis · Climate change · Temperature anomaly · Precipitation · Wet days' frequency

M. Ghosh (✉)
Fellow (ISCF) at National Institute of Urban Affairs, New Delhi, India

R. Bag
Doctoral Fellow (UGC-SRF) at Centre for the Study of Regional Development, School of Social Sciences, Jawaharlal Nehru University, New Delhi, India

S. Sharma et al. (eds.), *Climate Change Adaptation, Risk Management and Sustainable Practices in the Himalaya*,
https://doi.org/10.1007/978-3-031-24659-3_2

1 Introduction

The abundant scientific evidence indicates that the climate of the earth's troposphere is changing for a variety of natural and anthropogenic reasons[1] (Cicerone & Nurse, 2017). The Intergovernmental Panel on Climate Change (IPCC) made predictions in 2013 regarding the earth's average temperature, which is anticipated to increase sharply between 0.3 °C to 0.7 °C from the years 2016 to 2035, and the land surface temperature is expected to rise more. The analysis of historical data on land surface temperature shows that the average worldwide land surface temperature has risen by around 0.9 °C during the last century (Rohde et al., 2017). However, the Indian climate is predominantly characterized by the monsoon climatic systems, and the nature of the monsoon climate in India has been changing due to global climate change (Srinivasan, 2012). For example, the temperature of the land surface in India surged by 0.6 °C, while the temperature in the post-monsoon period significantly increased by 0.79 °C from 1901 to 2010. Moreover, North-East India, encompassing the study region and the country's western coast, is experiencing a decline in the monsoonal rainfall trend (Srinivasan, 2012). The frequency of warm days over the mainland of India has increased by 4–5 days during 1981–2015 (Krishnan & Sanjay, 2017). The IPCC (2014) predicts that in the near future, there will be an increasing trend of heavy precipitation events accompanied by the Indian summer monsoon. It was also found that the Hindu Kush Himalayan region has warmed up three times more rapidly than the worldwide average (Xu et al., 2009). Shrestha et al. (2015) have projected that extremely high temperatures have a growing trend over the northern part of the Brahmaputra river basin, and the glacial area in the Ganga river basin will be reduced by 35–45% of its present area by 2050. Since 1850, the last three decades have been warmer than the successive decades, and the number of excessive cold days are decreasing (Shrestha et al., 2015). Shrestha et al. (1999) have also rightly mentioned that global warming has been widely responsible for changing precipitation and glacial retreat in the Himalayan region (Annexure 1 shows yearly and season patterns of temperature and precipitation across several eco-regions, which emblematize that in Tarai-Dooars Savana Grassland eco-region climate is changing). Moreover, during the period of 1971–1994, the mean maximum temperature of the Siwalik and Terai Himalayas, located in the southern portion of the Himalayan Mountain range, increased by 0.03 °C/year. Shrestha et al. (2000) observed that the Himalayan mountains are warming at the rate of 0.09 °C annually and the Terai regions of West Bengal at the rate of 0.04 °C/annual, respectively. In other parts of West Bengal, the changing tendency of climate

[1] Recent changes in the earth's surface climate can also be attributed to human activities, including deforestation, shifting land uses, and expanding industry. Human effects have been the primary discernible force driving climate change in recent years, by interfering with the energy fluxes that occur naturally through modifications to the composition of the atmosphere, people primarily affect the global climate.

variables has also been widely recorded, for instance, the Sagar island's inter-annual rainfall deficit varied up to 44.8%, whereas the contribution of monsoon rainfall varied from 69% to 80%, and diurnal temperature variations increased during the years 1982–2017. In India's State Level Climate Change Report (2013), it was noted that West Bengal had a rising trend in mean maximum temperature, winter seasonal rainfall, and diurnal temperature range (+0.02 °C per year). In addition, the neighboring country, Bangladesh, had an upward trend in temperature, respectively, around 1 °C and 0.5 °C in May and November between 1985 and 1998. Furthermore, decadal rainfall anomalies have been above the long-term mean in Bangladesh since 1963 (Mizra & Dixit, 1997; Mizra, 2002). Rathore et al. (2013) show the rising trend of mean maximum temperature, winter season rainfall, and diurnal temperature range (+0.02 °C per year) over West Bengal. Nonetheless, literature on the pan-Indian level temperature and rainfall has been reviewed critically to understand the broad scenarios of climate change. The summer monsoons at different meteorological subdivisions in India at different time duration have also been critically reviewed. Table 2.1 illustrates the annual variation in precipitation for each of the 30 meteorological subdivisions, allowing the reader to comprehend the monsoon rainfall changes in India. The comparative trends of summer monsoon (JJAS) rainfall (mm/year) for 1871–2016 and 1981–2016 in the sub-Himalayan foothills of Sikkim and West Bengal subdivisions show that throughout the latter half, precipitation has declined dramatically at an annual rate of 8.49 mm (Table 2.1). Furthermore, Im et al. (2017) have projected the deadly heat wave[2] in India, and they considered this region to be most vulnerable to human habitat. The Ganges valley and Chotanagpur plateau will cross the survivability threshold by the turn of the century (2071–2100). Simultaneously, Indian states like Bihar, Odisha, Uttar Pradesh, and Andhra Pradesh would also cross the survivability threshold. In India, the lethal heat wave can put many people working as agricultural labors in the highly populated river valley at significant risk (Im et al., 2017). In addition, Shrestha et al. (1999) found that the average maximum temperature in the Siwalik and Terai Himalaya (the southern part of the Himalayan Mountain range) was increased by 0.03 °C/year from 1971 to 1994. Kapoor and Shaban (2014) observed a soaring annual precipitation fluctuation throughout the period of 1970–2014 and an upward trend of yearly mean temperature (maximum) by 0.5 °C annually in the Kullu region of Himalaya. Table 2.2 shows the current and projected pattern of climate change in India.

[2] At the global level, as the earth's temperature is rising, nearly one-third of the earth's population is under heat stress. The death rate due to heat wave has been on raising trend. By 2100, about 4 billion people will be at risk due to heat exposure, which will affect the global productive days (−2.2%) and indicates $2.4 trillion in economic losses. The construction workers, older, overweight, and poor people risk will increase 13 times and 35 times, respectively (Pennisi, 2020).

Table 2.1 Trend in summer monsoon (JJAS) rainfall (mm/year) for the period 1871–2016 and 1981–2016 for 30 meteorological subdivisions

Meteorological subdivisions	1871–2016 (146 years)	1981–2016 (37 years)	Meteorological subdivisions	1871–2016 (146 years)	1981–2016 (37 years)
Assam and Meghalaya	−0.74**	−5.95*	East Uttar Pradesh	−0.72**	−6.38***
Tripura, Naga-land, Mizoram, and Manipur	−1.64***	−1.71	West Uttar Pradesh plain	−0.53**	−3.53
Sikkim and sub-Himalayan West Bengal	−0.37	−8.49*	Haryana, Chan-digarh, and Delhi	0.13	−1.96
West Bengal (Gangetic)	0.40	1.71	Punjab	0.18	−2.07
Odisha	−0.24	2.06	Rajasthan (west)	0.21	1.99
Jharkhand	−0.44	0.37	Rajasthan (east)	−0.05	3.92*
Bihar	−0.55	6.43	Madhya Pradesh (west)	0.08	3.02
Chhattisgarh	−1.13***	2.96	Madhya Pradesh (east)	−1.07***	−0.62
Andhra Pradesh (coastal)	0.56***	0.56	Gujarat	−0.11	5.05
Telangana	0.37	−3.37	Kutch, Diu, and Saurashtra	0.4	8.26**
Rayalaseema	0.23	0.09	Goa and Konkan	2.96***	6.26
Pondicherry and Tamil Nadu	−0.1	−1.51	Maharashtra (central)	0.07	1.69
Karnataka (coastal)	1.72**	−8.49	Marathwada	−0.19	−2.05
Karnataka (north interior)	0.06	−2	Vidarbha	−0.35	1.78
Karnataka (south interior)	0.24	0.18	Kerala	−1.27*	−5.96

Where ***, **, and * are significant at the 1%, 5%, and 10% level, respectively
Source: Prepared by researchers based on the reports published by the Indian Institute of Tropical Meteorology (IITM), Pune, 2017

It is evident from Table 2.2 that the temperature increase rate is consistently more high during the winter months in most of the Himalayan regions, while a diminishing trend of rainfall was observed in the Eastern Himalayas. High warming rates have also been observed in the Western Himalayas (Kumar et al., 2006).

This study aims to assess the long-term anomalies as well as monthly and seasonal trends of climatic variables, such as temperature, precipitation, and the frequency of wet days, in the Jalpaiguri district from 1901–2000. This is due to the fact that there are several macrolevel research studies that discuss about the changing

Table 2.2 Existing and future scenario of climate change over the Indian Himalayas

Western Indian Himalaya	Central Indian Himalaya	Eastern Indian Himalaya
Das et al. (2018) observed that the temperature (maximum) went up by 0.09 °C from 1901 to 2003.	Diodato et al. (2012) showed that the Himalayan and Tibetan Plateau regions have warmed more recently than they did a century ago.	Dash et al. (2007) discovered that there was 1 °C increase in the maximum temperature over northeast India between the years 1901–2003.
Bhutiyani et al. (2007) anticipated an increase of temperature by 0.16 °C/ decade in the Western Indian Himalayas during the last century.	Shrestha et al. (1999) reported the central Himalayan region, particularly the Terai region, is experiencing a trend in the yearly average maximum temperature of 0.4–0.9 °C/decade.	Jhajharia and Singh (2011) observed that temperatures are rising in the eastern Indian Himalayas on average at the rate of 0.2–0.8 °C every 10 years and temperature (maximum) rises at the rate of 0.1–0.9 °C/decade.
Over the northwest Indian Himalayas, Singh et al. (2008) detected an increase in the maximum and seasonal average temperatures for all seasons except the monsoon.	Christensen et al. (2007) found that in the next 100 years, it was anticipated that the average temperature would climb faster in the Tibetan Plateau and the upper Himalayan regions, at the rate of 2.8 °C.	Pre-monsoon and winter months have seen more significant temperature increases over the eastern Himalayan region than the monsoon and post-monsoon months (Kumar et al., 2006).
The research also indicates that the western Himalayas will get a higher percentage of monsoon rainfall than the average precipitation rate during the years 2071–2100 (Kumar et al., 2006).		Between 1971 and 2000, the eastern trans-Himalaya and high mountains warmed by 0.9 and 1.2 °C/decade (Shrestha & Devkota, 2010).

dynamics of climate change, but very few studies accessed them at the district level in India. In addition, from an economic, agricultural, and disaster management perspective, it is crucial to evaluate how the precipitation and temperature have changed in each district over a considerable period.

This chapter has been organized as follows: The first section deals in detail with the location and general characteristics of the study region's climatic and meteorological components. The second section has discussed the data sources and methodology of the study. In the third section, the results have been analyzed and discussed. The last section, that is, the fourth section, concludes the study, where the impact and changing trend of the climatic element also has been discussed.

2 The Study Area

Jalpaiguri district, situated in the eastern Himalayan[3] foothills region and part of the sub-Himalayan[4] West Bengal climatic subdivision, has been selected for the present study. According to the Koppen climate scheme, the climate of this area is classified as Monsoon type and with typically dry winters (Cwg). The district has a regular seasonal pattern of rainfall, with 80–85% of the rain falling during the four summer months and a notable lack of rain throughout the winter (Roy, 2011). The average yearly rainfall in the area is 2347.82 mm, while the average temperature is around 24.33 °C throughout the year. The study area receives the highest rainfall in the entire sub-Himalayan West Bengal due to its location factor. The Bay of Bengal branch of the Southeast monsoon is blocked by the southern margin of the Himalayas, and heavy orographic rainfall occurs in this region. Rainfall distribution in Jalpaiguri is generally determined by the slope of the mountain and the direction of the Southwest monsoon, and it gradually decreases from east to west and north to south. Jalpaiguri is renowned for its frequent torrential downpours and floods[5] brought by the climate in the entire region. In 1993, Bharnobari and Chuapara tea estates, respectively, got a total of 2241 mm and 1988 mm of precipitation throughout the month of July. Extreme rainfall events happen regularly in the eastern part of the district.[6] It has been discovered that two-hour rainstorms that produce 200–300 mm of rain are not uncommon in this area (Roy, 2011). The date of arrival of monsoon in the study area varies between 9th to 15th of June each year. The Jalpaiguri district is dominated by cold and rainy seasons; the other two relatively short-spanned seasons are spring and autumn (Starkel et al., 2008). The location of the study region is depicted in Fig. 2.1, along with an illustration of the future monsoonal trend in Southeast Asia (2006–2050).

[3]It has been evident (through changing magnitude and acceleration) that the Eastern Himalayas are a hotspot and three times higher prone to glacial lake outburst than in any other Himalayan region, which emblematise the climatological change in recent past in this region (Veh et al., 2019). The ice loss is consistent and it is found that the average rate of ice loss become double during 2000–2016 in contrast to 1975–2000 (Maurer et al., 2019).

[4]India is divided into 36 homogenous monsoon climatic subregion; "sub-Himalayan West Bengal and Sikkim" is one of them in north-east India (IMD, 2017).

[5]The mentioned rainfall and temperature of Jalpaiguri is 100-year average from 1901 to 2000.

[6]Climate-induced disasters like flood and drought are two prime reasons for rural vulnerabilities, for example, 2017 flood in the northern part of West Bengal (including whole sub-Himalayan region) due to heavy rain over a week damaged 2,93,770.790 hectare of cropped area mainly paddy crop, and a total of 44,21,996 population have affected (West Bengal Disaster Management Report, 2017).

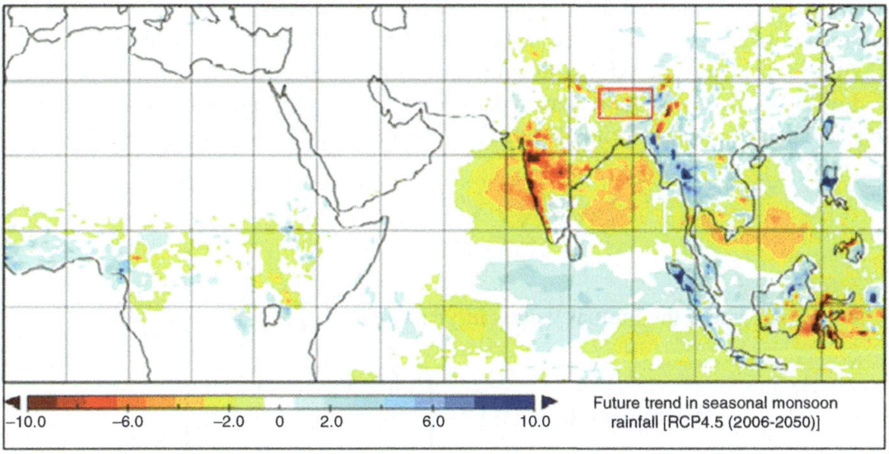

Fig. 2.1 Future trend of monsoon in South East Asia, 2006–2050 (*red rectangle* showing the study area). (Source: Centre for Climate Change Research, Indian Institute of Tropical Meteorology, Pune, 2017)

3 Data and Methods

The data of different climatic components, such as monthly mean precipitation, temperature (minimum and maximum temperature), and wet days' frequency of Jalpaiguri district over 100 years (1901–2000), have been downloaded from India Water Portal. In addition, the tea garden's stations data regarding heavy rainfall and rainstorm were collected during field surveys throughout the different months and years for a robust interpretation of the climatic trend. The Mann-Kendal (MK) trend statistics and Sen's slope estimation test was used for trend analysis of several climatic components (mean monthly precipitation, minimum and maximum temperature). The MK test statistics is a nonparametric robust statistical measure, widely used in atmospheric sciences to ascertain if a time series data has a monotonic trend/steady trend, especially for understanding the trend of temperature, precipitation, chemical particle duration, etc. (Chatterjee et al., 2016; Duhan & Pandey, 2013; Mandal et al., 2010). Therefore, the slope of the trend may be estimated by deploying Sen's technique, which employs a linear model. There is an absence of autocorrelation and seasonal component variation in the data series (Pohlert, 2017) when data are randomly distributed. The Mann-Kendall trend test outperforms the least-squares method because of its unique and desirable qualities. Instead of comparing the actual data values, the test primarily evaluates the sample data's relative magnitudes. The Mann-Kendall test can be used in situations when it is reasonable to anticipate that the time series will confirm the following model:

$$x_i = f\ (t) + \varepsilon_i$$

where *f(t)* is a continuous monotonic increasing or decreasing function of time and the residuals εi can be assumed to be from the same distribution with zero means. Therefore, it is considered that the variance of the distribution is constant in time.

Let $X_1, X_2, X_3 \ldots X_n$ represents n data points, where X_j represents the data point at time j. Then the Mann-Kendall statistic (S) is calculated using the formula:

$$S = \sum_{k=1}^{n-1} \sum_{j=k+1}^{n} sgn(x_j - x_k)$$

where x_j and x_k are the annual values in years j and k, $j > k$, respectively.

$$\text{Where Sign}\left(x_j - x_k\right) = \begin{Bmatrix} 1 & \text{if } x_j - x_k > 0 \\ 0 & \text{if } x_j - x_k = 0 \\ -1 & \text{if } x_j - x_k < 0 \end{Bmatrix}$$

The variance of S, VAR(S) is calculated by the following equation:

$$\text{VAR}(S) = \frac{1}{18}\left[n(n-1)(2n+5) - \sum_{p=1}^{q} t_p(t_p - 1)(2t_p + 5)\right]$$

Here q is the number of tied groups, and tP is the number of data values in the pth group.

The values of S and VAR(S) are used to compute the test statistic Z as follows:

$$Z = \begin{Bmatrix} \dfrac{S-1}{\sqrt{\text{VAR}(S)}} & \text{if } S > 0 \\ 0 & \text{if } S = 0 \\ \dfrac{S+1}{\sqrt{\text{VAR}(S)}} & \text{if } S < 0 \end{Bmatrix}$$

For time series with less than ten data points, the S test is used, and for time series with ten or more, the Z is used (Salmi et al., 2002), and the presence of a statistically significant trend is evaluated using the Z value. A positive or negative value of Z indicates an upward or downward trend. Either to test an upward or downward monotone trend (a two-tailed test), H_0: no trend is present, and H_a: the positive or negative trend is present and has been used in the Mann-Kendall trend statistics.

Sen's nonparametric statistics have been applied to determine the slope of the trend. Sen's method can be used in cases where the trend can be assumed linear. This means that:

$$f(t) = Q_t + \text{B}$$

where Q = slope and B = constant.

To get the slope estimate Q, we first calculate the slopes of all data value pairs (Sen, 1968):

$$Q_i = \frac{x_j - x_k}{j - k.}$$

where $j > k$

If there are n values Xj in the time series, we get as many as $N = n(n\text{-}1)/2$ slope estimates Q_i. Sen's slope estimator is the median of these N values of Qi. The N values of Q_i are ranked from the smallest to the largest, and Sen's estimator is:

$$Q = \begin{bmatrix} Q_{\{(N+1)/2\}} & \text{if N is odd} \\ \frac{1}{2}\left\{Q_{\left(\frac{N}{2}\right)} + Q_{(N+2)/2)}\right\} & \text{if N is Even} \end{bmatrix}$$

The positive value of Q indicates an increasing trend, and the negative value indicates the decreasing trend of the given observation.

Along with the trend analysis using Mann-Kendall statistical test, annual mean, minimum, and maximum temperature anomalies, pre-monsoon, monsoon, post-monsoon, and total rainfall anomalies for the district of Jalpaiguri from 1901 to 2000 have also been determined. At first, the average temperature of each year (mean, minimum, and maximum) was estimated, whereas the hundred-year anomalies of the climatic component were calculated using the following method: Initially average of each climatic component (mean temperature, minimum temperature, maximum temperature, total rainfall, monsoon, pre- and post-monsoon) for hundred years was calculated by the simple arithmetic mean. Finally, the hundred years of simple arithmetic mean have been deducted from each year's respective value.

4 Result and Discussion

The state action plan on climate change suggested that rainfall will decrease in January and February in almost all of the agroclimatic zone except Sundarban in West Bengal (Government of West Bengal, 2012). It is anticipated that there will be a significant rise in the precipitation amount throughout the winter season in the Terai, hilly, and new alluvial zone of the state (Government of West Bengal, 2012). As it was mentioned, the Jalpaiguri District in the sub-Himalayan West Bengal division is well known for its extreme rainfall events and experiencing high-impact climatic events; for example, extremely heavy rainfall $\geq$210 mm was reported during the 24 hrs ending at 08:30 hrs IST during the southwest monsoon season in 2017, the Kumargram tea gardens have received 250 mm on 22 June, and the Buxaduar and Alipurduar, respectively, 450 mm and 390 mm on 12 August in

2017 (Indian Meteorological Department, 2017). The result section has been designed into two parts for convenience. In the first part, the trend of temperature anomaly is explained in detail. Whereas in the second part, the results of descriptive statistics, findings of the Mann-Kendall test, and Sen's slope are explained.

4.1 Long-Term Temperature and Rainfall Anomalies

The scientific evidence of the temperature anomaly of a particular region supports the contemporary debate on global climate change. Temperature is considered a reliable indicator to figure out how the existing difference in the typical climate of a particular region, because it can be easily measured by the sophisticated thermometer, remote sensing technology, and its role in the earth's dynamic heat balance process (Barry & Chorley, 2003; Bhutiyani et al., 2007). The sub-Himalayan region is an ideal site to study the trend analysis of climatic components because of the varied climatic conditions throughout the Himalayan region (Bhutiyani et al., 2007). As mentioned in the methodology section, a long-term temperature anomaly has been calculated. Figure 2.2 illustrates the upward trend in average, minimum, and maximum temperatures during the last century.

From 1901 to 2000, the average temperature anomaly has risen. The linear trend line of the annual average temperature anomaly in Jalpaiguri has indicated that the temperature anomaly increased by about 0.006 °C per year (Fig. 2.2). It was highest in the year 1999 (+1.2 °C), whereas the maximum and minimum temperatures had gone up by 0.0058 °C and 0.0065 °C/year, respectively (Fig. 2.2). The temperature

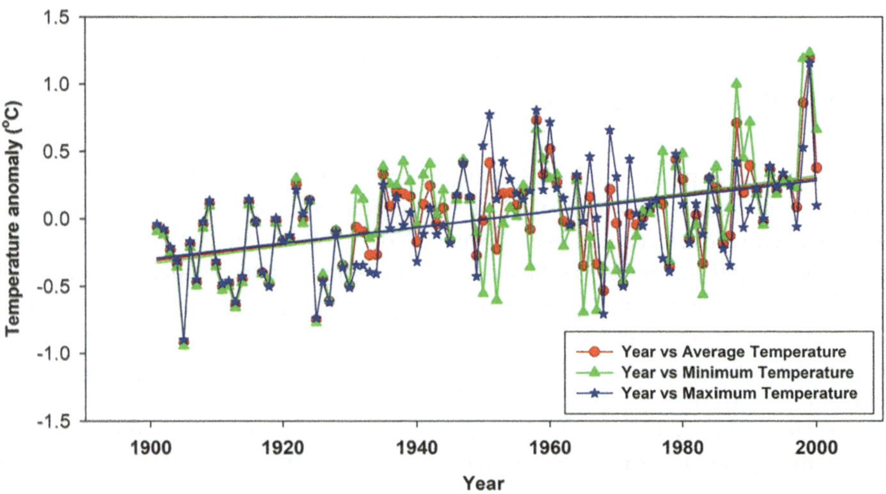

Fig. 2.2 Time series plot of yearly average, minimum, and maximum temperature anomalies (degree Celsius) of Jalpaiguri District

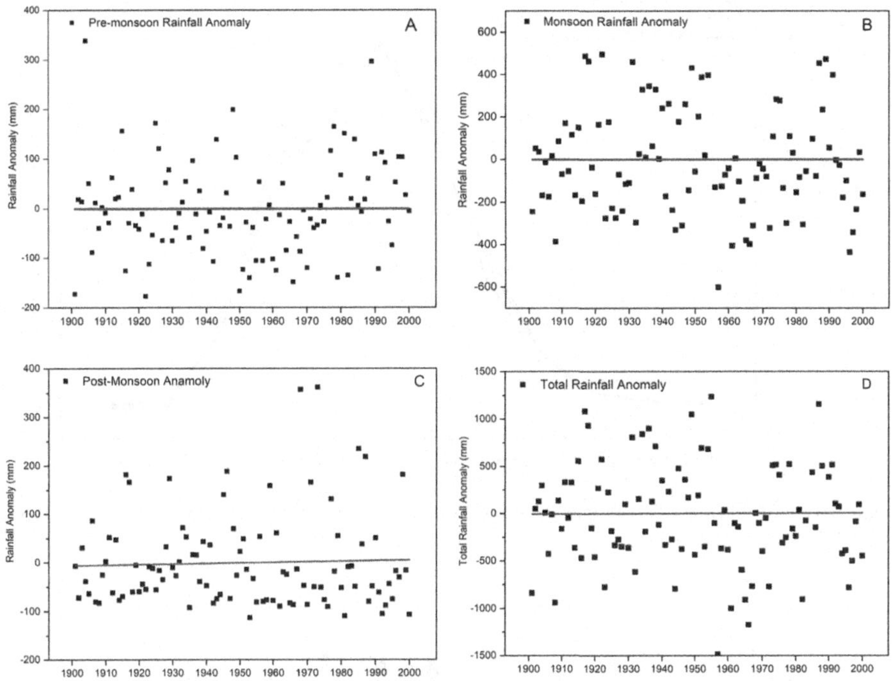

Fig. 2.3 Total rainfall anomalies in the Jalpaiguri District during the pre-monsoon, monsoon, and post-monsoon seasons

anomaly could affect different physical and socioeconomic sectors in the respective study region. For example, temperature anomaly could affect the physical nature of the atmosphere and the socioeconomic lives of rural people. It may influence the ecosystem, agricultural production, mortality, household welfare, etc. (Lee et al., 2016). Figure 2.3 displays the rainfall anomaly for various seasons, including pre-monsoon, monsoon, and post-monsoon, as well as total annual rainfall. It is evident from Fig. 2.3 that monsoon and post-monsoon rainfall anomalies have grown up, although the overall rainfall anomaly has increased at a nearly constant rate.

The average monthly precipitation and temperature distributions for the years 1901–1934, 1935–1964, and 1965–2002 have also been approximated, and it shows that the rainfall in June and August has comparatively varied over the period of 1901–1934 and 1965–2002 (Fig. 2.4). Figure 2.4 shows the normal distribution of rainfall and indicates that in September, the amount of rainfall has increased, which would have a reflecting influence on the extreme rainfall in the study area. In June, the opposite trend was observed, indicating a drought condition (see the black, red, and blue color dot symbols in Fig. 2.4).

Aside from the preceding analysis, Fig. 2.5 depicts the average monthly temperature distribution in three different periods of the study area, which reveal the changes in the mean temperature. Figure 2.5 indicates that in February, the mean

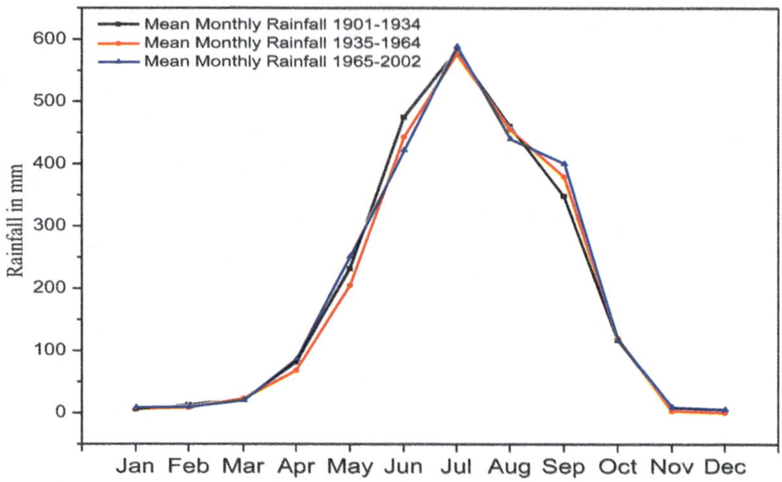

Fig. 2.4 Distribution of average monthly rainfall during the time periods of 1901–1934, 1935–1964, and 1965–2002

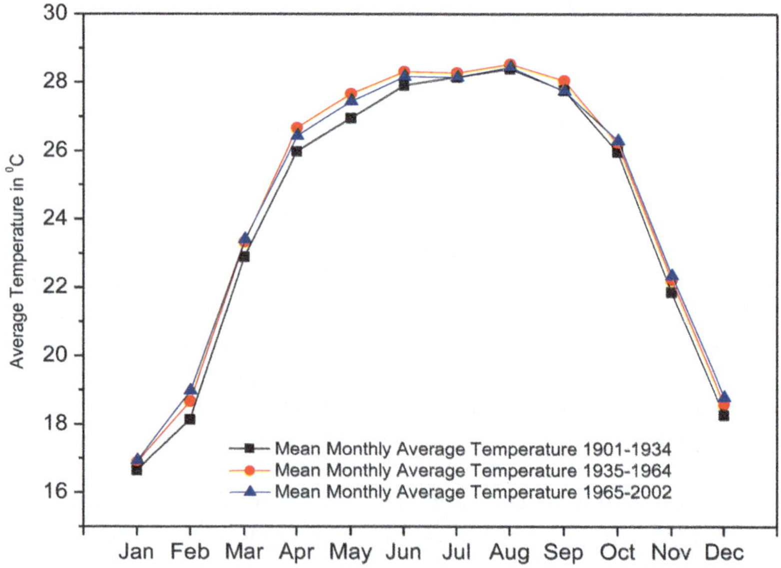

Fig. 2.5 Distribution of mean monthly temperature during 1901–1934, 1935–1964, and 1965–2002

monthly average temperature shows an increasing trend (see the black, red, and blue color dot symbols); besides, it's also indicating that in April, May, and June, the mean monthly temperature shows a changing trend (see the black, red, and blue color dot symbols in Fig. 2.5). Therefore, it would be an indication of the frequency of extremes, particularly the consequences of drought and heat wave in the area.

4.2 Mean Monthly Minimum Temperature Trend

The trend analysis of descriptive statistics of mean monthly minimum temperature shows that the minimum temperature was highest in July 1937 (26.57 °C) and lowest in January 1978 (7.55 °C) in the last century. If we consider the hundred-year average monthly minimum temperature, the month of August had the highest mean monthly minimum temperature (25.32 °C), and the month of January (10.12 °C) had the lowest. However, the Mann-Kendall statistical trend test exhibited a rising trend of mean monthly minimum temperature in April, October, and November at 0.007 °C/year (Table 2.3). Whereas, January and February showed high inconsistency in terms of mean monthly minimum temperature in the last century.

4.3 Mean Monthly Maximum Temperature Trend

The maximum temperature has risen more across the Himalayan and Tibetan plateau regions at a greater altitude than at lower elevations (Bhutiyani et al., 2007; Liu & Chen, 2000). In the district Jalpaiguri of sub-Himalayan West Bengal, April 1999 was the hottest month, with a mean monthly maximum temperature of 36.58 °C in the twentieth century, whereas in the year 1981, January recorded the lowest mean monthly maximum temperature (21.02 °C) among all the months in the last century. The Mann-Kendall test signifies that in the month of July, the maximum temperature is decreasing but not significantly. The month of August and September had the least amount of variations in maximum temperature, followed by July and June (Table 2.4). Similarly, during February and November, the temperature considerably rose by 0.01 °C annually in the last century. In May and October, the highest temperatures went up dramatically at the rate of 0.007 °C/year.

4.4 Mean Monthly Precipitation Trend

Since 1950, Indian summer monsoon rainfall has indicated a significant decreasing trend (Saha et al., 2014). One of the factors contributing to the lowering trend of precipitation in northeast India is the weakening of temperature gradient between the northern and southern hemispheres (Pachauri & Meyer, 2014). In contrast, the mid-latitude regions have seen a rise in precipitation since 1901 (Bollasina et al., 2011). Moreover, according to the Indian Parliament Special Report (2015), rainfall in West Bengal has been increasing at a rate of 3.63 mm/year during all the seasons, and it has also shown that the rainfall has been rising at the rates of 1.45 mm and 1.34 mm throughout the monsoon and post-monsoon seasons. In the study area, a

Table 2.3 Pattern of mean monthly minimum temperature of Jalpaiguri, 1901–2000 (Mann-Kendall test and Sen's slope)

Month	Minimum MMMT[a] temperature (°C) and year		Maximum temperature (°C) and year		Mean MMMT temperature (°C)	Std. deviation statistic	CV (%)	Skewness statistic	Kurtosis statistic	Mann- Kendal Z	Sen's slope
Jan	7.55	1978	13.01	1990	10.1257	0.96131	9.493	0.091	0.348	1.85+	0.007
Feb	8.40	1905	14.23	1988	11.7514	0.95876	8.158	0.088	0.924	4.07***	0.014
Mar	13.06	1952	18.28	1985	15.9677	1.10394	6.913	−0.421	−0.298	2.83**	0.011
Apr	17.76	1905	23.67	1999	20.2146	1.04400	5.164	0.373	0.561	2.16*	0.007
May	21.05	1986	24.67	1995	22.7374	0.79813	3.510	0.123	−0.486	2.60**	0.008
Jun	23.39	1913	25.91	1937	24.4724	0.53300	2.177	0.301	−0.114	1.85+	0.003
July	24.00	1956	26.57	1937	25.1230	0.46283	1.842	0.216	0.750	0.62	0.001
Aug	24.14	1972	26.30	1931	25.3226	0.45939	1.814	−0.162	−0.156	0.94	0.002
Sep	22.46	1972	25.26	1951	24.4310	0.55445	2.269	−0.297	0.375	0.37	0.001
Oct	19.73	1966	23.89	1998	21.6925	0.86542	3.989	0.300	−0.101	2.16*	0.007
Nov	14.16	1987	18.95	1979	16.0957	0.94587	5.876	0.544	0.516	2.08*	0.007
Dec	23.67	1972	27.13	1998	25.6816	0.58030	2.259	−0.115	1.000	1.12	0.002

[a]Mean monthly minimum temperature, *CV* coefficient of variation

Where *** if trend at = 0.001 level of significance, ** if trend at = 0.01 level of significance, * if trend at = 0.05 level of significance, + if trend at = 0.1 level of significance

Table 2.4 Trend of mean monthly maximum temperature of Jalpaiguri, 1901–2000 (Mann-Kendall test and Sen's slope)

Month	Minimum MMMT[a] temperature (°C) and year		Maximum MMMT temperature (°C) and year		Mean MMMT temperature (°C)	Std. deviation statistic	CV (%)	Skewness statistic	Kurtosis statistic	Mann-Kendal Z	Sen's slope
January	21.02	1981	27.12	1990	23.5795	0.98295	4.168	0.209	0.883	1.25	0.004
February	22.28	1905	28.25	1960	25.5479	1.04201	4.078	0.355	0.652	2.51*	0.010
March	27.31	1940	33.57	1969	30.5370	1.19124	3.900	−0.164	0.039	1.52	0.007
April	29.92	1949	36.58	1999	32.5603	1.26210	3.876	0.279	0.159	1.58	0.007
May	29.60	1977	34.11	1979	32.0310	0.95896	2.993	0.334	−0.284	2.11*	0.007
June	30.20	1929	33.04	1998	31.8480	0.61063	1.917	−0.156	−0.374	2.31*	0.005
July	29.88	1965	32.68	1972	31.3470	0.52204	1.665	0.067	0.660	−0.52	−0.001
August	30.59	1958	33.03	1983	31.6397	0.44542	1.407	−0.092	0.651	1.77+	0.002
September	30.23	1931	32.37	1982	31.3344	0.45026	1.436	0.159	−0.352	0.51	0.001
October	28.84	1968	32.66	1975	30.7045	0.83090	2.706	0.454	−0.170	2.24*	0.007
November	25.94	1967	30.75	1979	28.2946	0.96160	3.398	0.190	0.258	3.09**	0.010
December	30.86	1968	34.07	1958	32.3514	0.59929	1.852	0.124	0.114	2.23*	0.005

[a]MMMT mean monthly maximum temperature, *CV* coefficient of variation
Where *** if trend at = 0.001 level of significance, ** if trend at = 0.01 level of significance, * if trend at = 0.05 level of significance, + if trend at = 0.1 level of significance

total of 5 months (i.e., May to September) is the rainy season, although the district receives rainfall throughout the year, except few winter months. The trend analysis of mean monthly precipitation (Table 2.5) indicates the changing nature of precipitation in Jalpaiguri over the last century. The outcome shows that except for a few months (January, 1937, 1946, 1965, and February 1974), almost all the months had received rainfall in Jalpaiguri. The month of November and December are generally drier, because of low precipitation (mean precipitation 7.76 mm and 4.16 mm, respectively). If we divide the twentieth century into equal halves (1901–1950 and 1951–2000), then it was found that in the latter part of the twentieth century, 11 times out of 14 times the month of December did not get a single drop of rainfall. In the year 1955, July received 1105.93 mm of rainfall, recorded as the highest monthly rainfall in the last century. The district Jalpaiguri gets the height amount of rain among all the districts in West Bengal, whereas 79.48% of yearly rainfall occurred in the south-west monsoon season (June to September). Nevertheless, the Mann-Kendall test indicated that rainfall in June and August is observed to be decreasing at the rate of 0.93 mm and 0.18 mm/ year, respectively (Table 2.5).

4.5 Mean Monthly Wet Days' Frequency Trend

The regional precipitation behavior is an essential indicator for understanding climate change scenarios. In a large scale, there is much more evidence regarding the precipitation change and wet days' frequency; although, on a local or regional scale, the evidence of changing precipitation behavior of climate change is much less (Brunetti et al., 2002). The magnitude of the trend in wet day frequency over the past century was calculated using Sen's estimator, and it showed that in June, wet days' frequency was decreasing significantly in the study area (Table 2.6). The results also show that the month of July has the highest number of rainy days (17.39) and the month of December has the lowest number during the twentieth century, and there was a huge inconsistency in the number of rainy days in November and December (even some years reported no wet days in the above months). The July 1931 has the highest number of rainy days among all months; it has received 23 and a half-wet day, which is the highest of the last century. The Mann-Kendall test statistics show that the frequency of wet days has declined dramatically in June, by 0.016 days/year.

4.6 Trend of Seasonal Rainfall and Prevalence of Wet Days

Recently, it was discovered by Paul et al. (2016) that the Indian summer monsoon precipitation in central and north-east India is diminishing due to the changing land use and land cover pattern, particularly by agricultural intensification and plantation agriculture. Additionally, the Hindu Kush Himalaya is reportedly warming up

Table 2.5 Trend of mean monthly precipitation of Jalpaiguri, 1901–2000 (Mann-Kendall test and Sen's slope)

Month	Minimum precipitation (mm) and year		Maximum precipitation (mm) and year		Mean precipitation (mm)	Std. deviation statistic	CV (%)	Skewness statistic	Kurtosis statistic	Mann-Kendal Z	Sen's slope
Jan	000	–	37.601	1994	8.033	9.236	114.978	1.469	1.594	1.58	0.024
Feb	000	–	38.100	1938	11.285	9.289	82.309	0.936	0.150	−0.59	−0.014
Mar	0.199	1009	72.061	1926	22.839	15.805	69.2038	0.912	0.443	−0.25	−0.014
Apr	5.49	1999	179.480	1943	78.962	36.045	45.6484	0.371	−0.081	0.75	0.088
May	68.83	1968	564.352	1989	229.668	84.983	37.0025	1.109	2.919	1.16	0.302
Jun	172.83	1958	838.304	1984	448.974	143.764	32.0207	0.446	−0.173	−1.91*	−0.934
Jul	267.57	1944	1105.930	1955	584.758	160.409	27.4318	0.564	0.401	0.44	0.314
Aug	224.91	1975	733.556	1958	453.961	113.747	25.0565	0.489	−0.175	−0.47	−0.180
Sep	154.32	1912	688.251	1942	378.425	113.580	30.0139	0.367	−0.216	1.50	0.618
Oct	2.91	1981	483.832	1968	119.038	94.619	79.4862	1.709	3.120	−1.05	−0.209
Nov	000	–	38.263	1924	7.706	9.357	121.431	1.455	1.429	0.35	0.002
Dec	000	–	42.956	1979	4.169	7.904	189.586	2.836	8.586	−0.16	0.000

Where *** if trend at = 0.001 level of significance, ** if trend at = 0.01 level of significance, * if trend at = 0.05 level of significance, + if trend at = 0.1 level of significance, *CV* Coefficient of variation

Table 2.6 Trend of mean monthly wet day frequency of Jalpaiguri, 1901–2000 (Mann-Kendall test and Sen's slope)

Month	Minimum number wet days and year		Maximum number of wet days and year		Mean precipitation (mm)	Std. deviation statistic	CV (%)	Skewness statistic	Kurtosis statistic	Mann-Kendal Z	Sen's slope
Jan	0.00	–	2.52	1993	1.1166	0.64152	57.452	0.338	−0.455	1.28	0.001
Feb	0.00	–	2.19	1913	1.2266	0.42227	34.426	−0.026	0.505	−1.28	0.000
Mar	0.20	1909	3.71	1926	1.8993	0.67586	35.584	0.308	−0.332	−0.59	−0.001
Apr	1.14	1949	7.57	1977	4.4030	1.13104	25.687	−0.362	0.563	1.19	0.004
May	5.61	1969	15.34	1904	9.7258	1.79852	18.492	0.298	0.496	0.77	0.005
Jun	9.60	1905	20.06	1984	14.6691	2.28547	15.580	0.137	−0.390	−2.06*	−0.016
Jul	11.86	1967	23.50	1931	17.3912	2.14718	12.346	0.291	0.334	1.16	0.008
Aug	11.40	1975	19.81	1987	15.4096	1.77038	11.488	0.340	−0.211	−0.38	−0.002
Sep	8.36	1912	16.99	1942	12.6434	1.89725	15.005	0.021	−0.477	0.59	0.004
Oct	1.00	1914	7.53	1987	3.8546	1.38252	35.866	0.572	0.164	−0.99	−0.005
Nov	0.00	–	2.81	1995	1.0326	0.64374	62.341	0.196	−0.019	−0.81	0.000
Dec	0.00	–	1.00	1997	0.5567	0.42351	76.075	−0.132	−1.827	0.53	0.002

Where *** if trend at = 0.001 level of significance, ** if trend at = 0.01 level of significance, * if trend at = 0.05 level of significance, + if trend at = 0.1 level of significance, *CV* Coefficient of variation

Table 2.7 Trend of seasonal rainfall and wet day frequency of Jalpaiguri, 1901–2000 (Mann-Kendall test and Sen's slope)

	Rainfall in mm				Wet day frequency			
Season	Mean	SD	Mann-Kendal Z	Sen's slope	Mean	SD	Mann-Kendal Z	Sen's slope
Pre-monsoon (M, A, M)	331.47	95.58	1.03	0.304	16.0281	2.389	0.60	0.005
Monsoon (J, J, A, S)	1866.11	265.758	−0.47	−0.443	60.1133	4.115	−0.41	−0.006
Post-monsoon (O, N, D)	130.91	95.874	−0.65	−0.134	5.4439	1.547	−1.48	−0.008
Total annual	2347.82	302.868	−0.25	−246	83.9285	5.376	−0.57	−0.009

Where *** if trend at = 0.001 level of significance, ** if trend at = 0.01 level of significance, * if trend at = 0.05 level of significance, + if trend at = 0.1 level of significance, *CV* Coefficient of variation

around three times faster than the average world temperature (Shrestha et al., 2015), which will also affect the pattern of precipitation over the whole Himalayan region. Jain and Kumar (2012) showed that in the "sub-Himalayan West Bengal and Sikkim" and "Coastal Karnataka subdivision," rainfall increases by 0.87 mm/year in July and 0.40 mm/year in the post-monsoon season. However, the yearly rainfall fluctuations in the aforementioned locations are significantly lower in comparison to the rest of the country. Table 2.7 shows the results of hundred-year average and standard deviations of seasonal precipitation of the Jalpaiguri district in sub-Himalayan West Bengal. The hundred years average summer monsoon rainfall is about 1866.11 mm with 265.75 mm seasonal fluctuation. Table 2.7 also shows that the average fluctuation of precipitation is high during the last century, whereas pre-monsoon and post-monsoon fluctuation is comparatively low. Even though there is a tendency of declination in precipitation during monsoon seasons and post-monsoon seasons, although the pre-monsoon rainfall remains constant. In terms of wet day frequency, the district experienced an average of 84 wet days throughout the year, with January and February as an exception (mainly the dry winter season). There is also a high fluctuation of wet days' frequency in the monsoon period (Table 2.7).

5 Conclusion

In the nutshell, from the trend analysis of minimum temperature, average temperature, precipitation, and wet day frequency in the respective months in the Jalpaiguri district, we can argue that Himalayan foothills in West Bengal have been experiencing changes in the climatic components. The historical trend of the average monthly maximum temperature in Jalpaiguri has revealed a considerable upward tendency in February by 0.01 °C/year. In particular, the months of April, October, and November as well as the second half of the twentieth century exhibit an increasing trend in the mean monthly maximum temperature. The inconsistency in frequency and intensity of warm days in the mentioned month would be the prime cause of changes in other associated climatic components, because land surface temperature primarily determines the changes in thermodynamics of lower atmosphere. The changes in temperature over Jalpaiguri district during the twentieth century is also an indicator of the transforming hydroclimatic condition in the overall sub-Himalayan region. Nonetheless, the findings of relatively high fluctuation of monsoon rainfall compared to pre- and post-monsoon rainfall indicated toward the high rainy extreme events, which is also supported by high oscillation of wet day frequency in the monsoon period. The above mentioned changing climatic components signify toward a possible change in hydroclimatic dynamics in sub-Himalayan West Bengal. For example, the changing rainfall-induced severe flood of 2017 and 2021 emblematized a hydroclimatic change. Here, it also needs to mention that as the Jalpaiguri district is the hotspot of mosquito breeding due to humid climatic condition, the changing pattern of rainfall will also may intensify the mosquito-fetched health vulnerability. However, studies have shown that 1 °C rise in temperature during the growing period will reduce near about five million agricultural productions (Kumar et al., 2014). A large number of individuals in the densely populated sub-Himalayan West Bengal work as agricultural laborers, the increasing trend of mean monthly maximum temperature in April, May, and June, which will increase the deadly heat waves and cause high vulnerabilities among agricultural labors and common population. Moreover, it can also be argued that temporal changes in monthly rainfall and frequency of rainfall days directly influence the water flow of the river basin and groundwater level in the sub-Himalayan region, along with the whole natural ecosystem.

Annexure 1

Temperature and precipitation changes in the Himalayas. (a) Annual and season trends of temperature (°C/year). (b) Annual and season trends of precipitation (mm/year) shown in the bar graph in different eco-regions (Shrestha et al., 2012)

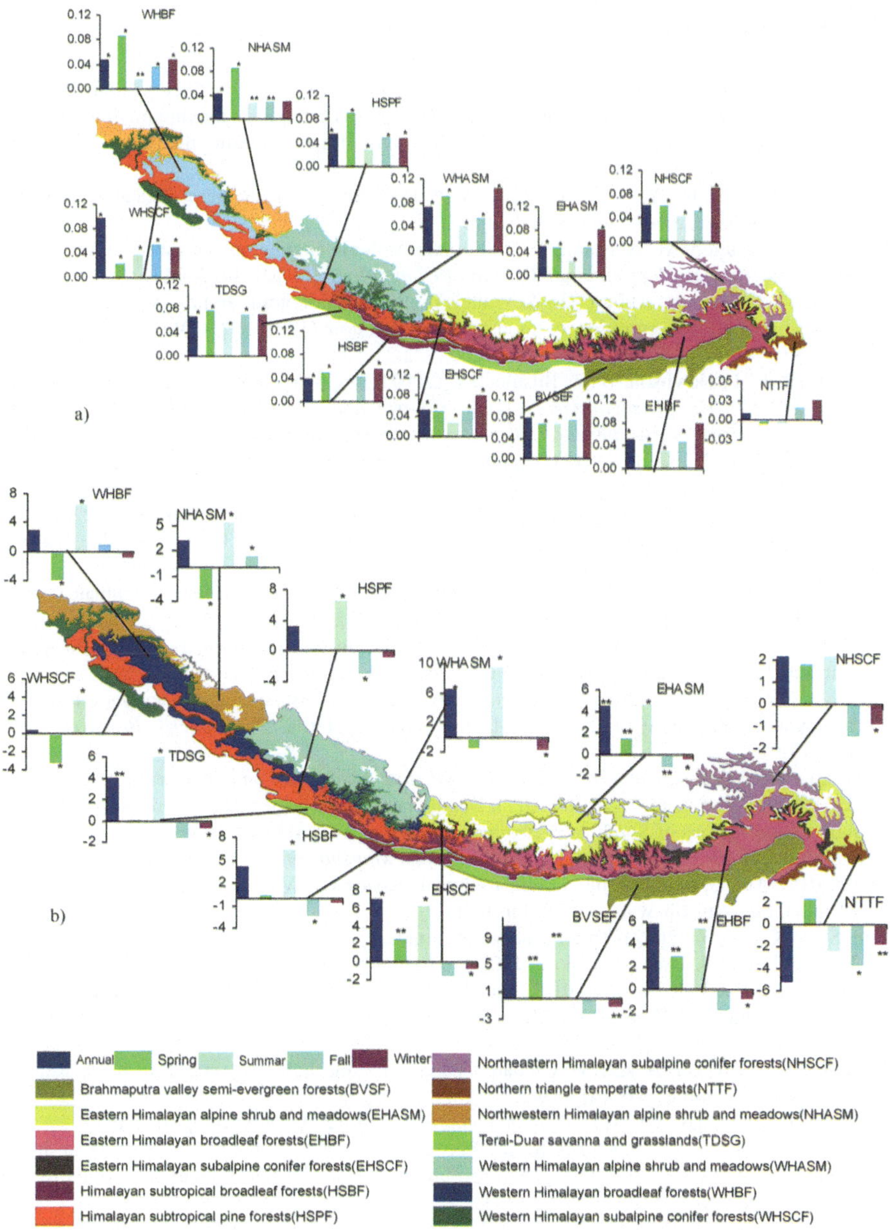

a)
WHBF
NHASM
HSPF
WHSCF
WHASM
EHASM
NHSCF
TDSG
HSBF
EHSCF
BVSEF
EHBF
NTTF
b)
WHBF
NHASM
HSPF
WHSCF
WHASM
EHASM
NHSCF
TDSG
HSBF
EHSCF
BVSEF
EHBF
NTTF
Annual
Spring
Summar
Fall
Winter
Brahmaputra valley semi-evergreen forests(BVSF)
Eastern Himalayan alpine shrub and meadows(EHASM)
Eastern Himalayan broadleaf forests(EHBF)
Eastern Himalayan subalpine conifer forests(EHSCF)
Himalayan subtropical broadleaf forests(HSBF)
Himalayan subtropical pine forests(HSPF)
Northeastern Himalayan subalpine conifer forests(NHSCF)
Northern triangle temperate forests(NTTF)
Northwestern Himalayan alpine shrub and meadows(NHASM)
Terai-Duar savanna and grasslands(TDSG)
Western Himalayan alpine shrub and meadows(WHASM)
Western Himalayan broadleaf forests(WHBF)
Western Himalayan subalpine conifer forests(WHSCF)

References

Barry, R. G., & Chorley, R. J. (2003). *Atmosphere, weather and climate* (8th ed.). Routledge.

Bhutiyani, M. R., Kale, V. S., & Pawar, N. J. (2007). Long-term trends in maximum, minimum and mean annual air temperatures across the Northwestern Himalaya during the twentieth century. *Climatic Change, 85*(1–2), 159–177.

Bollasina, M. A., Ming, Y., & Ramaswamy, V. (2011). Anthropogenic aerosols and the summer monsoon. *Science, 334*(October), 502–505.

Brunetti, M., Maugeri, M., & Nanni, T. (2002). Atmospheric circulation and precipitation in Italy for the last 50 years. *International Journal of Climatology, 22*(12), 1455–1471.

Chatterjee, S., Khan, A., Akbari, H., & Wang, Y. (2016). Monotonic trends in Spatio-temporal distribution and concentration of monsoon precipitation (1901-2002), West Bengal, India. *Atmospheric Research, 182*, 54–75.

Christensen, J. H., Hewitson, B., & Busuioc, A. (2007). Regional climate projections. In *Climate change 2007: The physical science basis. Contribution of working group 1 to the fourth assessment report of the intergovernmental panel on climate change.*

Cicerone, R. J., & Nurse, P. (2017). *Climate change evidence and causes.* Royal Society, U S National Academy.

Das, L., Dutta, M., Mezghani, A. & Benestad, R. E. (2018). Use of observed temperature statistics in ranking CMIP5 model performance over the Western Himalayan Region of India. *International Journal of Climatology, 38*, 554–570.

Dash, S. K., Jenamani, R. K., Kalsi, S. R., & Panda, S. K. (2007). Some evidence of climate change in twentieth-century India. *Climatic Change, 85*(April), 299–321.

Diodato, N., Bellocchi, G., & Tartari, G. (2012). How do Himalayan areas respond to global warming? *International Journal of Climatology, 32*(April 2011), 975–982.

Duhan, D., & Pandey, A. (2013). Statistical analysis of long term spatial and temporal trends of precipitation during 1901–2002 at Madhya Pradesh, India. *Atmospheric Research, 122*, 136–149.

Government of West Bengal. (2012). *West Bengal state action plan on climate change.*

Im, E., Pal, J. S., & Eltahir, E. A. B. (2017). Deadly heat waves projected in the densely populated agricultural regions of South Asia. *Science Advances, 2010*(August), 1–8.

Indian Meteorological Department. (2017). *Southwest Monsoon end of season report.*

IMD (2017). Annual Report of India Meteorological Department, Ministry of Earth Sciences, Govt. of India, Mausam Bhawan, New Delhi, India.

Jain, S., & Kumar, V. (2012). Trend analysis of rainfall and temperature data for India SK Jain, V Kumar Current Science (Bangalore) 102(1), 37–49. *Current Science, 102*(1), 37–49.

Jhajharia, D., & Singh, V. P. (2011). Trends in temperature, diurnal temperature range, and sunshine duration in Northeast India. *International Journal of Climatology, 31*(June 2010), 1353–1367.

Kapoor, M., & Shaban, A. (2014). Climate Change in Himachal: Evidence from Kullu Region. *Economic and Political Weekly*, 19–22.

Krishnan, R., & Sanjay, J. (2017). *Climate change over India: An interim report.* IITM.

Kumar K. R., Sahai A. K., Kumar K. K., Patwardhan S. K., Mishra P. K., Revadekar J. V., Kamala K., & Pant, G. B. (2006). High-resolution Climate Change Scenarios for India for the 21st Century. *Current Science 90*, 334–345.

Kumar, S. N., Aggarwal, P. K., Swaroopa Rani, D., Saxena, R., Chauhan, N., & Jain, S. (2014). Vulnerability of wheat production to climate change in India. *Climate Research, 59*, 173–187.

Lee, M., Villaruel, M. L., & Gaspar, R. (2016). *Effects of temperature shocks on economic growth and welfare in Asia.* Asian Development Bank.

Liu, X., & Chen, B. (2000). *Climatic warming in the Tibetan Plateau during, 1742, 1729–1742.*

Mandal, S., Choudhury, B. U., Mondal, M., & Bej, S. (2010). Trend analysis of weather variables in Sagar Island, West Bengal, India: A long-term perspective (1982–2010). *Current Science*, 1–7.

Maurer, J. M., Schaefer, J. M., Rupper, S., & Corley, A. (2019). Acceleration of ice loss across the Himalayas over the past 40 years. *Science Advances, 5*(6), 7266.

Mirza, M. M. Q., & Dixit, A. (1997). Climate Change and Water Resources in the GBM Basins. *Water Nepal, 5*, 71–100.

Mirza, M. M. Q. (2002). The Ganges water-sharing treaty: risk analysis of the negotiated discharge. *International Journal of Water, 2*(1), 57–74.

Pachauri, R. K., & Meyer, L. (2014). *Climate change 2014: Synthesis report. Contribution of working groups I, II, and III to the fifth assessment report of the intergovernmental panel on climate change*. IPCC.

Paul, S., Ghosh, S., Oglesby, R., Pathak, A., Chandrasekharan, A., & Ramsankaran, R. (2016). Weakening of Indian summer monsoon rainfall due to changes in land use land cover. *Scientific Reports, 6*(December 2015), 1–10.

Pennisi, E. (2020). Living with heat: Studies are probing how heat threatens health and how to lower. *Science, 370*(6518), 778–781.

Pohlert, T. (2017). *Non-parametric trend tests and change-point detection*, 1–18.

Rathore, L. S., Attri, S. D., & Jaswal, A. K. (2013). *State-level climate change trends*. India Meteorological Department, Ministry of Earth Sciences Government of India.

Rohde, N., Tang, K. K., Osberg, L., & Rao, D. S. P. (2017). Is it vulnerability or economic insecurity that matters for health? *Journal of Economic Behavior and Organization, 134*, 307–319.

Roy, S. (2011). *Flood Hazards in Jalpaiguri District*. Unpublished Ph.D. Thesis, Department of Applied Geography, University of North Bengal.

Saha, A., Ghosh, S., Sahana, A. S., & Rao, E. P. (2014). Failure of CMIP5 climate models in simulating Post-1950 decreasing trend of Indian monsoon. *Geophysical Research Letters, 41*(20), 7323–7330.

Salmi, T., Määttä, A., Anttila, P., Ruoho-Airola, T., & Amnell, T. (2002). *Detecting trends of annual values of atmospheric pollutants by the Mann-Kendall test and Sen's slope estimates -the excel template application makesens*. Finnish Meteorological Institute.

Sen, P. K. (1968). "Estimates of the regression coefficient based on Kendall's tau". *Journal of the American Statistical Association, 63*(324), 1379–1389.

Shrestha, A. B., & Devkota, L. (2010). *Climate change in the Eastern Himalayas: Observed trends and model projections*.

Shrestha, A. B., Wake, C. P., Mayewski, P. A., & Dibb, J. E. (1999). Maximum temperature trends in The Himalaya and its vicinity: An analysis based on temperature records from Nepal for the period 1971–94. *Journal of Climate, 12*, 2775–2786.

Shrestha, A. B., Wake, C. P., Dibb, J. E. & Mayewski, P. A. (2000). Precipitation fluctuations in the Nepal Himalaya and its vicinity and relationship with some large scale climatological parameters. *International Journal of Climatology, 20*, 317–327.

Shrestha, U. B., Gautam, S., & Bawa, K. S. (2012). Widespread climate change in the Himalayas and associated changes in local ecosystems. *PLoS One, 7*(5), 1–10.

Shrestha, A. B., Agrawal, N. K., Alfthan, B., Bajracharya, S. R., Maréchal, J., & Van Oort, B.. (2015). *HKH water atlas*. ICIMOD, GRID-Arendal, CICERO 2015.

Singh, P., Kumar, V., Thomas, T., & Arora, M. (2008). Changes in rainfall and relative humidity in river basins in northwest and central India. *Hydrological Processes, 22*, 2982–2992.

Srinivasan, J. (2012). Impact of climate change on India. In K. N. Dubash (Ed.), *Handbook of climate change and India: Development, politics, and governance* (pp. 29–40). Oxford University Press.

Starkel, L., Sarkar, S., Soja, R., & Prokop, P. (2008). Present-day evolution of the Sikkimese-Bhutanese Himalayan Piedmont. *Dział Wydawnictw Igipz PAN Ul. Twarda, 51*(55), 00–818.

Veh, G., Korup, O., & Walz, A. (2019). Hazard from Himalayan Glacier Lake outburst floods. *PNAS, 117*(2), 907–912.

West Bengal Disaster Management Report (2017). Annual Disaster Management Report of West Bengal, West Bengal Disaster Management & Civil Defence Department, Kolkata.

Xu, J., Grumbine, R. E., Shrestha, A., Eriksson, M., Yang, X., Wang, Y., & Wilkes, A. (2009). The melting Himalayas: Cascading effects of climate change on water, biodiversity, and livelihoods. *Conservation Biology, 23*(3), 520–530.

Chapter 3
Climate Change Impact on Major River Basins in the Indian Himalayan Region: Risk Assessment and Sustainable Management

K. Amrutha, Rasmi Patnaik, A. S. Sandeep, and Jitendra Kumar Pattanaik

Abstract Billions of people relay on water resources of the Himalayan region for drinking, irrigation, and other domestic purposes. Abundance of natural resources makes this region suitable for human settlements, despite the fact that the area experiences frequent natural hazards. Water resources including major rivers are one of the important components, responsible for high biodiversity of the Himalayas and its role in global atmospheric circulation. Recent climate changes have proved to affect the precipitation pattern and ice cover of the Himalayas, causing variations in the dynamics of rivers in the area. Climate change–induced variation in river flow quantity, timing, and unpredictability raises the danger of ecological changes and has a negative impact on aquatic life and the ecosystem depending on rivers. Agriculture is one important sector that is at highest risk due to climate change. This is a serious concern as the runoff patterns of the rivers are mainly determined by the precipitation pattern and ice cover in the upper reaches. Reduction in ice cover reduces the water storage capacity of the Himalayas, and fluctuations in the precipitation pattern cause floods and droughts. The increased frequency of natural hazards including floods and droughts affects the economy and is a threat to people's life. Climate change effects on water resources, namely, Himalayan snow and ice reservoirs and lake and river systems and the risk associated with it, can be monitored using different hydrological models. To cover vast geographical areas of the Himalayan region, adequate hydrological observatories need to be installed in order to monitor and record time series data of the hydrological parameters. Systematic monitoring will

K. Amrutha · J. K. Pattanaik (✉)
Department of Geology, School of Environment and Earth Sciences,
Central University of Punjab, Bathinda, Punjab, India

R. Patnaik
Independent Consultant, Bathinda, India

A. S. Sandeep
Department of Geography, School of Environment and Earth Sciences,
Central University of Punjab, Bathinda, Punjab, India

S. Sharma et al. (eds.), *Climate Change Adaptation, Risk Management and Sustainable Practices in the Himalaya*,
https://doi.org/10.1007/978-3-031-24659-3_3

help to predict how climate change will affect water resources in the future. Sustainable management of local resources based on suitable practices, adaptation strategies, and need-specific policies relevant to basin climate can further reduce frequent climate change-related impacts, risk, and vulnerability.

Keywords Climate change · Himalaya · Major rivers · Hydrological observatories · Risk assessment · Sustainable management

1 Introduction

Water resources in the Himalayan region include over ten of the largest rivers in Asia (Eriksson et al., 2009). Major rivers in the Indian Himalayan region (IHR) are distributed in the Indus basins, Ganga basins, and Brahmaputra (IGB) basins, and it possesses high water resource potential. Himalayas host large resources of snow, ice, and glaciers forming water tower of Asia (Xu et al., 2009). Large snow and ice cover are the source of fresh water to IGB rivers and their tributaries (Schild, 2008). The glacier coverage in Himalayas is estimated to be about 33,000 km^2, and it provides around 88×10^6 m^3 of water annually (NMSHE, 2010). Glacial melting impacts the water availability and seasonal flow of IGB basins. It provides many ecosystem services to people in the form of hydropower, water for agriculture, and household purposes and aid transport and tourism. Additionally, the biodiversity and global atmospheric circulation are significantly influenced by water resources. However, water-induced natural disasters including flash floods and riverine floods predominate in the Himalayan river basins. Increase in floods, droughts, landslides, endangered species, threat to biodiversity, food security, and people's livelihood are some of the impacts of climate change in the Himalayas (Xu et al., 2009; Mir et al., 2019). Climate change thereby affects the region socioeconomically, ecologically, and culturally. Although the Himalayas are hotspot for natural disasters due to frequent seismic activity and tectonic features like faults and thrusts, the frequency and severity of catastrophic disasters have grown due to recent climatic changes, posing a serious threat to people and their way of life, agriculture, infrastructure, and ecosystem. Understanding the behavior of IGB basins to climate change will help in studies related to disaster risk assessment and sustainable management.

Increase in air temperature, specific humidity, ocean heat, sea level rise, sea-surface temperature, decrease in extent of snow cover and glacier volume, and sea ice are some of the general climate change indicators (NOAA, 2009). Variations in temperature, precipitation patterns, water availability of IGB basins, fluctuations in streamflow, and extent and volume of glaciers are considered to be some of the important climate change indicators in the Himalayan region (IPCC, 2013).

In recent times, the IGB river basins have been affected by rising temperature, variation in precipitation trends, glacial retreat, and changes in runoff over time and space. Change in the strength and timing of the Monsson, winter westerlies, and high-pressure systems impact river flows and groundwater recharge, leading to natural hazards and damage to ecosystem and livelihood of people (ICIMOD, 2010). Reduction in water storage capacity due to decrease in the snow and ice is a major concern here (ICIMOD, 2010). However, the impacts may vary basin-wise in terms of rate, intensity, and direction. Rising temperature might affect the high altitude areas more than the low latitude areas as the rising temperature causes the melting of snow and ice in the region (Shrestha et al., 1999).

Despite the observed changes in the monsoon precipitation and temperature, the impact of future climate change on the hydrological regime of major river basins in the Himalayan region is not well studied (Gosain et al., 2011). One of the important challenges in understanding the hydrological response of IGB river basins is lack of availability of sufficient data on these basins as it remains classified (Mishra & Lilhare, 2016). Dam construction and river water diversion also complicate understanding climate change-induced hydrological variation in the river system. IGB river basins being one of the major basins with a lot of rivers in the Himalayan region are crucial to assess how susceptible their hydrological systems are to climate change and comprehend how these river systems are expected to react in the future.

2 Regional Setting of the Himalayan River System

The Indus Ganga and Brahmaputra (IGB) river systems broadly constitute the Himalayan river system. It supports ~700 million people in Asia (Nepal & Shrestha, 2015) and covers >50% area of the country (Khan et al., 2015). The water resource in Indus, Ganges, and Brahmaputra (IGB) river systems are the major source of water for drinking, irrigation, navigation, industry, and hydropower purposes (Mirza et al., 2003). The basin extends between 21°–37° N latitude and 66°–97° E longitude and covers India, Bangladesh, Pakistan, Nepal, and Bhutan. The Indus river flows into the Arabian Sea forming a delta of about 41,440 km^2 and Ganga river along with Brahmaputra falls into the Bay of Bengal (Patel et al., 2021). Distribution of these basins covers a large geomorphic area covering high elevated mountain areas, plains, and sea level. Indus and Brahmaputra river systems have a trans-Himalayan origin, and Ganga river system originates from the Himalayas. Indus river originates from Lake Ngangla rainfall-runoff (China); Ganga originates from Gangotri Glacier, Uttarakhand (India); and Brahmaputra river originates from Angsi Glacier (China) (India-WRIS, 2012). Jhelum, Ravi, Chenab, and Sutlej are the main tributaries of Indus; Yamuna, Rama Ganges, Gomti, Ghagra, Son, Gandak, Burhi Gandak, Koshi, and Mahananda are the major tributaries of Ganga; and the important tributaries of Brahmaputra are Dibang, Lohit, Dhansiri, Kolong, Kameng, Manas, Raidak, Jaldhaka, Teesta, and Subansiri (Shrestha et al., 2015).

The overall discharge from the IGB basins is affected differently by snowfall, glacial melt, surface runoff, and groundwater. Mean discharge of Brahmaputra river is higher than the Ganges, whereas Indus has the lowest mean discharge (Table 3.1). The total river discharge in the upper Indus basins is constituted by glacier meltwater

Table 3.1 Key geographical and hydrological features of the Indus-Ganga-Brahmaputra basins

Basin	India	Area (km^2)	Length (Km)	Annual discharge (m^3/s)	Annual basin precipitation (mm)	Glaciated area (Km^2)	Mean discharge (m^3/s)
Indus	39%	1,120,000	3180	5533	434	21,193	5533
Ganges	79%	1,087,300	2515	12,037	1094	9012	18,691
Brahmaputra	36%	543,400	2840	21,261	2143	14,020	19,824
Reference		Nepal & Shrestha et al. (2015) and references therein			Khan et al. (2015)	ICIMOD report (2005)	Mirza (2004)

(~41%), snowmelt (~22%), and rainfall runoff (~27%). However, the Ganga and Brahmaputra basin's streamflow is dominated by rainfall runoff (66% and 59%, respectively), followed by meltwater contribution of approximately 20% and 25% to the total runoff, respectively (Shrestha et al., 2015). Though the relative contribution of these sources is debatable, surface runoff controlled by precipitation including snowfall is the major contributor to the total discharge of the basin (Maurya et al., 2011). In Indus basins, the annual rainfall is higher in the mountain region (up to 2000 mm), and low lands experience a rainfall ranging from 100 mm to 500 mm (Aquastat, 2011). This is an important factor that contributes to mild weather in the southern parts and snowfall in the northern parts (Nepal & Shrestha, 2015).

The total discharge of the basin is the main source of water downstream, where only ~40% of water available can be utilized due to topographical constraints and uneven distribution of water resources (CWC, 2015). IGB river supports the agricultural sector in a vast geographical area; further, Indus basins alone constitute the largest irrigation network in the world, where the water is regulated by Tarbela Dam on the Indus river and Mangla Dam on the Jhelum (Laghari et al., 2012). About 144,900 ha of land is irrigated by Indus, 156,300 ha is irrigated by Ganges basins, and 6000 ha of land is irrigated in the Brahmaputra basins (Immerzeel et al., 2010) (Fig. 3.1).

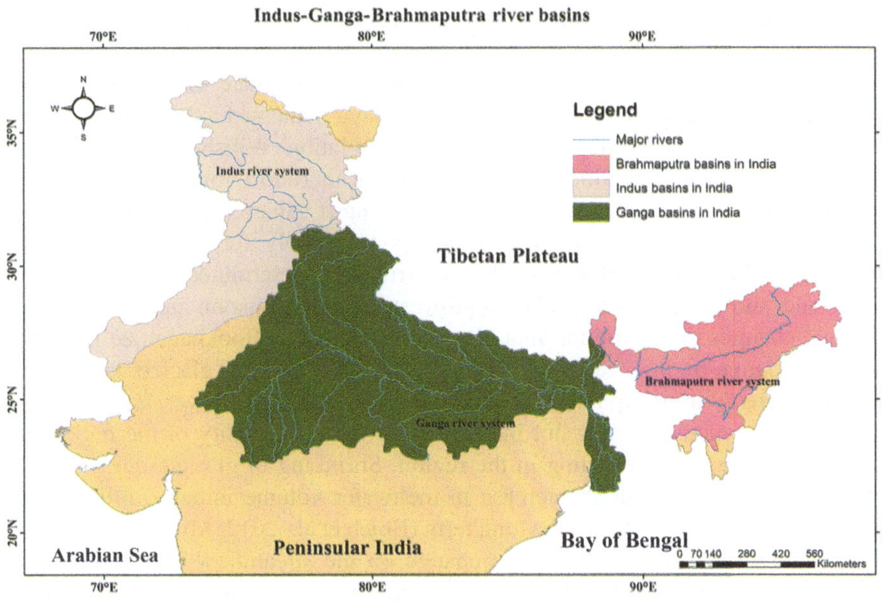

Fig. 3.1 Indus-Ganga-Brahmaputra river basins. (Modified after Patel et al., 2021)

3 Impact of Climate Change on Indus-Ganga-Brahmaputra River Basins

Temperature, precipitation, the amount of glacier/snow, and its melt are the common components studied to assess the impact of climate change on IGB river basins. Variations in precipitation, fluctuations in the amount of glacier/snow along with temperature changes, and seasonal extremes affect the life and chemistry of river basin (Bolch et al., 2012; Lutz et al., 2014). As temperatures rise, evapotranspiration increases, and glacial storage declines. This results in an increased glacier melt in short term and decreased amount of glacier melts in long term (Bolch et al., 2012; Milner et al., 2017). Fluctuating glacier volume leads to extreme weather events such as floods, droughts, flash floods, landslides, etc. (Nepal & Shrestha, 2015). Extreme weather conditions have become more common in India in recent years. For example, historical flood records from Alaknanda river in Uttarakhand, India, show an increase in extreme flood events (Goswami et al., 2006).

Himalayas being geologically dynamic, steep terrain with high seismicity and major thrust plains is highly vulnerable to natural hazards. Moderate to large magnitude earthquakes cause rock and ice avalanches and landslides, one of the major natural hazards of the region (Vaidya et al., 2019). Heavy rainfall with cloud bursts also activates landslides and debris flow. Unsustainable development and overuse of resources have a negative impact on slope stability and raise the risk of landslides. Sudden and unpredictable climatic change has made mountain ecosystems more susceptible to natural disasters, and it is one of the significant effects of climate change on the Himalayan river basins.

Depending on the basin location, the impact of climate change on glacier response and water supply varies. Eastern Himalaya witnessed a decrease in glacier volume due to decreased snowfall and increased ablation (Wiltshire, 2014). Continuous decrease in glacier volume causes increase in runoff for short term and a decrease in upstream water supply and annual runoff (Sharma et al., 2000). There can be up to 17% reduction in upstream water supply for Ganga river alone (Immerzeel et al., 2010). However, the river runoff is determined by both precipitation and snow/glacier melt. The southwest Indian Monsoon intensity and its seasonal changes determine the majority of precipitation (Bookhagen & Burbank, 2006). In Ganga basins, the reduced runoff due to decrease in glaciers is compensated by increase in precipitation (Immerzeel et al., 2013). However, an increase in rainfall during monsoon may not increase the water availability of the region but may result in increased flooding in the region. Shrinking of glaciers due to rise in temperature may result in an increase in meltwater volume initially, followed by water scarcity in some rivers in the long term (Bolch et al., 2012; Milner et al., 2017). Understanding the effects of climate change on the streamflow in the IGB river system is difficult, because of the spatial and temporal unpredictability of these changes, as well as practical constraints and a lack of monitoring of changes in river basins. As temperature and precipitation are primary driving factors, these data can be used for climate models to know the prospective impacts of climate change on the water supply in a river basin.

Temperature Even though the extreme topography and complex response of the Himalayas to greenhouse effect make climate change projections difficult, climate models predict that by 2100, the Indian subcontinent will heat up and the temperature will rise between 3.5 and 5.5 °C (Kumar et al., 2006). The glaciers in the Himalayas are retreating faster than the world average glacier retreat rate (Dyurgerov & Meier, 2005). Global warming is one of the important reasons for this. Global warming causes snowmelt to occur earlier causing shorter winter, and a large proportion of precipitation fall as rain (ICIMOD, 2010). This shows the interrelationship between temperature, glaciers, and runoff. Climate models based on different climate change scenario suggest that as a result of global warming, by 2050, 35% of the glaciers will retreat, and there will be a sudden increase in runoff between 2030 and 2050 (Qin, 2002). This might result in disappearance of smaller glaciers also. Increase in temperature can also affect the aquatic ecosystem by disturbing the environment suitable for their survival.

Precipitation Precipitation trends over the IGB basins have varied spatially and over time. Time period from 1998 to 2007 shows a decreasing trend in annual rainfall over IGB river basins where the mean annual rainfall has a heterogenous pattern at different locations (Patel et al., 2021). Interannual and interseasonal increase and decrease have been reported from the Himalayan region. Decreasing trend of both annual and seasonal rainfall is observed in the eastern Himalayan region (Brahmaputra basins), whereas Indus basins in the western region show an increasing trend. However, the middle part of the IGB basins shows complex pattern as both Indian summer Monsoon and westerlies contribute differently. This fluctuation in temperature and rainfall has an impact on the water storage in this region, causing water stress over time. Rainfall pattern and temperature can be correlated to the runoff of IGB basins (Patel et al., 2021). The intensity, amount, and distribution of total precipitation determine the river runoff and impact the downstream flow of river. Over the last years, overall rainfall in Indus basins shows an increasing trend, whereas Brahmaputra has a decreasing trend, and the Ganga basins experience both increasing and decreasing rainfall trend (Patel et al., 2021) (Table 3.2).

4 Impact of Future Climate Change on Himalayan River Basins

Himalayan region is very sensitive to climate change, which experiences more warming than the global average, very high variability in precipitation with increased frequency of precipitation events, and shrinking of glaciers (ICIMOD, 2010). Climate change might greatly impact the water supply of the IGB river basins causing potential threat to food security, energy security, environmental quality, livelihoods, and quality of life of people (Dahal et al., 2020). There is an increased occurrence of climate extremes in the past years, and it is likely to be more prominent in the future (Mishra et al., 2015; Easterling et al., 2000). One of the projected effects

Table 3.2 Overview of impact of climate change on water resources

Climate	Trend	Impacts on IGB river basins	Risk associated	Sustainable practices
Rainfall	Increase in rainfall intensity	Increased runoff with increased sediment load, flooding, and reduced ground water discharge	Groundwater exploitation, affect agriculture, human settlements, and infrastructure	Disaster preparedness and management, groundwater management, proper planning on urban drainage system and land use planning, soil and water conservation
	Decrease in rainfall intensity	Drying up of small tributaries and springs, reduction in seasonal flow of water, drought	Groundwater exploitation	Water harvesting
	Variation in rainfall pattern	Unpredictable river flow patterns	Affect agriculture	Efficient management of irrigation and water supply system
Temperature	Increase in temperature	Increase in temperature of water resources	Changes or imbalance in aquatic ecosystem	Proper monitoring of temperature changes
Glacier volume	Decrease in glacier volume	Unpredictable river flow patterns	Affects perennial water sources	Efficient management of irrigation and water supply system

Source: Climate change adaptation in Himachal Pradesh; Sustainable strategies for water resources (2010)

of climate change is the decrease in availability and access to freshwater by 2050 (ICIMOD, 2010). The "National Commission for Integrated Water Resources Development" (NCIWRD) reports that by the year 2050, the demand for water is expected to be 973–1180 billion cubic meters (BCM). Agriculture sector will dominate the demand share (70%) followed by households (9%) and industries (7%) (NCIWRD, 1999). The mismatch between the present availability and projected demand for water in the coming years explains the need to study the major river basins and factors controlling their discharge and/ recharge. Regular water scarcity and flooding, deteriorating water quality, and excessive dependence on groundwater in past years have increased the concern that water resources may be vulnerable to global climate change (IPCC, 2007). The hydrological regime of IGB can be affected by factors such as seasonal changes in rainfall onset, its intensity, amount, and duration causing fluctuations in the precipitation pattern over IGB basins (Anja du, 2019; Berger et al., 2019). Rising temperature has resulted in overall increase of total runoff in IGB basins, where the runoff changed by −5% to +12% for Upper Indus, 1–27% for Upper Ganga, and 0–13% for the Upper Brahmaputra basins by 2050 (Lutz et al., 2014). Increase in runoff also predicts the possibility of future flood events (Gain et al., 2011). Global circulation models (GCMs) predict an increase in mean peak discharge in Brahmaputra river (Mirza, 2002), and along with the precipitation, accelerated melt runoff causes increase in runoff up to 2050 (Lutz et al., 2014). There will be ~20% reduction in upstream water supply in Brahmaputra river, and the glacier ice melt will accelerate up to 2040, followed by a decrease (Immerzeel et al., 2010; Prasch et al., 2011). Overall, the river basins experience a decrease in the amount of water availability in upstream, higher peak flow, and reduced volume and area of glaciers and snow. Along with surface runoff, soil erosion and sediment yield of a river basin also get affected by climate change (Lilhare et al., 2015). According to Lutz et al. (2014), the Indus basin's annual discharge may rise by 7–12% by 2050 as a result of increased precipitation and rapid melt. Reduction in snowfall and a decrease in annual snowmelt runoff were associated with an increase in temperature, and a decrease in glacier area might decrease the river discharge drastically (Singh & Bengtsson, 2004; Akhtar et al., 2008). Projected impact of climate change on the hydrology of Indus basins shows that by 2100, glacier area might reduce by 33% and glacier volume by 50%, and there will be glacier melt peak in 2044 or 2065 followed by a decline in glacier volume (Immerzeel et al., 2013).

5 Risk Assessment

Natural hazards like flash floods and landslides are caused by an increase in the frequency of high-intensity rainfall (ICIMOD, 2002), which adversely affects human settlements. River runoff, which is dependent on the total precipitation, variation in intensity, and quantity and type of precipitation, will disturb the high latitude wet lands, water flow, and sediment transport in the rivers. Disaster risk assessment is an interplay between hazard, exposure, and vulnerability (Wester et al., 2019). Risk assessment starts with awareness about risks such as the intensity of existing and anticipated risk. It is important to identify, quantify, and characterize the hazards toward people's life and the environment for disaster risk assessment. This information forms the base of practices and policies related to disaster risk reduction. While risk analysis has both political and scientific dimensions, public perception on risk also has an important role in disaster risk analysis (Slovic, 1999). Impact of climate change on the hydrologic sensitivity of rivers can be monitored through precipitation trends, temperature changes, surface runoff data, rate of evapotranspiration, and streamflow data (Mishra & Lilhare, 2016). The Himalayan mountain system is very susceptible to global warming (Bandyopadhyay & Gyawali, 1994). Even though climate change is affecting the whole IGB basins, the effect is not uniformly distributed. Since the Western Himalayas are dominated by the westerlies and the eastern end is dominated by the southwest monsoon, global warming affects differently from west to east of the Himalayas. Each basin has a different and /unique response to climate change; hence, its impact varies within a basin from upstream to downstream and also in different catchment areas.

Monitoring of climatic variables such as temperature and rainfall and their influence on water availability over the IGB basins shows a decrease in terrestrial water storage (-12.6 mm year^{-1}) with increase in mean annual temperature (0.02 °C year^{-1}), and there is a positive correlation between terrestrial water storage ($r = 15.8$) (Patel et al., 2021). Variations in temperature and precipitation cause water stress over different part of IGB basins. Glaciers in Ganga river basins are sensitive to global warming; the contribution of glacier to water resources in the Indus basins is higher than that in the Ganges Basins (Immerzeel et al., 2012). Sensitivity analysis in Ganges Basins shows that the effect of warming may be higher in the northern, upstream region of Ganga river basins due to the presence of snow and glaciers (Mishra & Lilhare, 2016). Hence, the overall streamflow in the basin is controlled by the monsoon system. This shows that even within the Ganges Basins, the risk associated with future climate change has to be done locally to develop a robust mitigation program. Reduced river flow adversely affects the agricultural production, hydroenergy generation and agricultural production, and physical infrastructure of the area (Mirza 2011; Rupper et al., 2012).

Glacier retreat can destabilize the surrounding slopes causing landslides (Dadson & Church, 2005). Outbreak floods, flash floods, and debris flows are also some of the consequences of glacier retreat and excessive rain. Excessive meltwater formed as a result of glacial melt formed glacial lakes in the Eastern and central Himalayas. Most of these high-altitude lakes are moraine-dammed, which are weak and prone to sudden breach (ICIMOD, 2010). Floods due to this glacial outburst can cause serious damage to life, property, and ecosystem downstream. However, the impacts of glacial retreat are pronounced in basins, which are highly glaciated and receive less precipitation (Rees & Collins, 2006). Warming climate is the major reason for precipitation extremes. Human-induced global warming also contributes to the intensification of extreme precipitation events (Min et al., 2011).

Global warming intensifies the extreme precipitation events causing variability in occurrences of droughts and floods (Sheffield & Wood, 2008) and hazards like landslides. Large-scale landslides can lead to damming of river flow upstream resulting in the formation of landslide dam, which can subsequently result in outburst flood (Reynolds, 2014). Intense rainfall or/and breaching of natural dams cause flash floods. Though riverine floods and flash floods are common in IGB river basins, flash floods often occur rapidly without giving much time to provide warning.

Risk associated with natural hazards can be assessed in different dimensions such as physical, social, environmental, and economic dimensions. Damage to physical assets such as infrastructure, suffering of human life, and economic loss come under physical, social, and economic dimensions, respectively. While natural hazards disrupt the balance of ecosystem, poor environmental management, overexploitation of natural resources, ecosystem degradation, and climate change make people more susceptible to disasters and cause more damage (ISDR & UNEP, 2007). Climate change increases the vulnerability to disaster and leads to greater destruction. Environmental degradation can also bring on anthropogenic interventions like massive construction projects. Extreme temperature changes such as heat wave and cold wave itself are hazardous to human and natural systems and can be classified as climatological hazards (IPCC, 2012).

An important part of risk assessment is identifying the triggers for natural hazards and assessing its exposure to formulate a disaster risk management strategy according to geographical and geopolitical scales (Reynolds, 2014). Risk assessment is reliable when it is used to support practical risk management, the gathering of historical damage data to support evidence-based risk management, land use and urban planning, risk-informed decision, and policy making. Regular risk assessment can reduce the vulnerability to natural hazards. Lack of risk assessment, unplanned and unscientific developmental works, and lack of disaster mitigation programs worsen the impact of disaster. Sharing information related to scientific and indigenous knowledge, infrastructure, institutions, and insurance can enhance resilience, and government needs to adopt a regionally standardized, multihazard risk assessment approach (Wester et al., 2019). It is crucial to include watershed features such as land cover and land use, hydrological parameter, soil type, landscape, anthropogenic activities, and agricultural practices while assessing the implications of future climate change. Coupling the principles of climate change, hydrology, and cryosphere science with economic analysis might help in identifying economic impacts of climate change, which is one of the important risks associated with climate change (Mishra et al., 2020). Though these areas have been reported to have experienced increase in the frequency of glacial lake outburst floods, the distinction between climate-induced events and anthropogenic influences is still not clear (Sud et al., 2015).

6 Sustainable Practices, Adaptation Strategies, and Policies

Considering the long-term impact, risk, and vulnerability of climate change on societies, it calls for more planned adaptations along the river basins. This includes strengthening of the local knowledge, practices, and functioning of the institutions, which are trying to cope with the climate change. Studies show that there are many sustainable adaptation methods that have been practicing in the Himalayan region. In the early 1960s, local villagers residing near the upper Ganga region collectively prepared a traditional recharge structure "Chahal" to maintain the springs of the region. In the downstream area, villagers are mostly cultivating potato, maize, chilly, and pumpkin on sandy soil, which is a popular adaptation practice. Sharing livestock or Adhi system is also an important adaptation mechanism followed by poor people in this region. In Indus basins, people use drip irrigation and sprinkler irrigation along with several off-farm activities, which are very effective adaptation strategies. The spring shed management by Sikkim people is also a good example of adaptation according to climate changes (Dilshad et al., 2019). Toward efficient planning of climate change adaptation, a holistic understanding of the different components of the river systems focusing on sustainable management alternatives for ecosystems and healthy rivers is essential. Impacts of climate change at all three levels (local, downstream, and global level) should be thoroughly studied in connection to each other, and accordingly, adaptive strategies should be developed from each of these perspectives. Local adaptive strategies can be planned by taking into account of local effects, whereas adaptation at downstream level needs to evaluate the downstream effects and model accordingly. Adaptive methods should be based on future projections, on the global level, and on the possible feedback mechanism of the environmental changes in the Himalayas to global warming (Eriksson et al., 2009). It would be quite useful if the trends of all possible hydrological and hydrometeorological variables are assessed for concurrent years to draw meaningful inferences and cause-effect relationships between key variables, and based on the inferences, adaptive strategies should be designed. Since the Himalayan basins are international basins and some of their data are in classified domain, integration of all the relevant information for effective planning and management of climate change adaptation for this region would require a transboundary approach and multilateral collaboration across the regions. Policies for such river basin should be based on an interdisciplinary method and an integrated governance policy.

To facilitate such policy integration, climate change adaptation should be mainstreamed into integrated river basin plans rather than treated as an isolated activity. Successful river basin management requires participatory and equitable decision-making and an effective coordinating institution with adequate authority and mandate to act. It should be accountable to all relevant stakeholder groups (Oglethorpe et al., 2015) and should be able to coordinate across geographical boundaries. Policies should be coordinated to avoid contradictions and overlaps among prevailing plans that already exist. The need of putting science-based policies in the conservation of the whole riverine ecosystems is important. The policy interventions should be based on an integrated knowledge base. The key to sustainability and sustainable management of resources for climate change adaptation lies in effective utilization of multidisciplinary scientific knowledge. As a starting point for policymaking, flows in rivers need to be studied in a holistic manner. Volume of flow per unit time, its constituents and composition, energy content of the river flow, biodiversity, and sediment budget are a few parameters that need to be considered other than simple hydraulic quantities. The occurrence of floods in the river basins should be duly considered at the time of policy development. The longitudinal, lateral, and vertical river flows must be integrated and should be the basis of any new policy related to this area. A holistic analysis of the upstream-downstream properties in the IHR river basins is important in planning for watershed management and cost assessment for ecosystem services. Basin-level governance should be a priority during policymaking, which urgently needs to be put in practice. Plans for adaptation should include built-in flexibility and scenario planning that takes future uncertainties into account. Water management strategies need to be developed to utilize the predicted water availability during monsoon season in future, and it is crucial to have policies and practices that can manage the uncertainty in precipitation pattern. In light of growing climatic unpredictability, uncertainty, and extreme events, coordinating institutions must be adaptable enough to allocate water and natural resources while also being able to adopt novel strategies as and when necessary. Policy and institutional solutions must adequately address the structural and underlying causes of climate vulnerability. By establishing monitoring programs for snow, ice, and water, downscaling climate models, using hydrological models to forecast water availability in various regions and localities, and creating basin-wide scenarios that take socioeconomic development and water demand into account, the related knowledge gap can be reduced (Eriksson et al., 2009). Policy pushes to research on developing informed data instruments is essential in this context.

Water from rivers is used by a variety of parties and sectors as it flows from upstream to downstream. Climate change-related risks like landslides, floods, and droughts would affect significantly those individuals who reside within the basin area and/or live further downstream. The unique needs and goals of these vulnerable groups should therefore be taken into account in adaptation policies and programs. An integrated management system along the basins with the goals of catastrophe risk management, resilience building, and fair benefit sharing is crucial for the long-term sustainability of the river ecosystem and its communities. In this regard, managing urban and industrial growth along the basin areas with respect to water supplies, efficient use of catchment area, discharge management, identifying key environmental assets (protected areas, critical habitats such as wetlands and floodplains, sensitive ecosystems) and some of the crucial elements in basin planning to enable sustainable management, which must be included in IHR river basin policy, include identifying the special demands of basin populations and appreciating the value of the eco-services they give to people and their livelihoods (Eriksson et al., 2009; Dilshad et al., 2019; Pradhan et al., 2021). Effective public participation and regional and cross-border cooperation hold the key to successful policy implementation. To do this, cross-sectoral coordination between the government, local communities, community-based organizations, NGOs, and research agencies is required in the formulation of adaptation and livelihood measures. Effective utilization of traditional knowledge locally can significantly reduce community's vulnerability and will contribute socially inclusive economic development as per Sustainable Development Goals. One such strategy is integrated water resources management (IWRM), which encourages the coordinated development and management of water, land, and related resources in order to maximize the resulting economic and social benefits in a fair way without risking the sustainability of crucial ecosystems (Global Water Partnership, 2000). This might be helpful in creating adaptive management strategies across geographic ranges in the Himalayan region by deliberately focusing on the connections between upstream and downstream regions at the macro (river basin), meso (catchment), and micro (local) scales (Pradhan et al., 2021). Recognizing the climate change threat to these basins, different countries have adopted policy measures in their respective government as National Communications and National Adaptation Programs of Action. These policies place a strong emphasis on utilizing the human and infrastructure resources at hand to increase people's capacity for adaptation and decrease vulnerabilities (Sud et al., 2015).

7 Climate Change Adaptation Policies Associated with IGB River Basins

IGB river basins have a geographical spread in India, Bangladesh, Pakistan, Nepal, and Bhutan. These countries have formulated regulations, organizations, and networks to achieve climate change policies. India has National Ganga River Basin Authority (NGRBA) and Brahmaputra River Valley Authority (BRVA) focused on integrated water resource development and Draft National Water Policy 2012, which proposes a legal framework for water governance, giving emphasis on climate change impacts (Sud et al., 2015 and references therein). India's first National Action Plan on Climate Change (NAPCC) was released on June 2008. In this action plan, policies and programs related to climate mitigation and adaptation were addressed. The national action plan has been divided into eight national missions, namely, (1) National Solar Mission, (2) National Mission for Enhanced Energy Efficiency, (3) National Mission on Sustainable Habitat, (4) National Water Mission, (5) National Mission for Sustaining the Himalayan Ecosystem (NMSHE), (6) National Mission for a "Green India," (7) National Mission for Sustainable Agriculture, and (8) National Mission on Strategic Knowledge for Climate Change. There are two national missions, NMSHE and "National Water Mission" under NAPCC specifically deal with the Himalayan region and river climate. NMSHE targets the sustenance of the Himalayan ecosystem by developing different task forces dedicated for different themes such as natural and geological wealth, water, ice, snow resources, forest resources and plant biodiversity, microflora and fauna, wildlife and animal population, Himalayan agriculture, and traditional knowledge system of the Himalayas. These task forces concentrate on establishing databases, design monitoring systems, and do modelling and simulation and apply it in vulnerability assessment and policy research.

Under the National Mission on Himalayan Studies (NMHS), water resource management is a broad thematic area covering the states of Jammu and Kashmir, Himachal Pradesh, Uttarakhand, Arunachal Pradesh, Sikkim, Manipur, and Tripura. The Ministry of Environment, Forest and Climate Change (MoEF & CC) implemented this mission and have nearly 25 ongoing projects targeted on water resources in the Indian Himalayan Region. The objectives of projects range from spring rejuvenation in different states to development of low-cost water purification systems. Understanding climate change scenarios up to year 2100 in IHR states, groundwater augmentation, quantification of hydrological processes, policy recommendation on land use practices and water use, capacity, and awareness building of stakeholders are some of the overall objectives of the mission (NMSHE). The Mission targets to develop inventories of mountain spring in selected states of IHR, monitor and model prediction of water-induced disasters, and monitor water quality and ecological integrity of selected springs, lakes, and rivers. Under the mission, a decision support system is developed to support water demand in Jammu and Kashmir and Uttarakhand, low-cost landslide early warning system in

Uttarakhand, and natural hazard prediction models in Uttarakhand, and with the help of geospatial techniques and hydrological models, hydropower potential zones in Himachal Pradesh have been identified.

Similarly, Bangladesh has instituted different centers and national committee such as National Steering Committee on Climate change, Flood Forecasting and Warning Centre (FFWC), and National Disaster Management Council (NDMC), to address the issue related to water resources and impact of climate changes. In Afghanistan, National Climate Change Committee and Supreme Council for Water Resources Management coordinate and function toward the steering and coordination of water resource management. A small country like Bhutan has a government body called "Bhutan Water partnership" (BhWP), which is to coordinate programs related to water resource protection and development, and committee like "National Climate Change Committee," a technical-level task force for project implementation. In Pakistan, Indus River System Authority (IRSA) and Prime Minister's Committee on Climate Change focus on knowledge generation and management. Muti-Stakeholder Climate Change Initiatives Coordination Committee (MCCICC) in Nepal coordinates among government agencies and stakeholders to work toward climate change.

8 Conclusion

Urbanization, agriculture expansion, population explosion, and recent industrial and economic developments have affected the health of water resources. Changes in land use and cropping pattern and overexploitation of groundwater along with recent climate changes have significantly affected the freshwater resources in India. Himalayan river systems are reportedly affected by recent climate changes. Rapid reduction in glaciers and associated changes in the downstream of Himalayan river basins indicate the effect of climate change in the Himalayan river basins. Unpredicted and accidental disasters in the mountain region is one of the consequences of these changes. Hence, mountain settlements are more susceptible to natural disasters like earthquakes, floods, landslides, avalanches, and severe weather. The availability and accessibility of water resources in the downstream area are significantly impacted by climate change indicators like rising temperatures, abrupt changes in precipitation, glacial retreat, and related risks. However, due to complex topography of the Himalayas and lack of sufficient snow, ice, and water monitoring system, the information about impact of short- and long-term climate change on water resources in the Himalayan region is limited. Since climate change-induced natural hazards are difficult to predict, it is important to have more hydrological units to gauge the river flow, precipitation, and climate models required to be developed. These models need to be evaluated and updated regularly to predict the effect of climate change on Himalayan river basins. Society must adopt sustainable practices as part of disaster preparedness. Sharing meteorological information regionally and seasonal migration and restricting construction activities in landslide areas may reduce the risk associated. Even though there are several studies conducted on IHR, the policy development for climate change adaptation needs to be dealt with in a comprehensive manner and for different levels such as a general policy for the entire IHR river basin and also as separate policies for each basin.

Acknowledgments The authors are thankful to the Vice-Chancellor, Central University of Punjab, for his administrative support for this work. K. Amrutha would like to acknowledge DST for providing INSPIRE fellowship toward a PhD.

References

Akhtar, M., Ahmad, N., & Booij, M. J. (2008). The impact of climate change on the water resources of Hindukush–Karakorum–Himalaya region under different glacier coverage scenarios. *Journal of Hydrology, 355*(1–4), 148–163.

Anja du, P. (2019). *Climate change and freshwater resources: Current observations, impacts, vulnerabilities and future risks*. Springer. https://doi.org/10.1007/978-3-030-03186-2

Aquastat, F. A. O. (2011). *FAO's information system on water and agriculture*. Food and Agriculture Organization of the United Nations (FAO).

Bandyopadhyay, J., & Gyawali, D. (1994). Himalayan water resources: Ecological and political aspects of management. *Mountain Research and Development, 14*, 1–24.

Berger, S., Bliefernicht, J., Linstädter, A., Canak, K., Guug, S., Heinzeller, D., Hingerl, L., Mauder, M., Neidl, F., Quansah, E., & Salack, S. (2019). The impact of rain events on CO_2 emissions from contrasting land use systems in semi-arid West African savannas. *Science of the Total Environment, 647*, 1478–1489.

Bolch, T., Kulkarni, A., Kääb, A., Huggel, C., Paul, F., Cogley, J. G., Frey, H., Kargel, J. S., Fujita, K., Scheel, M., & Bajracharya, S. (2012). The state and fate of Himalayan glaciers. *Science, 336*(6079), 310–314.

Bookhagen, B., & Burbank, D. W. (2006). Topography, relief, and TRMM-derived rainfall variations along the Himalaya. *Geophysical Research Letters, 33*(8).

Climate change adaptation in Himachal Pradesh: Sustainable strategies for water resources 2010. Asian Development Bank.

CWC. (2015). *Central Water Commission, Annual Report 2015*. Government of India Ministry of Water Resources, River Development & Ganga Rejuvenation.

Dadson, S. J., & Church, M. (2005). Postglacial topographic evolution of glaciated valleys: A stochastic landscape evolution model. *Earth Surface Processes and Landforms: The Journal of the British Geomorphological Research Group, 30*(11), 1387–1403.

Dahal, P., Shrestha, M. L., Panthi, J., & Pradhananga, D. (2020). Modeling the future impacts of climate change on water availability in the Karnali River Basin of Nepal Himalaya. *Environmental Research, 185*, 109430.

Dilshad, T., Mallick, D., Udas, P. B., Goodrich, C. G., Prakash, A., Gorti, G., Bhadwal, S., Anwar, M. Z., Khandekar, N., Hassan, S. T., & Habib, N. (2019). Growing social vulnerability in the river basins: Evidence from the Hindu Kush Himalaya (HKH) region. *Environmental Development, 31*, 19–33.

Dyurgerov, M. B., & Meier, M. F. (2005). *Glaciers and the changing Earth system: A 2004 snapshot* (Vol. 58). Institute of Arctic and Alpine Research, University of Colorado.

Easterling, D. R., Evans, J. L., Groisman, P. Y., Karl, T. R., Kunkel, K. E., & Ambenje, P. (2000). Observed variability and trends in extreme climate events: A brief review. *Bulletin of the American Meteorological Society, 81*(3), 417–426.

Eriksson, M., Jianchu, X., Shrestha, A. B., Vaidya, R. A., Nepal, S., & Sandström, K. (2009). *The changing Himalayas – Impact of climate change on water resources and livelihoods in the greater Himalayas*. International Centre for Integrated Mountain Development. ISBN 978 92 9115 111 0 (printed) 978 92 9115 114 1 (electronic).

Gain, A. K., Immerzeel, W. W., Sperna Weiland, F. C., & Bierkens, M. F. P. (2011). Impact of climate change on the stream flow of the lower Brahmaputra: Trends in high and low flows based on discharge-weighted ensemble modelling. *Hydrology and Earth System Sciences, 15*(5), 1537–1545.

Gosain, A. K., Rao, S., & Mani, A. (2011). *Hydrological modelling: A case study of the Kosi Himalayan basin using SWAT*. CABI Publishing.

Goswami, B. N., Venugopal, V., Sengupta, D., Madhusoodanan, M. S., & Xavier, P. K. (2006). Increasing trend of extreme rain events over India in a warming environment. *Science, 314*(5804), 1442–1445.

ICIMOD, (2002). *Hazard and RISK Mapping (Internal Report)*. Kathmandu: Participatory Disaster Management Programme (Nep 99/014) and ICIMOD.

ICIMOD (2005). Reports of APN 2004-03-CMY Project "Inventory of Glaciers and Glacial Lakes and the Identification of Potential Glacial Lake Outburst Floods (GLOFs) Affected by Global Warming in the Mountains of India, Pakistan and China/Tibet Autonomous Region". ICIMOD.

ICIMOD. (2010). *Climate change vulnerability of mountain ecosystems in the Eastern Himalayas*. International Centre for Integrated Mountain Development. 104 p.

Immerzeel, W. W., Van Beek, L. P., & Bierkens, M. F. (2010). Climate change will affect the Asian water towers. *Science, 328*(5984), 1382–1385.

Immerzeel, W. W., van Beek, L. P. H., Konz, M., Shrestha, A. B., Bierkens, M. F. P. (2012). Hydrological response to climate change in a glacierized catchment in the Himalayas. *Climatic Change 10*(3–4), 721–736.

Immerzeel, W. W., Pellicciotti, F., & Bierkens, M. F. P. (2013). Rising river flows throughout the twenty-first century in two Himalayan glacierized watersheds. *Nature Geoscience, 6*(9), 742–745.

India-WRIS. (2012). *River basin atlas of India*. RRSC-West, NRSC, ISRO.

IPCC. (2007). *Summary for policymakers of climate change 2007: The physical science basic. Contribution of working group to the fourth assessment report of the intergovernmental panel on climate change*. Cambridge University Press.

IPCC. (2012). *Managing the risks of extreme events and disasters to advance climate change adaptation. A special report of working groups I and II of the intergovernmental panel on climate change*. Cambridge University Press.

IPCC. (2013). *Climate change. The physical science basis, working group I. Contribution to the fifth assessment report of the intergovernmental panel on climate change, WMO/UNEP*. Cambridge University Press.

ISDR, & UNEP. (2007). *International strategy for disaster reduction and United Nations environment programme annual report*.

Khan, A. A., Pant, N. C., Goswami, A., Lal, R., & Joshi, R. (2015). Critical evaluation and assessment of average annual precipitation in the Indus, the Ganges and the Brahmaputra basins, Northern India. In R. Joshi, K. Kumar, & L. M. S. Palni (Eds.), *Dynamics of climate change and water resources of Northwestern Himalaya* (pp. 67–84). Springer.

Kumar, K. R., Sahai, A. K., Kumar, K. K., Patwardhan, S. K., Mishra, P. K., Revadekar, J. V., Kamala, K., & Pant, G. B. (2006). High-resolution climate change scenarios for India for the 21st century. *Current Science*, 334–345.

Laghari, A. N., Vanham, D., & Rauch, W. (2012). The Indus basin in the framework of current and future water resources management. *Hydrology and Earth System Sciences, 16*(4), 1063–1083.

Lilhare, R., Garg, V., & Nikam, B. R. (2015). Application of GIS-coupled modified MMF model to estimate sediment yield on a watershed scale. *Journal of Hydrologic Engineering, 20*(6), C5014002.

Lutz, A. F., Immerzeel, W. W., Shrestha, A. B., & Bierkens, M. F. P. (2014). Consistent increase in High Asia's runoff due to increasing glacier melt and precipitation. *Nature Climate Change, 4*(7), 587–592.

Maurya, A. S., Shah, M., Deshpande, R. D., Bhardwaj, R. M., Prasad, A., & Gupta, S. K. (2011). Hydrograph separation and precipitation source identification using stable water isotopes and conductivity: River Ganga at Himalayan foothills. *Hydrological Processes, 25*(10), 1521–1530.

Milner, A. M., Khamis, K., Battin, T. J., Brittain, J. E., Barrand, N. E., Füreder, L., et al. (2017). Glacier shrinkage driving global changes in downstream systems. *Proceedings of the National Academy of Sciences, 114*(37), 9770–9778.

Min, S. K., Zhang, X., Zwiers, F. W., & Hegerl, G. C. (2011). Human contribution to more-intense precipitation extremes. *Nature, 470*(7334), 378–381.

Mir, B. H., Kumar, R., & Rather, N. A. (2019). Water resources scenario of Himalayan region. In R. Kumar et al. (Eds.), *Applied agricultural practices for mitigating climate change* (Vol. 2, 1st ed.). CRC Press/Taylor & Francis Group.

Mirza, M. M. Q. (2002). Global warming and changes in the probability of occurrence of floods in Bangladesh and implications. *Global Environmental Change, 12*(2), 127–138.

Mirza MMQ (2011) Climate change, flooding in South Asia and implications. *Regional Environmental Change 11*, S95–S107.

Mirza, M. M. Q. (Ed.). (2004). *The Ganges water diversion: Environmental effects and implications* (Vol. 49). Springer.

Mirza, M. M. Q., Warrick, R. A., & Ericksen, N. J. (2003). The implications of climate change on floods of the Ganges, Brahmaputra and Meghna rivers in Bangladesh. *Climatic Change, 57*(3), 287–318.

Mishra, V., & Lilhare, R. (2016). Hydrologic sensitivity of Indian sub-continental river basins to climate change. *Global and Planetary Change, 139*, 78–96.

Mishra, V., Ganguly, A. R., Nijssen, B., & Lettenmaier, D. P. (2015). Changes in observed climate extremes in global urban areas. *Environmental Research Letters, 10*(2), 024005.

Mishra, V., Shah, H., López, M. R. R., Lobanova, A., & Krysanova, V. (2020). Does comprehensive evaluation of hydrological models influence projected changes of mean and high flows in the Godavari River basin?. *Climatic Change, 163*, 1187–1205.

NCIWRD. (1999). *Integrated water resource development: A plan for action* (Report of the National Commission for Integrated Water Resource Development (NCIWRD)) (Vol. I). Ministry of Water Resources, Government of India.

Nepal, S., & Shrestha, A. B. (2015). Impact of climate change on the hydrological regime of the Indus, Ganges and Brahmaputra River basins: A review of the literature. *International Journal of Water Resources Development, 31*(2), 201–218.

NMSHE. (2010). *National Mission for sustaining the Himalayan ecosystem*. Government of India.

NOAA, U. S. (2009). *National oceanic and atmospheric administration*. NOAA.

Oglethorpe, J., Regmi, S., Bartlett, R., Dongol, B. S., Wikramanayake, E., & Freeman, S. (2015). *The value of a river basin approach in climate adaptation* (p. 57). Organizers.

Patel, A., Goswami, A., Dharpure, J. K., & Thamban, M. (2021). Rainfall variability over the Indus, Ganga, and Brahmaputra River basins: A spatio-temporal characterisation. *Quaternary International, 575*, 280–294.

Pradhan, N. S., Das, P. J., Gupta, N., & Shrestha, A. B. (2021). Sustainable management options for healthy rivers in South Asia: The case of Brahmaputra. *Sustainability, 13*(3), 1087.

Prasch, M., Marke, T., Strasser, U., & Mauser, W. (2011). Large scale integrated hydrological modelling of the impact of climate change on the water balance with DANUBIA. *Advances in Science and Research, 7*(1), 61–70.

Qin, D. H. (2002). Assessment of environment change in western China, 2nd Volume, Prediction of environment change in western China. *Beijing: Science Press, p64, 73*, 115–132.

Rees, H. G., & Collins, D. N. (2006). Regional differences in response of flow in glacier-fed Himalayan rivers to climatic warming. *Hydrological Processes: An International Journal, 20*(10), 2157–2169.

Reynolds, J. M. (2014). Natural disaster preparedness for hydropower projects in high mountain environments. In *HYDRO 2014 conference proceedings, Cernobbio, Italy, 13th–15th October 2014.*

Rupper, S., Schaefer, J. M., Burgener, L. K., Koenig, L. S., Tsering, K., Cook, E. R., (2012). Sensitivity and response of Bhutanese glaciers to atmospheric warming. *Geophysical Research Letters 39*, L19503. https://doi.org/10.1029/2012GL053010

Schild, A. (2008). ICIMOD's position on climate change and mountain systems: The case of the Hindu Kush-Himalayas. *Mountain Research and Development, 28*, 328–331.

Sharma, K. P., Vorosmarty, C. J., & Moore, B. (2000). Sensitivity of the Himalayan hydrology to land-use and climatic changes. *Climatic Change, 47*(1), 117–139.

Sheffield, J., & Wood, E. F. (2008). Projected changes in drought occurrence under future global warming from multi-model, multi-scenario, IPCC AR4 simulations. *Climate Dynamics, 31*(1), 79–105.

Shrestha, A. B., Wake, C. P., Mayewski, P. A., & Dibb, J. E. (1999). Maximum temperature trends in the Himalaya and its vicinity: An analysis based on temperature records from Nepal for the period 1971–94. *Journal of Climate, 12*(9), 2775–2786.

Shrestha, A. B., Agrawal, N. K., Alfthan, B., Bajracharya, S. R., Maréchal, J., & Oort, B. V. (2015). *The Himalayan climate and water atlas: Impact of climate change on water resources in five of Asia's major river basins.*

Singh, P., & Bengtsson, L. (2004). Hydrological sensitivity of a large Himalayan basin to climate change. *Hydrological Processes, 18*(13), 2363–2385.

Slovic, P. (1999). Trust, emotion, sex, politics, and science: Surveying the risk-assessment battlefield. *Risk Analysis, 19*(4), 689–701.

Sud, R., Mishra, A., Varma, N., & Bhadwal, S. (2015). Adaptation policy and practice in densely populated glacier-fed river basins of South Asia: a systematic review. *Regional Environmental Change, 15*, 825–836.

Vaidya, R. A., Shrestha, M. S., Nasab, N., Gurung, D. R., Kozo, N., Pradhan, N. S., & Wasson, R. J. (2019). Disaster risk reduction and building resilience in the Hindu Kush Himalaya. In P. Wester, A. Mishra, A. Mukherji, & A. Shrestha (Eds.), *The Hindu Kush Himalaya assessment* (pp. 389–419). Springer.

Wester, P., Mishra, A., Mukherji, A., & Shrestha, A. B. (2019). *The Hindu Kush Himalaya assessment: Mountains, climate change, sustainability and people* (p. 627). Springer Nature.

Wiltshire, A. J. (2014). Climate change implications for the glaciers of the Hindu Kush, Karakoram and Himalayan region. *The Cryosphere, 8*(3), 941–958.

Xu, J., Grumbine, R. E., Shrestha, A., Eriksson, M., Yang, X., Wang, Y. U. N., & Wilkes, A. (2009). The melting Himalayas: Cascading effects of climate change on water, biodiversity, and livelihoods. *Conservation Biology, 23*(3), 520–530.

Chapter 4
Forestry Policies and Practices to Promote Climate Change Adaptation in the Indian Western Himalayan States

Manoj Kumar

Abstract Indian Western Himalayas (IWH) represent one of the regions of the world where climate change is likely to be rapid and pose a serious threat to the rich biodiversity and ecosystem services. The IWH region is one of the mega-biodiversity centers where many initiatives have been taken to conserve its biodiversity. The region serves as the water catchment area for several rivers, providing water to the downstream states for agriculture and consumption and indirectly feeding and employing a significant portion of the country's population. Climate change brings with it new difficulties to undermine basic ecosystem resilience. The protection and sustainable management of ecosystems of the IWH, on the other hand, may provide a variety of socioecological advantages for long-term adaptation against both present and future climate change. Forests are essential for climate change mitigation efforts; however, they must adapt to climate change to maintain their own life as well as to provide support for the people who rely on them. Adaptation actions are performed to mitigate or minimize the detrimental effects of climate change by taking advantage of good possibilities and building resistance to the negative ones that arise. Forests' ability to adapt to climate change varies based on their geographical location and species. To deal with the problems of climate change, it is necessary to develop an adaptation framework that considers the roles of various sectors in forestry development and management. A framework may be created by combining the perspective of the community with the policies and plans of the government. This chapter illustrates the various existing forestry policies prevalent in the western Himalayan states aiming at climate change adaptation. The findings would be useful for policymakers to further integrate and seek possibilities of synergizing actions to develop a better adaptation framework for the region.

Keywords Adaptation · Forest management · Climate change · Ecosystem services · Biodiversity

M. Kumar (✉)
GIS Centre, Forest Research Institute (FRI), Dehradun, India

S. Sharma et al. (eds.), *Climate Change Adaptation, Risk Management and Sustainable Practices in the Himalaya*,
https://doi.org/10.1007/978-3-031-24659-3_4

1 Introduction

There are just a few regions on Earth where climate change might be as fast as the Himalayan region. Climate change would impact biodiversity, ecological services, and human well-being in this region much significantly (Chaturvedi et al., 2011). The Indian Western Himalaya (IWH) is a mega-biodiversity center and has been selected for several biodiversity conservation projects (IIRS, 2002; NMSHE, 2010) and forestry-based measures to combat the impacts of climate change. The region is the water catchment area for many rivers that provide water to the downstream states for irrigation and personal as well as industrial consumption (MoEF, 2010). The main consequences seen in the area include floods in snow-fed rivers, glacier retreat, water unpredictability, and soil, forest, and grassland degradation (Pachauri et al., 2014). The region is highly vulnerable to climate change and is facing the increasing incidences of forest fires, drying springs, loss of biodiversity, forest land diversion, increasing human-wildlife conflict, and illegal trade of forest products (Kumar et al., 2021a).

Adaptation is defined as "adjustment in natural or anthropogenic systems in response to definite or expected climatic stimuli or their effects, which moderate destruction or exploit advantageous opportunities" (IPCC, 2007). Adaptation measures are taken to nullify or reduce the adverse impact of climate change by exploiting beneficial opportunities and developing immunity against the negative ones. Modulation for sustenance is carried out keeping in view the present risk and anticipating further diversification in climatic conditions. To counteract the rising human population density, equilibrating processes are pulled from a shrinking region of global natural habitats. Earth ecosystems supply multiple resources that are shared by numerous countries and have an impact that goes beyond the borders of a particular nation. To prevent a "tragedy of the commons," when a common resource is overexploited, governments and other stakeholders must resist this temptation to exploit (Galvani et al., 2016).

Adaptation for *natural forests* involves "conservation, protection, and restoration strategies," whereas options for adapting to climate change in *managed forests* include "sustainable forest management, adjusting tree species compositions to build resilience, and coping with the increased risks of pests and diseases and fire" (IPCC, 2022). Restoring natural forests and improving the sustainability of managed forests, in general, will make carbon stocks and sinks more resilient (Haq et al., 2021). Policies must be framed in a way to work together with local communities and indigenous peoples to ensure forest adaptation. Mountain ecosystems have already hit their hard adaptation limitations as a result of rising global temperature. With the rise of global warming by 1.5 °C, mountain ecosystems are anticipated to lose their ability to adapt (IPCC, 2022).

In the absence of adequate adaptation measures, climate change in addition to the nonclimatic drivers will harm the world's forests (McCarthy, 2001; Smit & Wandel, 2006). Government policies for various developmental plans and measures to combat the adverse impacts of climate change demand consideration of factors such as improving awareness among the community about climate risks and its mitigation options and initiatives to lower the cost of adaptations, improving the farm-household assets and alternative livelihood, and improving access to services (Ali & Erenstein, 2017; Dhyani et al., 2021; Kalra & Kumar, 2018; Kumar et al., 2021b). There is a need to work together with the government and the local community to make sure that policies are properly implemented, especially those that could minimize the impacts of climate change with an assurance that people adapt to it (Kumar et al., 2019b). A study in the Indian Himalayas on community forestry by Gupta and Koontz (2019) demonstrated how the government can work together with nongovernment partners (NGOs) for the effective implementation of policies of forest conservation. The government could provide technical and financial assistance, while on the other, NGOs could drive the communities to those resources and compensate for each other's weaknesses. This will make it easier for communities to participate in forest governance.

The adaptation of forests to climate change varies to various degrees depending on their region and type. To cope with the challenges of climate change, the requirement is to design a framework for adaptation considering the role of different sectors. The framework can be developed by synergizing community perception and various governmental policies and programs (Kumar et al., 2021b). This chapter aims to describe the numerous climate change adaptation programs that have been launched and implemented in the IWH region, with a focus on forestry. The major findings of the policy document on the State Action Plan on Climate Change (SAPCC) for the respective IWH states have been summarized with a focus on presenting information that directly relates to the actions oriented for the forestry sector that would assist in climate change adaptation. In the end, other major policies that could be integrated for climate change adaptation measures involving forestry-related initiatives have been summarized. The chapter will assist planners to have a ready source of information on existing policies for the IWH region. At the same time, this will supplement information for people who intend to formulate any policies advocating forestry-based adaptation measures to combat the impacts of climate change.

2 Biophysical and Socioeconomic Characteristics of the Study Region

The IWH is a complex geographical region with varying altitudes and orientations. This makes the region complex and would witness the effects of climate change in a different manner, which will have spatial variations (Padma, 2014). The long-term trends in temperatures across the northwestern Himalayas imply a significant increase in air temperature, with winter warming happening at a faster rate (Bhutiyani et al., 2007). The region may witness minimum temperature rise by 1–4.5 °C and maximum temperatures rise by 0.5–2.5 °C (MoEF, 2010). The varying

Table 4.1 Biophysical and socio-economic characteristics of the Indian Western Himalayas (IWH)

Parameter	J & K (includes Ladakh)	Himachal Pradesh	Uttarakhand	IWH
General				
Geographical area (km^2)	2,22,236	55,673	53,483	3,31,392
Population (million)[a]	12.50	6.86	10.12	29.58
Number of districts[a]	22	12	13	47
Number of villages[a]	4939	20,690	16,793	42,422
Land use pattern[a] (hectare)				
Permanent pastures and other grazing land	33.07	1,507,522	192,077	16,99,632.07
Fallow land (current and other than current)	111.94	78,791	143,619	2,22,521.94
Land under misc. tree crop and groves not included in net area sown	1096.62	64,905	387,817	4,53,818.62
Net area sown	7,57,450.70	543,365	700,171	20,00,987.70
Climatic parameters				
Annual temperature (min.-max., °C)	Subzero to 40	−15 to 43	−2.4 to 41.5	−2.4 to 43
Average rainfall (mm)	600–800	1800	1550	600–1800

Source: ISFR (2011) and [a]Census of India (2011)

altitude of the region produces unique pattern of vegetation that varies spatially having subtropical, conifer, grasslands, and alpine meadows. The changing climate has started influeincing the flora and fauna of the region with a shifting of their habitat (Singh et al., 2020). The region is represented by two union territories of Jammu and Kashmir (J & K) and Ladakh and two states, namely, Himachal Pradesh (HP) and Uttarakhand (UK). *Before the creation of union territories in 2019, the J & K and Ladakh represented the common state of Jammu and Kashmir. Therefore, most of the statistics and policies reviewed and presented in this chapter are common for the combined UT of present J & K and Ladakh mentioned as* J & K *or* JK. The biophysical and socioeconomic characteristics of the region are highlighted in Table 4.1.

The IWH can be divided into three zones based on geology and topography, namely, higher Himalayas or Himadri, lesser Himalayas or Himachal, and outer Himalayas or Shiwaliks. The climate of the IWH is quite temperate – often referred to as the extratropical mountain climate. The altitude of the IWH region varies from 97 to 7144 m above the mean sea level (Fig. 4.1). There is also a strong influence of the continental effects as reflected in the large annual and diurnal range of temperature (IIRS, 2002). In the valleys of Kashmir, Kullu, and Dehradun, predominantly alluvial soil is found, whereas brown soils are found in the hills.

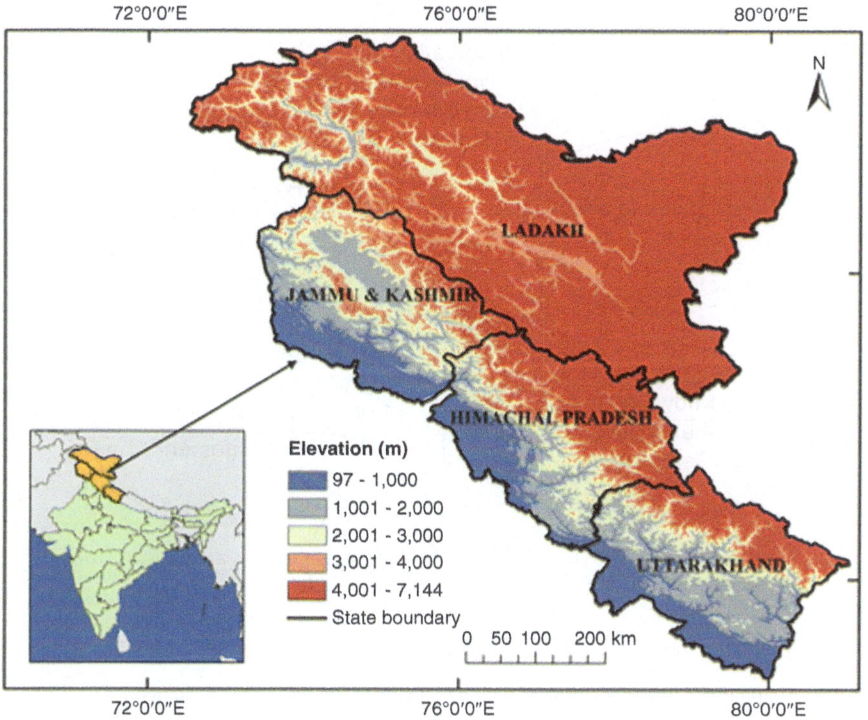

Fig. 4.1 Elevation ranges in the Indian Western Himalayan region

3 Adaptation Initiatives in the IWH Region

Adaptation initiatives aim to address the risks and effects of climate change. Well-framed adaptation strategies will drastically reduce the stress of climate change. Sustainable forest management (SFM) practices are a dynamic and growing concept that aims to continue and improve the socioecological and economic importance of forests, for the benefit of present and coming generations (Dhyani et al., 2021; Kumar et al., 2021b). However, well-planned target-oriented adaptation initiatives will ensure better preparedness to withstand the ill effects of climate change and extreme events. For climate-sensitive areas such as IWH, climate change adaptation should be an integral part of good governance and development. The forest ecosystem needs the longest response time to adapt to climate change, which varies for forest types and regions (Kumar et al., 2018; Rawat et al., 2020).

Adaptation plans are needed to deal with the effects of climate change and the harm it causes. It should be ensured to incorporate adaptation strategies in developmental and mitigation projects. Challenges of connecting adaptation research to policymaking include insufficient information about existing indigenous knowledge,

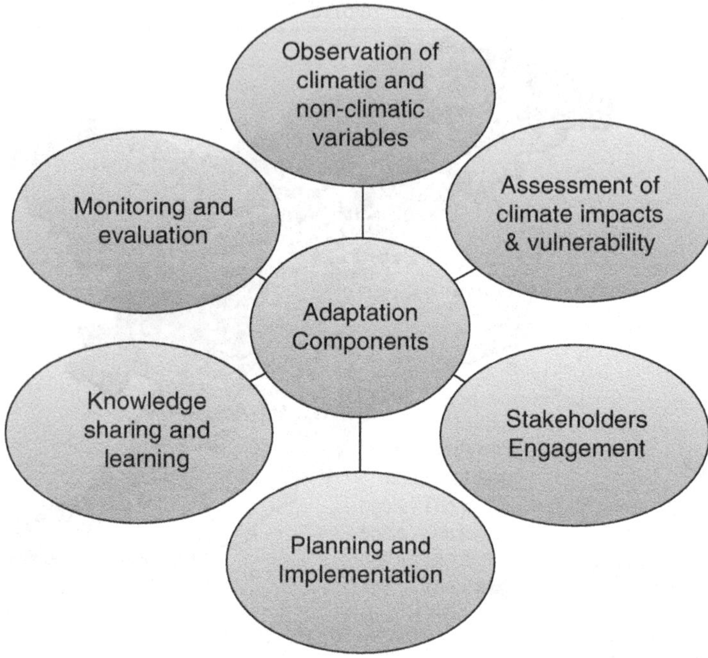

Fig. 4.2 Adaptation components linked with forest ecosystem and climate change

poor participation of local communities in decision-making processes, lack of awareness and education among the forest-dependent communities, poor forecasting and early warning systems, etc. (Murthy et al., 2011). Various adaptation components that can be linked with the forest ecosystems are shown in Fig. 4.2.

3.1 Integration of Adaptation Initiatives in the State Action Plan on Climate Change

The Indian Government launched the National Action Plan on Climate Change (NAPCC) on June 30, 2008, which focuses on understanding climate change-related issues. All the states and union territories were directed by the Central Government to prepare and adopt the State Action Plan on Climate Change (SAPCC) in 2009. The design and guidance for preparations were provided by the Union Ministry of Environment, Forest and Climate Change. States act as focal points for respective geographic areas. The development of the State Action Plan has been a significant effort in delineating regional climate vulnerability, exploring future projections, and formulating actionable strategies. The SAPCC adheres to the objectives and strategies of the national action plan on climate change to safeguard the poor and vulnerable sections of the societies through sustainable development strategy, sensitivity to climate change, enhancing ecological sustainability and mitigating greenhouse gas emissions, using appropriate technologies for adaptation and mitigation of greenhouse gas emissions, and embracing international cooperation.

The SAPCC is in accordance with eight national missions forming the core of the National Action Plan, which are "National Solar Mission, National Mission for Enhanced Energy Efficiency, National Water Mission, National Mission on Sustainable Habitat, National Mission for Sustaining the Himalayan Ecosystem, National Mission for a Green India, National Mission for Sustainable Agriculture, and National Mission on Strategic Knowledge for Climate Change" (https://dst.gov.in/climate-change-programme). SAPCC outlines the policies for addressing climate change–related issues. The action plan is based on state sectoral policies and themes. The strategy focuses on impact mitigation and adaptation methods to preserve the states' unique environment. Table 4.2 depicts the objectives, thrust areas, and initiation of the state action plan in the IWH.

3.1.1 Major Proposed Actions Under Jammu and Kashmir State Action Plan on Climate Change

The key priorities proposed under the J & K SAPCC include (i) implementation of J & K State Forest Policy 2011 for forest conservation and improvement of degraded forests; (ii) quantify per capita annual fuelwood availability and consumption; (iii) generation of weather and climate data and to ensure its easy access; (iv) preparation of an eco-restoration plan; (v) to study the phenology of important tree species by monitoring flowering and fruiting pattern; (vi) to conduct study to estimate the soil organic matter (SOC) pool under different forest covers, serving as a benchmark for future investigations of changes in SOC pool; (vii) to study the feasibility for developing mechanism under the UNFCCC to REDD+; (viii) collect data on climate effects, development of framework for the assessment of impacts, decision support tools, and successful adaption methods; (ix) to track carbon influxes/outflows from diverse forest types/trees and their contribution to carbon sequestration; (x) capacity building through the specific training modules and the dissemination of information via regional workshops; and (xi) annually convene educational seminar, forum, workshop and to disseminate knowledge-based products. Suggested actions focused on forestry in combating climate change is summarized in Table 4.3.

3.1.2 Major Proposed Actions Under Himachal Pradesh State Action Plan on Climate Change

The key priorities proposed under the HP SAPCC include: (i) Sanjhi Van Yojna Scheme by the involvement of grass root level institutions such as Gram Panchayats, Mahila Mandals, Yuvak Mandals, Village Forest Development Societies (VFDSs), Community-Based Organizations (CBOs) and NGOs in sustainable management of forest resources; (ii) Jan-Jan Sanjeevani Van Abhiyan initiated in 2008, which ensured the distribution of more than 1.50 million of medicinal plants to the rural

Table 4.2 Initiation, objectives, and thrust areas of Jammu and Kashmir (includes Ladakh), Himachal Pradesh, and Uttarakhand State Action Plan on Climate Change

State	Enacted date	Objectives	Thrust areas
Jammu and Kashmir (includes Ladakh)	August 9, 2011	Construct adaptation and mitigation strategies toward stabilization of emissions, increase ecosystem resilience, diversify the dependency on natural resources	Management of forests Soil and water conservation Estimation of soil carbon Stabilizing emissions Enhancing ecosystem resilience Eco-restoration plan Improving the livelihood of forest-based communities Wetland development Forest protection
Himachal Pradesh	November 14, 2011	Improving the societal awareness and preparedness Increasing the robustness of infrastructural design and long-term investments Increasing the flexibility of vulnerable managed systems	Reducing forest degradation Decreasing livestock pressure on forests Encouraging community participation through JFM Promoting alternatives to fuelwood Monitoring forest encroachment Regeneration of crops Networking the fragmented protected area Improving the production of natural resources and income of rural households Sustainable development Agrarian economy Water sources for drinking and irrigation Sustainability of hydro economy
Uttarakhand	January 19, 2011	Creating framework with respect to response strategies and sustainable development	Conservation of natural resources Enhancing forest area Conservation of soil and water Capacity building Management of fire Encouraging sustainable development Securing livelihood Buffering ecosystem services

Table 4.3 Actions and programs for the forestry sector focusing adaptation measures in Jammu and Kashmir (JK & Ladakh)

S. no.	Major thrust area	Proposed activities
1.	Implementation of J & K State Forest Policy 2011	A comprehensive plan for the conservation of existing forests Restoring degraded forests
2.	Capacity building and awareness for all levels of stakeholders	Improving the skills of forest department personnel to recognize the impacts of climate change on forests Training includes legal aspects of forest and environmental issues
3.	Gene bank development for climate-adaptable species	Identification of vulnerable species and species that have wider adaptability over larger geographical area to ensure their genetic conservation
4.	Eco-restoration through afforestation and climate oriented eco-restoration plan	Preparation of an eco-restoration plan for climate change mitigation Influence the State's carbon budget
6.	Flora and Fauna vulnerability study	To undertake a study to see the changes in floral and faunal composition Understand and manage horticulture, which is a major source of livelihood and revenue.
7.	Studies of carbon influxes/out fluxes of various forest types/trees and their role in carbon sequestration	Carbon stock estimation and assessment of sequestration potential for prominent forest species To ensure improved estimates based on micro-meteorological and inventory methods
8.	Study on per capita firewood consumption and alternative livelihood	Quantify per capita annual fuelwood availability and consumption Livelihood improvement of forest-dependent communities Encouraging communities for forest conservation
9.	Study on SOC of forest area	To conduct studies for the estimation of the SOC pool
10.	Climate impact study in undisturbed/protected forest areas	To conduct climate impact studies in undisturbed and protected forests to study the possible climate change effects without interference from other disturbances, including anthropogenic causes of forest degradation
11.	Climate change impacts on undisturbed forest areas like national parks	To study the impact of climate change on national parks such as the efficiency of carbon sequestration, changes in feeding patterns, reproduction and migration pattern, impact of tourism, etc.

(continued)

Table 4.3 (continued)

S. no.	Major thrust area	Proposed activities
12.	E-green portal with geo-reference	To conduct a field survey with modern instruments /standalone GPS Developing an E-green portal to provide thematic information
13.	Nursery development for climate-adaptable species	To conduct programs for the production of high-quality climate-adaptable seedlings Establishment of clonal nurseries and root trainers The seedlings to be utilized by the forest department, while surplus seedlings to be sold to the public
14.	REDD+ feasibility study for carbon sequestration	To study the feasibility for developing mechanism under the UNFCCC to REDD+
15.	Studies on sustained water availability	To stop decline in groundwater level, conserve surface water run-off, ensure soil conservation, reduce soil erosion, water conservation measures through new and improved technologies To study the rate of glacier melting and its impact in associated areas To address critical issues of water sustainability influenced by increasing population and climate change

Source: http://moef.gov.in/wp-content/uploads/2017/08/Jammu-Kashmir.pdf

and urban households through 5000 distribution points; (iii) in 2013, Department of Environment, Science and Technology (DEST) developed an Environment Master Plan that established baseline data for the identification of ecological sensitive zones and the critical issues that impact them; (iv) BioCarbon sub project as an additional component of the mid Himalayan watershed program with an objective to engage small and marginal farmers in tree planting initiatives to add value to ongoing watershed operations; (v) Herbal gardens for the cultivation of medicinal plants; (vi) Community-led Assessment, Awareness, Advocacy and Action Plan for Environment Protection, and carbon neutrality to evolve Himachal as a sustainable and climate resilient state; (vii) Organic Himachal initiative to take into account a complete range of organic farming such as vision, mission strategy about policy, awareness, technology and support to the farmers, quality assurance of the state produces and to develop policies to make Himachal an organic compost rich state; (viii) As a part of national bamboo mission, planting bamboo species in areas of Kangra, Nahan, Mandi, Hamirpur along with marketing of bamboo.

The government of Himachal Pradesh took upon the Environment Master Plan (EMP) to ensure a long-term outlook on attaining sustainable development. It was established to safeguard the long-term viability of natural resources and cultural legacy. The EMP has been envisaged as a guide tool to provide strategies regarding environmental issues. The EMP of the state for the forestry sector has been summarized in Table 4.4.

Table 4.4 Actions and programs for the forestry sector focusing adaptation measures in Himachal Pradesh

S. no.	Major thrust area	Proposed actions
1.	Forest degradation	Encouraging community participation through joint Forest Management (JFM) Implementation of state plans for the reforestation and regeneration of forests Promoting natural regeneration, afforestation and plantation Eco-development programs Encouraging wild fruit trees plantation Development of watersheds in order to conserve soil and water Monitoring of project programs like JFM, eco-development program, watershed Strengthening of catchment area treatment plan
2.	Increase in anthropogenic pressure	Increase in forest cover per capita through JFM in wastelands or degraded lands. Developing plans for managing fallow lands and wastelands Developing livelihood programs and educating the migratory Gujjar grazers to adopt agro-economic activities Suggest alternatives of fuel wood, such as utilizing renewable resources and also promoting smokeless *chulha* by providing subsidy for better kitchen hygiene
3.	Increasing livestock pressure on forests	Plantation of community orchards and fodder crops Identification of common land for pasture Upgrading breeds and reducing nonproductive livestock Developing stall feeding and growing grasses to reduce grazing pressure on forests
4.	Timber distribution rights	Update and amendment in the regulation of timber distribution Assessment of timber needs and monitoring of felling of marked trees Promoting agroforestry to meet the need for timber. Strict actions against defaulters
5.	Forest encroachment	Demarcation of the forest land with pillars Periodic monitoring of the area Imposing strict laws like the Forest act, town and planning act
6.	Unsustainable harvesting of the forest produce	Monitoring at the time of harvest of forest produce Preserving and documenting the germplasm of aromatic as well as medicinal plants Use of biotechnology for propagation Enforcing regulatory measures Conducting programs and surveys for spreading awareness among people

(continued)

Table 4.4 (continued)

S. no.	Major thrust area	Proposed actions
7.	Ban on green felling leading to inhibited natural regeneration	To remove the blockage and to induce regeneration of crops, thinning can be carried out Setting up permanent plots for long-term monitoring to understand the forest ecology
8.	Not taking into account diversification of species	Encouraging mixed plantations of willow, oak, fir, bamboo, wild fruit species Planting deodar and Kail along with Chil and fir/spruce Use of pine needles for making of briquettes so as to reduce pressure on fuelwood Monitoring afforestation, plantation, and nursery works
9.	Natural hazards and forest fire	Cooperation of villagers to protect forests from fire as well as help the forest department Zila Parishad, Gram Panchayat to motivate community involvement in firefighting, protection, and afforestation
10.	Forest diversion for non-forestry purposes	Alternative solutions like bio-engineering under catchment area treatment plans A strict application of policy guidelines
11.	Fragmented protected area network	Reviewing the existing protected area network in respect to bio-geographic zones and forest types at regular intervals
12.	Crop damage by wild animals	Introducing crop insurance schemes Vegetative barriers Providing compensation Creating buffer zones
13.	Stress on wildlife population	Monitoring and reporting illicit cases to the concerned authority Carrying out awareness programs to educate the people about the ill effects about poaching
14.	Alien and invasive species leading to loss of native vegetation	Monitoring and eradication of alien and exotic species from forest area Preparation of suitable work plans to identify infested areas with invasive species Establishing a techno-economic feasibility assessment and encouraging the use of lantana bush products

Source: http://www.indiaenvironmentportal.org.in/files/file/ExecutiveSummaryEnvironmentMasterPlan.pdf

3.1.3 Major Proposed Actions Under Uttarakhand State Action Plan on Climate Change

The key priorities proposed under the UK SAPCC include (i) improving silvicultural techniques with climate change considerations; (ii) research on removal of invasive species and ex situ conservation; (iii) surveying the population dynamics and movement of wildlife; (iv) estimation of total carbon stock and annual increment for Uttarakhand; (v) assessing biodiversity of various ecosystems; (vi) documenting the traditional knowledge related to biodiversity; (vi) evaluating the impact of climate change on high-altitude wetlands, alpine meadows, and moraines; (vii) forest fire management and prioritizing corridors; (viii) enhancing the area under forests and trees and also increasing the quality and density of degraded forests by natural regeneration in moderately dense forests, planting climate-resilient species, and managing interference in scrub forests; (ix) recognizing vulnerable areas with expert assistance; and (x) revamping local institutions. Suggested actions focused on forestry in combating climate change is summarized in Table 4.5.

4 Developmental Policies, Programs, and Practices for Adaptation focusing Forestry in the IWH Region

Forests contribute to the global atmospheric gas concentration by influencing the carbon cycle by the capturing and releasing of carbon through its various processes. Realising the importance of forests for the global climate change scenario, the UN adopted Agenda 21, which is a nonbinding, voluntarily implemented action plan with a focus on climate change, conservation of biodiversity, and forests.

The carbon potential of India's forests changes due to various promotional initiatives for enhancing the forest cover as well as various pressures, which leads to forest degradation and deforestation (Savita et al., 2018a) leading to loss of biomass and thus carbon. India's various acts and legislations are focusing on the management and conservation of its forests such as the Forest Conservation Act (1980) to limit arbitrary forest land diversion and regulate land-use changes in existing forest areas, thereby reducing deforestation. National Forest Policy, 1988, was the backbone of major policies adopted in India, which prioritized forests for maintaining environmental steadiness and ecological equilibrium, motivating participatory forest management, and the stimulus to farm forestry. It provided the first accusation on forest produce to local communities. Important initiatives by various agencies of India for forest conservation and management are presented in Table 4.6.

Among all initiatives, the JFM initiated in 1990 by the Ministry of Environment and Forests (MoEF) serves in a better way that involves the local community in the management actions for the management of forest resources. It ensures the involvement of local communities for the protection and management of forests. Among the most important aspects of this program are increasing the capacity at various levels, the establishment of institutions, and sharing of benefits between the state and the communities (Ravindranath & Sudha, 2004). The local communities and government manage the forests resources and share the cost and benefits equally. The 2000 guidelines on JFM by the MoEF, Government of India, strengthened the original mission, included women's involvement, and brought structural challenges all under one roof. The JFM is envisioned as a powerful instrument for preventing additional

Table 4.5 Actions and programs of the forestry sector focusing adaptation measures in Uttarakhand

S. no.	Major thrust area	Proposed activities
1.	Increasing the existing area under forests and improving open and degraded forests	Management interventions in scrub forests (1 km^2/year) Site-specific activities for the maintenance of existing forests and new plantation Plantation of climate-resilient species that would benefit local communities Implementation of assisted natural regeneration activities
2.	Enhancing natural resources and livelihood options of the vulnerable sections	Providing livelihood options to local communities Rangeland management Agroforestry/farm forestry Mapping of important rangelands using remote sensing approach Alternate livelihood options such as NTFPS collection, biomass briquetting, eco-tourism, establishment of nurseries, and cultivation of medicinal plants Establishment of quality planting material production centers Conservation of bugyals (2500 ha per annum) Protection and management initiative toward regulated grazing
3.	Soil and water conservation	Close watch on glaciers Maintenance of soil moisture regime Raising large-scale rainwater harvesting structures for water conservation Vegetation approach to soil and moisture conservation Introduction of three-tier forest plantations
4.	Fire management	Management of fuel load Maintenance of fire lines to control forest fire Ensure alternate energy sources Making firefighting quick response team equipped with advance firefighting tools and early warning systems Regular monitoring of fire threats Controlled and cool burning
5.	Research and capacity building	Monitoring and evaluation of carbon stock Monitoring the forest carbon flux Annual increment of total carbon stock in Uttarakhand
6.	Short-term research projects	Climate change–related impacts and adaptation strategies Ecological and physiological studies in important forest types Adapt and develop vegetation models for climate change impact studies

Source: https://moef.gov.in/wp-content/uploads/2017/08/Uttarakhand-SAPCC.pdf

Table 4.6 Developmental policies, programs, and practices in the forestry sector relevant to climate change adaptation

Social- and climate-related initiatives	Objectives	Date of initiation
Joint Forest Management (JFM)	To safeguard forest with collaboration between the forest department and local communities	1990
Compensatory Afforestation Fund Management & Planning Authority (CAMPA)	Utilizing fund for afforestation in efficient manner	2002
National Afforestation Programme (NAP)	Rehabilitation of degraded forest through JFM & Forest Development Agency	2000
National Afforestation & Eco-development Board	Promote afforestation, ecological, restoration and eco-development activities	1992
Pradhan Mantri Krishi Sinchayee Yojana (PMKSY-WC)	More crop per drop-extending irrigation coverage and improving water use efficiency	2015
Reducing Emissions from Deforestation and Forest Degradation (REDD+)	Reducing gas emissions, halting reversing forest loss	2015/2017
Forest PLUS	Promote scientific and technical collaboration in the forestry sector and accelerate India's transition to a low carbon economy	2010
Green India Mission (GIM)	Protecting, restoring, and enhancing forest cover	2008
National Rural Employment Guarantee Act (NREGA)	Guarantees employment to unemployed people	2006
Rainfed Area Development	Enhancing agriculture productivity and minimizing the risk associated with climatic variabilities	2011–2012
Namami Ganga	Abatement of pollution, conservation and rejuvenation of national river Ganga	2014

forest degradation. The logic behind JFM is to put a collective responsibility on local communities and forest staff to protect the forest meticulously. The JFM can fulfil local subsistence requirements of fuelwood, fodder, NTFPs, small timber, etc., while avoiding degradation of forests that give local, national, and international environmental advantages.

Forests are an important source of varieties of ecological services, to compensate for these losses; many afforestation programs have been carried out in the IWH region to divert nonforest land into forest land. Since afforestation doesn't start giving services overnight, there is still the loss of goods and services that would have been provided by the forests that had been laid down for completing the demands of industries and other developmental activities. To compensate for the loss in the interim, the net present value (NPV) of the diverted forest is calculated for 50 years, and the same is recovered from the agency that is diverting the forest land. To manage and utilize the amount collected sum, Compensatory Afforestation Fund

Management and Planning Authority (CAMPA) was formed. The State CAMPA is meant to speed up efforts related to forest preservation, wildlife management, sector infrastructure development, and other related activities. These efforts accelerate the annual rate of plantations. The plantations raised are mainly of eucalyptus, teak, acacia, poplar, casuarinas, pine, bamboo, and other miscellaneous species.

The National Afforestation and Eco-development Board (NAEB) works in the direction of promoting afforestation and eco-development activities across the country, with special attention toward degraded forest areas, protected areas like national parks and sanctuaries, as well as ecologically fragile areas like IWH. Different schemes operated by NAEB include afforestation and eco-development schemes, fodder and fuelwood schemes, NTFPs, and medicinal plant-related schemes. To assist the state forest departments in the effective implementation of technologies and to provide other aids, the board has seven regional centers. These centers evaluate NAEB's program in the field and organize training programs and workshops. They also act as a platform for exchanging ideas and experiences among different regions. The regional centers' work programs are meant to promote the sustainability of different afforestation and forest management programs. These programs cover training for forest-based microenterprises, studies on improved silvicultural practices, and interdepartmental linkage workshops at the district level for synergy of various forest conservation schemes of the government. Based on these programs, it is expected that forest-based microenterprises could scale up as a channel of promoting sustainable livelihoods for forest fringe communities. According to the principle of eliciting and cultivating people's involvement, the National Association for Forest Healing (NAEB) offers suitable financial support to organizations working in the cause of forest healing.

The World Bank takes into account various issues like climate change, finance, trade, agriculture, education, and food security intending to reduce poverty and share prosperity in developing countries. Various World Bank-assisted programs being enacted in the IWH region include (i) Sustainable Land and Ecosystem Management (SLEM), (ii) Uttarakhand Decentralized Watershed Development Project, (iii) Himachal Pradesh Horticulture Development Program, and (iv) Mid-Himalayan Watershed Program implemented in the HP.

The SLEM is a bilateral program initiated by the Government of India and the Global Environmental Facility (GEF) under the Country Partnership Programme. The SLEM program aimed at maintaining the capacity of the ecosystem to deliver goods and services, while keeping climate change in mind, promoting sustainable land management, and conserving and developing natural resources. The program has been initiated in the states of Madhya Pradesh, Nagaland, Rajasthan, and Uttarakhand. In the IWH region, the program has been carried out in Uttarakhand only. The main objective of the SLEM project in Uttarakhand was to rehabilitate and preserve the ecosystem's functions and biodiversity, also increasing income and improving livelihood. It aims to achieve zero net land degradation, that is, aiming at practices that lead to a balance of carbon and also mitigating climate change by sequestering carbon dioxide.

India's national REDD-plus (Reducing Emissions from Deforestation and Forest Degradation) strategy focuses on increasing and improving the country's forest and tree cover, hence increasing the quantity of forest ecosystem benefits that come to local communities. Development and implementation of REDD+ at the national level require close coordination and strong linkages between all stakeholders of the forest sector, which will need to be guided by the MoEF. The Ministry of Environment, Forests, and Climate Change has formed a REDD+ Cell, which is tasked with the coordination and guidance of REDD+-related activities on a national and subnational level, as well as at regional, state, and local levels. The REDD+ Cell would provide guidance on the concept, planning, financing, execution, supervision, and assessment of REDD+ programs in various jurisdictions. The Cell works closely with the Climate Change Division and the Forest Policy Division within MoEF for analytical application, implementation, and reporting of REDD-plus actions and also actively facilitates nationwide coordination.

Almost 2,00,000 villages are classified as forest fringe villages in India, which represents the dependence of communities on forest resources (FRI, 2017). This provides enough scope and opportunity in the Indian communities for integrating REDD+ initiative activities (Kumar et al., 2020b). However, this suggests evolving methodologies and strategies for a procedural framework addressing REDD+ objectives for ensuring people's participation and sharing of the benefits accruing from REDD+ incentives.

Forest PLUS is a joint program between India and the United States for developing different methods for forest management and also for the sustainable use of forests in India. Forest PLUS program stimulates India's conversion to a low carbon economy and also formulates India to implement REDD+ successfully. Forest Plus is associated with the Green India Mission, REDD+, and National Action Plan on Climate Change and also works with MoEF, local governments, NGOs, and State Forest Department. It is a USAID (United States Agency for International Development)-funded project focusing on the reduction of greenhouse gas emissions from the forested landscapes of India. The main objectives of the program are (i) sustainable use of forests, (ii) low carbon economy, (iii) development of systems for monitoring forest carbon, (iv) improving the land use planning and reducing deforestation, (v) conducting inventories for greenhouse gases, and (vi) cost-effective management of forests. Forest PLUS brings together Indian and American experts to strengthen, enhance, and improve various technologies, techniques, and strategies of forest management in meeting the challenges of managing and maintaining forests for ecological sustainability, carbon storage, species diversity, and livelihood opportunities. It associates with the local communities and the Indian forestry institutions in the areas of Karnataka, Madhya Pradesh, Sikkim, and Himachal Pradesh (Kumar et al., 2021b).

The Green India Mission (2010) was considered an inclusive participatory mission for greening India through the decentralized governance structures mandating community engagement in traditional afforestation programs. The mission ensures participation of grass-root groups in the preparation, decision-making, execution, and evaluation phases. The key objectives of the Mission are to "increase forest/tree cover on 5 m ha of forest/non-forest lands and improved quality of forest cover on another 5 m ha (a total of 10 m ha); improve ecosystem services including biodiversity, hydrological services, and carbon sequestration as a result of treatment of 10 m ha; increase forest-based livelihood income for 3 million forest-dependent households; enhance annual CO_2 sequestration of 50–60 million tons by the year 2020." Green India Mission directly or indirectly is associated with the climate change adaptation strategy.

Rainfed area development has a very crucial role to play in the Indian economy as well as agriculture (Savita et al., 2018a). Rainfed regions are the regions where crop production is dependent on rainfall. India grades one among rainfed countries but least in rainfed yields (<1 t/ha). Sixty-four percent of the country's sown area is rainfed (https://nraa.gov.in/). These areas are occupied by the majority of India's rural poor and marginal farmers experiencing difficulties and risks in biophysical and socioeconomic conditions. Poverty, water scarcity, low yields, malnutrition, and lack of infrastructure are few difficulties faced by the communities depending on these areas. Rainfed areas are susceptible to changing climate due to their poor potential to cope with severe water and weather stress. According to Central Research Institute for Dryland Agriculture, rainfed areas are those that receive an annual rainfall of 750–800 mm and has <30% of irrigated land. The National Rainfed Authority (NRAA) is now serving as an advisory body for policy formulation and monitoring schemes to resolve agricultural challenges across the vast rainfed system of the nation, whereas the National Mission for Sustainable agriculture aims at promoting sustainable agriculture.

River Ganga originates from the IWH, which directly and indirectly is correlated with the forest ecology of the Himalayan region. Forest-dependent communities (FDC) are dependent on the Ganga river for several livelihood activities. Namami Gange Programme is a "flagship program" of government of India, initiated in 2014 with a budget outlay of 20,000 crore to abate pollution, and to ensure conservation and rejuvenation of National River Ganga within a time frame of 5 years for medium-term activities and 10 years of long-term activities (https://nmcg.nic.in/). The vision for Ganga Rejuvenation is to ensure "Aviral Dhara," that is, continuous flow, and "Nirmal Dhara," that is, unpolluted flow. One of the envisaged major activities for Ganga rejuvenation is forestry intervention to enhance the productivity and diversity of forests in headwater areas all along the river (Savita et al., 2018b). Multi-tier plantations including trees, shrubs, grasses, medicinal plants, etc. can be raised. Deterioration in the form of biodiversity decline, degradation of land and water resources, pollution, etc. can be controlled by developing a natural corridor along the river. A holistic approach through amalgamating various schemes and programs like National Afforestation Programme, Green India Mission, NREGA, etc. can help attain the objectives effectively.

Being in the lap of the great Indian Western Himalayas, enriched with large forest cover and glacier source of various rivers, states like Jammu and Kashmir, Himachal Pradesh, and Uttarakhand are vulnerable to flash floods, forest fire, and cloud bursts along with other natural disasters. Disasters are disturbances to a community's existence and livelihood caused by hazard effects, resulting in loss of life and environmental damage beyond its capability to manage alone. Disaster management provides aid to cope with calamity and act as a core component of sustainable development. Disaster management plans are aimed to build resilience in the socioeconomic resources and functions to reduce vulnerability and mitigate risks. With a vision of building a safer and disaster-resilient India, the Disaster Management Act, 2005, was framed. Under this act, various national and state-level institutions were established for reducing risks, preventing losses, and preparing, responding, and recovering through hazardous conditions.

The regions of the Indian Western Himalayas are rich in diversity, where the valleys receive a good amount of rainfall and also have rich soils. But the temperate region of Ladakh faces climatic stress against agricultural growth. These regions also face soil erosion, thus affecting the soil fertility as well as the crop produce. Wheat, maize, and rice are the major crops produce, whereas apples, plums, and apricots are the major fruits produced in these regions. Many schemes and programs have been implemented aiming toward sustainable agriculture. Few schemes have been discussed in this section. The National Mission for Sustainable Agriculture (NMSA) obtains its acceptance from Sustainable Agricultural Mission, which is one of the eight missions in the National Action Plan on Climate Change. The mission (NMSA) is designed to make Indian agriculture resilient to climate change. NMSA has been prepared to improve agricultural productivity, especially in rainfed areas with an emphasis on the efficiency of water use, resource conservation, and soil health management. The components of NMSA are rainfed area development, soil health management, climate change, and sustainable agriculture. NMSA emphasizes sustainable agriculture through a course of adaptation measures such as refined seed crops, the efficiency of water use, improved livestock and fish cultures, pest and nutrient management, livelihood diversification, markets and agricultural insurance, and access to credit support and information.

5 Mainstreaming Adaptation in Forest Planning and Management

Adaptation addresses the risks and effects of climate change. Well-framed adaptation strategies will drastically reduce the stress of climate change. Strategies such as REDD+, Sustainable Forest Management (SFM), and conservation and enhancement of carbon stocks already exist. The SFM strives to maintain and increase the economic, social, and environmental value of all forest types for current and future generations. Effective adaptation measures will improve preparation for climate-related changes, especially severe occurrences. Climate change adaptation should be part of effective governance and development in climate-sensitive places like IWH. The forest ecosystem needs the longest response time to adapt to climate change.

Adaptation practices vary for different regions and forest types. A few adaptation strategies including the suggestion of Murthy et al. (2011) are indicated below.

- Promoting in situ and ex situ conservation
- Planned afforestation and adopting silvicultural practices
- Reducing forest fragmentation in order to facilitate species migration
- Promoting natural regeneration and mixed-species forestry
- Linking and expanding protected areas to promote the migration of species
- Developing sustainable forest management practices and reducing deforestation
- Promoting and developing temperature-tolerant and pest-resistant species
- Alternate techniques to fuelwood to decrease pressure on forests, promoting the use of bio-briquettes
- Building capacity to create and execute climate change adaption methods
- Promoting conservation of biodiversity-rich forests
- Increasing the amount of land already covered by forests while also enhancing the quality and density of degraded forests
- Promoting agroforestry as a means of increasing biophysical and socioeconomic resilience
- Establishing quality planting material production centers
- Conservation of soil and water by rejuvenating traditional *chals*
- Fire management involving quick response teams, regular monitoring of fire threats using satellite imagery and information technology
- Education and training programs for forest-dependent communities
- Infrastructure development in the rural areas
- Involving the active participation of local communities in forest conservation schemes
- Promoting regeneration of native species and short-rotation species
- Monitoring and management of forest areas
- Promoting bio-energy plantations and adaptive co-management between the forest authorities and local communities

There is a definite need for effective adaptation strategies involving the above-suggested measures to cope with climate change and damages caused due to climate change.

6 Conclusion and Way Forward

Climate change is posing new challenges and has the potential scope to severely impact various ecosystems. The unprecedented speed and type of change threaten the fundamental ecosystem resilience (IPCC, 2007). On the other hand, the conservation and sustainable management of ecosystems can facilitate multiple socioecological benefits and support long-term approaches to climate change adaptation against both current climate change and future climate change (Kumar et al., 2019a; Schmitt et al., 2009).

Forests are critical for mitigative measures for climate change (Kumar et al., 2020a). However, forests also need to adopt the changes in the climate to sustain their existence as well as to support the dependent communities. Broadly, the linkage between forests and climate adaptation is twofold: firstly, forests are important due to their role in delivering ecosystem services; and secondly, reducing the vulnerability of the society to the ill effects of climate change. Hence, there is a need to develop and implement policies that have two components of "forests for adaptation" and "adaptation for forests."

The varying levels of uncertainty, limited information, and risk, against the impacts of climate change, are posing problems for the conservation of the forests. Moreover, it is difficult to extrapolate the appropriate knowledge about the forest response to adapt against the disturbances in the forests due to climate change (Füssel, 2007; Giupponi & Biscaro, 2015; Hinkel, 2011; Kumar et al., 2020a; Smit & Wandel, 2006). The segregated evaluation of forest ecosystems across the world through models shows that there will be impacts on the forests from cell level to ecosystem levels such as species composition, the productivity of the forests, and phenology, resulting in changing the deliverables of the forests to the communities. These changes in forests will be detrimental to the forest itself and communities too, specifically for the adaptation aspect. Therefore, by neutralizing and nullifying these impacts and conserving the system, it becomes capable to modulate itself according to the prevailing conditions. For this, an appropriate mechanism has to be devised and implemented. Implementing the earlier approaches of forest management may be unreceptive and can be ecologically inappropriate and socially undesirable. Therefore, modified approaches inclusive of various resources and all stakeholders' involvement with cross-scale and cross-sectoral linkages will be essential to capture the dreadful impacts leading to better adapt the changes (Locatelli et al., 2008).

In the above context, under the changing climate, there is a need to implement measures for reducing the climate change impacts on the forests. The mechanism for supporting the adaptability of the forests revolves around strengthening the intrinsic characteristics of the forest ecosystem as well as supporting extrinsically by various means leading to the growth and development of the forest ecosystem. The internal components of the forests include the topography and edaphic factors besides the micro-climatic condition and hydrology, the important components for enriching the food preparation, and assimilation of the plants. Biodiversity is the most important component for measuring the strength of adaptability of the forest ecosystem. Balancing these factors through any means ranging from engineering works to anthropogenic activities may facilitate forest ecosystem growth.

The external support may broadly include those mechanisms that are either directly or indirectly supporting the growth of the forest ecosystem during the changes occurring in the climate. These factors include governance to silvicultural management and facilitating the communities with resources that support climate-smart farming so that a balance can be made between the social and ecological systems. The other includes the role of the community and support from the industrial sector as well as the inflow of research inputs and right information. A framework addressing the major prominent issues of the IWH region should be in place to support the policy need for a better adaptation plan.

References

Ali, A., & Erenstein, O. (2017). Assessing farmer use of climate change adaptation practices and impacts on food security and poverty in Pakistan. *Climate Risk Management, 16*, 183–194.

Bhutiyani, M. R., Kale, V. S., & Pawar, N. J. (2007). Long-term trends in maximum, minimum and mean annual air temperatures across the Northwestern Himalaya during the twentieth century. *Climatic Change, 85*, 159–177.

Census of India. (2011). *Office of the Registrar general & Census Commisioner, India*. Ministry of Home Affairs, Government of India, New Delhi. Office of the Registrar General & Census Commisioner, India. Ministry of Home Affairs, Government of India, New Delhi.

Chaturvedi, R. K., Gopalakrishnan, R., Jayaraman, M., Bala, G., Joshi, N. V., Sukumar, R., & Ravindranath, N. H. (2011). Impact of climate change on Indian forests: A dynamic vegetation modeling approach. *Mitigation and Adaptation Strategies for Global Change, 16*, 119–142.

Dhyani, S., Murthy, I. K., Kadaverugu, R., Dasgupta, R., Kumar, M., & Gadpayle, K. A. (2021). Agroforestry to achieve global climate adaptation and mitigation targets: Are South Asian countries sufficiently prepared? *Forests, 12*, 303.

FRI. (2017). *Forest resource dependence and ecological assessment of forest fringes in rainfed districts of India*. Forest Research Institute. Ministry of Environment, Forest and Climate Change, Government of India, Dehradun.

Füssel, H.-M. (2007). Vulnerability: A generally applicable conceptual framework for climate change research. *Global Environmental Change, 17*, 155–167.

Galvani, A. P., Bauch, C. T., Anand, M., Singer, B. H., & Levin, S. A. (2016). Human–environment interactions in population and ecosystem health. *Proceedings of the National Academy of Sciences, 113*, 14502–14506.

Giupponi, C., & Biscaro, C. (2015). Vulnerabilities – Bibliometric analysis and literature review of evolving concepts. *Environmental Research Letters, 10*, 123002.

Gupta, D., & Koontz, T. M. (2019). Working together? Synergies in government and NGO roles for community forestry in the Indian Himalayas. *World Development, 114*, 326–340.

Haq, S. M., Calixto, E. S., & Kumar, M. (2021). Assessing biodiversity and productivity over a small-scale gradient in the protected forests of Indian Western Himalayas. *Journal of Sustainable Forestry, 40*, 675–694.

Hinkel, J. (2011). "Indicators of vulnerability and adaptive capacity": Towards a clarification of the science–policy interface. *Global Environmental Change, 21*, 198–208.

IIRS. (2002). *Biodiversity characterisation at landscape level in Western Himalayas, India using satellite remote sensing and geographic information system*. Indian Institute of Remote Sensing, Dehradun.

IPCC. (2007). *The physical science basis: Contribution of Working Group I to the Fourth Assessment Report of the Intergovernmental Panel on Climate Change, Intergovernmental Panel on Climate Change (IPCC), Climate change 2007*. Cambridge University Press.

IPCC. (2022). *Climate change 2022. Impacts, adaptation and vulnerability. Summary for policymakers. Working Group II Contribution to the Sixth Assessment Report of the Intergovernmental panel on Climate Change*. Cambridge University Press.

ISFR. (2011). *India State of Forest Report*. Forest Survey of India, Ministry of Environment, Forest and Climate Change, Government of India, Dehradun.

Kalra, N., & Kumar, M. (2018). Simulating the impact of climate change and its variability on agriculture. In S. Sheraz Mahdi (Ed.), *Climate change and agriculture in India: Impact and adaptation* (pp. 21–28). Springer.

Kumar, M., Rawat, S. P. S., Singh, H., Ravindranath, N. H., & Kalra, N. (2018). Dynamic forest vegetation models for predicting impacts of climate change on forests: An Indian perspective. *Indian Journal of Forestry, 41*, 1–12.

Kumar, M., Savita, Singh, H., Pandey, R., Singh, M. P., Ravindranath, N. H., & Kalra, N. (2019a). Assessing vulnerability of forest ecosystem in the Indian Western Himalayan region using trends of net primary productivity. *Biodiversity and Conservation, 28*, 2163–2182.

Kumar, M., Singh, M. P., Singh, H., Dhakate, P. M., & Ravindranath, N. H. (2019b). Forest working plan for the sustainable management of forest and biodiversity in India. *Journal of Sustainable Forestry*, 1–22.

Kumar, M., Kalra, N., & Ravindranath, N. H. (2020a). Assessing the response of forests to environmental variables using a dynamic global vegetation model: An Indian perspective. *Current Science, 118*, 700–701.

Kumar, M., Savita, & Kushwaha, S. P. S. (2020b). Managing the forest fringes of India: A national perspective for meeting the sustainable development goals. In: *Sustainability perspectives: Science, policy and practice, strategies for sustainability* (p. 331). Springer.

Kumar, M., Kalra, N., Singh, H., Sharma, S., Rawat, P. S., Singh, R. K., Gupta, A. K., Kumar, P., & Ravindranath, N. H. (2021a). Indicator-based vulnerability assessment of forest ecosystem in the Indian Western Himalayas: An analytical hierarchy process integrated approach. *Ecological Indicators, 125*, 107568.

Kumar, M., Phukon, S. N., & Singh, H. (2021b). The role of communities in sustainable land and forest management. In: *Forest resources resilience and conflicts* (pp. 305–318). Elsevier.

Locatelli, B., Kanninen, M., Brockhaus, M., Colfer, C. P., Murdiyarso, D., & Santoso, H., (2008). *Facing an uncertain future: How forests and people can adapt to climate change.*

McCarthy, J. J. (2001). *Climate change 2001: Impacts, adaptation, and vulnerability: Contribution of Working Group II to the Third Assessment Report of the Intergovernmental Panel on Climate Change*. Cambridge University Press.

MoEF. (2010). *INCCA: Indian Network for Climate Change Assessment*. Climate change and India: A 4x4 assessment.

Murthy, I., Tiwari, R., & Ravindranath, N. H. (2011). Climate change and forests in India: Adaptation opportunities and challenges. *Mitigation and Adaptation Strategies for Global Change, 16*, 161–175.

NMSHE. (2010). *National Mission for Sustaining the Himalayan Ecosystem under National Action Plan on Climate Change.*

Pachauri, R. K., Meyer, L., Van Ypersele, J. P., Brinkman, S., Van Kesteren, L., Leprince-Ringuet, N., & Van Boxmeer, F. (2014). *Climate change 2014 synthesis report the Core Writing Team technical support unit for the synthesis report. Russ. Fed. Hoesung Lee (Republic Korea) Scott. B. Power NH Ravindranath.* Available online https//www. ipcc. ch/pdf/assessment-report/ar5/syr/SYR_AR5_FINAL_full_wcover. pdf. Accessed 28 Jan 2020.

Padma, T. V. (2014). Himalayan plants seek cooler climes: Race is on to record mountain biodiversity before it is lost. *Nature, 512*, 359–360.

Ravindranath, N. H., & Sudha, P. (2004). *Joint forest management in India: Spread, performance and impact*. Universities Press.

Rawat, A. S., Kalra, N., Singh, H., & Kumar, M. (2020). Application of vegetation models in India for understanding the forest ecosystem processes. *Indian Forester, 146*, 99–100.

Savita, Kumar, M., & Kushwaha, S. P. S. (2018a). Forest resource dependence and ecological assessment of forest fringes in rainfed districts of India. *Indian Forester, 144*, 211–220.

Savita, Mathur, P. K., Sharma, L. K., & Kumar, M. (2018b). Forestry interventions for Ganga rejuvenation: A geospatial analysis for prioritising sites. *Indian Forester, 144*, 1127–1135.

Schmitt, C. B., Burgess, N. D., Coad, L., Belokurov, A., Besançon, C., Boisrobert, L., Campbell, A., Fish, L., Gliddon, D., & Humphries, K. (2009). Global analysis of the protection status of the world's forests. *Biological Conservation, 142*, 2122–2130.

Singh, H., Kumar, N., Kumar, M., & Singh, R. (2020). Modelling habitat suitability of western tragopan (Tragopan melanocephalus) a range-restricted vulnerable bird species of the Himalayan region, in response to climate change. *Climate Risk Management, 29*, 100241.

Smit, B., & Wandel, J. (2006). Adaptation, adaptive capacity and vulnerability. *Global Environmental Change, 16*, 282–292.

Chapter 5
Climate-Induced and Geophysical Hazards and Risk Reduction Financing in Mountain Regions

Kamleshan Pillay, Hari Ballabh, and Srinivasan Pillay

Abstract Geophysical hazards are ubiquitous phenomena occurring in the Himalayan Mountains. These hazards are directly linked to the unique natural physical characteristics of the region, climate-related phenomena such as cloudbursts and storms that lead to flash flooding, and the anthropogenic influence of human activities and developmental structures such as road infrastructure, hydroelectric power plants, settlements, and agricultural activities. Widespread devastation, loss of life, and developmental setbacks ensue in the immediate wake of the event, but economic and livelihood setbacks may take years, if at all. Both direct and indirect economic impacts are important at the community level. Direct economic impacts refer to the loss of physical assets such as homes and damaged infrastructure for information and communication technology (ICT), transport, agriculture, and energy. Alternatively, indirect economic impacts refer to loss of utility as a result of direct impacts such as the inability to travel to places of employment and supply chain disruptions. To manage the impacts of the hazards, communities may integrate climate adaptation measures into their dwellings to reduce their risk exposure. Risk reduction measures must be coupled with financial resilience actions to ensure adequate risk management. Greater financial resilience can be delivered from public and private finance sources, both of which are discussed in this research.

Keywords Himalayas · Geophysical hazards · Risk reduction · Climate finance

K. Pillay (✉)
Global Change Institute, University of Witwatersrand, Johannesburg, South Africa

H. Ballabh
District Disaster Management Authority, Haridwar, Uttarkhand, India

S. Pillay
Department of Environmental Sciences, University of KwaZulu-Natal, Durban, South Africa
e-mail: pillays2@ukzn.ac.za

S. Sharma et al. (eds.), *Climate Change Adaptation, Risk Management and Sustainable Practices in the Himalaya*,
https://doi.org/10.1007/978-3-031-24659-3_5

1 Introduction

Changes to the Earth-atmosphere interactions generating climate change are not a new phenomenon. Paleoclimatic records have shown that the Earth's climate has varied over geologic timescales due to natural forcing such as orbital variations, volcanism, etc. and amplified by feedback mechanisms in the Earth-atmosphere system (Krishnan et al., 2020). However, it is recognized that in recent times, the Earth's atmospheric composition has been so significantly altered by such human activities as, inter alia, fossil fuel combustion, deforestation, air pollution, and land use changes, that humans have become the catalysts of accelerated climatic changes with knock-on impacts to ecological systems.

The Fifth Assessment Report by the Intergovernmental Panel on Climate Change (IPCC AR5) concluded that the Global Mean Surface Temperature (GMST), comprising of both global land surface, and global sea surface temperature (LSAT and SST, respectively) have increased since the mid-nineteenth century (Krishnan et al., 2020). In this regard, Hartmann et al. (2013), for instance, presented similar evidence based on Earth-atmosphere temperature changes as well as changes to the cryosphere, sea-level rise, ocean acidification, and the increase in extreme climate phenomena. Global average temperature has increased by 1 °C since pre-industrial times, while the average warming over India since the beginning of the last century (1901–2018) is around 0.7 °C (Krishnan et al., 2020). However, CMIP5 projections based on the high concentration radiative forcing (RCP8.5) for the next century (to 2100) show average temperatures over India to increase by approximately 4.4 °C. In like vein, CMIP5 models project increases in mean precipitation as well as its variability together with increases in daily precipitation extremes (Krishnan et al., 2020) over India.

The relative warming of the atmosphere has been experienced more acutely in mountain regions, most notably in the Hindu Kush Himalayas (HKH), which has experienced rapid warming of approximately 1.3 °C over the period 1951–2014 (Krishnan et al., 2020). This phenomenon has been accompanied by a significant decline in the Himalayan cryosphere together with changes in precipitation characteristics. These changes, coupled with an increase in climatically induced extreme events, have led to an increase in the severity, frequency, and magnitude of hazards such as floods and landslides. Geophysical hazards are not new or uncommon to the Himalayan Mountains (Ballabh et al., 2014; Krishnan et al., 2020). This tectonically active region in which the northward-moving Indian Plate continually drives into the Eurasian Plate has resulted in crustal deformation that has produced numerous faults, deformed, and sheared rocks with low cohesive stability, readily susceptible to mass movement. The towering, rugged, and awe-inspiring terrain is therefore a distinctly fragile environment (Ballabh et al., 2014).

In undisturbed terrain, the steep slopes with tectonically folded rock structures make landslides and slumps an ever-present danger but more so where human activities and developmental structures such as road infrastructure, hydroelectric power plants, settlements, and agricultural activities have significantly weakened the already sensitive environment. Geophysical hazards in the Himalayan context are inextricably interlinked with climate-related phenomena such as cloudbursts and storms that lead to flash flooding and pose a grave danger to inhabitants of the region. For instance, glacial retreat due to climate warming in the region has led to the infilling of glacial lakes while leaving behind unstable glacial till—all with the potential to trigger off a myriad of hazards such as floods and mudflows, landslides, and the slumping of large sections of mountain slopes. Over 11,000 natural incidents were reported by the Uttarakhand State Disaster Management Authority (USDMA) for the period 2015 to November 2021 with the majority being landslides (USDMA, 2021).

The most recent such incident occurred in February 2021 in Uttarakhand State, India, when a massive slab of frozen debris broke off a retreating glacier causing flash flooding, mudslides, and landslides down in the valley with considerable damage to property, infrastructure, and loss of life. The infrastructure damage to the Tapovan Vishnugad HEP alone is estimated at over 1500 crore rupees; 72 people lost their lives, while 135 persons were missing or unaccounted for (Thakkar, 2021). The devastation caused by this and other such natural disasters from an environmental standpoint is enormous, as is the loss of life and widespread economic setback for the inhabitants and the region.

The existing vulnerability of the region to hazards also often results in lost development gains. Repeated losses require repeated resourcing and re-financing. Both direct and indirect economic impacts are important at the community level (Pillay et al., 2017). Direct economic impacts refer to the loss of physical assets such as homes and damaged infrastructure for information and communication technology (ICT), transport, agriculture, and energy (Pillay, 2016). Alternatively, indirect economic impacts refer to loss of utility as a result of direct impacts such as the inability to travel to places of employment and supply chain disruptions.

To manage the impacts of the hazards, communities may integrate climate adaptation measures into their dwellings to reduce their risk exposure. Risk reduction measures must be coupled with financial resilience actions to ensure adequate risk management. Greater financial resilience can be delivered from public and private finance sources, both of which will be assessed in this study.

2 Focus Area

The Hindu Kush Himalaya (HKH) constitutes the most extensive mountain region on Earth, spanning over 4.2 million km^2 (Bajracharya et al., 2015) and is known to contain the largest area of glaciation outside the polar regions (Shrestha et al., 2012). This vast, biodiverse region contains four global biodiversity hotspots and is the source of numerous fluvial systems including ten of the largest river systems of southeastern Asia (Sharma et al., 2019) (Fig. 5.1).

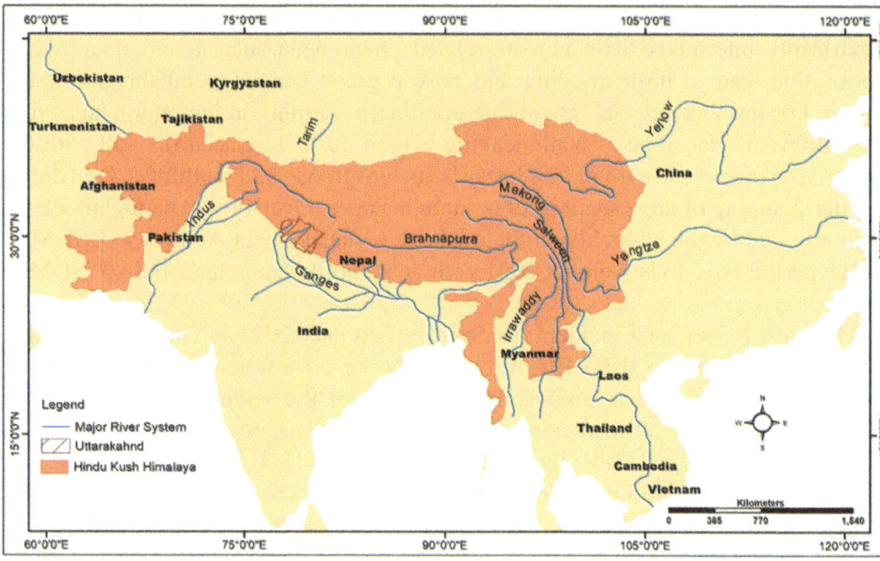

Fig. 5.1 The Hindu Kush Himalaya with the major river systems, bordering countries and the state of Uttarakhand in India

The 2017 population of the HKH was estimated at 240 million. This number scales up to an estimated 1.9 billion people if the population of the 10 major drainage basins containing rivers that emanate from, and are part of or peripheral to, the HKH are taken into consideration (Rasul et al., 2017). This study is based on a portion of the HKH, known as the Western Himalayas (WH) comprising dominantly of the Himalaya biodiversity hotspot and numerous protected areas. Three of the states of India that make up a large part of the WH are Jammu and Kashmir (including recently separated Ladakh), Himachal Pradesh, and Uttarakhand.

This chapter is focused principally on the latter, also known as Uttarakhand (UK). The UK region covers an area of close to 42,000 km^2 and comprises of 11 districts (Uttarkashi, Chamoli, Rudraprayag, Champawat, Tehri Garhwal, Pauri Garhwal, Almora, Pithoragarh, Nainital, Dehradun, and Bageshwar) (Apollo, 2017). Apollo (2017), using 2011 statistics, estimated the population of the Uttarakhand Himalaya at just over 6 million, with a population density of approximately 145 persons per km^2. This represents a 3.8-fold increase in population over a century, placing immense strain on resources and concomitant developmental pressures on an already fragile, natural hazard-prone environment. The changes to global climate have further exacerbated the situation by the increased propensity of hazards due to the rapidly changing mountain climate regime.

In this chapter, the main geophysical hazards of the Uttarakhand Himalayan part of the Western Himalayan region are discussed within the context of climate change and the subsequent economic implications of the impacts. The key objectives of this chapter are as follows:

- Outlining the main geophysical hazards of the region.
- Evaluating the changes to these hazards under future climate scenarios in terms of altered magnitudes and frequencies.
- Estimating economic losses from climate hazards at the community level.
- Identification of risk reduction measures that can be used to reduce the exposure of communities.
- Evaluating the financial options to increase resilience among communities.

3 Climate Drivers of Hazards in the Himalayas

3.1 Temperature

Since the 1850s to early 2020, concentrations of CO_2 have increased from 280 to 416 ppm, modulating radiative fluxes and inducing near-surface climatic changes (Krishnan et al., 2020). It is not surprising therefore that several studies and IPCC assessment reports implicate anthropogenic forcing for driving atmospheric warming by just over 1 °C since the 1850s. Krishnan and Dhara (2021) (in Krishnan et al., 2020) point out that since the mid-1950s, India has witnessed broad climatic changes including, inter alia, a rise in average temperatures, increases in extreme temperature and rainfall events, and a decrease in monsoon precipitation. However, they caution that the complexity of these changes is enhanced due to internal variability of climatic phenomena at local scales. Over the past half-century or more (50–60 years), significant temperature and precipitation shifts from the norm have been the general trend in the HKH region, while extremes of both parameters have become more frequent (Ren et al., 2017, in AR6). The mean decadal surface air temperature increase of 0.1 °C over the Himalayas over the period 2001–2014 has been shown to be more rapid than the global average (Ren et al., 2017, in AR6).

Shrestha et al. (2012) contend that while global mean surface temperature has increased on average by 0.8 °C in the last century, results show that the temperature increase in the Himalayas from 1982 to 2006 is 1.5 °C, considerably higher than the global average. This situation was previously mentioned in AR5 where the HKH was projected to continue the warming trend and at a rate higher than the global mean. Shrestha et al. (2012) reported CMIP5 results show temperature increases of 1–2 °C across the HKH with localized variability over the next three decades compared to 1961–1990. CMIP6 models report that the Western Himalayas are projected to experience temperature increases exceeding 6 °C by the end of the twenty-first century under SSP5–8.5 relative to 1995–2014 (Almazroui et al., 2020).

A further compounding factor is the rising trend of extreme warm events and a reduction of extreme cold events (observed over the five decades preceding 2015) (Krishnan et al., 2019; Wester et al., 2019). Added to this is the phenomenon of elevation-dependent warming (EDW) in which higher elevations have been observed to show a steeper rise in mean temperatures compared to low (coastal) elevations although this phenomenon is still to be better quantified.

Significantly, these climatic changes have resulted in the glacial recession and a reduction in glacial mass. These phenomena have been accompanied by the accumulation of meltwater in glacially carved depressions behind unconsolidated moraine deposits presenting potentially disastrous flood hazards. Glacial lakes so formed are often supplemented by melting permafrost and snow, enhancing lake volumes while simultaneously reducing soil and substrate cohesiveness.

3.2 Precipitation

CMIP5 models project increased precipitation over the HKH in the twenty-first century (Palazzi et al., 2015; Kitoh & Arakawa, 2016) although projections are subject to large uncertainties in CMIP5 and CORDEX 25 (Hasson et al., 2013, 2017; Mishra, 2015; Sanjay et al., 2017). CMIP6 projects an increase in winter precipitation over the western Himalayas, with a corresponding decrease in the east (Almazroui et al., 2020). IPCC (2018) stated that heavy precipitation risk in high-elevation regions is projected to be higher at 2 °C compared to 1.5 °C of global warming (medium confidence).

According to AR6, there is medium confidence that HKH precipitation will increase in the coming decades, while the SROCC predicts substantial glacial mass loss and degradation (with melting) of permafrost over the HKH and other mountain regions. Of course, the size of the area together with regional differences in warming and precipitation projections and glacier properties will cause considerable differences in glacier response (Kraaijenbrink et al., 2017).

Critical to the temperature and precipitation changes is the acceleration of glacial mass loss of the "third pole" due to decreased snowfall and a longer season of melting. This loss is anticipated to continue through the twenty-first century, increasing with RCP after 2030 (Marzeion et al., 2014; Section 9.5.1.3).

4 Climate Change Impacts in the Himalayas

A large number of hazards are experienced in the HKH, and the important climate change-related physical hazards impacting (or likely to impact) the UK are flood events caused by heavy rainfall, cloudbursts, or glacial lake outbursts (GLOFs), while landslides are frequent and often directly or indirectly linked to climate phenomena.

One of the more alarming consequences of warming over the Himalayas is glacial retreat with the concomitant formation of glacial lakes (King et al., 2017, 2018; Dubey & Goyal, 2020) as meltwater accumulates behind terminal moraine deposits. Sudden flood outbursts are the obvious hazard should the unconsolidated retaining moraine material give way, quickly releasing massive volumes of water into the laterally confined downstream valleys. Such glacial lake outburst floods (GLOF) may be triggered by a number of causes: masses of detached higher elevation material suddenly crashing into a glacial lake as rockfall, debris flows, landslides, or avalanches, instability of the supporting moraine, or sudden discharge of water into the lake from GLOF events further upslope (Carrivick & Rushmer, 2006; Carrivick & Tweed, 2013; Rounce et al., 2016). Devastation from GLOFs is of such magnitude as to completely wipe out entire settlements, farmlands, livestock, infrastructure, and buildings, together with the likelihood of loss of human lives. With regard to major developments, Schwanghart et al. (2016) note that 68% of all HEPs are located on potential GLOF pathways. It is therefore critically important to estimate risks associated with GLOF events and the potential financial implications of such loss.

Floods may also be caused by cloudbursts, an intense, short-duration storm occurring over an area of up to 30 km^2 and in which rainfall intensities may exceed 100 mm/hour (Indian Meteorological Department (IMD), 2013). Deoja et al. (1991) cite a downpour range of 200–1000 mm/day or, in a shorter duration, 100–250 mm/hour. Chaudhri and Sing (2012) point out that massive cloud formation is facilitated by an abundant supply of condensation nuclei, originating from the weathering and transformation of the dominant gneiss, biotite schist, and garnet mica schist into clays such as illite and kaolinite. Khandudi (2020) citing Das (2015) describes the cloudburst as a collapsed balloon of water falling over a relatively small area. Such large volumes of water, suddenly discharged into steep-sided, confined valleys, have the potential to cause major destruction as the water mass surges down in the valley. Over the past 6 years (2015 to November 2021), a total of 42 cloudburst incidents, averaging seven per year, were reported in Uttarakhand State (USDMA, 2021).

This phenomenon in the Himalayan region typically occurs during the monsoons (late June to early September) when thermodynamic instability results in rapid, dynamic uplift over the steep Himalayan orography (Das et al., 2006). The intense instability leads to rapid convection and the generation of dense cumulonimbus clouds, which gives rise to the cloudburst. Flash flooding, with valley and downstream inundation, ensues and may be accompanied by mass wasting phenomena such as debris flows, rock falls, and the more common landslides. The saturation of soils on the steep mountain slopes together with undercutting of riverbanks and lower slope support material presents ideal scenarios for the initiation of landslides.

A landslide is a general term for the downward movement of surface materials (including rock, soil, earth, or debris) (Gill and Denton, 2015). While landslides can be categorized into varying types, generally based on the type of movement involved, for the purposes of this paper, the relative frequency and magnitude of landslides as a consequence of climate drivers (such as intensity, frequency, and volume of rainfall, as well as floods and human activities) are considered. Events such as cloudbursts are characterized by heavy rainfall over short time periods (commonly exceeding 100 mm/hour). Heightened infiltration leads to soil/rock saturation and a rapid decrease of interparticle effective stress and resistive forces. Soil and rock material therefore experience decrease in their ability to stay rigid and solid and begin to move in response to gravity. Loss of soil and rock debris stability at high altitudes may be further exacerbated at high altitudes due to melting of permafrost and through glacial recession, increasing the risk of landslides. While landslides themselves are pervasive throughout the HKH region and therefore one of the major hazards experienced, they may also lead to secondary hazards such as increasing sediment yield of river systems, or the temporary damming of rivers caused by the deposition of landslide debris across river channels.

5 Consequences of Hazards in the Uttarakhand Himalayas

Flooding in the UK and in the greater HKH is a relatively common event following heavy rainfalls. However, extreme flood events are particularly destructive, sweeping away all before it, causing extensive erosion, transfer, and deposition of massive amounts of sediment, with loss of lives, homes, farmlands, livelihoods, infrastructure, and other developments.

Khandrudi (2020) reported on 50 individual cloudburst events that occurred in Uttarakhand over an approximate 50-year period from July 20, 1970, to September 7, 2019. This author provides detail of losses incurred as a consequence of each event, but overall, it is estimated that 1323 persons lost their lives as a direct consequence, while the number of missing persons totals 4026 (presumed dead). Of the latter, a large number of unaccounted persons (4021) are attributed to the Kedarnath Disaster that occurred on June 16–17, 2013. The cloudbursts that have occurred during the 50-year period also caused damage to homesteads (3299), livestock (with 2696 cattle lost), and a massive loss to infrastructure facilities, including roads and bridges. The financial losses run into hundreds of crore rupees.

In the Himalayas, as with similar terrain worldwide, landslides are one of the most pervasive recurring type hazards. Most of the landslides are triggered by excessive rainfall, but the cause of landslide generation can be attributable to one or more of a number of contingent causes.

The lower Himalayan region possesses immense potential for hydropower generation but is also highly susceptible to mass wasting, especially landslides. Susceptibility to landslides increases markedly with human activity, especially such large-scale developmental projects (Ballabh et al., 2014). The predicted increase in extreme climatic events portends disastrous consequences for the generation of more destructive landslides in the future. Khandrudi (2020) (citing Pradhan et al., 2020) notes that in the context of the Indian Himalayas, during the 50-year period mentioned above, 580 landslides occurred with 477 triggered by rainfall. Additionally, partial or complete destruction to road and other infrastructure is a concomitant hazard for every landslide that occurs where some development is situated.

The process of reconstruction, rehabilitation, and restoration of the devastation in the aftermath of extreme events in the UK and the entire HKH in general has resulted in significant financial losses. In many cases, restoration to prior status is often not possible with restoration goals having to be revised to more achievable targets. While the destruction of lives and livelihoods is difficult to overcome, more tangible physically based risk reduction, reconstruction, and restoration are highly dependent on institutional arrangements that govern disaster finance access and distribution. It is therefore important to consider the financing of disaster risk reduction as a precursor to aid the mitigation of hazards in the Himalayas.

6 Disaster Risk Reduction Financing in the Himalayas

6.1 Increases in Disaster Hazards Under Climate Change

Globally, disasters have increased in frequency from 991 in 2000 to 1100 to 2010 (Ray et al., 2019). In the last two decades, India has experienced a significant increase in frequency and magnitude of disasters. Floods are the highest source of annual disaster losses in India and are estimated to be approximately USD 7 bn every year (World Bank, 2021) as shown in Fig. 5.2. There is a likelihood that India will also experience more flooding as extreme precipitation events become more common under future climate scenarios. According to the INFORM risk index, India faces severe disaster risk (IMF, 2021). More specifically, India faces exposure to flooding (fluvial, pluvial, and coastal), tropical cyclones, and drought. Disaster risk in India is heightened by social vulnerability, which reduces the ability of the country to be resilient to disaster risk.

The World Bank estimates that India loses up to 2% of its GDP and 12% of government revenue to direct losses from natural disasters (Gulati, 2015). Therefore, there is a critical need to invest in disaster risk reduction (DRR) to reduce the

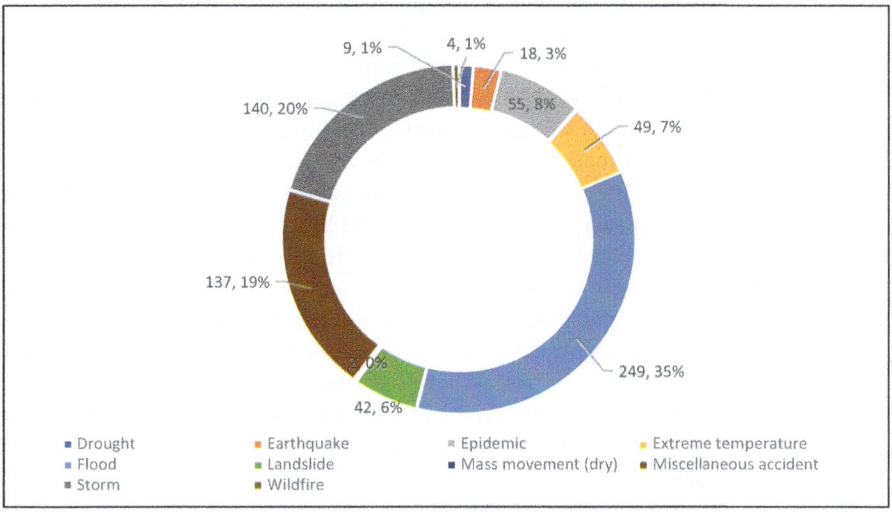

Fig. 5.2 Average annual natural hazard occurrence for 1980–2020. (Data Source: World Bank, 2021)

economic losses resulting from natural disasters. Investing in DRR in a timely manner not only reduces economic losses in the relief and reconstruction phases but also ensures that communities are able to rebound as quickly as possible. Investing in DRR has also been shown to be more cost-effective as opposed to managing post-disaster costs (Price, 2018).

7 India's Policy Framework for Disaster Management

The highest authority of disaster management in India is the National Disaster Management Authority (NDMA) with the nodal ministry being the Ministry of Home Affairs. The NDMA is assisted by the National Executive Committee (NEC) (Gulati, 2013). The NEC is chaired by the Home Secretary and brings together different ministries and government departments involved in DRM. The NDMA acts as the implementing arm of the NDMA enforcing approved policies, plans, and guidelines (Carty & Pozarny, 2016). The National Institute of Disaster Management is responsible for training, research and development, and capacity building. India has a specialized unit that is responsible for disaster relief—the National Disaster Response Force (Gulati, 2013). National-level disaster institutions are required to coordinate and interact with State and District level governments. DRR has not been explicitly defined in the Indian Disaster Management Act (2005) (Gulati, 2013).

The institutional landscape at the national level is similar at both the district and state levels, where a disaster management authority and executive committee are the governing bodies that are responsible for disaster management (Ogra et al., 2021). The Disaster Management Act of 2005 is the guiding legislation that sets out the disaster management architecture at the central, state, and district level in India. This ensures that disaster management plans are implemented and that they integrate measures focused on all phases of the disaster risk cycle, including prevention and risk reduction measures. Other policy levers that integrate DRR mainstreaming include the Ministry of Finance's stipulation for development plans of one billion rupees to undertake a natural disaster impact assessment (Chakrabarti & Prabodh, 2012). There is also coordination between the NDMA and the Planning Commission to ensure that technological innovation can facilitate greater integration of disaster risk reduction into planning processes. Nodal ministries and departments that possess DRR-related plans and programs include the Ministry of Environment and Forest, Ministry of Water Resources, Ministry of Earth Sciences, Ministry of Health and Family Welfare, Ministry of Rural Development, and Department of Space.

The Disaster Management Act also provides for the allocation of disaster funds at different levels of government (Bindal et al., 2021). More specifically, the Act provides for the development of a National Disaster Mitigation Fund that focuses on funding for disaster mitigation. The Fund is envisioned to cover funding gaps that may occur owing to different funding mandates at the central, state, and district levels (Gulati, 2013). The Act also encourages the development of state-level funds.

In terms of disaster relief, costs are funded through the State Disaster Relief Fund (SDRF). If disasters are high-magnitude, funds may be supplemented through the National Disaster Relief Fund. It must be noted that funds may only be used for relief with certain disasters not being covered by these funds. Consequently, there is a funding gap for the reconstruction phase, post-disaster, with affected parties often reallocating internal funds or relying on external aid. The lack of funding for reconstruction is also exacerbated by the speed in which these funds are available. This has implications for the ability of a region to regain development losses.

The primary objective of the National Policy on Disaster Management 2009 is to ensure efficient response and relief to vulnerability in society. The National Mission for Sustaining the Himalayan Ecosystem (NMSHE) is a mission that has been initiated under the National Action Plan on Climate Change (NAPCC) (NMSHE, 2021). NMSHE focuses on enhancing the understanding of climate change, impacts, and adaptation initiatives required for the Himalayas.

8 Financing Disaster Risk Reduction and Climate Adaptation in HKH

Finance is the most significant challenge to achieving holistic disaster risk management in the HKH (Mishra et al., 2019). For most disasters in the recent past, long-term disaster recovery funds have not been available. More severe natural disasters such as the 2004 tsunami event have received international support (mostly from the World Bank and the Asian Development Bank (ADB)) (World Bank, 2005). There is a significant gap in funding between the economic losses incurred in India and financial support gained through the SDRF/NDRF and state development funds. This highlights the need to explore risk transfer approaches. The focus of disaster relief by the Indian government has resulted in disaster mitigation and resilience-building efforts remaining dispersed, untargeted, and uncoordinated.

Enhancing disaster risk reduction in the HKH will require substantial capital flows for implementation efforts. Furthermore, owing to the lack of tangibility of disaster risk reduction benefits, private financing to support adaptation has not materialized (Pillay et al., 2017). Success of these models has already been demonstrated by funding mechanisms in operation in Bhutan, Bangladesh, and Nepal. Bangladesh established a climate change trust fund that has mobilized USD600 million, with USD400 million being mobilized from internal government resources and a further USD200 million from development financial institutions (Mishra et al., 2019). Another example is the Bhutan Trust Fund for Environmental Conservation, which supports small-scale activities at the local and community level (Mishra et al., 2019).

Enhanced participation in adaptation financing by the private sector is crucial for the scaling of funding flows ensuring that national and regional adaptation targets are reached (Pauw et al., 2016). Adaptation finance from global climate finance sources

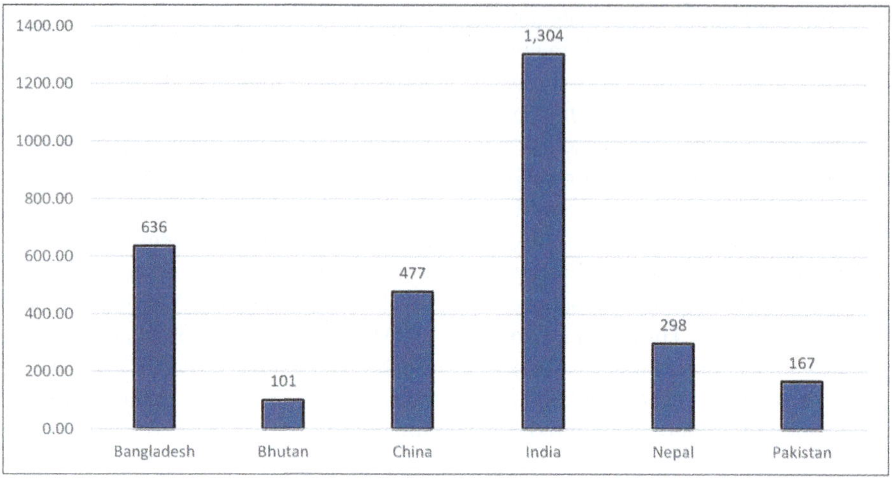

Fig. 5.3 Total amount of funding approved and received from climate finance sources by HKH countries (USD millions). (Climate Funds Update, 2021)

has been slow to reach local government and community levels with mobilized funds being insufficient to meet the financing needs for the HKH countries (DCF Alliance, 2019). The costing of disaster risk reduction and adaptation needs is a difficult undertaking (Pillay et al., 2017). This is owing to uncertainties of future climate scenarios and the appropriate initiatives to significantly reduce risk. Therefore, a comprehensive assessment of disaster risk reduction requirements is needed for the HKH. Attempts have been made to quantify the high-level financing needs for the HKH region. It is unlikely that international climate finance flows will be sufficient to meet the disaster risk reduction needs of HKH countries. Therefore, it is critical that domestic and private sector sources be mobilized.

According to the Climate Funds Update (2021), HKH countries, as shown in Fig. 5.3, have managed to mobilize approximately USD 2.9 billion from different sources of climate finance. In the HKH region, proactive disaster risk reduction undertaken by formal businesses has been absent. Furthermore, there has been little quantification of the possible impacts that could result from increasing climate risk on business operations. This is likely to change with the Task Force on Climate-Related Financial Disclosures (TCFD) regulations should they become mandatory. TCFD regulations will necessitate that businesses evaluate and report their exposure to climate risks (TCFD, 2017). India's disaster risk exposure is exacerbated by rapid urbanization. For example, the population of Delhi has increased from 2 to 16 million people between 1950 and 2008, thereby increasing flood risk exposure (UNFCCC, 2009). The Government of India has estimated that it would require USD 230 million over the next 5 years to support the National Mission for Sustaining the Himalayan Ecosystem (Mishra et al., 2019). A significant portion of India's National Adaptation Fund for Climate Change (NAFCC) has supported projects in the Himalayan region

(~ USD 68 million). Most project types have been focused on sustainable agriculture (Gupta et al., 2018).

Coordination between business associations and communities is critical to foster greater adaptation implementation. This is demonstrated by stakeholder engagements with the Association of Bhutanese Industries and the Bhutan Chamber of Commerce and Industry (Mishra et al., 2019). In the case of Bhutan, the Punatsangchhu Hydro Projects I and II provided for financial support for the installation of early-warning systems (UNDP, 2012). Private sector buy-in has also been garnered in Bangladesh mostly for agriculture and climate change agriculture (Fayolle et al., 2019). Other examples that are noteworthy in the HKH are the pilot program of climate resilience in Nepal delivered by the IFC where agribusinesses provided training to farmers to reduce crop losses and increase productivity under changing climatic conditions (IFC, 2016). In Uttarakhand, private companies are experimenting with ICT-based business models allowing for weather forecasting for farmers.

9 Comprehensive Disaster Risk Financing

An appropriate disaster risk financing strategy must be based on a comprehensive risk analysis with a needs and gaps assessment. Within the risk analysis, an understanding of the appropriate balance between risk retention and risk transfer must be attained within a layering approach as outlined in Fig. 5.4 (Warner et al., 2013). In addition to assessing financing needs, technological innovation requirements must also be assessed, thereby creating an enabling environment for DRR implementation. The engagement of the private sector through the development of risk transfer

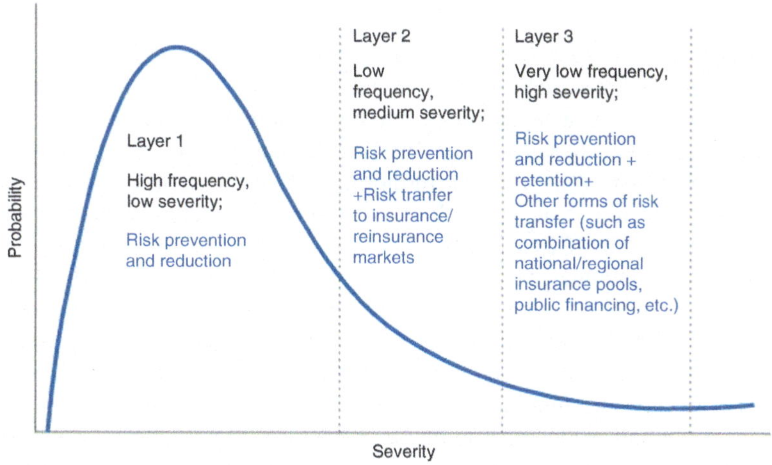

Fig. 5.4 Layering approach to disaster risk financing. (Warner et al. 2013)

markets is critical to ensure that their support can be leveraged, thus ensuring that funding gaps are closed and that long-term financial sustainability is prioritized.

Risk transfer instruments such as insurance require consideration as they would be used to cover the finance gap for relief efforts that are not covered by SDRF/NDRF scheme and to fund non-immediate rehabilitation and reconstruction (Gulati, 2015). More generally, insurance will also reduce the reliance of the Indian government on contingent reserves and funds—self-insurance. Insurance would typically be used for low-frequency, high-severity disasters. Insurance may also be able to initiate more risk-averse behavior through well-design incentives that allow for premium cost reductions if risk reduction implementation is undertaken by the policyholder (Pillay, 2016). Disaster insurance offerings would need to bear low-income groups in mind and the subsequent affordability of products targeted at these groups. Microinsurance will be discussed in detail in the next section. It is likely that government would need to support premiums of insurance products for low-income groups for a particular period (Pillay et al., 2017). Donor financing may also be a possible option to support premiums of low-income groups.

Beyond the availability of insurance policies for the government, a well-developed insurance sector is critical for ensuring that economic impacts of natural disasters are limited. At present, NatCat insurance offerings are not common in the Indian market (Ray et al., 2019). Furthermore, the reinsurance market for natural catastrophe coverage is also underdeveloped in India (Ray et al., 2019). A well-developed insurance and reinsurance sector also limits the opportunity for partnerships. Partnerships between government and the insurance sector have been used to jointly support for low-income groups directly or their participation in microinsurance schemes.

10 Microinsurance in India

There are several social protection schemes that have been implemented by the Government of India, which have leveraged insurance solutions. An example of such a scheme is the Ayushman Bharat Pradhan Mantri Jan Arogya Yojana scheme (Bindal et al., 2021). Microinsurance service offerings are well-developed in India. However, the primary issue limiting further growth of the market is not the affordability of microinsurance but the lack of a robust administration system that allows for ease of access (UNFCCC, 2009).

Typically, low-income groups are significantly affected by natural disasters. This is owing to their small asset bases, exposure, and their lack of savings. According to the UNFCCC (2009), less than 5% of annual income is contributed to savings in rural northern India with many farmers in debt. Consequently, there is a high dependence of rural communities on the government for both protection and support during the relief and reconstruction phase of the disaster risk cycle. From a regulatory perspective, policies that have been implemented that require insurance companies to grow their quota of low-income clients over time have influenced the growth of microinsurance products (Roth et al., 2007). Technological innovation has also enhanced the interest of insurers in the low-income segment. More specifically, technological innovation has assisted insurers in creating improved channels to access low-income clients. This has traditionally been a significant issue that has limited insurance penetration.

For communities that are affected by natural disasters, the speed of insurance payout is instrumental in minimizing the impact and facilitating recovery. Consequently, microinsurance schemes that operate using parametric policies may be able to facilitate faster payouts as they are based on exceedances of thresholds as opposed to on-the-ground post-disaster economic assessments. In the case of the sovereign risk pools focused on disaster risk, payouts occurred between 7 and 14 days after a natural disaster occurred (Pillay, 2016). Parametric insurance policies can be challenging as there may be a mismatch between economic damages and payouts (basis risk). This is especially the case if the quality of historical data is poor. There is also a possibility that economic damages may not result in a payout as a certain threshold may not be exceeded. This can reduce trust in microinsurance policies. Parametric policies are envisioned to self-correct over time with the payout-threshold relationship being more accurate as more data is integrated into policy design. Payouts may be used for any purpose—damages to assets owned by the policyholder or losses experienced as a result of lost income.

Based on the current market, weather index-based crop insurance is a focus area of the microinsurance segment. For example, in the state of Himachal Pradesh, microinsurance is available for a multitude of crops including ginger, potato, tomato, and pea crops (Mishra et al., 2019). Pilot schemes are also being implemented in Uttarakhand and Assam States by private insurers (UNFCCC, 2009). Beyond crop microinsurance, there are also pilots being developed in Bangladesh and Nepal focused on flooding and hailstorms, respectively (GFDRR, 2009; Swiss Re, 2018). For microinsurance based on parametric policies in particular, significant investments must be made in weather stations to create indices on which the policies are based. Often insurers are hesitant to make these investments without a detailed understanding of the potential market size and demand.

11 Concluding Remarks

This chapter outlines the main climate-related geophysical hazards that impact the Uttarakhand State of the Western Himalayan region. Temperature and precipitation variations have been shown in prior research to be expected outcomes of climate change. This paper provides a synopsis of these changes including an overview of the consequences of geophysical hazards experienced. The process of reconstruction, rehabilitation, and restoration of the devastation post the extreme events in Uttarakhand and the entire HKH in general has resulted in significant financial losses. While lives and livelihoods lost are difficult to overcome, physically based risk reduction, reconstruction, and restoration is possible but dependent on institutional arrangements that govern disaster finance access and distribution. Finance is the most significant challenge to achieving holistic disaster risk management in the HKH. Enhancing disaster risk reduction in the HKH will require substantial capital flows for implementation. It is essential that decision-makers and the wider financial sector consider options that move beyond the reliance on the public sector such that comprehensive disaster risk management can be enabled. Comprehensive disaster risk financing could include the greater use of insurance, microinsurance, risk pooling, catastrophe bonds, and countercyclical financial instruments. It is essential that disaster risk financing not only consider post-disaster financing but also mechanisms that promote risk reduction and adaptation investments. This will ensure that development gains are protected and maintained, especially in contexts where post-disaster financing may be slow.

References

Almazroui, M., Saeed, S., Saeed, F. Islam, M. N., & Ismail M., (2020). Projections of Precipitation and Temperature over the South Asian Countries in CMIP6. *Earth Systems and Environment, 4*, 297–320. https://doi.org/10.1007/s41748-020-00157-7

Appolo, M. (2017). *Contemporary studies in environment and tourism. Chapter 9: The population of Himalayan regions—By the numbers: Past, present and future*. Cambridge Scholars Publishing.

Bajracharya, S. R., Maharjan, S. B., Shrestha, F., Guo, W., Liu, S. Immerzeel W., and B. Shrestha, B., (2015). The glaciers of the Hindu Kush Himalayas: current status and observed changes from the 1980s to 2010. *International Journal of Water Resources Development, 31*(2), 161–173. https://doi.org/10.1080/07900627.2015.1005731

Ballabh, H., Pillay, S., Negi, G. C. S., & Pillay, K. (2014). Relationship between selected physiographic features and landslide occurrence around four hydropower projects in Bhagirathi Valley of Uttarakhand, Western Himalaya, India. *International Journal of Geosciences, 5*, 1088–1099.

Bindal, M. K., Kumar, S., Thomas, G., & Gupta, A. (2021). *Working group report: Disaster risk financing, insurance and risk transfer*. National Institute of Disaster Management.

Carrivick, J. L., & Rushmer, E. L. (2006). Understanding high-magnitude outburst floods. *Geology Today, 22*(2), 60–65. https://doi.org/10.1111/j.1365-2451.2006.00554.x

Carrivick, J. L., & Tweed, F. S. (2013). Proglacial lakes: Character, behaviour and geological importance. *Quaternary Science Reviews, 78*, 34–52. https://doi.org/10.1016/j.quascirev.2013.07.028

Carter, B., & Pozarny, P. (2016). *National Disaster Management Authorities (NDMA)* (pp. 1–14). Helpdesk Research Report.

Chakrabarthi, D., & Prabodh, G. (2012). *Understanding existing methodologies for allocations and tracking DRR resources in India* (pp. 1–94). UNISDR.

Chaudhri, A., & Sing, M. (2012). Clay minerals as climate change indicators- A case study. *American Journal of Climate Change, 1*, 231–239.

Climate Funds Update. (2021). Available online https://climatefundsupdate.org/

Das, S., Ashrit, R., & Moncrieff, M. W. (2006). Simulation of a Himalayan cloudburst event. *Journal of Earth System Science, 115* (3), 299–313.

Das, P.K., 2015. Global warming, glacial lakes and cloud burst events in Garhwal-Kumaon Himalaya: A hypothetical analysis. International Journal of Environmental Sciences, 5(4), 697-708.

DCF Alliance. (2019). *The devolved climate finance mechanism: Principles, implementation and lessons from four semi-arid countries* (pp. 1–40). IIED.

Deoja, B., Dhital, M., Thapa, B., & Wagner, A. (1991). *Mountain risk engineering handbook, Part I & II* (p. 875). ICIMOD.

Dubey, S., & Goyal, M. K. (2020). Glacial lake outburst flood hazard, downstream impact, and risk over the Indian Himalayas. *Water Resources Research, 56*. https://doi.org/10.1029/2019WR026533

Fayolle, V., Fouvet, C., Soundarajan, V., Nath, V., Acharya, S., Gupta, N., & Petrarulo, L. (2019). *Engaging the private sector in financing adaptation to climate change: Learning from practice* (pp. 1–36). ACT.

Gill, J. R., & Denton, T. P. (2015). Landslides. In J. Gill & R. Tostevin (Eds.), *Hazards in the Himalayas* (Geology for global development). UK School of Seismology Project.

Global Facility for Disaster Reduction and Recovery. (2009). *Agricultural insurance feasibility study for Nepal* (pp. 1–114). World Bank.

Gulati, A. G. (2013). *Financing disaster risk reduction* (pp. 1–9). Presentation to Expert Meeting Group on Effective Strategies for Mainstreaming Disaster Risk Reduction in Asia and the Pacific.

Gulati, A. G. (2015). Financing disaster risk reduction: The Indian context. *Society for Policy Studies*, 1–6.

Gupta, N., Mishra, A., Agrawal, N. K., & Satapathy, S. (2018). Inter-state cooperation to address climate change adaptation in the Indian Himalayan region—Need of the hour. *Economic and Political Weekly, 12*, 35–40.

Hasson S., Böhner, J., & Lucarini. V. (2017). Prevailing climatic trends and runoff response from Hindukush–Karakoram–Himalaya, upper Indus Basin. *Earth System Dynamics, 8*, 337–355. https://www.earth-syst-dynam.net/8/337/2017/. https://doi.org/10.5194/esd-8-337-2017

Hasson, S., Lucarini, V., & Pascale, S. (2013). Hydrological cycle over South and Southeast Asian river basins as simulated by PCMDI/CMIP3 experiments. *Earth System Dynamics, 4*(2), 199–217. https://doi.org/10.5194/esd-4-199-2013

Hartmann, D. L., Klein, A. M. G., Tank, M. Rusticucci, L. V., Alexander, S., Brönnimann, Y., Charabi, F. J., Dentener, E. J., Dlugokencky, D.R., Easterling, A., Kaplan, B. J., Soden, P. W., Thorne, M., Wild & P.M. Zhai. (2013). Observations: Atmosphere and Surface. In Stocker, T. F., D. Qin, G.-K. Plattner, M. Tignor, S. K. Allen, J. Boschung, A. Nauels, Y. Xia, V. Bex, & P. M. Midgley (Eds.), *Climate change 2013: The Physical Science Basis. Contribution of Working Group I to the Fifth Assessment Report of the Intergovernmental Panel on Climate Change*. Cambridge University Press, Cambridge, United Kingdom and New York, NY, USA.

IMD. (2013, July). *A preliminary report on heavy rainfall over Uttarakhand during 16–18 June 2013*. Government of India, Ministry of Earth Science, Indian Meteorological Department, New Delhi.

IPCC. (2018). *Global Warming of 1.5°C. An IPCC Special Report on the impacts of global warming of 1.5°C above pre-industrial levels and related global greenhouse gas emission pathways, in the context of strengthening the global response to the threat of climate change, Sustainable development, and efforts to eradicate poverty* (eds.) [Masson-Delmotte, V., P. Zhai, H.-O. Pörtner, D. Roberts, J. Skea, P.R. Shukla, A. Pirani, W. Moufouma-Okia, C. Péan, R. Pidcock, S. Connors, J.B.R. Matthews, Y. Chen, X. Zhou, M.I. Gomis, E. Lonnoy, T. Maycock, M. Tignor, and T. Waterfield].

International Finance Corporation. (2016). *Planting a more resilient future for farmers in Nepal*. https://www.ifc.org/wps/wcm/connect/news_ext_content/ifc_external_corporate_site/news+and+events/news/impact-stories/planting-a-more-resilient-future-for-farmers-in-nepal

International Monetary Fund (IMF). (2021). *IMF climate change dashboard*. Available online https://climatedata.imf.org/pages/fi-indicators

Khandrudi, S. (2020). Cloudbursts over Indian sub-continent of Uttarakhand Himalaya: A traditional habitation input from Bansoli, District-Chamoli, India. *International Journal of Earth Sciences Knowledge and Applications, 2*(2), 48–63.

King, O., Dehecq, A., Quincey, D., & Carrivick, J. (2018). Contrasting geometric and dynamic evolution of lake and land-terminating glaciers in the central Himalaya. *Global and Planetary Change, 167*, 46–60. https://doi.org/10.1016/j.gloplacha.2018.05.006

King, O., Quincey, D. J., Carrivick, J. L., & Rowan, A. V. (2017). Spatial variability in mass loss of glaciers in the Everest region, Central Himalayas, between 2000 and 2015. *The Cryosphere, 11*(1), 407–426. https://doi.org/10.5194/tc-11-407-2017

Kitoh, A., & Arakawa, O. (2016). Reduction in the east–west contrast in water budget over the Tibetan Plateau under a future climate. *Hydrological Research Letters, 10*(4), 113–118. Published online in J-STAGE https://www.jstage.jst.go.jp/browse/hrl). https://doi.org/10.3178/hrl.10.113

Kraaijenbrink, P., Bierkens, M., Lutz, A. & Immerzeel, W. W., (2017) Impact of a global temperature rise of 1.5 degrees Celsius on Asia's glaciers. *Nature 549*, 257–260. https://doi.org/10.1038/nature23878

Krishnan & Dhara (2021), Executive Summary in R. Krishnan, J. Sanjay, C. Gnanaseelan, M. Mujumdar, A. Kulkarni & S. Chakraborty: Editors: *Assessment of Climate Change over the Indian Region. A Report of the Ministry of Earth Sciences (MoES), Government of India.* (eBook). https://doi.org/10.1007/978-981-15-4327-2

Krishnan, R., Sanjay, J., Gnanaseelan, C., Mujumdar, M., Kulkarni, A., & Chakraborty, S. (Eds.). (2020). *2020: Assessment of climate change over the Indian region: A report of the Ministry of Earth Sciences (MoES), Government of India* (p. 226). Springer. https://doi.org/10.1007/978-981-15-4327-2. Cited sections: Introduction and Chapter 11.

Krishnan, R., Shrestha, A. B., Ren, G., Rajbhandari, R., Saeed, S., Sanjay, J., Syed, M. A., Vellore, R., Xu, Y., You, Q., & Ren, Y. (2019). Unravelling climate change in the Hindu Kush Himalaya: rapid warming in the mountains and increasing extremes. In: Wester P, Mishra A, Mukerji A, Shrestha AB (eds) *The Hindu Kush Himalaya assessment: mountains, climate change, sustainability and people*. Springer Nature, Cham, Switzerland.

Marzeion B., Cogley, J. G., Richter K., & Parkes D. (2014) Attribution of global glacier mass loss to anthropogenic and natural causes. *Science, 345*, (6199), 919–921. https://doi.org/10.1126/science.1254702

Mishra, A., Appadurai, N., Regmi, B., & Alam, M. (2019). Adaptation to climate change in the Hindu Kush Himalaya. In *The Hindu Kush Himalaya assessment: Mountains, climate change, sustainability and people* (pp. 1–35). Springer.

Mishra, V. (2015). *Climatic uncertainty in Himalayan water towers. Journal of Geophysical Research-Atmospheres, 120*(7), 2689–2705.

National Mission for Sustaining the Himalayan Ecosystem. (2021). *About NMSHE.* Available online http://www.knowledgeportal-nmshe.in/

Ogra, A., Donovan, A., Adamson, G., Viswanathan, K. R., & Budimir, M. (2021). Exploring the gap between policy and action in disaster risk reduction: A case study of India. *International Journal of Disaster Risk Reduction, 63*, 102428.

Palazzi, E., Hardenberg, J., Terzago, S. & Provenzale, A. (2015). Precipitation in the Karakoram-Himalaya: a CMIP5 view. *Climate Dynamics, 45*, 21–45. https://doi.org/10.1007/s00382-014-2341-z

Pauw, W. P., Klein, R. J. T., Vellinga, P., & Biermann, F. (2016). Private finance for adaptation: Do private realities meet public ambitions. *Climate Change, 134*, 489–503.

Pillay, K. (2016). *Climate risk management: Opportunities and challenges for risk pooling* (pp. 1–32). Commonwealth Small States Digest.

Pillay, K., Aakre, S., & Asbjorn, T. (2017). *Mobilizing adaptation finance in developing countries* (pp. 1–42). CICERO.

Pradhan, B., Dikshit, A., Sarkar, R., Segoni, S., & Alamri, A. M. (2020). Rainfall induced landslide studies in Indian Himalayan region: A critical review. *Applied Sciences, 10*, 1–24.

Price. (2018). *Cost-effectiveness of disaster risk reduction and adaptation to climate change* (pp. 1–21). Institute of Development Studies.

Rasul, G., Hussain, A., Mahapatraa, B., & Dangola, N. (2017). Food and nutrition security in the Hindu Kush Himalayan region. *Journal of the Science of Food and Agriculture, 98*(2). https://doi.org/10.1002/jsfa.8530

Ray, S., Jain, S., & Thakur, V. (2019, September). *Financing resilience against natural disasters*. Report. Indian Council for Research on International Economic Relations, New Delhi.

Ren, Y. Y., Ren, G. Y., Sun, X. B., Shrestha, A. B., You, Q. L., Zhan, Y. J., et al. (2017). Observed changes in surface air temperature and precipitation in the Hindu Kush Himalayan region during 1901–2014. *Advances in Climate Change Research, 8*(3). https://doi.org/10.1016/j.accre.2017.08.001

Roth, J., McCord, M., & Liber, D. (2007). *The landscape of microinsurance in the world's poorest 100 countries*. Microinsurance Center.

Rounce, D. R., McKinney, D. C., Lala, J. M., Byers, A. C., & Watson, C. S. (2016). A new remote hazard and risk assessment framework for glacial lakes in the Nepal Himalaya. *Hydrology and Earth System Sciences, 20*, 3455–3475. https://doi.org/10.5194/hess-20-3455-2016

Sanjay, J., Krishnan, R., Shrestha, A. B., Rajbhandari, R., & Ren, G-Y. (2017). Downscaled climate change projections for the Hindu Kush Himalayan region using CORDEX South Asia regional climate models. *Advances in Climate Change Research 8*(3):185–198. https://doi.org/10.1016/j.accre.2017.08.003

Schwanghart, W., Worni, R., Huggel, C., Stoffel, M., & Korup, O. (2016). Uncertainty in the Himalayan energy-water nexus: Estimating regional exposure to glacial lake outburst floods. *Environmental Research Letters, 11*(7), 074005. https://doi.org/10.1088/1748-9326/11/7/074005

Sharma, E., D. Molden, A. Rahman, Y. R. Khatiwada, L. Zhang, S. P. Singh, T. Yao, et al., (2019). Introduction to the Hindu Kush Himalaya Assessment. In: *The Hindu Kush Himalaya Assessment: mountains, climate change, sustainability and people* (eds. Wester, A., A. Mukherji, and A.B. Shrestha). Pp. 1–16. Cham: Springer International Publishing.

Shrestha, U. B., Gautam, S., Bawa, K. S. (2012). Widespread climate change in the Himalayas and associated changes in local ecosystems. *PLoS ONE 7*(5), 36741. https://doi.org/10.1371/journal.pone.0036741

Swiss Re. (2018). *Closing the gap: Flood protection for Bangladesh*. Available online https://www.swissre.com/dam/jcr:9d44ab7c-8087-469c-88f4-5d9e23148a1a/Case_study_Bangladesh.pdf

TCFD. (2017). *Recommendations of the task force on climate-related financial disclosures*.

Thakkar, H. (2021, March 6). *The force multipliers of disasters in Uttarakhand Economic & Political Weekly*, p. 13.

United Nations Development Programme. (2012). *Technical and social impact assessment: Reducing climate change-induced risk and vulnerabilities from glacial lake outburst floods in Punakha, Wangdue and Chamkhar Valleys*.

United Nations Framework Convention on Climate Change (UNFCCC). (2009). *Mechanisms to manage financial risks from direct impacts of climate change in development countries* (pp. 1–114). UNFCCC.

USDMA. (2021). *Uttarakhand State Disaster Management Authority online incident logging system*. http://usdma.uk.gov.in/sachet/DashBoard/Default.aspx. Accessed on 19/10/2021.

Warner, K., Yuzva, K., Zissener, M., Gille, S., Voss, J., & Wanczeck, S. (2013). *Innovative insurance solutions for climate change: How to integrate climate risk insurance into a comprehensive climate risk management approach*. Munich Climate Insurance Initiative and GIZ.

Wester, P., Mishra, A., Mukherji, A., & Shrestha, A. B. (Eds.). (2019). *The Hindu Kush Himalaya assessment: Mountains, climate change, sustainability and people*. ICIMOD, HIMAP. https://doi.org/10.1007/978-3-319-92288-1

World Bank. (2005). *World Bank committed to rebuilding tsunami devasted region*. Available online https://reliefweb.int/report/india/world-bank-committed-rebuilding-tsunami-devastated-region

World Bank. (2021). *Climate change knowledge portal: India*. Available online https://climateknowledgeportal.worldbank.org/country/india/vulnerability

Chapter 6
Assessment of Existing Himalayan Glacier Inventories for Glacier Studies: A Case Study from the Ravi Basin of North-Western Himalaya (India)

Ishtiaq Ahmed, Vikram Sharma, Rinku Kumar, Devi Lal, Rajan Bhandari, and Pritam Chand

Abstract Outside of the polar regions, the Hindukush-Karakoram-Himalaya (HKH) has the highest cluster of snow cover and glaciers, which offer various ecosystem services, including water, to the billions of people who live across this region. A glacier inventory is a vital prerequisite for researching a wide range of diverse phenomena, processes, and effects of such glacier changes across these regions. In recent years, several glacier inventories are available for the HKH region, namely, the Geological Survey of India (GSI) Glacier Inventory, the Space Application Center (SAC) Glacier Inventory, the International Centre for Integrated Mountain Development (ICIMOD) Glacier Inventory, Randolph Glacier Inventory (RGI), and Glacier Area Mapping for Discharge from the Asian Mountains (GAMDAM) Glacier Inventory (GGI). Prior to being used for any glacier investigations, it's critical to evaluate the quality and consistency of these inventory datasets. Thus, the current study provides a detailed quality assessment of all these available glacier inventories by comparing them with the detailed Ravi basin glacier inventory (RBGI). The comprehensive RBGI was created using the Landsat Enhanced Thematic Mapper (ETM+) images (2002) with a supplement of medium- to high-resolution imagery and field validations. The RBGI consists of 285 glaciers in 2002 with a mapped area of 164.5 ± 7.5 km^2. There are 71 glaciers out of the total glaciers that have debris-covered parts, which occupy 36.1 ± 2.1 km^2 (~22% of the whole area covered by glaciers). Large variations were found in the glacial area

I. Ahmed · R. Kumar · D. Lal · P. Chand (✉)
Department of Geography, School of Environment and Earth Sciences, Central University of Punjab, Bathinda, Punjab, India
e-mail: pritam.chand@cup.edu.in

V. Sharma
Department of Geography, Institute of Science, Banaras Hindu University, Varanasi, India

R. Bhandari
Department of Geography, DAV College, Panjab University, Chandigarh, India

S. Sharma et al. (eds.), *Climate Change Adaptation, Risk Management and Sustainable Practices in the Himalaya*,
https://doi.org/10.1007/978-3-031-24659-3_6

(ranging from 202 to 112.7 km^2) and a total number of glaciers (ranging from 299 to 192) mapped within the Ravi basin among these available glacier inventories. With few spatial differences in the total number of the glacier, their extent, and median elevation, it was found that the recently updated GGI inventory, which is incorporated into the revised version of RGI V6 for the Himalayan region, is most comparable to our RBGI inventory. Likely causes of the significant difference among these inventories include standard glacier definition (minimum area of glacier mapping and headwall definition), misinterpretation of the seasonal snow cover, demarcation of debris-covered areas, and consequences of excluding glacier sections in the shaded regions.

Keywords Glacier inventory · Ravi Basin Glacier Inventory (RBGI) · Digital Elevation Model · Glacier mapping · Satellite images · Debris covered

1 Introduction

Outside of the polar regions, the Hindukush-Karakoram-Himalaya (HKH) region has the largest cluster of snow cover and glaciers. The widespread melting and receding of the Himalayan glacier are currently one of the most manifest effects of reported climate warming (Mayewski & Jeschke, 1979; Bolch et al., 2012). The life of heavily inhabited areas downstream is dependent on river discharge, which is continuously altered by shrinking glaciers and shorter snow cover periods (Barnett et al. 2005); Immerzeel et al., 2010). Moreover, the formation of moraine-dammed lakes, which are vulnerable to catastrophic outburst floods and pose major risks to mountain communities and infrastructure, is commonly associated with such glacier retreat across these regions (Worni et al., 2013). Therefore, more thorough long-term monitoring of the Himalayan glaciers is crucial and necessitates datasets providing information on its precise extent and topographic characteristics, or glacier inventories. During the last decade, a number of organizations and institutes (or groups/consortiums) provided glacier inventories, for example, Geological Survey of India (GSI) Glacier Inventory (Raina & Srivastava, 2008), the Space Application Center (SAC) Glacier Inventory (SAC, 2010), the International Centre for Integrated Mountain Development (ICIMOD) Glacier Inventory (Bajracharya & Shrestha, 2011), Randolph Glacier Inventory (RGI) (Pfeffer et al., 2014), and Glacier Area Mapping for Discharge from the Asian Mountains (GAMDAM) Glacier Inventory (GGI) (Nuimura et al., 2015) for the Himalayan region. These inventories have been used to evaluate recent glacier change (Bajracharya et al., 2014), regional scale glacier volume, ice thickness, mass balance (Maurer et al., 2016), and glacier lake evolution (Gardelle et al., 2011) as well as to make correlations between the distribution of glacier altitudes and precipitation (Nuimura et al., 2015). However, there has been a reported variance in these glacier inventories throughout the Himalayan region (Nuimura et al., 2015). Such variances and uncertainties can be ascribed to glacier characterization (i.e., selection of glacier minimum area for mapping), selection of dataset, flaws in implemented glacier mapping semiautomatic methods particularly for the debris-covered area, co-registration between multi-resolution images and aerial photography, categorization criteria, etc. Additional factors that distort the results and cause large-scale variances in such inventories include void zones, noise, and the different used digital elevation models (DEMs), for example, Shuttle Radar Topography Mission (SRTM) and ASTER GDEM, which are used for analyzing topographic details. As a result, any changes in the glacier area and its related regional characteristics particularly three-dimensional (3-D) characteristics that these inventories may suggest may be more fictitious than real (Racoviteanu et al., 2009). Thus, prior to being used for any glacier investigations, it is essential to evaluate the quality and consistency of these inventory datasets. With this consideration, the present study generates a precise and robust glacier inventory for the Ravi basin of Himachal Himalaya using Landsat Enhanced Thematic Mapper (ETM+) (2002) imageries (aided by ASTER image of 2002, Google Earth, and field inputs) and to compare it with available glacier inventories to assess their quality and accuracy. The implication of this study is to identify the appropriate glacier inventories among available inventories for precise glacier resource assessment, more accurately simulate sea-level rise projections, and other applied applications across the Himalayan region.

2 Regional Setting and Climate of the Study Region

The Ravi basin includes the Siul, Budhil, and upper Ravi sub-basins and is situated in the eastern most parts of the districts of Chamba and Kangra in Himachal Pradesh, India (Fig. 6.1). This area is geographically situated in the Himachal Himalaya, a regional portion of the northwestern Himalayas. The Raigarh Glacier, which has a snout at 4050 meters above sea level (m a.s.l.) and second-largest valley glacier in the Pir-Panjal range (with an area of 8.8 km^2), is the origin point of the Ravi River that runs into the Indus River. The majority of Ravi's course is north-westerly and follows the intersection of two corresponding mountain ranges: the Pir-Panjal to the north and the Dhaula-Dhar to the south, which divides the Ravi basin from Chenab and Beas basins, respectively (Marh, 1986). The study area has a basin outflow at the junction of the Siul and Ravi rivers, is roughly 4900 km^2 in size, and has a range of elevation of 765–6125 m a.s.l.The basin is located between Himachal Pradesh's maximum (Dharamshala) and lowest (Lahaul) precipitation areas in terms of annual precipitation (Chand & Sharma, 2015b). As a result of being in a rain shadow, the portion of the Dhauladhar range in the north experiences less precipitation than the southern part of the Pir Panjal range. For instance, during the past 20–25 years, the rain-shadowed tiny town of Holi (1830 m a.s.l.) averaged 816 mm of annual precipitation, while Bharmour (2150 m a.s.l.) in the Pir-Panjal range averaged 1180 mm of annual precipitation (Bhagat et al., 2004). According to data from the years 1990 to 2013, November through March, the mid-latitude westerlies contribute 32.4% (278.5 mm a^{-1}) of the mean annual rainfall, while the Indian Summer Monsoon (ISM) mainly contributes 56.9% (486.4 mm a^{-1}) of it (Pareta & Pareta,

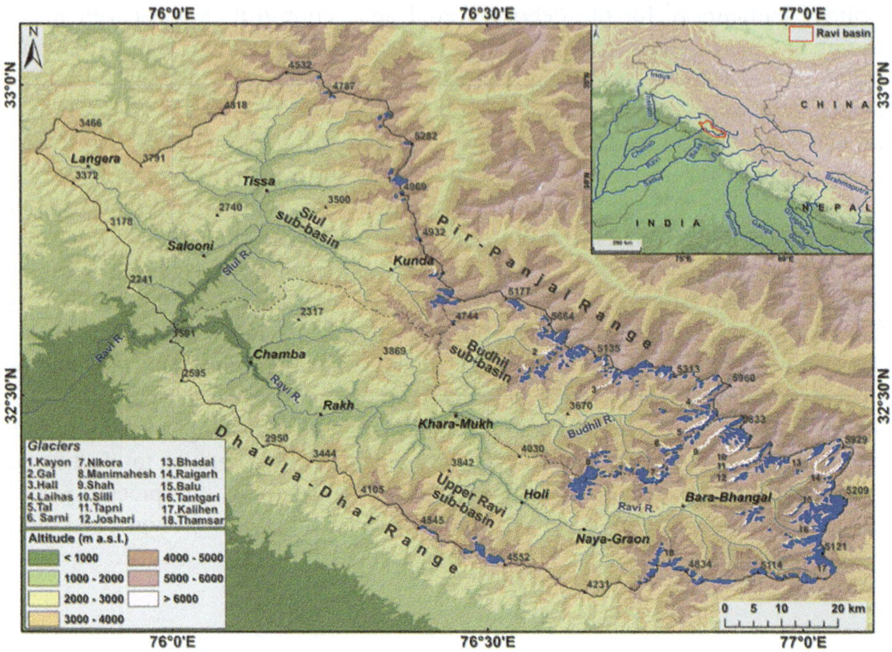

Fig. 6.1 The location and extent of the Ravi basin. The locations of the study region within the northwestern Himalayas are depicted in the inset figure

2014). These precipitation patterns show that the ISM and the winter mid-latitude westerlies have a significant impact on the glaciers in the Ravi basin (Chand & Sharma, 2015b). Moreover, the spring snowmelt and the ISM rainfall are the primary factors controlling the regional water balance and discharge of the Ravi basin (HPSEB, 2004).

3 Existing Himalayan Glacier Inventories

The substantial work on the Himalayan glacier inventory, comprising the Himachal Himalaya, has been accomplished by a number of different organizations and consortiums, including the GSI, SAC, ICIMOD, RGI, and GGI. Additionally, the GLOBGlacier and a team of researchers provide certain regional to sub-basin inventory of glaciers, for example, Frey et al. (2012) and Bhambri et al. (2013). RGI database includes the majority of these local/regional inventories. The RGI is a global database of glacier outlines that first went public in 2012 and most recently released version 6 in 2017. Each upgrade to the RGI versions introduces modifications to the definition of unsubdivided ice masses and region outlines and incorporated the refined regional inventories created by other groups or institutes. The latest

RGI V6 database provides the almost entire world's glacier digital outline. It includes descriptions of over 215, 547 glaciers having a total glacierized area of 705,738.793 km^2, with almost 13% of the glaciers located in High Mountain Asia (South Asia East as well as West and Central Asia). A national glacier inventory with 9575 glaciers totaling an area of 37,466 km^2 for Indian states was released by GSI in 2008 and 2009, and it includes earlier inventories conducted for India's Himalayan region in 1977 (e.g., Vohra, 1980; Kaul & Puri, 1999). The glacier outlines of this inventory were mainly obtained either using the Survey of India (SoI) topographic maps or aerial photographs/satellite imageries. Moreover, the SAC created a glacier inventory on a scale of 1: 50000 using IRS LISS-III data in 2010 for the main Himalayan river basins, that is, Brahmaputra, Ganges, and Indus. It inventoried around 32,392 glaciers with 37 characteristics and a total area of 71,182.1 km^2. In 2011, ICIMOD provides a revised glacier inventory almost for the entire HKH region with some additional prior respective or availability of digital outlines through collaboration with Cold and Arid Regions Environmental and Engineering Research Institute for China region (CAREERI) (Bajracharya & Shrestha, 2011). This inventory mainly mapped the glacier outlines from Landsat TM/ETM+ satellite images (2005 $\pm$ 3) and identified 54,000 glaciers having a total area of 60,000 km^2, across the ten river basins of the HKH region. Recently, GGI, a new glacier inventory designed for high mountain Asia, is created using more than 200 Landsat ETM+ scenes between 1999 and 2003, which includes 87,084 glaciers with a total area of 91,263 $\pm$ 13,689 km^2. The GGI inventory latterly incorporated in a revised version of RGI V5, and the same continues in the latest version of RGI, that is, RGI V6 (RGI Consortium, 2017). Therefore, the present study has taken RGI V4 and GGI digital outlines separately to compare these with available glacier inventories and RBGI. The digital outlines of ICIMOD, RGI, and GGI inventories are publicly available, whereas the GSI inventory is published in a tabular format. SAC inventory is not accessible neither tabular nor digitally and therefore couldn't include in the accuracy evaluation.

4 Methods and Datasets

4.1 Data Source

The present study used Landsat 7 ETM+ multispectral and medium spatial resolution satellite images (e.g., ASTER) to map and inventory glaciers. Since the 1970s, Landsat data have proven useful for studies of the global cryosphere, particularly for generating glacier inventories and changes in related spatial-temporal patterns because of its repeated and synoptic data coverage and multispectral resolution capability to distinguish snow from clouds and glaciers from other terrains (Roy et al., 2014).

The United States Geological Survey's (USGS) Earth explorer website (http://glovis.usgs.org) was accessed to choose the appropriate images at the ending of the ablation period with less seasonal snow cover and cloud cover (Paul & Svoboda,

2009) (Table 6.1). Accordingly, the multispectral Landsat ETM+ level 1 terrain-corrected (L1T) images for the year 2002 were chosen and utilized as a baseline for glacier inventory. The Landsat (L1T) data product offers orderly radiometric accuracy, geometric precision incorporating ground control points (GCPs), and topographic correctness through the use of a DEM. Additionally, the data selected near the year 2000 standard, which is advised for a worldwide glacier inventory as the foundational data to support worldwide cryosphere applications and to ensure consistency when comparing data with other glacier inventories that are currently accessible (Paul et al., 2009). Moreover, the ASTER (2002) image with better spatial resolution as compared to Landsat ETM+ with a limited swath and areal coverage was also utilized to further support accurately mapping the glacier boundary. Besides, the ASTER GDEM V2 (30 m) from Japan Space Systems was utilized as reference DEM for evaluating accuracy assessments among other available open-source DEMs, that is, SRTM and CartoDEM, for this region and further used for semiautomatic demarcation of the watershed basin and to extract glacier's topographic variables. The glacier digital outline for all available glacier inventories was procured from the respective organization's website or upon request.

4.2 Digital Elevation Models (DEM) and Quality Assessment

Separating distinct glaciers across its catchment divides and determining inventory allied topographic criteria like mean, medium, minimum, and maximum elevation, aspect, and mean slope require a DEM of the proper quality and resolution (Paul et al., 2009). Several studies have used open-source SRTM3v4 DEM (e.g., Bhambri et al., 2013) and the ASTER GDEM V2 (e.g., Frey et al., 2012) to extract glacier topographic inventory parameters in different parts of HKH region. Recently, at the Indian national level, Cartosat-1 DEM (CartoDEM) having a spatial resolution of ~30 m (CartoDEM V1 for the India region and CartoDEM V1.1 R1 for particular western Himalaya regions) are available. These public domain DEMs are provided at the Bhuvan web portal (http://bhuvan.nrsc.gov.in). The CartoDEM is generated from the Cartosat stereo images for all of India with an absolute vertical and planimetric accuracy of 8 m at LE90 (Linear Error of 90%) and 15 m at CE90 (circular error of 90%), respectively (ISRO, 2011). The accuracy of all of the abovementioned open-source DEMs has been established prior to their use for further analysis of glacier mapping and characteristics. Absolute vertical accuracies of all the DEMs were measured by comparison with 88 GCPs collected from non-differential GPS, that is, Garmin GPSMAP 76Cx 3–10 m (44 points; elevation range 2097–4418 m a.s.l.) in different field campaigns (2010–2013). Benchmark (BM)/spot height data (44 points, elevation range 2339–5857 m a.s.l.) were taken from SoI toposheet no. 52 D, 52 H, and 53 A. GPS control points were collected along the main trail routes and at sites along and/or across the glacial valley during geomorphological mapping within a range of elevation between 2060 m and 4382 m a.s.l. (accuracy within ± <2–8 m). The GCPs (e.g., Benchmark and Spot Heights)

Table 6.1 The detail of datasets used in the present study

Satellite/sensor	Date of acquisition	Spatial resolution (m)	Scene/product ID	RMS x,y (m)
Satellite data				
Landsat 7 ETM+	August 2, 2002	15,30	LE71470372002214SGS00	<15
	October 28, 2002	15,30	LE71480372002301SGS00	<15
ASTER	October 28, 2002	15,30,90	AST_L1A_003_10282002054845_11112002172853	<15

based on the SoI maps are randomly distributed and cover most of the significant peaks and stable geomorphological features. The vertical datum of CartoDEM, SRTM, ASTER GDEM, GPS, and SoI topographic sheets is WGS84, EGM96, EGM96, WGS84, and MSL, respectively, and shows variation. Besides, the horizontal datum (WGS 84) of all these datasets is the same. To standardize the vertical datum, CartoDEM and GPS height were shifted from WGS84 to EGM 96 datum using the mathematical model as suggested by Mukherjee et al. (2013). Since the vertical datum of the SoI toposheet is set to mean sea level (MSL) and the EGM96 surface is an extremely accurate approximation of MSL (Sun et al., 2003), the height obtained from SoI topographic map is directly compared with elevations acquired from DEMs (Mukherjee et al., 2013).

4.3 Glacier Identification, Mapping, and Inventory

The current work adhered to the Global Land Ice Measurements from Space (GLIMS) program guidelines (http://www.glims.org) to delineate the glacier boundaries from satellite images. When mapping the glacier outlines, the region above the bergschrund was left out since it occasionally contributes additional ice and snow to the glacier surface by avalanches and creeps flow (Bhambri et al., 2011). In addition, the defined glacier polygon region was cleared of internal rocks and nunataks. Prior research has demonstrated that the normalized difference snow index (NDSI) and band ratio methods are reliable and time-effective methods for mapping clean glacier ice (Racoviteanu et al., 2009). Additionally, it has been noted that the band ratio near-infrared (NIR)/shortwave infrared (SWIR) outperforms RED/SWIR and NDSI for shaded locations with thin debris cover (Andreassen et al., 2008). In order to map clean ice cover areas from Landsat images, the present study investigated a number of image band ratios (RED/NIR; RED/SWIR), NDSI, and classification methods (supervised and unsupervised). When compared to other tested band ratio methods, the band ratio RED/SWIR was found to be the most suitable one because it takes usage of the spectral heterogeneity of ice/snow in the SWIR region of the spectrum. To remove isolated pixels and fill in small gaps in areas covered in glaciers, the binary images were created using a threshold value of ~2 after which a 3 × 3 median filter was applied (Paul et al., 2002; Chand & Sharma, 2015a). In order to create a vector glacier boundary, these binary images were transformed. The resulting individual glacier vector polygon was manually corrected to remove incorrectly classified features such as turbid water surfaces (wide rivers and lakes), misclassified pro-glacial lakes, shadow areas, clouds, and remaining seasonal snow cover patches and snow aprons, as well as to correct any incorrect morphological delineation of debris-free area using the aforementioned band rationing method. Debris covers many Himalayan glaciers in the ablation zones, and no promising method for automated mapping of debris-covered glaciers has yet been established (Benn & Owen, 2002; Racoviteanu et al., 2009). The majority of suggested debris cover mapping techniques are either region-specific, or their universal applicability is not established; hence, manual digitization is preferred. Manual delineation requires careful observation and is favored to map the debris-covered glacier areas by interpreting the surface thermokarst features, for example, uneven less-relief supra-glacial ponds, ice cliffs, etc. (Iwata et al., 1980). Additionally, mapping the termini and boundary of debris-covered glaciers proved fairly challenging using only Landsat ETM+ multispectral imagery. In order to map the debris-covered glacier boundary more accurately, the multispectral bands of Landsat 7 ETM+ with its additional comparatively better spatial resolution PAN band was merged using a Brovey transform image fusion approach (Bahuguna et al., 2004). The goal was to improve the spatial resolution to 15 m from 30 m in order to better identify the glacier terminal and its form. In addition, medium-resolution datasets from ASTER (2002) and high-resolution datasets from IRS-1D Panchromatic (PAN) (2004), Bhuvan 2D/3D, and Google Earth (GE) were employed as supplementary sources to enhance the glacier outlines of debris-covered glaciers (e.g., Paul et al., 2013). The most likely location of the glacier termini was also determined based on the existence of small meltwater ponds, incipient meltwater streams near the termini, fractures in the surface slope, spectrum color differences, and indicators of movement (recognized by overlays of multi-temporal images) (Chand & Sharma, 2015a). The determination of past and present glacier termini has also been aided by field investigations of a few glaciers, including Shah, Tal, Thamsar, Manimahesh, Laihas, Tapni, and Sarni. These investigations have used handheld GPS, field mapping of recessional moraines, and field photographs. As suggested by Bolch et al. (2010), the continuous masses of ice were divided into its respective drainage basins (i.e., hydrological divides) using a semiautomated method utilizing ASTER GDEM V2 (2010) and manually improved wherever required. We use 0.02 km^2 as the minimum size to map glaciers because the majority of the prior inventory used a similar size criteria, for example, GLOBGlacier (Frey et al., 2012) and ICIMOD (Bajracharya & Shrestha, 2011). In order to maintain consistency when comparing our glacier inventory with other glacier inventories, certain rock glaciers (identified in the study area) were excluded (Frey et al., 2012). To assess the disparity among the inventories, the glacier area and median elevation were compared for each 3×3 km grid cell.

4.4 Mapping Uncertainty

In order to ensure the precision and importance of the results, it is imperative to properly analyze the problem of uncertainty and how it manifests itself in glacier mapping based on remote sensing. For clean-ice glaciers, previous research showed a mapping uncertainty of between 2% and 4% (Paul et al., 2002), and even under ideal circumstances, some studies managed to attain an accuracy of less than half-pixel (Bolch et al., 2010). Using the 2002 Landsat ETM+ PAN merged image (15 m) as the base for the inventory, we evaluated the mapping variance for each glacier using a buffer around the glacier edges, as proposed by Granshaw and Fountain

(2006). Since only one side can be impacted by the shift (Bolch et al., 2010), the buffer size was determined to be half of the predicted shift induced by misregistration, leading to the usage of a 7.5 m buffer for the Landsat 7 ETM+ outline. Due to a small glacier's larger abundance of edge pixels, this technique accounts for the relatively higher inaccuracy of small polygons (Bolch et al., 2010).

5 Results

5.1 DEM Quality Assessment

The ASTER GDEM V2 ($r^2 = 0.996$) has slightly better accuracy as compared to ASTER GDEM V1 ($r^2 = 0.995$), CartoDEM V1 ($r^2 = 0.913$), CartoDEM V1.1 R1 ($r^2 = 0.992$), and SRTM3v4 ($r^2 = 0.911$) (Fig. 6.2). Therefore, ASTER GDEM V2 was utilized to partially automatically delineate drainage basins and extraction of topographic glacier parameters. Although several studies reported that the SRTM DEM is marginally better than the photogrammetrically derived ASTER GDEM, nevertheless, it is the opposite for the SRTM void regions. Though, for the Himalayan region, both datasets have been used for the compilation of glacier inventories,

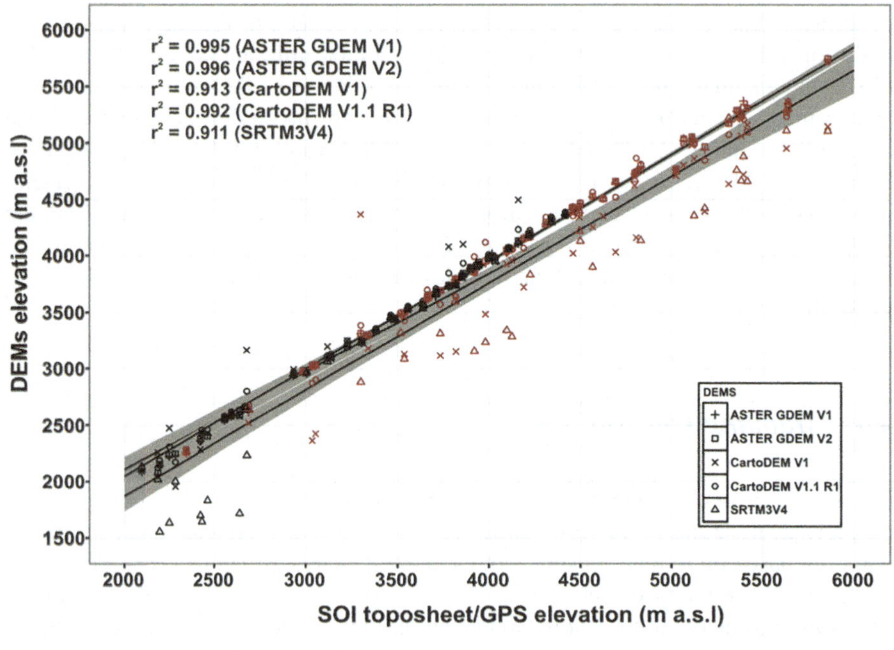

Fig. 6.2 Accuracy assessment of open-source DEMs, that is, ASTER, SRTM, and CartoDEM, with GCPs collected from the field using non-differential GPS (2010–2014) and benchmark/spot heights taken from SoI toposheet

in this study, ASTER GDEM V2 was preferred over SRTM3v4 DEM since (a) 27% of the mapped glacier area existed in the interpolated portions of the SRTM3v4 data gaps, which are having consistently too low accuracy and reported higher uncertainty for these gaps (Frey et al., 2012); (b) the gross errors would degrade the accuracy of the topographic parameters and further misidentify where catchment divides are located, which would affect the overall extent of the glacier (Frey et al., 2012); and (c) according to Singh et al. (2015), preliminary analysis demonstrates that ASTER GDEM V2 has comparatively higher accuracy to CartoDEM V1/V1.1 R1 for the current study area and its adjacent basin area, namely, the Beas. Additionally, CartoDEM V1 especially V1.1 R1 (or recently available CartoDEM V3) can be used in the future for studies on glaciers in the Himalayan region, but because it is still in its early stages of development, the study favored ASTER GDEM V2 to CartoDEM. Early analyses of the ASTER GDEM V1 indicated several high-frequency biases, including sensor jitter and a lack of co-registration of the stacked individual DEMs, as opposed to the ASTER GDEM V2, which has superior co-registration (Kääb et al., 2012).

5.2 *Ravi Basin Glacier Inventory (RBGI) and Topographical Characteristics*

Based on Landsat ETM+ imagery from 2002, the RBGI consists of 285 glaciers with a mapped area of 164.5 ± 7.5 km^2 (Table 6.2). There are 71 glaciers out of the total glaciers that have debris-covered parts, which occupy 36.1 ± 2.1 km^2 (~22% of the total area covered by glaciers). The basin contains a variety of small to large valley glaciers, with an average size of 0.6 km^2 and having an area range from 0.02 to 10.3 km^2. In comparison to other glacierized basins in the HKH region, such as Brahmaputra (1.2 km^2), Ganga (1.1 km^2), Alaknanda/Saraswati basin (3.7 km^2), Bhagirathi (1.3 km^2), Chenab (1.1 km^2), and Shyok (1.4 km^2), the average size of the glaciers in the studied area is comparatively less (Bajracharya & Shrestha, 2011; Frey et al., 2012; Bhambri et al., 2013). The relationship between the number of glaciers and their total area shows that glaciers having a size of less than 1.0 km^2 comprise 87.4% (249) of the total number but contribute only 39.2% of the total mapped area (Fig. 6.3). Moreover, the comparatively large glacier >5 km^2 size class has only five glaciers (i.e., 1.8% of the total count) but covers 21.7% (35.6 ± 1 km^2) of the area of total glaciers. This follows the relationship between the number of glaciers and their total area with the other glaciated regions of the HKH (Bajracharya & Shrestha, 2011; Frey et al., 2012; Bhambri et al., 2013).

The increase in size and number of glaciers from west to east (from south to north) indicate that glaciers and glaciation occurred mainly in the eastern, north-eastern, and south-eastern upper part of the Ravi basin, encompassing Budhil, Tunda basins, and Bara-Bhangal region (Shah, Raigarh, Tantgari basins) in the Pir-Panjal range and Thamsar, Kalihen glacial valleys in the upper eastern part of the

Table 6.2 Gl Ravi basin's glacier parameters (2002) and those of its sub-basins (2002)

Parameters	Ravi	Sub-basins (2002) Siul	Budhil	Upper-Ravi
Minimum mean elevation (m)	4623	4369	4642	4648
Maximum mean elevation (m)	5038	4765	5072	5054
Mean elevation mean (m)	4828	4576	4854	4846
Median elevation mean(m)	4828	4579	4854	4847
Mean elevation range (m)	415	395	430	406
Minimum elevation (m)	3473	3754	3473	3636
Maximum elevation (m)	6024	5465	5732	6024
Mean slope (°)	27.33	27.04	28.50	26.39
Mean aspect	SW (207.8)	W (252.9)	SW (206.8)	S (201.4)
Debris-covered area (km^2)	36.1 (22%)	0.7 (7.6%)	13.7 (23.6%)	21.8 (22.3%)
Clean-ice area (km^2)	128.3 (78%)	8.4 (92.4%)	44.2 (76.4%)	75.8 (77.7%)
Area by glacier size (km^2)				
<1	64.41	4.21	26.93	33.27
1–2	24.23	4.89	8.58	10.76
2–5	40.15		22.30	17.85
5–11	35.66			35.66
Total glacierized area (km^2)	164.45	9.1 (5.5%)	57.8 (35.2%)	97.5 (59.3%)

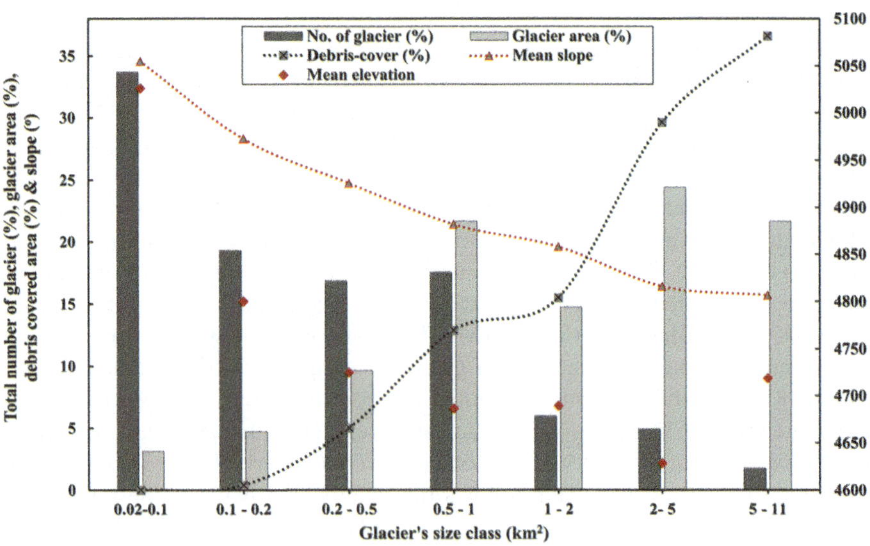

Fig. 6.3 Relationship of glacier characteristics according to the glacier size class, for example, glacier number, glacier area, debris covered area, average glacier elevation, and average slope in different glacier size classes

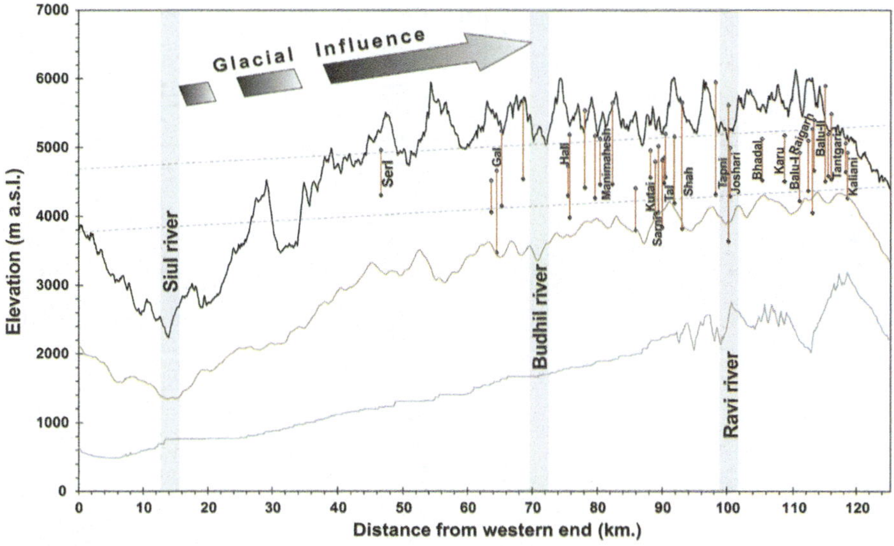

Fig. 6.4 Altitude-distance profile from west to east in Ravi basin along the Pir-Panjal and Dhauladhar ranges, showing summit elevation and elevation range of major valley and the glaciers (red lines). Topographical control on the distribution of glaciers is represented by the concentration of glaciers toward the eastern part of the basin

Dhauladhar range (Fig. 6.4). Besides, the increase in size and number of glaciers from south to north reveals the influence of glaciation in the northern portion of the basin, that is, Pir-Panjal range. Thus, the Pir-Panjal range of the Upper-Ravi sub-basin contains the vast majority of the largest glaciers (>5 km^2). The presence of large glaciers in this region further can be ascribed to topographical factors such as higher relief with maximum elevation range, wide catchment morphology, and changing gradient of precipitation as the rain turns to snow at higher altitudes and feeds the glacier's accumulation areas.

In addition, the general profile of most of the valley glaciers is concave, with steeper slopes in the upper part and gentler slopes in the lower part with some variations due to influenced by bedrock lithology and structure in the basin (Fig. 6.5). According to some of the glacial profiles, the glaciers altered the terrain by polishing and exhuming the bedrock and creating a knobby topography that is distinctive of the higher portions of glacial valleys (Anderson & Anderson, 2006; Castillo & Muñoz-Salinas, 2013). The primary differentiating feature of valley glaciers is the existence of ice-free slopes overlooking the glacier surface. Such steep slopes are significant contributors to the buildup of ice and snow in the form of avalanches as well as the rock debris that falls over the surface of glaciers (Benn & Evans, 2010). The majority of the glaciers are classified as mountain glaciers in the basin, where a primary source of nourishment are direct snowfall and drift snow.

The glacier's mean elevation is 4828 meters a.s.l., with a range of 3941–5822 meters. Additionally, the glaciers' median elevation, which is frequently used to

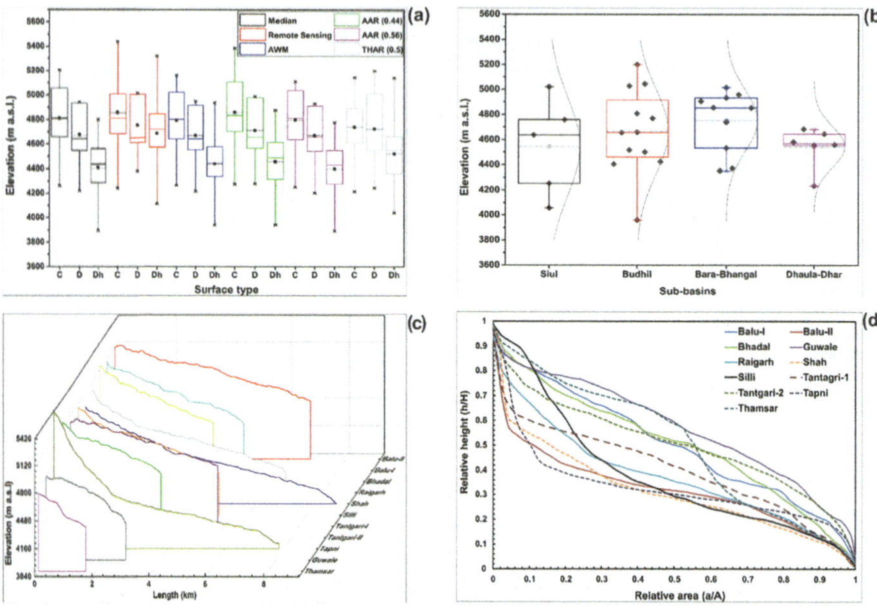

Fig. 6.5 ELA and profiles, (**a**) Average ELA calculated for clean, debris and highly debris-covered glaciers using different methods, (**b**) average glacier ELA for the glaciers in different sub-basins, (**c**–**d**) 3D and 2D representation of glaciers longitudinal profiles

approximate long-term equilibrium line altitude (ELA) based on topographic (Braithwaite & Raper, 2009), is alike to the mean elevation, with ranges from 3896 to 5859 m a.s.l. and a mean elevation of 4828 m a.s.l (Fig. 6.5). Moreover, the mean elevation for the glaciers (~4878 m a.s.l.) situated in the Pir-Panjal is comparatively higher than glaciers (~4604 m a.s.l.) located in the Dhauladhar range. This variance in average mean elevation of ~280 m between the glaciers in the Pir-Panjal range and the Dhauladhar range is in approximation with the difference of mean relief of ~350 m between these mountain ranges (Fig. 6.5). However, the lowest elevations are reached by debris-covered valley glaciers of the Pir-Panjal range as compared to the Dhauladhar range mainly due to the existence of large size and fully formed glaciated basins in the Pir-Panjal range. Additionally, the mean elevation changes depending on the size of the glacier, being lower for larger glaciers than for smaller ones. Moreover, the ELA for the clean ice glacier is comparatively higher than the glacier with less and extensive debris cover on their surface (Fig. 6.5). It suggests that the debris cover makes a significant role in the morphology characteristics of the glacier and further in glacier dynamics.

Figure 6.6a shows the hypsometry (glacier's area distribution by the elevation) of the total glacier, debris-covered, and clean-ice-covered surface as per the different classified glacier size classes. The majority of the glaciers in each size class are located between 4600 and 4800 m a.s.l., with lesser size-class glaciers often located at a higher elevation than large-size glaciers. Additionally, due to debris-covered

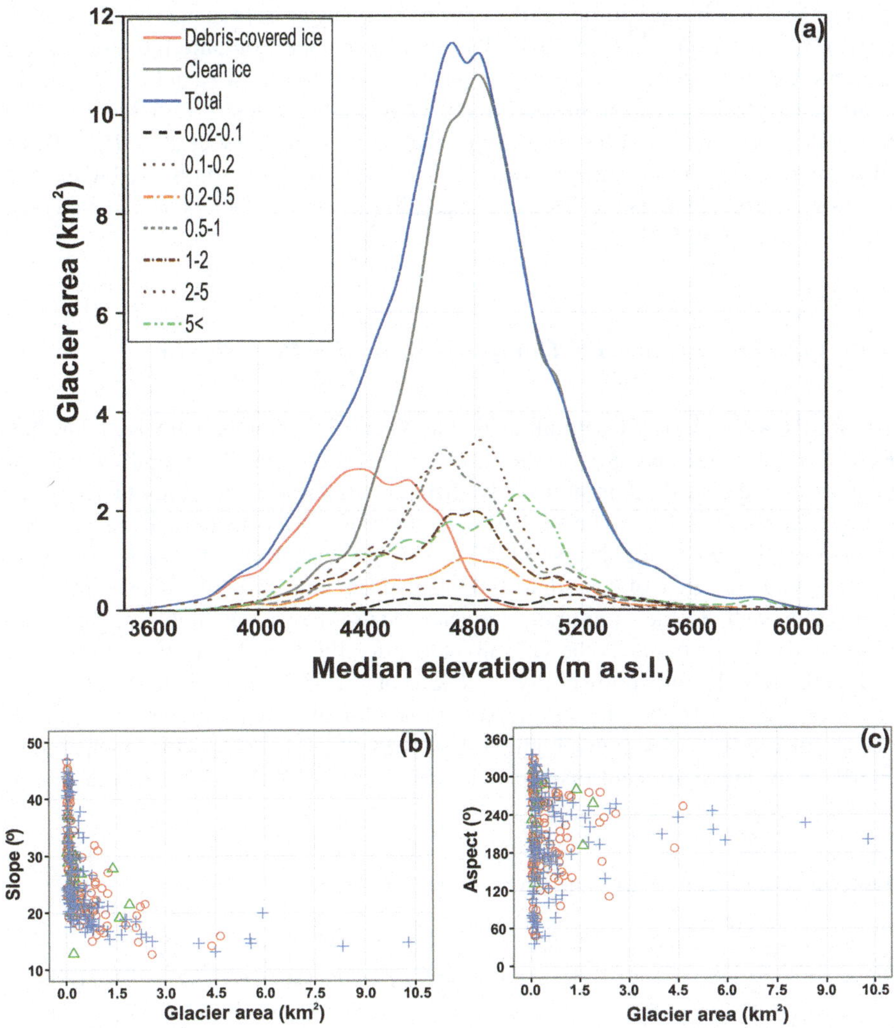

Fig. 6.6 Hypsometry (elevation-area relationship) of debris-covered, clean ice, total glaciers for different glaciers size classes (**a**). Glacier characteristics in 2002 (**b**) glacier area relative to aspect, (**c**) glacier area versus slope. The glaciers of Siul, Budhil, and the Upper Ravi sub-basin are represented by the triangle, circle, and cross, respectively

portions that make up 37.1% of the glacier area in this size class, glaciers in the size class >5 km^2 have an additional peak in the hypsometry curve at various elevations (<~4500 m) (Fig. 6.6a). Additionally, it has been noted that a heavy debris layer lowers the steady-state AAR, one of the governing elements in glacier dynamics, necessitates a higher ablation area to counteract accumulation, and this additional affects the glacier hypsometry (Benn & Evans, 2010). The mapped glaciers are situated on moderate slopes with an average slope of 27.3°; similar mean slopes for glaciers in other basins in the north-western Himalayas were stated by Frey et al. (2012), ranging from 23.9° to 30.9°. The smaller glaciers compared to large valley glaciers are on a steeper slope in the basin (Fig. 6.6c). It was indicated that, in contrast to glaciers with a steeper slope, large valley glaciers (>5 km^2) with a moderate slope contained the majority of the ice volume (Frey et al., 2012, 2014). Most of the glaciers are facing west (24.7%), followed by south (~24.4%), and south-west (16.6%); nonetheless, on average, 51.6% of the glaciers are situated on a southerly side (Fig. 6.6b).

5.3 Glacier Inventories Comparison in the Ravi Basin

The RGBI was compared with all available glacier inventories, namely, ICIMOD, RGI, GGI, and GSI inventory in the Ravi basin (Table 6.3). The overall number of glaciers and their extent is noticeably different in these inventories (Table 6.3). Comparing the RBGI with the RGI v4 inventory reveals a considerable discrepancy in the overall number of glaciers (39.7% or 188 glaciers). However, we removed the glaciers with an area of less than 0.05 km^2 from our subsequent evaluation to uniformly define the minimum glacier size across available inventories (Table 6.4). Comparable to the GGI inventory for the Ravi basin, the total glacier area in the RBGI is much lower (by −2.2 km^2 or −1.35%) than the ICIMOD (by −49.86 km^2 or −30.67%). However, as compared to the RGBI inventory, the total glacier area in RGI v4 is much higher (by 40.1 km^2 or 24.64%). When compared to RGBI, there are significant discrepancies between the total number of glaciers in

Table 6.3 Comparison of glacier inventories in the Ravi basin

Inventory	Number of glaciers	Area (km^2)	Mean size	Minimum glacier size (>km^2)	Data used (year)	References
GSI (2008)	172	193	–	–	SOI Toposheet and Aerial Photograph	Raina and Srivasatva (2008)
ICIMOD (2004)	198	235.2	–	–	IRS-1D LISS-III (2000–2002)	Bhagat et al. (2004)
ICIMOD (2011)	217	113.6	0.52	0.02	Landsat ETM+ (2005 ± 3)	Bajracharya and Shrestha (2011)
RGI v4	473	213	0.44	0.02	Landsat ETM + (2002) and ALOS PALSAR (2007)	Frey et al. (2012)
GGI	267	160.4	0.60	0.05	Landsat ETM + (1999–2003)	Nuimura et al. (2015)
RBGI	285	164.5 ± 7.5	0.58	0.02	Landsat ETM + (2002)	Present study

Table 6.4 Summary of the glaciers mapped in the RBGI, GGI, RGI, and ICIMOD inventory

Inventory	Number of glaciers	Area (km^2)	Mean size	Selected minimum glacier size ($>km^2$)	Difference (number)	Difference (area) (km^2)	(%)
RBGI	233	162.57	0.70	0.05	–	–	–
ICIMOD	192	112.71	0.59	0.05	−41	– 49.86	– 30.67
RGI v4	299	202.63	1.80	0.05	66	40.06	24.64
GGI	267	160.37	0.60	0.05	34	– 2.20	– 1.35

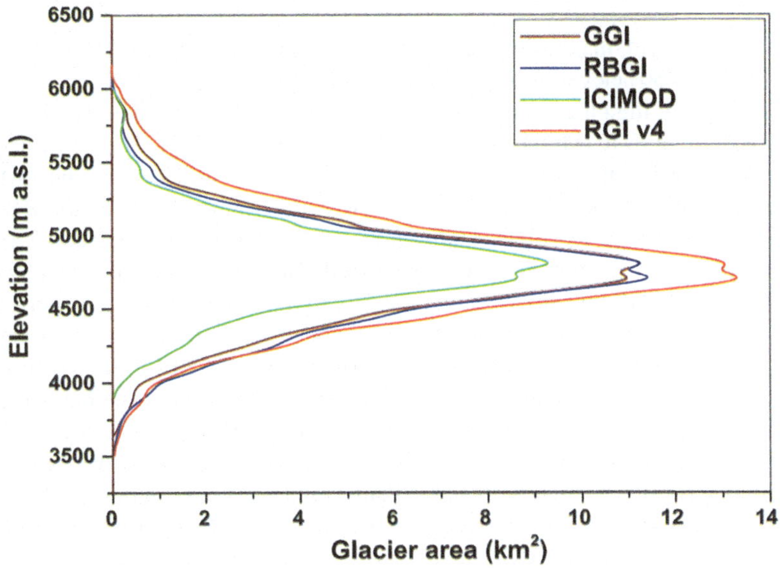

Fig. 6.7 Glacier hypsometry for the Ravi basin from RBGI, GGI, RGI, and ICIMOD inventory. Note for each inventory, only glaciers larger than >0.05 km^2 in the area are taken into consideration

RGIv4 (by 66 glaciers), ICIMOD (by −41 glaciers), and GGI (by 34 glaciers) (Table 6.4).

Moreover, we analyzed how much glacier area was distributed according to elevation in each inventory (i.e., the hypsometry) as depicted in Fig. 6.7. RGI and ICIMOD inventorie′s glacier area distribution by elevation zone clearly shows the difference in glacier size than those in the RGBI. For instance, glaciers in the RGI are mapped more extensively than those in the RBGI as the distribution of glacier area in almost all elevation zones, particularly between 4250 and 6000 m, is significantly larger in the RGI hypsometry than in the RBGI (Fig. 6.7). However, glacier size in the ICIMOD inventory is comparatively lower to the RGBI as glacier hypsometry for the Ravi basin is significantly lower in the ICIMOD than in the RBGI inventory for elevation between 3900 and 5000 m (Fig. 6.7). The GGI inventory shows similarity in glacier hypsometry on comparison with RGBI inventory in nearly all

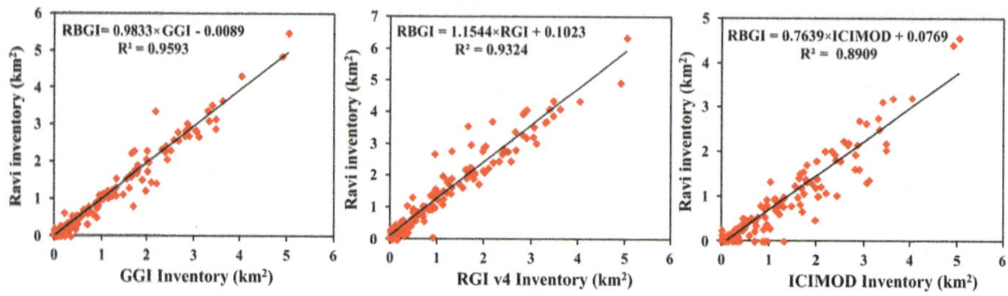

Fig. 6.8 Scatterplots of glacier area in each 3 × 3 km grid cell of the different inventories, for example, GGI, RGIv4, and ICIMOD, plotted against the RBGI in the Ravi basin

elevation zones with maximum and minimum glacier area in upper (5300–5800) and lower (4250–3750) elevation zones, respectively (Fig. 6.7).

We evaluated the glacier area for each 3 × 3 km grid cell to identify differences among RBGI, GGI, ICIMOD, and RGIv4 inventories (Fig. 6.8). The figure shows a scatter diagram of the glacier area in each 3 × 3 km grid cell of the different inventories, for example, GGI, RGIv4, and ICIMOD, plotted against the RBGI in the Ravi basin. The scatter plots clearly show the large difference between ICIMOD and RGI inventory as plotted against the RBGI. However, GGI inventory is comparable with RBGI inventory as the same can be inferred Pearson correlation coefficient, which is close to 1, and the behavior of the line of best fit. In order to find variations among all available inventories and with RBGI, we additionally assessed the spatial distributions of glacier area and the area-weighted mean of median elevation for each 3 × 3 km grid cell (Figs. 6.9 and 6.10). It is observed that glacier area is higher in the RGI in most of the grids across the Ravi basin, mainly along the upper reaches of the Pir-Panjal range (Figs. 6.9 and 6.10). However, the total glacier area is comparatively lesser in the ICIMOD than in the RBGIRavi Basin Glacier Inventory (RBGI) inventory, as the same trends of smaller glacier areas are found in the GGI inventory in comparison with RBGIRavi Basin Glacier Inventory (RBGI). The difference in grid-wise total glacier area is found minimum on a comparison of GGI with RBGIRavi Basin Glacier Inventory (RBGI).

6 Discussion

The present study provides the comprehensive glacier inventory for the Ravi basin based on visual interpretation, specifically the exclusion of seasonal snow patches from Landsat ETM+ images with the addition of medium to high-resolution images from Bhuvan 2D/3D, IRS-1D PAN, and Google Earth, as well as field inspections. The present RBGI identified 285 glaciers with a total area of 164.5 ± 7.5 km^2 for the year 2002. The overall extent and distribution of glaciers explicitly reflect the interplay between topography (e.g., altitude, aspect, relief, and distance from the moisture sources), geomorphic history (e.g., upliftment history of the Ravi basin),

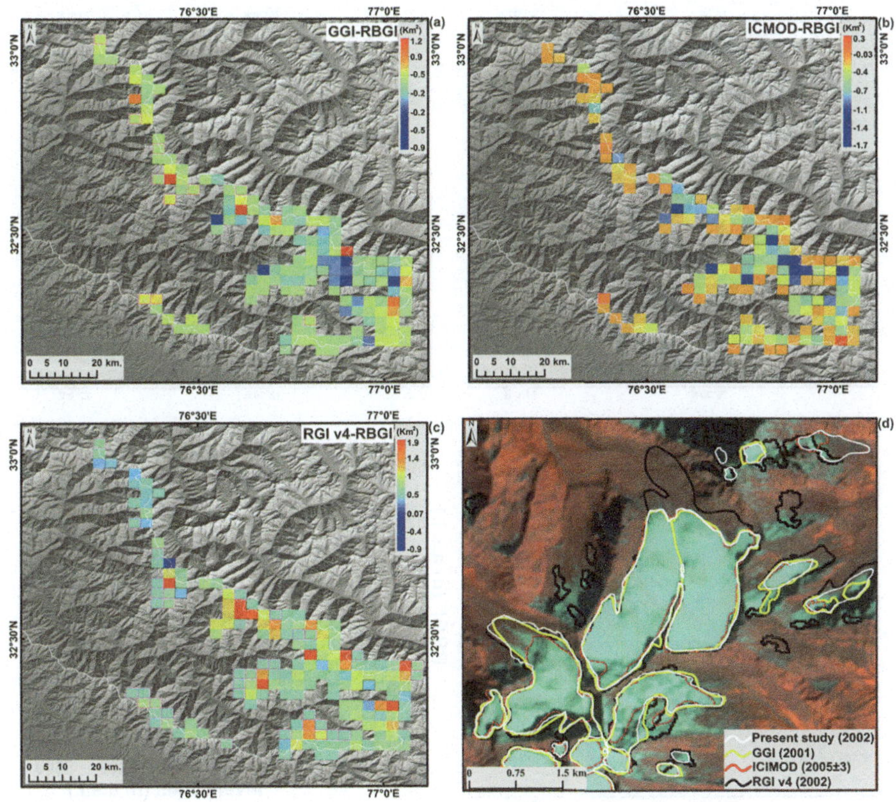

Fig. 6.9 Differences of glacier area in the different inventories for each 3 × 3 km grid cell in the Ravi basin (**a–c**). (**d**) Misinterpretation of the seasonal snow-covered area and glacier boundary in different inventories

and climatic factors (e.g., orographic control precipitation gradient and temperature) that systematically vary around one sub-basins to the other (Benn & Evans, 2010). The Ravi basin's high percentage of debris cover is crucial because the presence of debris, which either slows or fasts the surface melting, is concentrated on low-lying tongues, where more melting is anticipated, and therefore further study is required to understand the response of debris-covered on glacier dynamics in the basin (Scherler et al., 2011). As mentioned that a number of organizations are providing glacier inventories, for example, GSI, SAC, ICIMOD, RGI, and GGI for the Himalayan region. These inventories provide a valued input for many cryospheric and glacio-hydrological studies. However, the noticeable difference has been reported for these inventories, and the same has been observed when compared with RBGI. Additionally, when these inventories were compared on a regional level, a sizable discrepancy was discovered in the overall glacier area for the Himachal Himalaya (Fig. 6.11 and Table 6.5). Thus, to further identified the appropriate glacier inventories among available inventories for precise glacier resource assessment, the study in detail compared these inventories with RBGI. For instance, we investigated the differences

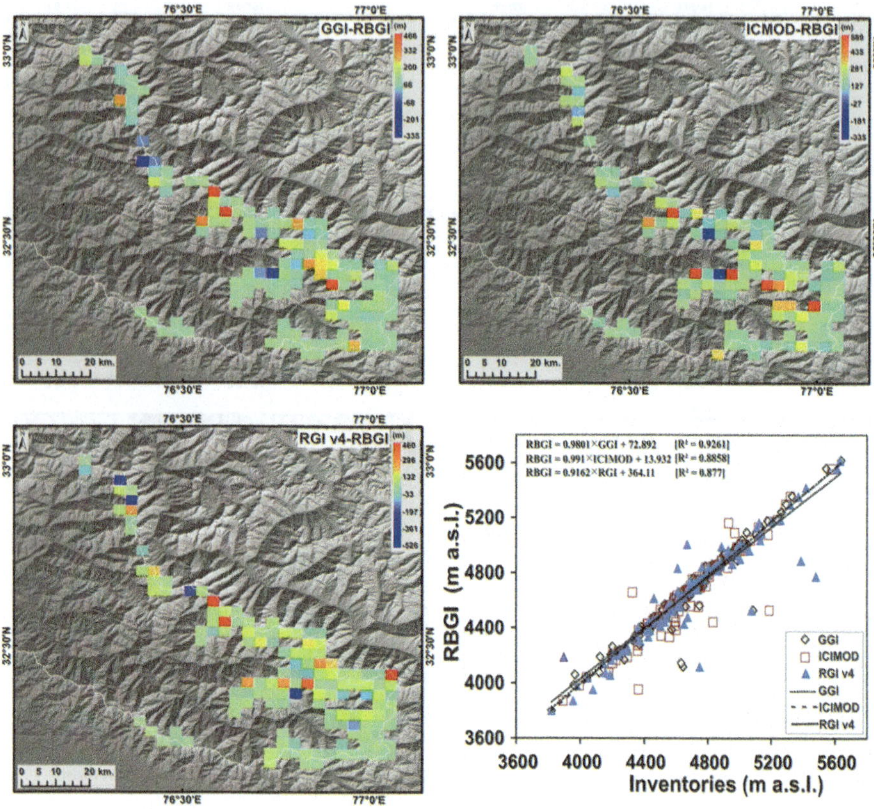

Fig. 6.10 Area-weighted average median elevation variations among various inventories for each 3 × 3 km grid cell in the Ravi basin (**a–c**). (**d**) Scatter plots of the area-weighted average of a median glacier in each 3 × 3 km grid cell of the different inventories, for example, GGI, RGIv4, and ICIMOD, plotted against the RBGI in the Ravi basin

in glacier area and median elevation by comparison of all available inventory datasets with RBGI inventory and found significant differences during the comparison of these inventory datasets.

Since the underlying Landsat imagery's acquisition dates are nearly the same for all of these inventories except GSI, the difference in the total glacier area and elevation between inventories cannot be linked to long-term changes in the Ravi basin's glacier extent. Thus, we propose that the following plausible variables may be to account for these variations in inventories: (i) the incidental inclusion of periodic snow patches or aprons at the bottom of steep/avalanche slopes as well as periodic snow on paleo-cirques in the RGI v4, resulting from automatic mapping; (ii) variations in interpretation of debris-covered glaciers; (iii) variations in classification/definitions of glacier areas/boundary, particularly exclusion/inclusion of steep

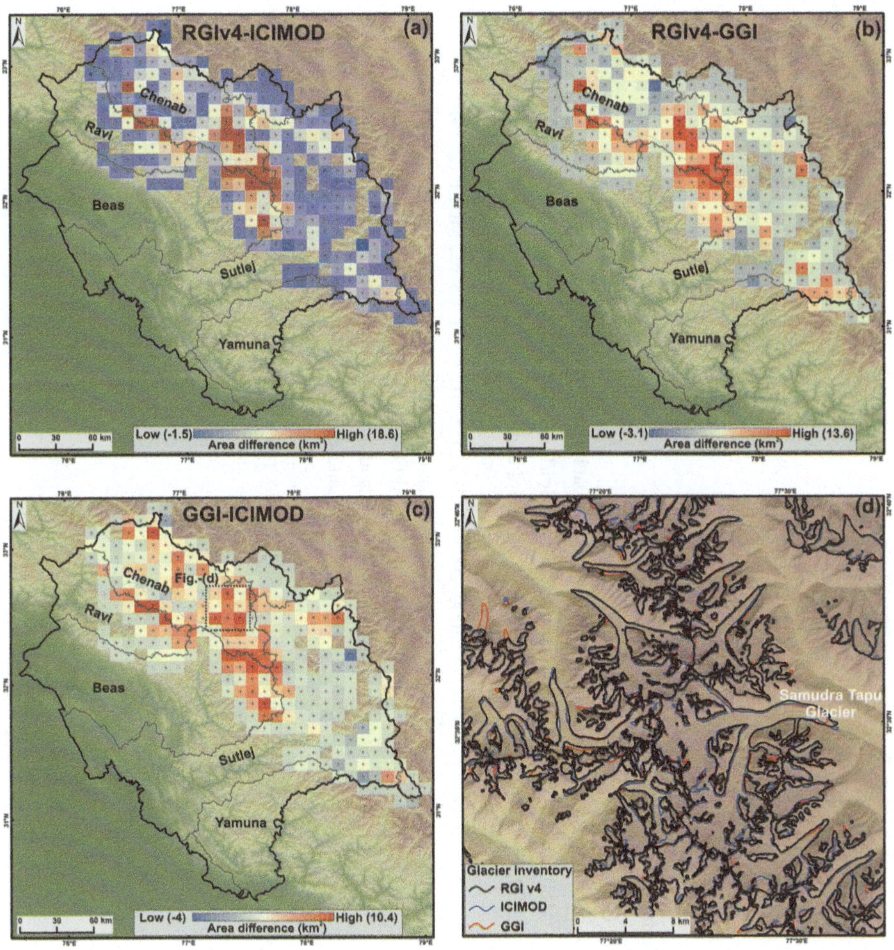

Fig. 6.11 Inconsistencies in glacier area among different available glacier inventory. Note that despite using the same datasets (Landsat TM/ETM), all of these glacier inventories show a significant variance in the area. Grid boundaries for all glacier inventories are represented by the plus sign (+)

headwalls; and (iv) depending on the goal of the inventory, connecting ice masses may be divided into numerous glaciers or considered as a single unit.

Furthermore, the GSI glacier inventory had fewer glaciers as compared to RBGI, most likely as a result of the topographic map's limited scale (1:50,000) and the incorrect interpretation of glacier boundaries from aerial images caused by periodic snow and debris cover (Bhambri & Bolch, 2009). The GGI inventory shows almost similarity in the total area in comparison with RGBI with significant variation in the total number of glaciers mapped, which is attributed to the omission of shaded glacier areas in the GGI (Nuimura et al., 2015).

Table 6.5 Inventories of glaciers for Himachal Pradesh (India)

Inventory	Year	Min. area mapped (km^2)	Number of glacier	Glacier area (km^2)	Area % of the total area	Year of data sources	Data sources	References
ICIMOD[a]	2004	N.A.	2554	4160.6	7.5	1999–2001	SoI/IRS LISS-III	Bhagat et al. (2004)
GSI	2008	N.A.	2100	3799.1	6.8	1970s	SoI and Aerial Photos	Sangewar (2005) and Raina and Srivastava (2008)
SAC	2009	N.A.	–	–	–	2000–2005	IRS LISS-III	SAC, MOEF (2010)
RGI v4[a]	2013	0.02	4926	3751.7	6.8	2000–2002/2007	Landsat ETM+/ALOS PALSAR	Pfeffer et al. (2014)
GAMDAM[a]	2013	0.05	3347	3281.2	5.9	1999–2003	Landsat TM/ETM	Nuimura et al. (2015)
ICIMOD[a]	2011	0.02	3239	2809.4	5.1	2005 ± 3	Landsat TM/ETM+	Bajracharya and Shrestha (2011)

Note: [a]The total number of glaciers and their area were determined using the Himachal Pradesh boundary polygon from the Census of India's clipped glacier polygon

7 Conclusions

The study presents a detailed glacier inventory (RBGI) for the Ravi basin by utilizing Landsat ETM+ (2002) scenes with a supplement of high- to medium-resolution images (i.e., IRS-1D PAN, ASTER), DEM, and field inputs. The RBGI consists of 285 glaciers with a mapped area of 164.5 ± 7.5 km^2 in 2002 (Table 6.2). There are 71 glaciers out of the total glaciers that have debris-covered parts, which occupy 36.1 ± 2.1 km^2 (~22% of the total area covered by glaciers). The overall glacier area mapped by the RBGI is comparable to the GGI inventory. The absence of shaded glacier areas from the GGI inventory is the primary cause of the spatial disparities in the total number of mapped glaciers, their area, and median elevations between these inventories, which nevertheless show substantial regional inconsistency. A comparison of RBGI with RGI and ICIMOD in the Ravi basin revealed a significant difference in total glacier area and the total number of glaciers mapped. The total mapped glacier area is considerably higher (24.6%) in the RGI as compared to the RBGI. It is primarily attributed to the unintentional inclusion of periodic snow patches or aprons at the bottom of steep/avalanche slopes as well as deposited periodic snow in the paleo-cirques in the RGI, resulting from automated mapping and the variations in the interpretation of debris-covered glaciers. Furthermore, the ICIMOD inventory's total mapped glacier area is considerably less than the RBGI inventory's (30.7%) due to differences in how glacier areas are classified and defined, particularly the inclusion or exclusion of vertical headwalls and connecting ice masses may be divided into numerous glaciers or considered as a single unit depending on the goal of the inventory. Given that glaciers have seen rapid climatically induced changes over the past decades, the glacier inventories need to be revised and updated utilizing the most recent high- to medium-resolution satellite data, field validation, and widely accepted international standards. Additionally, it would be advantageous and essential to include the root mean square (RMS) error for each digitized or mapped glacier feature, source data, and characteristics that are sensitive to climate change in the inventory's list of parameters in the updated global or regional glacier inventories.

Acknowledgments The authors acknowledge ICIMOD, GGI, and RGI to provide digital outlines for the research community. We appreciate the anonymous reviewer's constructive comments, which helped us to enhance the manuscript's content. The corresponding author thanks the SERB, New Delhi, for its financial assistance with SERB-DST Project No. PDF/2017/002717 [NPDF Scheme].

References

Anderson, R. S., & Anderson, S. P. (2006). *Geomorphology: The mechanics and chemistry of landscapes: Great Britain*. Cambridge University Press.

Andreassen, L. M., Paul, F., Kääb, A., & Hausberg, J. E. (2008). Landsat-derived glacier inventory for Jotunheimen, Norway, and deduced glacier changes since the 1930s. *The Cryosphere, 2*, 131–145. https://doi.org/10.5194/tc-2-131-2008

Bahuguna, I. M., Kulkarni, A. V., & Nayak, S. (2004). DEM from IRS-1C PAN stereo coverages over Himalayan glaciated region – Accuracy and its utility. *International Journal of Remote Sensing, 25*, 4029–4041. https://doi.org/10.1080/01431160310001652376

Bajracharya, S. R., & Shrestha, B. (2011). *The status of glaciers in the Hindu Kush-Himalayan Region*.

Bajracharya, S. R., Maharjan, S. B., Shrestha, F., et al. (2014). *Glacier status in Nepal and decadal change from 1980 to 2010 based on landsat data – HIMALDOC*.

Barnett, T. P., Adam, J. C., & Lettenmaier, D. P. (2005). Potential impacts of a warming climate on water availability in snow-dominated regions. *Nature, 438*, 303–309. https://doi.org/10.1038/nature04141

Benn, D. I., & Evans, D. J. A. (2010). *Glaciers and glaciation* (2nd ed.). Hodder Arnold Publication.

Benn, D. I., & Owen, L. A. (2002). Himalayan glacial sedimentary environments: A framework for reconstructing and dating the former extent of glaciers in high mountains. *Quaternary International, 97–98*, 3–25. https://doi.org/10.1016/S1040-6182(02)00048-4

Bhagat, R., Kalia, V., Sood, C., et al. (2004). *Inventory of glaciers, glacial lakes and identification of potential Glacial Lake Outburst Floods (GLOFs) affected by global warming in the mountains of Himalayan Region*.

Bhambri, R., & Bolch, T. (2009). Glacier mapping: A review with special reference to the Indian Himalayas. *Progress in Physical Geography, 33*, 672–704. https://doi.org/10.1177/0309133309348112

Bhambri, R., Bolch, T., Chaujar, R. K., & Kulshreshtha, S. C. (2011). Glacier changes in the Garhwal Himalaya, India, from 1968 to 2006 based on remote sensing. *Journal of Glaciology, 57*, 543–556. https://doi.org/10.3189/002214311796905604

Bhambri, R., Bolch, T., Kawishwar, P., et al. (2013). Heterogeneity in glacier response in the upper Shyok valley, Northeast Karakoram. *The Cryosphere, 7*, 1385–1398. https://doi.org/10.5194/tc-7-1385-2013

Bolch, T., Menounos, B., & Wheate, R. (2010). Landsat-based inventory of glaciers in western Canada, 1985–2005. *Remote Sensing of Environment, 114*, 127–137. https://doi.org/10.1016/j.rse.2009.08.015

Bolch, T., Kulkarni, A., Kääb, A., et al. (2012). The state and fate of Himalayan glaciers. *Science (80-), 336*, 310–314. https://doi.org/10.1126/science.1215828

Braithwaite, R. J., & Raper, S. C. B. (2009). Estimating equilibrium-line altitude (ELA) from glacier inventory data. *Annals of Glaciology, 50*, 127–132.

Castillo, M., & Muñoz-Salinas, E. (2013). Assessing the impact of quaternary glaciations on the topography of Sierra de Gredos (Central Spain). *The Revista Mexicana de Ciencias Geológicas, 30*, 312–323.

Chand, P., & Sharma, M. C. (2015a). Glacier changes in the Ravi basin, north-western Himalaya (India) during the last four decades (1971–2010/13). *Global and Planetary Change, 135*, 133–147. https://doi.org/10.1016/j.gloplacha.2015.10.013

Chand, P., & Sharma, M. C. (2015b). Frontal changes in the Manimahesh and Tal Glaciers in the Ravi basin, Himachal Pradesh, northwestern Himalaya (India), between 1971 and 2013. *International Journal of Remote Sensing, 36*, 4095–4113.

Frey, H., Paul, F., & Strozzi, T. (2012). Compilation of a glacier inventory for the western Himalayas from satellite data: Methods, challenges, and results. *Remote Sensing of Environment, 124*, 832–843. https://doi.org/10.1016/j.rse.2012.06.020

Frey, H., Machguth, H., Huss, M., et al. (2014). Estimating the volume of glaciers in the Himalayan–Karakoram region using different methods. *The Cryosphere, 8*, 2313–2333. https://doi.org/10.5194/tc-8-2313-2014

Gardelle, J., Arnaud, Y., & Berthier, E. (2011). Contrasted evolution of glacial lakes along the Hindu Kush Himalaya mountain range between 1990 and 2009. *Global and Planetary Change, 75*, 47–55. https://doi.org/10.1016/j.gloplacha.2010.10.003

Granshaw, F. D., Fountain, A. G. (2006). Glacier change (1958–1998) in the North cascades national park complex, Washington, USA. *Journal of Glaciology, 52*(177), 251–256. https://www.cambridge.org/core/journals/journal-of-glaciology/article/glacier-change-19581998-in-the-north-cascadesnational-park-complex-washington-usa/3E8AD445283FA72CD3C2A95B2EF56309

HPSEB. (2004). *Chamba Hydro-electric project Himachal Pradesh.*

Immerzeel, W. W., van Beek, L. P. H., & Bierkens, M. F. P. (2010). Climate change will affect the Asian water towers. *Science, 328*, 1382–1385. https://doi.org/10.1126/science.1183188

ISRO. (2011). *Evaluation of Indian National DEM from Cartosat-1 data.*

Iwata, S., Watanabe, O., & Fushimi, H. (1980). Surface morphology in the ablation area of the Khumbu Glacier. *Journal of the Japanese Society of Snow and Ice, 41*, 9–17. https://doi.org/10.5331/seppyo.41.Special_9

Kääb, A., Berthier, E., Nuth, C., et al. (2012). Contrasting patterns of early twenty-first-century glacier mass change in the Himalayas. *Nature, 488*, 495–498. https://doi.org/10.1038/nature11324

Kaul, M. K., & Puri, V. M. K. (1999). *Inventory of the Himalayan glaciers: A contribution to the International Hydrological Programme.*

Marh, B. S. (1986). *Geomorphology of the Ravi River.* Inter-India Publications.

Maurer, J. M., Rupper, S. B., & Schaefer, J. M. (2016). Quantifying ice loss in the eastern Himalayas since 1974 using declassified spy satellite imagery. *The Cryosphere, 10*, 2203–2215. https://doi.org/10.5194/tc-10-2203-2016

Mayewski, P. A., & Jeschke, P. A. (1979). Himalayan and Trans-Himalayan glacier fluctuations since AD 1812. *Arctic and Alpine Research, 11*, 267–287.

Mukherjee, S., Joshi, P. K., Mukherjee, S., et al. (2013). Evaluation of vertical accuracy of open source Digital Elevation Model (DEM). *International Journal of Applied Earth Observation and Geoinformation, 21*, 205–217. https://doi.org/10.1016/j.jag.2012.09.004

Nuimura, T., Sakai, A., Taniguchi, K., et al. (2015). The GAMDAM glacier inventory: A quality-controlled inventory of Asian glaciers. *The Cryosphere, 9*, 849–864. https://doi.org/10.5194/tc-9-849-2015

Pareta, K., & Pareta, U. (2014). Climate change impact on land and natural resource in Chamba Tehsil of Himachal Pradesh State, India. *The International Journal of Science & Technoledge, 2*, 38–48.

Paul, F., & Svoboda, F. (2009). A new glacier inventory on southern Baffin Island, Canada, from ASTER data: II. Data analysis, glacier change and applications. *Annals of Glaciology, 50*, 22–31.

Paul, F., Kääb, A., Maisch, M., et al. (2002). The new remote-sensing-derived Swiss glacier inventory: I. Methods. *Annals of Glaciology, 34*, 355–361. https://doi.org/10.3189/172756402781817941

Paul, F., Barry, R. G., Cogley, J. G., et al. (2009). Recommendations for the compilation of glacier inventory data from digital sources. *Annals of Glaciology, 50*, 119–126.

Paul, F., Barrand, N. E., Baumann, S., et al. (2013). On the accuracy of glacier outlines derived from remote-sensing data. *Annals of Glaciology, 54*, 171–182. https://doi.org/10.3189/2013AoG63A296

Pfeffer, W. T., Arendt, A. A., Bliss, A., et al. (2014). The Randolph glacier inventory: A globally complete inventory of glaciers. *Journal of Glaciology, 60*, 537–552.

Racoviteanu, A. E., Paul, F., Raup, B., et al. (2009). Challenges and recommendations in mapping of glacier parameters from space: Results of the 2008 Global Land Ice Measurements from Space (GLIMS) workshop, Boulder, Colorado, USA. *Annals of Glaciology, 50*, 53–69.

Raina, V., & Srivastava, D. (2008). *Glacier Atlas of India.*

RGI Consortium. (2017). *Randolph glacier inventory – A dataset of global glacier outlines: Version 6.0: Technical report.* Global Land Ice Measurements from Space, Colorado, USA.

Roy, D. P., Wulder, M. A., Loveland, T. R., et al. (2014). Landsat-8: Science and product vision for terrestrial global change research. *Remote Sensing of Environment, 145*, 154–172. https://doi.org/10.1016/j.rse.2014.02.001

SAC. (2010). *Snow and glaciers of the Himalayas: Inventory and monitoring.*

Sangewar, C. V. (2005). *Thematic compilation of glacier inventory of Himachal Himalaya.*

Scherler, D., Bookhagen, B., & Strecker, M. R. (2011). Spatially variable response of Himalayan glaciers to climate change affected by debris cover. *Nature Geoscience, 4*, 156–159. https://doi.org/10.1038/ngeo1068

Singh, M. K., Gupta, R. D., Snehmani, et al. (2015). Scenario-based validation of moderate resolution DEMs freely available for complex Himalayan Terrain. *Pure and Applied Geophysics*, 1–23. https://doi.org/10.1007/s00024-015-1119-5

Sun, G, Ranson K., Kharuk, V., & Kovacs, K (2003). Validation of surface height from shuttle radar topography mission using shuttle laser altimeter. *Remote Sensing of Environment, 88,* 401–411. https://doi.org/10.1016/j.rse.2003.09.001. https://www.sciencedirect.com/science/article/abs/pii/S0034425703002244

Vohra, C. P. (1980). Some problems of glacier inventory in the Himalayas. In *World glacier inventory – Inventaire mondial des glacier* (pp. 67–74). International Association of Hydrological Sciences.

Worni, R., Huggel, C., & Stoffel, M. (2013). Glacial lakes in the Indian Himalayas – From an area-wide glacial lake inventory to on-site and modeling based risk assessment of critical glacial lakes. *Science of the Total Environment, 468–469*(Suppl), S71–S84. https://doi.org/10.1016/j.scitotenv.2012.11.043

Chapter 7
Landslide Susceptibility Mapping of Tehri Reservoir Region Using Geospatial Approach

Gaurav Tripathi, Achala Shakya, Ritambhara Kumari Upadhyay, Suraj Kumar Singh, Shruti Kanga, and Sandeep Kumar Pandey

Abstract Uttarakhand is one of the most landslide-susceptible states because of its geographical setting, which consists of 86% of the Himalayan terrain. However, in recent years, landslides have increased dramatically due to the large number of settlements, farms, road buildings, and a wide variety of hydroelectric projects. Therefore, this is a need to study the landslides scrupulously at a regional scale to rein the future developmental planning models. In the current work, a comprehensive study has been undertaken for the assessment of landslide susceptibility zones using the weight of evidence (WOE) and risk assessment for the Tehri region, specifically around the Tehri reservoir. Landslides are derived through remote-sensing techniques and other sources such as slope, geology, aspect, geomorphology, land use/land cover, drainage, lineaments, and more. After that, the WOE method is applied to integrate causative factors for the mapping of susceptible landslide zones, where the weights have been assigned to each layer according to

G. Tripathi
Department of Geoinformatics, Central University of Jharkhand, Ranchi, India

A. Shakya
School of Computer Science, University of Petroleum and Energy Studies, Dehradun, India

R. K. Upadhyay
Centre of Advanced Studies in Geology, Panjab University, Chandigarh, India

S. K. Singh
Centre for Climate Change and Water Research, Suresh Gyan Vihar University, Jaipur, India
e-mail: suraj.kumar@mygyanvihar.com

S. Kanga (✉)
Department of Geography, School of Environment and Earth Sciences, Central University of Punjab, Bathinda, India
e-mail: shruti.kanga@cup.edu.in

S. K. Pandey
Gujarat Institute of Disaster Management, Gandhinagar, Gujarat, India
e-mail: apspm-gidm@gujarat.gov.in

S. Sharma et al. (eds.), *Climate Change Adaptation, Risk Management and Sustainable Practices in the Himalaya*,
https://doi.org/10.1007/978-3-031-24659-3_7

available literatures. Subsequently, vulnerability is prepared for the area by integrating layers through the weighted sum technique. Finally, a risk map was prepared by integrating a susceptibility and vulnerability map. All three maps, namely, vulnerability, landslide susceptibility, and risk maps, were classified into five zones: very low, low, moderate, high, and very high. The results obtained from final maps and plots indicate that approximately 8% of the area is in a high susceptible zone, 50% is in a moderate susceptible zone, 54% is in a very low-risk zone, 23% is in a moderate-risk zone, and 14% is in a very high-risk zone. This study identified and illustrated the causative factors, combined into a GIS environment to identify landslide-prone locations. Then, depending upon the potency of an element, suitable and effective preventive measures may be taken to reduce the impact of the disaster. The concerned government agencies can use the same map while mapping disaster management, developing future strategies, implementing rehabilitation programs, and environmental planning.

Keywords Tehri Reservoir · Landslide Hazard Zonation · Weights · Remote Sensing · GIS

1 Introduction

1.1 Background

There are a variety of geo-environmental factors related to geology and meteorology and that are human-induced that cause landslides. These factors can be categorized as intrinsic and extrinsic factors (Hutchinson, 1995; Varnes, 1984). Some of the most significant intrinsic factors include geology of the area (type of lithology, inclination of the bedrock presence of linear structures and discontinuities such as lineaments, faults, folds, etc.), geomorphology (slope gradient, aspect, and relative relief in the region), soil (type, thickness of soil cover, permeability, porosity, soil moisture, etc.), land use/land cover, and other hydrologic conditions such as presence of streams that cause down-cutting and eventually landslides. Some of the extrinsic factors can be attributed to the amount of rainfall in the region, seismic activities, blasting, drilling, cloudbursts, flash floods, and other human interferences or disturbances like drilling and blasting in the name of development like cutting for roads, highways, buildings, tunnels, etc.

The highly rugged lesser Himalayan terrain hosts the Tehri dam and reservoir at the confluence of rivers Bhagirathi and Bhilangana. The dam is 260.5 m high, whereas the reservoir is 67 km long. Several studies have supported that the Tehri Dam has induced severe adverse effects on the geo-environment in the region (Gupta & Anbalagan, 1997). Needless to say that owing to the vastness of the project, a number of villages have been negatively impacted with innumerable landslides leading to loss of farmland. The slope instability can be greatly attributed to the slope, relative relief, lithology, geology, and structural discontinuity that characterize the Himalayan zone and make them dynamic with earthquakes, landslides, etc.

As is well known, the Himalayas are young-fold mountains, and the height of the ranges is increasing at a rate of 5 cm per year.

The most important factor in the formulation of mitigation strategies for landslides is a comprehensive probability mapping that is needed along with characterization of the various factors that are responsible for inducing landslides in the region. The probable zones of landslides can be mapped with the help of the factors that cause landslides in the region. In fact, the landslide probable zones are the areas where landslides are expected to occur in the future based on previous instability studies in the region (Carrara et al., 1995; Guzzetti et al., 1999; Kanungo et al., 2006; Varnes, 1984).

Landslide susceptibility zonation (LSZ), landslide hazard zonation (LHZ), and landslide susceptibility mapping (LSM) are interrelated and are based on similar principles. The landslide zones are the regions that have a high probability of landslides (Varnes, 1984). The term landslide susceptibility was introduced by Brabb in the year 1984 to describe the spatial probability of landslide depending upon a variety of geo-environmental factors.

Remote sensing and geospatial techniques are highly useful, especially in regions that are inaccessible. Remote sensing data at temporal resolution facilitates the study of landslide-prone with reference to rainfall, that is, before rainfall and after rainfall and also in mapping the factors leading to landslides. With the help of GIS, spatial data extraction, integration, and analysis are greatly facilitated (Shakya et al., 2021). The integration of remote sensing and GIS has been found to be extremely useful in landslide susceptibility studies. There are a number of attributes and factors that are assigned weight or ratings for landslide susceptibility mapping, and this requires extensive data integration.

1.2 Landslide Inventory

For making landslide susceptibility maps, landslide inventory is a prerequisite. Landslide inventory map gives a quick reference to distribution of previous landslides in the region, the type of mass movements, location of previous landslides, pattern of re-occurrence of slope failures in the region, evolution of landscapes, processes of mass wasting, types of mass movements in the region, etc. (Guzzetti, 2000; Guzzetti et al., 2012; Hansen et al., 1984; Pašek, 1975; Wieczorek, 1984). Landslide maps can be prepared using visual interpretation techniques for identifying landslides using high-resolution satellite images such as LISS IV, Quickbird, GeoEye, etc. Field studies or investigations are also needed for the ground validation of the landslides identified through satellite image interpretation. With the help of remote sensing and Geographical Information System, landslide inventory maps can be prepared, which can be further validated through ground truthing to assess accuracy.

1.3 Susceptibility Mapping

The spatial probable mapping of the landslide and various causative agents is the backbone of landslide susceptibility in a specific time frame (Varnes, 1984). There are a number of intrinsic causative factors. In landslide susceptibility zonation (LSZ), the terrain is segmented based on homogeneity and ranked or weighted based on the likelihood that a given element will cause a landslide.

In the LHZ map, the terrain is divided into a number of zones such as least probability, medium probability, high probability, etc. depending upon the degree of probability of landslide in the region. The zones are identified based on a methodology that incorporates various factors such as parameters to be considered, nature of terrain, geology of the region, soil type, degree of slope, amount of rainfall, seismicity, etc. Usually, there are three different methodologies adopted, namely, heuristic (knowledge-based), deterministic (geotechnical/geo-engineering-based), and statistical (based on landslide inventories) (Clerici et al., 2006). Deterministic is most useful for mapping of landslides on large scale and is based on stability models and includes proper understanding of the physical laws that govern slope instability (Dietrich et al., 1995; Dunne et al., 1991; Mantovani et al., 1996; Montgomery & Dietrich, 1994; Terlien et al., 1995). It includes safety factors and gives the probability of slope failures in the region. In the heuristic methods, rank and weightage are provided to causative factors for slope failure and require a high degree of subjectivity. For mapping landslide susceptibility zones on a medium scale, the statistical method is best suited and based on the conditions that caused landslides in the past (Clerici et al., 2006).

1.4 Risk Mapping

Landslide risk mapping is a comprehensive approach involving exposure (or elements at risk), vulnerability, and susceptibility/hazard (Carreño et al., 2007). Risk or vulnerability mapping is imperative for hazard analysis and comprises both components, namely, susceptibility and vulnerability assessment of various elements (like settlement, road, and land use) taken under consideration for risk analysis. There are a number of methods for landslide risk or vulnerability assessment (Bunce et al., 1997; Guzzetti, 2000; Kanga et al., 2017; Yao et al., 2008). Risk zonation on a small scale can be applied for large-scale studies so that proper mitigation strategies can be formulated (Abella & Van Westen, 2007; Carreño et al., 2007; Kanga et al., 2017).

Major research questions that followed up throughout the research are as follows: (a) What are the important parameters for landslide susceptibility mapping? (b) How is the susceptibility and vulnerability map useful for risk analysis? (c) How accurately can the landslide inventory map be prepared with the visual interpretation technique through high-resolution satellite images? (d) How is the risk map useful for analysis for the future?

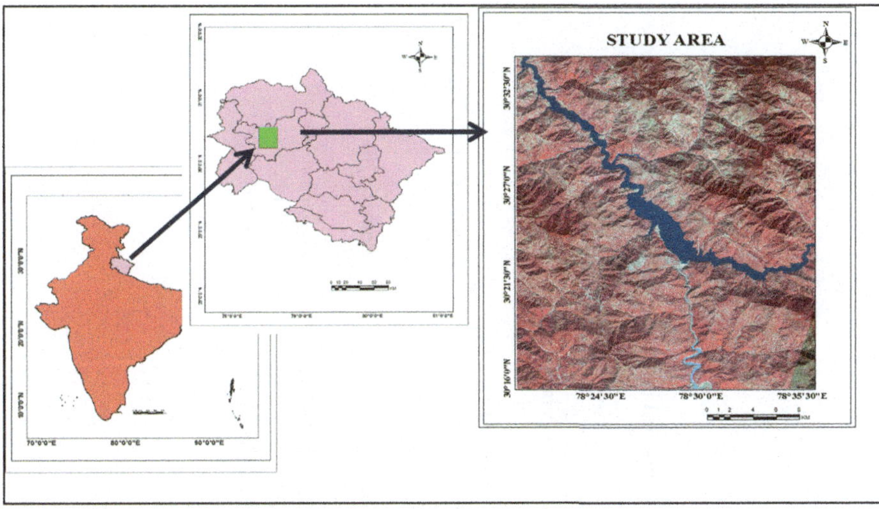

Fig. 7.1 Location of the study

In this proposed study, the following are the primary and minor goals to be achieved: (a) to prepare landslide hazard/ susceptibility zonation map by extracting essential factors from remote sensing data using digital and visual interpretation techniques and establishing a relationship between the selected parameters and landslides, (b) to prepare a vulnerability map, and (c) to prepare risk map.

2 Study Area

The area is located in the Tehri Garhwal and Uttarkashi districts of Uttarakhand, India, with central longitude and latitude of 78.5E and 30.5 N, respectively. It is also mentioned on Survey of India topographic sheet no. 53 J/10 (1985), 53 J/6 (1960), 53 J/7 (1960), 53 J/11 (1960) (Fig. 7.1).

3 Data Used and Methodology

To achieve the intended result, certain dataset are required in all scientific investigations. Therefore, this research involves satellite data, maps, and other auxiliary data involving ground-truth data (field information) as well. Different topic layers were produced to provide parameters for the inventory of landslides, susceptibility mapping, and risk mapping using software such as Erdas Imagine 2014, ArcGIS 10.3. The following table gives details of these materials (Table 7.1 and Fig. 7.2).

Table 7.1 Characteristics of the satellite and secondary data used in this study. The satellite data were procured in March 2015

Sl. no	Name of data	Temporal resolution	Spatial resolution	Acquisition date	Purpose	Source
1	IRS-Liss 3	24 days	23.5 m	March 8, 2012, March 11, 2015	NDVI map, lineament map	Bhuvan
2	ASTER GDEM	–	30 m	–	Aspect map, slope map	USGS
3	Drainage map	–	1:25,000	–	Drainage delineation	Survey of India Toposheet
4	Geomorphology map	–	–	–	Geomorphology map preparation	Bhuvan (2005–2006)
5	Land use/land cover map	–	–	–	Land use/land cover map preparation	Bhuvan (2011–2012)
6	Geology map, tec-tonic map	–	–	–	Geology map, tectonic map preparation	Geological Map (K S Valdia 1980)
7	Soil map	–	–	–	Soil map preparation	NBSS & LUP

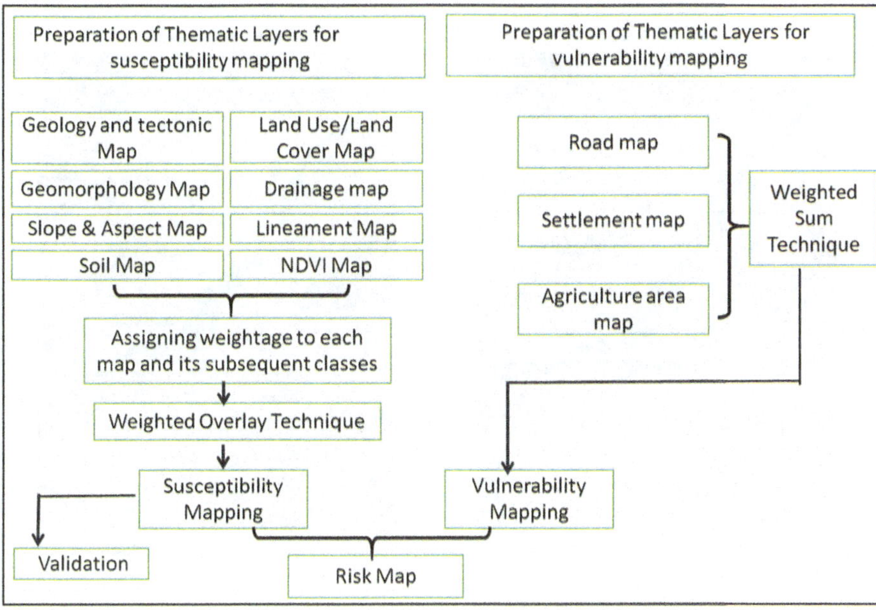

Fig. 7.2 Methodology flow chart

3.1 *Landslide Inventory Mapping*

The main emphasis is on high-resolution optical imaging for the identification and mapping of landslides to prepare the landslide inventory map. The landslide mapping is carried out with a high-resolution LISS 3 (23.5 m) multispectral satellite data of different periods to locate landslides. The result of this work was further validated using Google Earth photos. Since the region of investigation is near the Tehri reservoir, this visual interpretation approach produced good results (Fig. 7.3).

3.2 *Landslide Susceptibility Mapping*

This is an indirect method that describes how susceptibility the terrain is, based on multiple factors, which indicate the occurrence of landslide, which is also experienced by several scientists considering various factors, including geology, slope classes, soil depth or land use, etc. (Mantovani et al., 1996). The main aim is to identify landslides in a region by keeping a view on several internal causative elements. This is basically called LSZ, which can be described by dividing a geographical area into almost invariant sections and classifying them by the current or expected risk levels (Abella & Van Westen, 2007; Kanungo et al., 2006). The primary objective is to build a landslide susceptibility map of the Alaknanda valley for sufficient catastrophe risk mitigation actions by delineating high-risk zones.

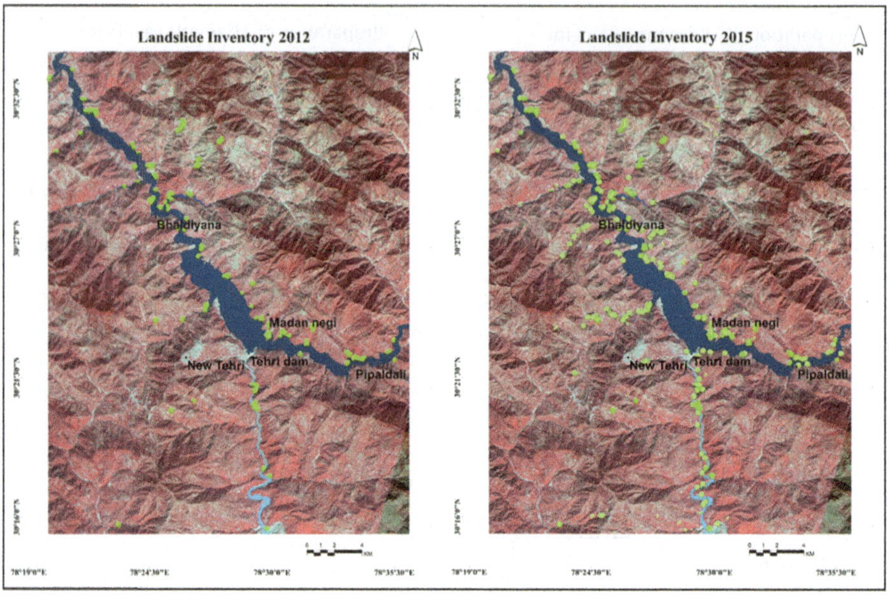

Fig. 7.3 Landslide inventory maps of Tehri region using LISS-3 data of March 8 and 11 of 2012 and 2015, respectively

3.2.1 Preparation of Database for Susceptibility Analysis Using a Weighted Overlay Technique (Figs. 7.4 and 7.5)

Slope maps are an important factor in landslide susceptibility mapping because they provide information on the steepness of the terrain. Landslides are more likely to occur on steep slopes than on gentle ones. Slope maps can be used to identify areas that have a high risk of landslides due to the steepness of the terrain. The slope map can be used to classify the area into different categories of the slope, such as low, moderate, high, and very high, and then each category can be assigned a weight according to the risk of a landslide it poses. Additionally, slope maps can be used to identify areas that are vulnerable to shallow landslides, deep-seated landslides, rock falls, and debris flows, which are all caused by different types of slopes. By identifying the type of slope, the susceptibility mapping can be made more accurate. Furthermore, a slope map can be used to identify the direction of the slope and hence the direction of potential landslides (Fig. 7.4). This information can be used to predict the path of a landslide and the potential damage it may cause.

Aspect maps, like slope maps, are important factors in landslide susceptibility mapping because they provide information on the orientation of the terrain. Aspect refers to the direction that a slope is facing, typically measured in degrees clockwise from true north. Aspect maps can be used to identify areas that have a high risk of landslides due to certain aspects, as certain aspect can influence the susceptibility to landslides (Fig. 7.5). For example, north-facing slopes are usually wetter and cooler

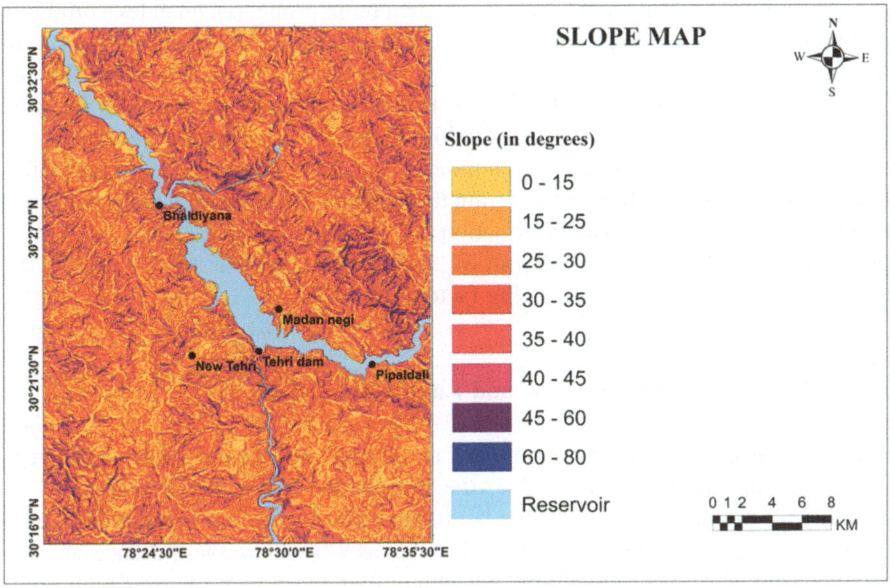

Fig. 7.4 Slope map of Tehri region using ASTER GDEM data

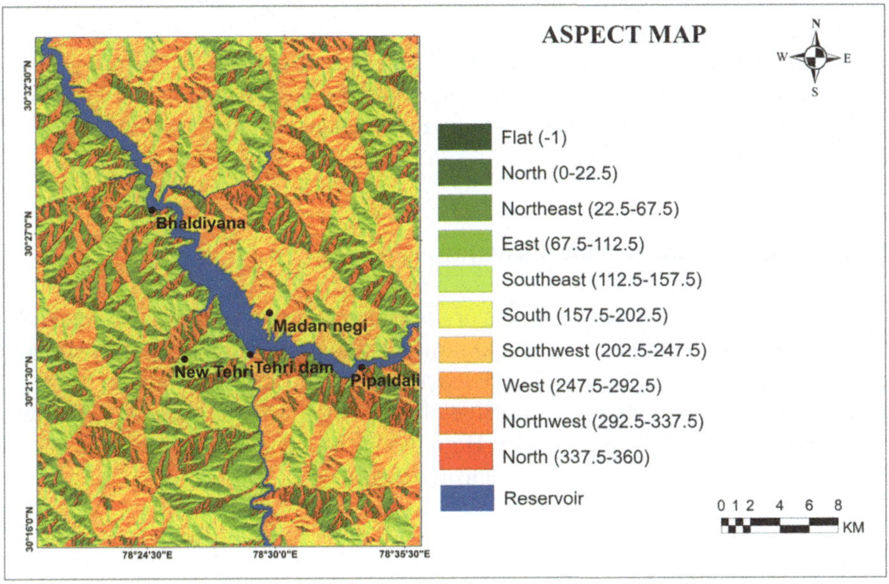

Fig. 7.5 Aspect map of Tehri region using ASTER GDEM data

than south-facing slopes, which can make them more prone to landslides. Similarly, east-facing slopes can be more susceptible to landslides during monsoon season, as they receive more rainfall. Aspect maps can also be used to identify areas that are vulnerable to different types of landslides. For example, shallow landslides are more likely to occur on south-facing slopes, while deep-seated landslides are more likely to occur on north-facing slopes. In summary, aspect maps provide important information about the orientation of the terrain that can be used to identify areas that are at a high risk of landslides. The information can also be used to make more accurate predictions about the potential impact of landslides on the area, by considering the aspect of the slope and its influencing factors.

3.2.2 Vegetation and Land Use and Land Cover Map

The NDVI map shows the vegetation type in the area. NDVI value stretches between −1 to +1, and the higher the value, the greener the space. More availability of greenery means there will be less number of landslide occurrence because vegetation binds and holds the surface soil. Thus, those areas were having less chance of slope failure. Moreover, most slopes in higher altitudes seem to be barren with some shrubs, rhododendrons, moose, lichen, and certain windflowers, which led to the occurrence of landslides. Terrace cultivation is prevalent in those areas where primary crops are potatoes, pulse, and barley; hence, slope failures are prominent in these region (Fig. 7.6).

As per the LULC map, available classes are forest, agriculture, barren ground, fallow land, grass/grazing, urban areas, and water bodies. Occurrence of landslides in forest and water body classes is often less spontaneous compared to the fallow land and cropland (Fig. 7.7).

3.2.3 Linear Features (Lineaments)

Lineaments indicate areas of weakness or disruption; hence, increasing the weakness may lead toward the failure. Faults and thrusts are the tectonic formations that were observed and are associated with extensive cracks and sharp variations in relief. Lineament map are categorized into five classes based on their density (low, very low, moderate, high, and very high). Here, it can be concluded that occurrence of landslides increases as per the class density (Fig. 7.8).

3.2.4 Geomorphology

The morphological arrangement of any terrain can be shown using geomorphology of that area, which also could provide insight about landslide susceptibility. Geomorphic land formations like dissected hills are more landslide-prone; however, areas like the piedmont zone are less susceptible in comparison to dissected hills

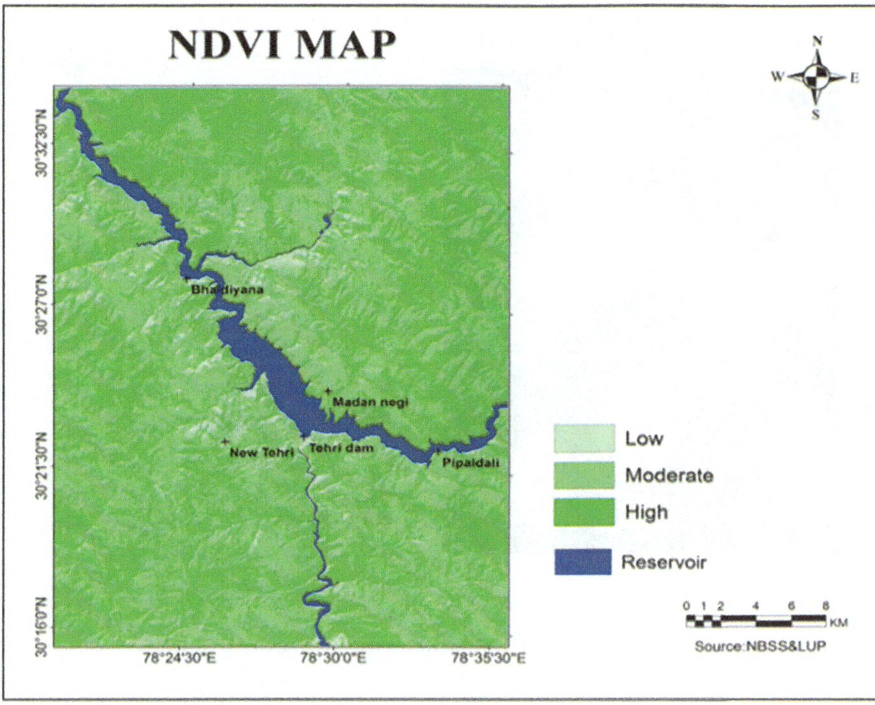

Fig. 7.6 NDVI Map of Tehri region using IRS LISS-3 data

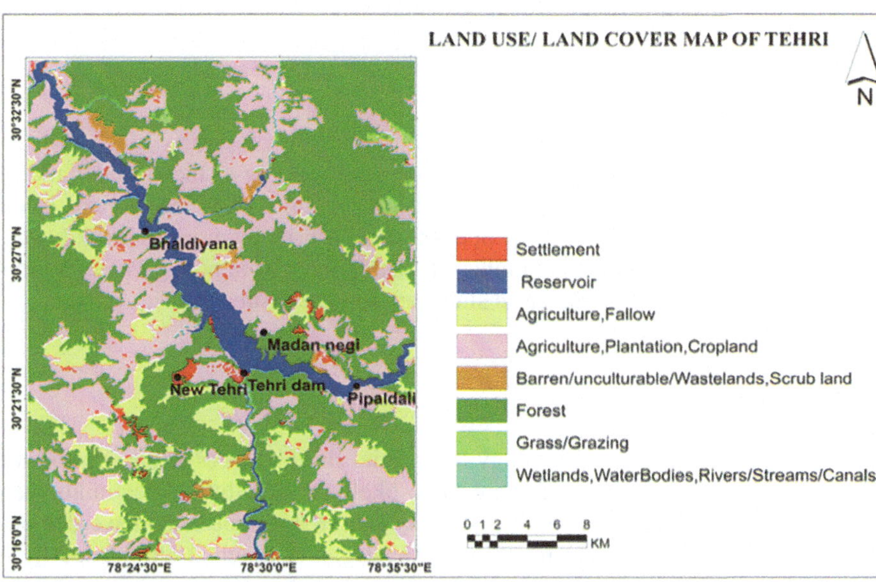

Fig. 7.7 Land use/land cover map of Tehri region based on Bhuvan (2011–2012) repository

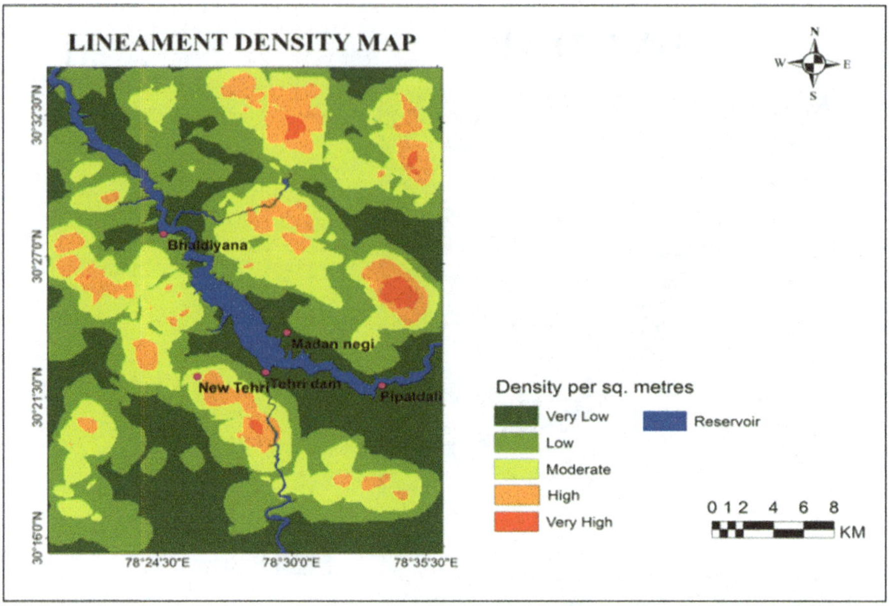

Fig. 7.8 Lineament map of Tehri region using LISS-3 data of March 11, 2015

(Anbalagan & Singh, 1996). The research area includes highly dissected hills and valleys, moderately dissected hills and valleys, an alluvial piedmont plain, an active flood plain, and a younger plain. Eighty percent of the region is occupied with very dissected hills, which is of major concern (Fig. 7.9).

3.2.5 Drainage Map

The drainage map was obtained from the topo sheet Survey of India. The drainage network has a unique relationship with landslides. In hilly locations, streams erode their banks regularly and form steep slopes that are especially prone to breakdowns. During the field investigation, a number of such failures were noted. These failures are likewise gradual. More terrestrial slides are seen near streams (Fig. 7.10).

3.2.6 Lithology

When it comes to landslides, lithology is essential since it serves as the slip surface for the parent material. Landslides are more likely in some lithological units than others. Several hard and large rocks, such as granite and limestone, have demonstrated a high level of erosion resistance (Anbalagan & Singh, 1996). Aside from that, sandstone rock is more sensitive to erosion, making it more susceptible to landslides (Fig. 7.11).

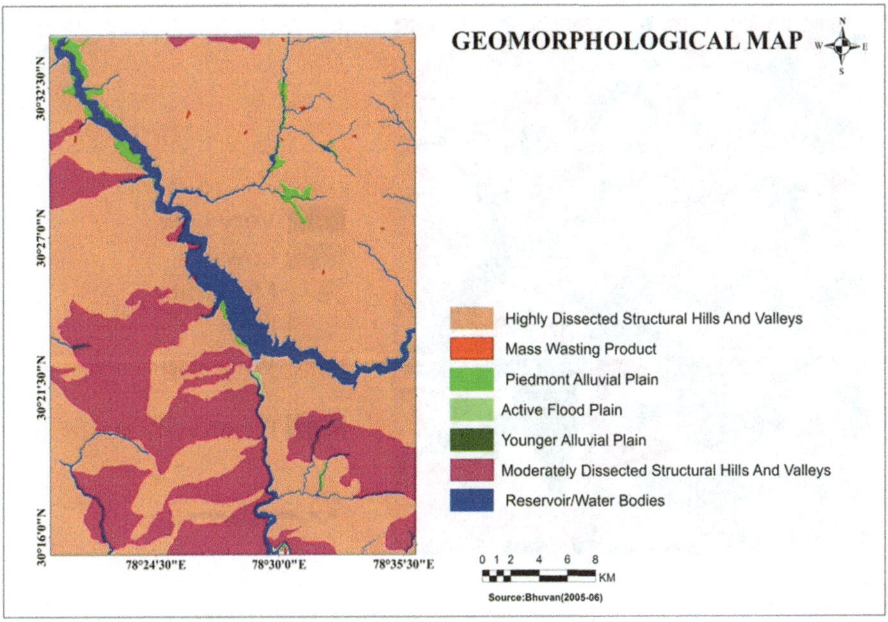

Fig. 7.9 Geomorphological Map of Tehri region based on Geological Map (K S Valdia, 1980)

3.2.7 Soil Type

It is composed of organic stuff and is found at the top of any terrain. Consequently, the movement of rocks or sediments is governed by the type of soil present. According to research, landslides are more likely to occur in loamy silty soils than in clay-rich soils, which are more stable (Fig. 7.12).

3.3 Methods

3.3.1 Weighted Overlay Technique

The weighted overlay technique is a qualitative map combination approach for landslide hazard zonation. Several steps are involved in the map combination approach for LSZ mapping (Mantovani et al., 1996):

(i) To select and map the causative factors.
(ii) To prepare thematic data layer with relevant classes.
(iii) Assigning weights to the respective classes.
(iv) To integrate thematic data layers.
(v) To prepare LSZ map with different zones.

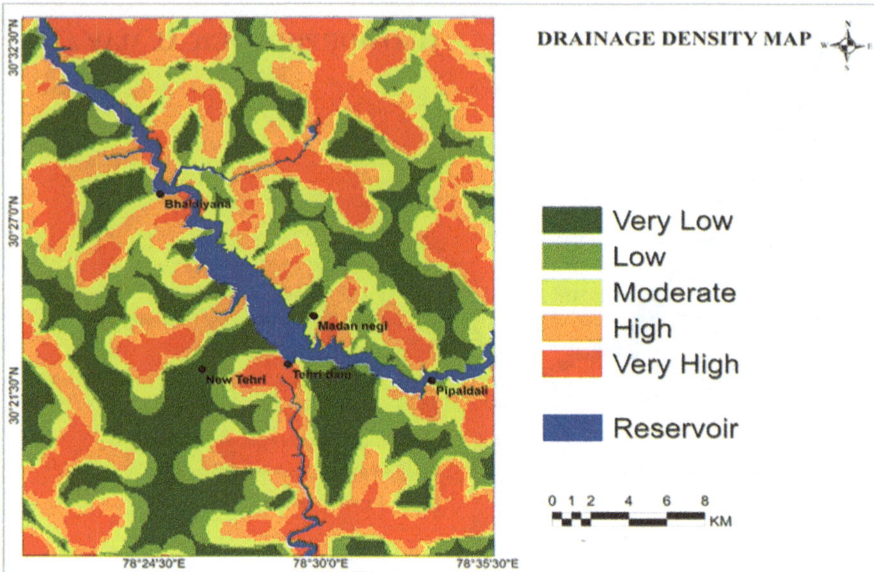

Fig. 7.10 Drainage Map of Tehri region based on SOI Topo sheet (1:25,000)

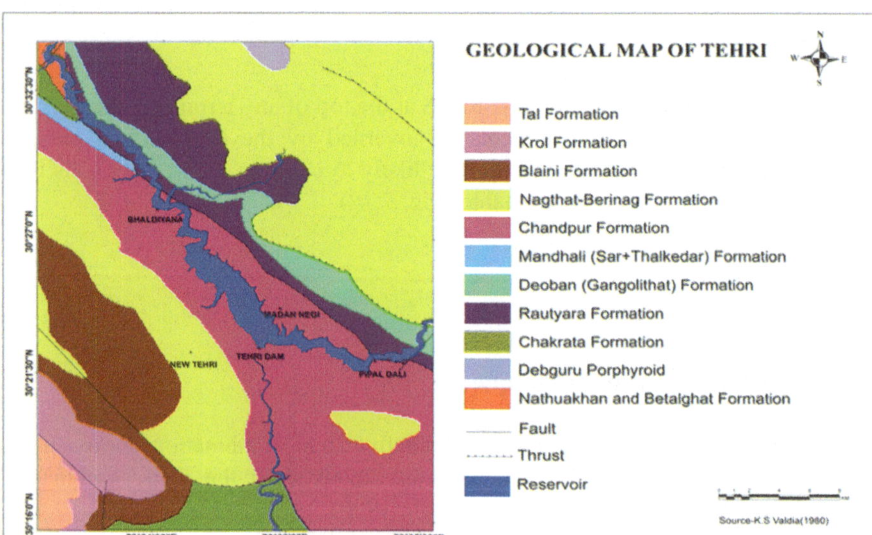

Fig. 7.11 Lithology map of Tehri region based on Geological Map (K S Valdia, 1980)

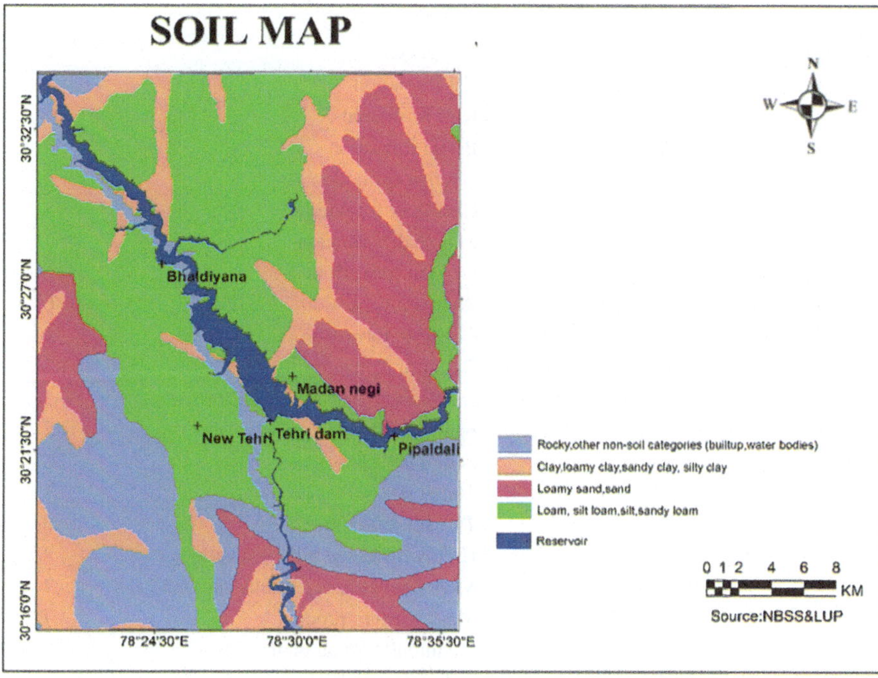

Fig. 7.12 Soil Map of Tehri region based on NBSS and LUP dataset

The preparation of thematic data layers pertaining to various causative factors is required for LSZ mapping. These factors commonly include slope, drainage, lineament, land use, land cover, and lithology (Kanungo et al., 2006). The weights assigned to the map layer in terms of landslide occurrence and the study area are highly dependent upon knowledge about the study region and the causative elements of landslides in the study area. The experts' opinion regarding this issue over a particular study area was taken into consideration while assigning weights. The weights may differ from one expert to another and from one geographical area to another. The most notable limitation of this method is the bias with which weights were allocated for each thematic data layer and its classes. The susceptibility map, on the other hand, generically categorizes based on their susceptibility to landslides into five important sections, which are as follows: very low, low, moderate, high, and very high susceptible areas (Table 7.2).

3.3.2 Risk Map Preparation

To perform risk analysis for any hazard, one must have to consider vulnerability factor. In this study, vulnerability layers considered for risk analysis are settlements, roadways, and agricultural fields. Vector layers were converted in raster and

Table 7.2 Weight and influence for thematic layers

Parameter	Weight	Category	Rank
Geology	13	Tal formation	5
		Krol formation	7
		Blaini formation	5
		Nagthat formation	5
		Chandpur formation	8
		Mandhali formation	4
		Deoban formation	6
		Rautyara formation	6
		Chakrata formation	5
		Debguru formation	3
		Nathuakhan and Betalghat formation	3
Structure	9	Fault (0–100 m)	7
		Fault (100–200 m)	6
		Thrust (0–100 m)	8
		Thrust (100–200 m)	6
Geomorphology	12	Highly dissected hills and valleys	8
		Moderately dissected hills and valleys	7
		Mass wasting product	7
		Piedmont alluvial plain	3
		Active flood plain	2
		Reservoir/waterbodies	4
		Younger flood plain	2
Land use/land cover	11	Settlement	5
		Reservoir	2
		Agriculture, fallow	6
		Agriculture, plantation	4
		Barren/waste land	6
		Forest	2
		Grass/grazing	5
		Wetlands, waterbodies	5
Slope	14	0–15	1
		15–25	2
		25–30	2
		30–35	2
		35–40	3
		40–45	7
		45–60	7
		60–80	9
Aspect	4	Flat	1
		North	1
		Northeast	2
		East	4
		Southeast	5
		South	6

(continued)

Table 7.2 (continued)

Parameter	Weight	Category	Rank
		Southwest	5
		West	3
		Northwest	1
		North	1
NDVI	10	Low	6
		Medium	4
		High	2
Soil	7	Rocky, other non-soil categories	2
		Clay, loamy clay, sandy clay	4
		Loamy sand, sand	5
		Loamy, silty loam, silty, sandy loam	6
Lineament density	13	Very low	1
		Low	1
		Moderate	3
		High	5
		Very high	7
Drainage density	7	Very low	1
		Low	1
		Moderate	3
		High	4
		Very high	5

reclassified. Moreover, spatial resolution was made same for all layers by using upscaling/downscaling technique. By combining these layers, ARC's vulnerability map can be created using the weighted overlay tool. Last but not least, the risk map is created by multiplying the susceptibility and vulnerability maps together.

4 Results and Discussion

A susceptibility map is a visual representation of an area divided into several classifications based on how prone it is to landslides. The susceptibility map produced by the weighted overlay technique divides the Tehri region into five zones based on landslide susceptibility. There are five zones of vulnerability: very low, low, moderate, high, and very high. Analyzing the map reveals that the high susceptibility zones in the study area are primarily concentrated in high slope areas and drainage corridors (Fig. 7.13).

The loss of life, infrastructure, and property as a result of the landslide is referred to as vulnerability. Adding together the three thematic layers such as settlement, road, and agricultural area resulted in the creation of a vulnerability map. Vulnerability levels range from very low to moderate majorly in the study region. A risk map is prepared by multiplying susceptibility and vulnerability together (Fig. 7.14).

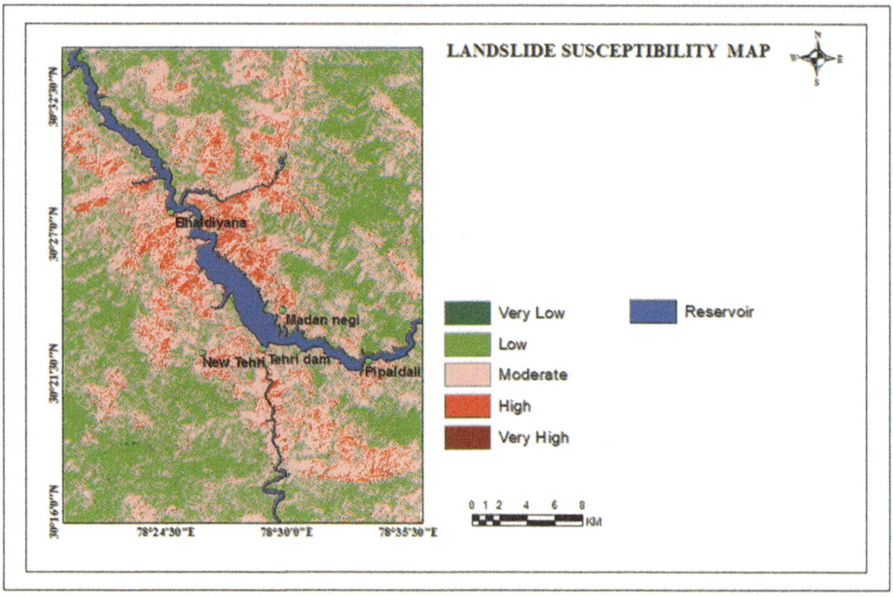

Fig. 7.13 Landslide susceptibility map of the study area

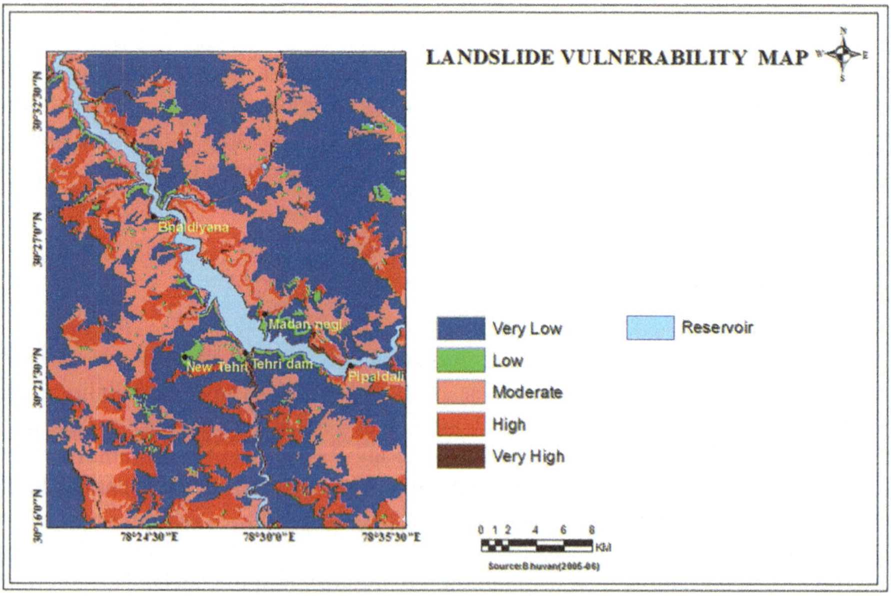

Fig. 7.14 Landslide vulnerability map of the study area

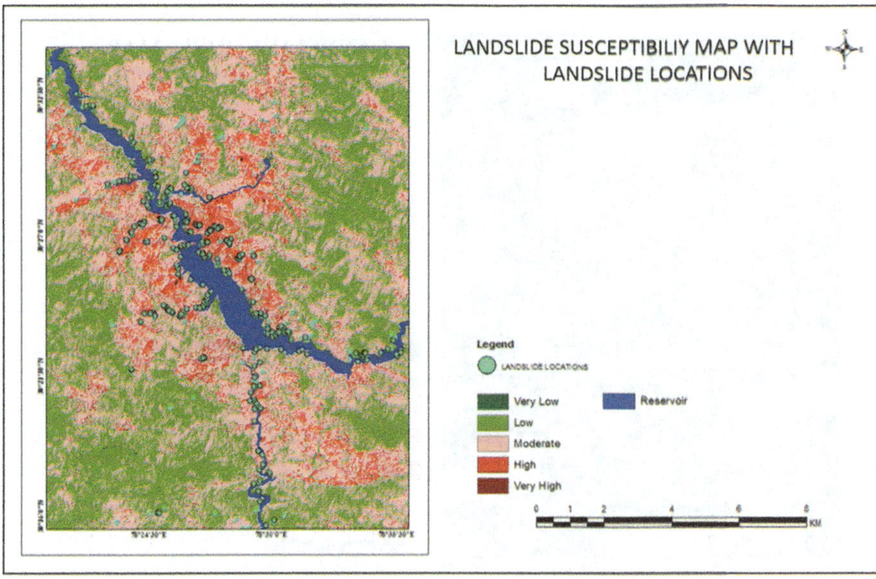

Fig. 7.15 Landslide susceptibility map with landslide locations over study area. Landslide locations were shown using cyan color circles, and Tehri reservoir area was shown in blue color

Integrating various thematic layers using the weight of evidences (WoE) method, the three maps, namely, susceptibility, vulnerability, and risk, have been prepared for the study area. All the three maps have been classified into five zones, namely, "very low," "low," "moderate," "high," and "very high." The overall analysis indicates that most of the area lies in low to moderate susceptible as well as risk zone. It is observed that about 50% and 42% area lie in moderate and low susceptibility zone, whereas the area that lies in high to very high susceptible zone is only 8%. The vulnerability map categorizes 53% and 42% area under very low and high vulnerability zone. The final risk map produced by integrating the vulnerability and susceptibility maps presents 54% area in very low, 23% in moderate, and 14% in high-risk zone. The susceptibility map also shows that the area to the left of Bhagirathi is more vulnerable due to mass wasting in comparison to the rest of the region. This finding has been corroborated by the field survey and spot and world-view images obtained from Google Earth. The susceptibility map is validated by a field survey. The susceptible areas have moderately to high steep slopes, somewhere in excess of 60° and do not have any in situ rocks as such. They are mostly covered by weathered river-borne material (RBM) or highly weathered phyllites (Figs. 7.15 and 7.16).

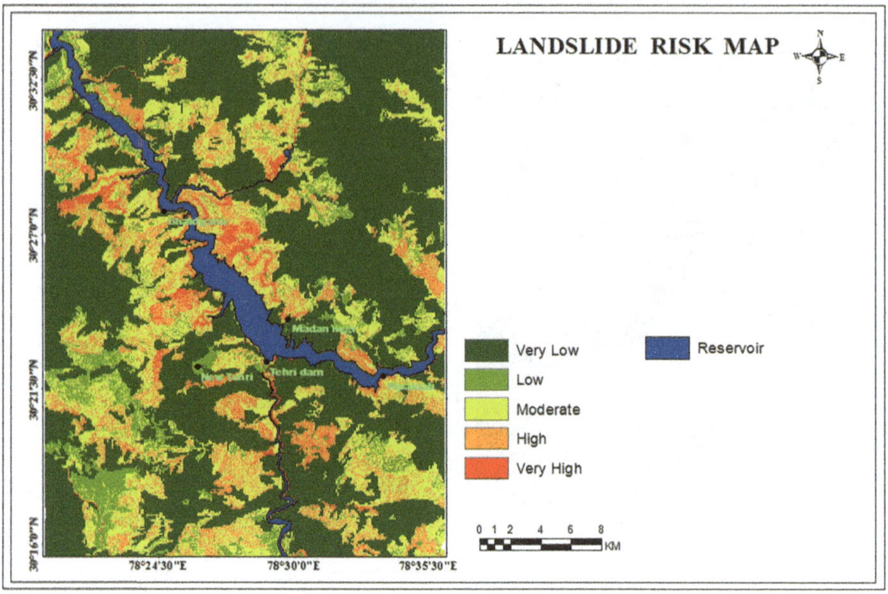

Fig. 7.16 Landslide risk map over study area. Tehri reservoir area is shown in blue color

5 Conclusion

Being a young-fold mountain range, the Himalayan terrain has always been susceptible to mass wasting activities (Anbalagan & Singh, 1996). However, the vulnerability has increased significantly in the past two decades or so due to unrestricted and unchecked human interventions in previously undisturbed terrain (Bunce et al., 1997; Gupta & Anbalagan, 1997; Hutchinson, 1995; Kanungo et al., 2006; Montgomery & Dietrich, 1994). Experiences not only from India but across the globe have only corroborated the idea that mass wasting and interference in the form of agriculture, farming, settlements, infrastructure, hydroelectric projects, etc. are intricately and persistently correlated. All such developments have actually decreased the threshold for any disaster and made the already stressed Himalayan terrain more prone to mass wasting activities (Kanga et al., 2017; Mantovani et al., 1996; Montgomery & Dietrich, 1994). Consequently, even a little excess rainfall or moderate earthquake shocks trigger landslide in some nook or corner of the state, which largely affects towns and cities, communication systems, and large structures, including dams and bridges. The current study not only shows how landslides might occur as a result of infrastructure development but also shows how remote sensing and GIS tools can be used to map landslide hazards and risks at a regional scale. Each causative factor's potential is enumerated and depicted in the current study, which is as follows: drainage map, NDVI map, geomorphology map, slope map, aspect map, land use/land cover map, soil map, thrust (buffer) map, lineament density map, and geological map, which merge them under GIS environment to delineate landslide-prone areas. Then, depending upon the potency of a factor, suitable and effective preventive measures may be taken to cushion the impact of the factor. The concerned government authorities can use such maps to build disaster management and preparedness plans, carry out rehabilitation programs, plan for the environment, and frame other future development strategies. Identifying vulnerable locations on a regional scale will help with planning ahead of time in the event of a disaster and may help to reduce the loss of life and property. Risk in landslides can be minimized and prevented in a variety of ways. When the entire terrain is vulnerable, one can adopt several conventional procedures to stabilize the slopes by planting shrubs, bushes, and other trees (Guzzetti et al., 1999; Hansen et al., 1984). Additionally, depending on the vulnerability and danger, specialized measures such as grouting may be implemented.

References

Abella, E. A. C., & Van Westen, C. J. (2007). Generation of a landslide risk index map for Cuba using spatial multi-criteria evaluation. *Landslides, 4*(4), 311–325. https://doi.org/10.1007/s10346-007-0087-y

Anbalagan, R., & Singh, B. (1996). Landslide hazard and risk assessment mapping of mountainous terrains – A case study from Kumaun Himalaya, India. *Engineering Geology, 43*(4), 237–246. https://doi.org/10.1016/S0013-7952(96)00033-6

Bunce, C. M., Cruden, D. M., & Morgenstern, N. R. (1997). Assessment of the hazard from rock fall on a highway. *Canadian Geotechnical Journal, 34*(3), 344–356. https://doi.org/10.1139/t97-009

Carrara, A., Cardinali, M., Guzzetti, F., & Reichenbach, P. (1995). Gis technology in mapping landslide hazard. In A. Carrara & F. Guzzetti (Eds.), *Geographical information systems in assessing natural hazards* (Vol. 5, pp. 135–175). Springer. https://doi.org/10.1007/978-94-015-8404-3_8

Carreño, M. L., Cardona, O. D., & Barbat, A. H. (2007). A disaster risk management performance index. *Natural Hazards, 41*(1), 1–20. https://doi.org/10.1007/s11069-006-9008-y

Clerici, A., Perego, S., Tellini, C., & Vescovi, P. (2006). A GIS-based automated procedure for landslide susceptibility mapping by the conditional analysis method: The Baganza valley case study (Italian Northern Apennines). *Environmental Geology, 50*(7), 941–961. https://doi.org/10.1007/s00254-006-0264-7

Dietrich, W. E., Reiss, R., Hsu, M.-L., & Montgomery, D. R. (1995). A process-based model for colluvial soil depth and shallow landsliding using digital elevation data. *Hydrological Processes, 9*(3–4), 383–400. https://doi.org/10.1002/hyp.3360090311

Dunne, T., Zhang, W., & Aubry, B. F. (1991). Effects of rainfall, vegetation, and microtopography on infiltration and runoff. *Water Resources Research, 27*(9), 2271–2285. https://doi.org/10.1029/91WR01585

Gupta, P., & Anbalagan, R. (1997). Slope stability of Tehri Dam Reservoir Area, India, using landslide hazard zonation (LHZ) mapping. *Quarterly Journal of Engineering Geology and Hydrogeology, 30*(1), 27–36. https://doi.org/10.1144/GSL.QJEGH.1997.030.P1.03

Guzzetti, F. (2000). Landslide fatalities and the evaluation of landslide risk in Italy. *Engineering Geology, 58*(2), 89–107. https://doi.org/10.1016/S0013-7952(00)00047-8

Guzzetti, F., Carrara, A., Cardinali, M., & Reichenbach, P. (1999). Landslide hazard evaluation: A review of current techniques and their application in a multi-scale study, Central Italy. *Geomorphology, 31*(1–4), 181–216. https://doi.org/10.1016/S0169-555X(99)00078-1

Guzzetti, F., Mondini, A. C., Cardinali, M., Fiorucci, F., Santangelo, M., & Chang, K.-T. (2012). Landslide inventory maps: New tools for an old problem. *Earth-Science Reviews, 112*(1–2), 42–66. https://doi.org/10.1016/j.earscirev.2012.02.001

Hansen, A. D. A., Rosen, H., & Novakov, T. (1984). The aethalometer – An instrument for the real-time measurement of optical absorption by aerosol particles. *Science of the Total Environment, 36*, 191–196. https://doi.org/10.1016/0048-9697(84)90265-1

Hutchinson, J. N. (1995). *Keynote paper: Landslide hazard assessment* (pp. 1805–1841). Presented at the international symposium on landslides in Imperial College of Science Technology and Medicine, London. Retrieved from http://pascal-francis.inist.fr/vibad/index.php?action=getRecordDetail&idt=6312114

Kanga, S., Tripathi, G., & Singh, S. K. (2017). Forest fire hazards vulnerability and risk assessment in Bhajji Forest Range of Himachal Pradesh (India): A geospatial approach. *Journal of Remote Sensing & GIS, 8*(1), 25–40.

Kanungo, D. P., Arora, M. K., Sarkar, S., & Gupta, R. P. (2006). A comparative study of conventional, ANN black box, fuzzy and combined neural and fuzzy weighting procedures for landslide susceptibility zonation in Darjeeling Himalayas. *Engineering Geology, 85*(3–4), 347–366. https://doi.org/10.1016/j.enggeo.2006.03.004

Mantovani, F., Soeters, R., & Van Westen, C. J. (1996). Remote sensing techniques for landslide studies and hazard zonation in Europe. *Geomorphology, 15*(3–4), 213–225. https://doi.org/10.1016/0169-555X(95)00071-C

Montgomery, D. R., & Dietrich, W. E. (1994). A physically based model for the topographic control on shallow landsliding. *Water Resources Research, 30*(4), 1153–1171. https://doi.org/10.1029/93WR02979

Pašek, J. (1975). Landslides inventory. *Bulletin of the International Association of Engineering Geology, 12*(1), 73–74. https://doi.org/10.1007/BF02635432

Shakya, A., Biswas, M., & Pal, M. (2021). Parametric study of convolutional neural network based remote sensing image classification. *International Journal of Remote Sensing, 42*(7), 2663–2685. https://doi.org/10.1080/01431161.2020.1857877

Terlien, M. T. J., Van Westen, C. J., & van Asch, T. W. J. (1995). Deterministic modelling in Gis-based landslide hazard assessment. In A. Carrara & F. Guzzetti (Eds.), *Geographical information systems in assessing natural hazards* (Vol. 5, pp. 57–77). Springer Netherlands. https://doi.org/10.1007/978-94-015-8404-3_4

Varnes, D. J. (1984). *Landslide hazard zonation: A review of principles and practice*. United Nations.

Wieczorek, G. F. (1984). Preparing a detailed landslide-inventory map for hazard evaluation and reduction. *Environmental & Engineering Geoscience, xxi*(3), 337–342. https://doi.org/10.2113/gseegeosci.xxi.3.337

Yao, X., Tham, L. G., & Dai, F. C. (2008). Landslide susceptibility mapping based on support vector machine: A case study on natural slopes of Hong Kong, China. *Geomorphology, 101*(4), 572–582. https://doi.org/10.1016/j.geomorph.2008.02.011

Chapter 8
Socioeconomic, Livelihood, and Ecological Transformation in the Sikkim Himalayan Region

Lak Tsheden Theengh, Priyadarshinee Shrestha, Passang D. Lepcha, Rajarshi Chakraborty, and Udai Gurung

Abstract The Himalayan landscape is host to rich bioresources, which fulfil diverse needs of the communities living in it. The region is also rich in cultural diversity, and the people living in this region possess enormous wisdom on the resources they live with. The region hosts a growing number of rural poor who are dependent on forest and ecosystem services for their livelihood, which primarily comprises of agriculture, livestock, horticulture, forest, pastures, etc., and there is a huge linkage between people and nature. Though the region is vital to ecological security in terms of providing forest cover, feeding perennial rivers are the source of drinking water and biodiversity-rich but are vulnerable to anthropogenic pressures and climate change impacts. This chapter highlights socioeconomic situation of villages in Sikkim drawing from the village-level community discussions, PRA exercises, and data from household baseline. It also includes the current livelihood options that communities in the village are engaged, along with various challenges, strengths, weaknesses, and opportunities. Various programs that are ongoing in the villages have also been mapped along with the presence of government departments and other organizations at a cluster level.

Keywords Himalayas · Cultural diversity · Climate change · Ecosystem services · Livelihoods · Sustainable development · Sikkim

---

L. T. Theengh (✉) · P. Shrestha (✉)
WWF-India, Khangchendzonga Landscape Program, Sikkim Office, Gangtok, India
e-mail: ltheengh@wwfindia.net; pshrestha@wwfindia.net

P. D. Lepcha
Darjeeling, West Bengal, India

R. Chakraborty
UNDP, SECURE Himalaya, New Delhi, India

U. Gurung
Forest Department, Government of Sikkim, Sikkim, India

S. Sharma et al. (eds.), *Climate Change Adaptation, Risk Management and Sustainable Practices in the Himalaya*,
https://doi.org/10.1007/978-3-031-24659-3_8

1 Introduction

Indian Himalayan region stretches over a length of about 3000 km of northern Pakistan, Nepal, and Bhutan and over 2500 km and width of 220–330 km in northwestern and northeastern states of India; the Himalaya hotspot includes all of the world's mountain peaks higher than 8000 m, which includes the world's highest mountain, the Mt. Everest. This immense mountain range, which covers nearly 750,000 km^2, has been divided into two regions: the Eastern Himalaya, which covers parts of Nepal, Bhutan, the northeast Indian states of West Bengal, Sikkim, Assam, and Arunachal Pradesh, southeast Tibet (Autonomous Region of China), and northern Myanmar, and the Western Himalaya, covering the Kumaon-Garhwal, northwest Kashmir, and northern Pakistan (Fig. 8.1).

Further, within the country, the hilly region of Jammu and Kashmir and Himachal Pradesh are mentioned as Western Himalaya, Uttarakhand as Central Himalaya, Sikkim and Arunachal Pradesh as Eastern Himalaya and of these two states together with Meghalaya, Assam, Manipur, Mizoram, Nagaland and Tripura, as north-eastern hill region of India.

The Indian Himalayan region has also been recognized as a World Heritage Site, including the famous Khangchendzonga National Park, also a Biosphere Reserve located in Sikkim, India, which was inscribed to the UNESCO World Heritage Sites list in July 2016, becoming the first "Mixed Heritage." The site stretches across both the West and North Districts of Sikkim.

Though immensely rich in natural resources, the region is marginal in terms of socioeconomic indicators of development like per capita income, road density, modern technologies, industrialization, and advance education/health care systems; adding to it, the climate change impacts are also visible, which influence lives and livelihoods variedly.

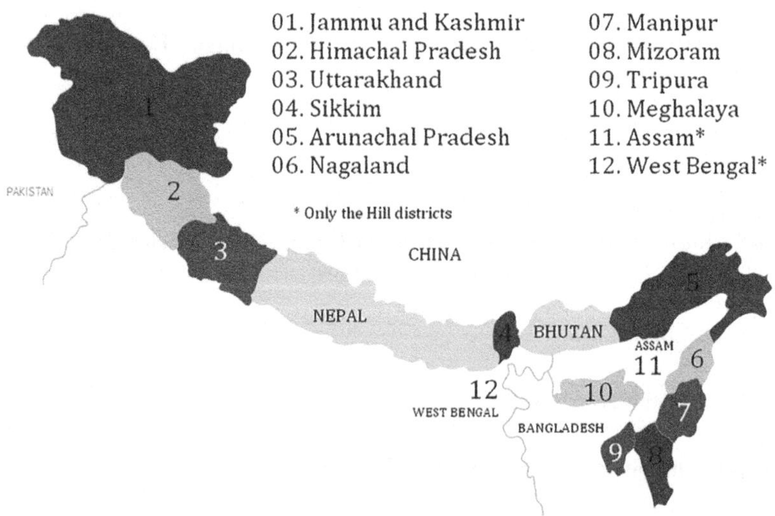

Fig. 8.1 The Himalayan States of India. (Source: G-SHE, 2009)

Study by Bawa and Ingty (2012) and Basnett et al. (2013) mentions North Sikkim faces increasing minimum and average temperature, shift in timing of snowfall, decline in snowfall amount, expansion of old lakes, and formation of new lakes and glacier retreat.

Findings from the community consultation mention that glaciers have significantly receded, and there is less snow in the winter months. Previously, more than 3–4 feet of snow were recorded; currently, it is barely 1–2 feet. Most mountains surrounding the villages that were snow-bound throughout the year are now without snow cover from June to August or early September. Communities also mention the drying of wetlands significantly.

Fragmentation of the upper broadleaf forests due to development activities is already happening; however, no studies on biodiversity or genetic diversity loss have been undertaken. Changes in species composition and decreasing productivity of alpine grasslands have been reported by Sharma et al. (2009) in certain Himalayan alpine regions.

This was corroborated by the interaction with the locals, who also pointed out that the grazing pastures are rapidly drying up. The quality of grass has decreased, mainly due to less snowfall, expanding border patrol infrastructure, and presence.

A report by Kothari et al. (2017), mentions of certain phenological changes in rhododendron blooms and shift of certain birds, insects, mammals, and plants from their geographic distribution to higher altitudes in response to the changing climatic conditions in Sikkim.

2 Study Area

2.1 Sikkim (Overview)

To assist in understanding socioeconomic, livelihood, and ecological transformation in the Himalayan region, a case study-based approach has been followed, from Sikkim Himalayan region. Sikkim is a small mountainous state in northeast India that is situated between 27°00′ 46″ to 28°07′ 8″ N latitude and 88° 00′ 58″ to 88° 55′ 25″ E longitude with an area of 7096 km^2. It constitutes only 0.22% of the total area of India. The state had four districts: North Sikkim, South Sikkim, East Sikkim, and West Sikkim, but recently on December 21, 2021, the Government of Sikkim announced the formation of two new districts as well as renaming of existing four districts, North District as Mangan, West District as Gyalsing, East Sikkim as Gangtok District, and South District as Namchi District with additions of two more new districts, Pakyong and Soreng, of which Mangan is the largest district with an area of 4226 and least population.

Sikkim shares its international borders with the Tibet Autonomous Region of China, Nepal, and Bhutan (Fig. 8.2). The highest point is the summit of Mount Khangchendzonga at 8586 m, and the landscape is the major catchment of river Teesta that originates from Tso Lhamo Lake in the North Sikkim.

Fig. 8.2 Map of Administrative Districts of Sikkim. (Source ENVIS Centre, MoEF, GOI)

2.1.1 Rainfall

Sikkim receives heavy rainfall, and monsoon takes the form of incessant rain during the months from May to September, while July is considered to be the wettest of all, as per the reports from IMD, which has 15 rain gauge stations in the state and around 600–700 mm of rainfall takes place here each month, and the temperature lies between 17 and 22 °C.

Rainfall Characteristics Pattern and Distribution analysis at Tadong East Sikkim by Das et al. (2017) mentions that from the year 1983 to 2015, year 2003 received the highest amount of annual rainfall (3740.6 mm), and the year 2010 received the highest number of rainy days (205), and the year 2009 received the lowest number of rainy days (128). The average number of rainy days in Sikkim (days with rain of 2.5 mm) ranges from 100 at in places like Thangu to 184 at Gangtok.

2.1.2 Forest Resources

Within an area of 7096 km^2, Sikkim has 30.77% of forest as protected area comprising of seven wildlife sanctuaries and one national park, which is the highest in the country. As per the Forest Environment and Wildlife Management Department, the state has the largest recorded forest land area covering 82.31% of the geographical area.

There are mainly five types of forest; subtropical, moist mix deciduous, wet temperate, conifer, and subalpine forest in Sikkim.

The ecosystem provided by the forest in Sikkim is indispensable. Reserve Forests including Khasmal and Goucharan are among the three categories of forest in the state of which Khasmal forest is a forest land set aside by the Government of Sikkim for meeting the domestic needs like timber, firewood, and fodder of the communities residing in the adjoining areas (Fig. 8.3). (https://sikkim.gov.in/departments/forest-environment-and-wildlife-department).

2.1.3 Biodiversity

Sikkim covers 0.2% of the geographical area of the country. Though small in size, it has rich biodiversity and is identified as a hotspot in the Eastern Himalayan region.

There are ten biogeographic zones and 25 biotic provinces, which have 16 major forests types and >200 subtypes (Champion & Seth, 1968).

Sikkim has over 4500 flowering plants, 550 orchids, 36 *Rhododendrons*, 16 conifers, 28 bamboos, 362 ferns and its allies, 9 tree ferns, 30 *Primulas*, 11 oaks, and over 424 medicinal plants, and the faunal wealth of Sikkim consists of about 144+ species of mammals, 550 species of birds, 600 species of butterflies, 33 species of reptiles, 16 species of amphibians, and 48 species of freshwater fishes (Hajra & Verma, 1996; Ganguli-Lachungpa and Rahmani, 2003; Tambe, 2007).

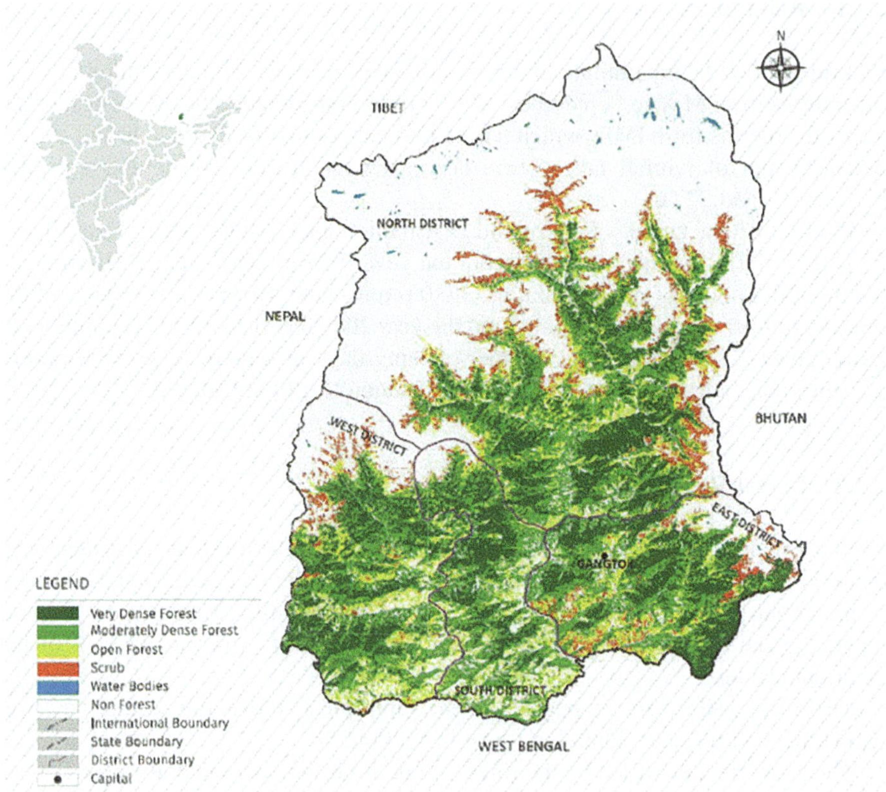

Fig. 8.3 Map of Forest Resource, Sikkim. (Source: Indian State of Forest Report, 2015)

The medicinal plants are found mostly in high-altitude areas and are rare and endangered. Some of these medicinal plants are used as the age-old traditions for various treatments by the local communities.

The small state also has 28 mountains/peaks, more than 80 glaciers, 300+ high-altitude wetlands, and over 104 river and streams of which Teesta and Rangit are the two main rivers and lifelines for the state.

2.2 *Study Sites*

SECURE Himalaya landscape spans across 27 villages in the Khangchendzonga Biosphere Reserve fringes in both North Sikkim and West Sikkim. There are 16 villages selected under the SECURE Himalaya Project in North Sikkim and 11 villages selected in West Sikkim. The villages have been grouped into clusters based on geographic location and contiguity with an indication of the number of villages under each.

2.2.1 Overview of the Villages

North Sikkim, the high-altitude area of Lachen, Thangu, Muguthang, and Lachung toward the extreme north of the district and situated along the upper reaches of the Teesta River are inhabited by the Lachenpa and Lachungpa community, which are a race of Tibetan lineage. Higher up in the cold desert plateau, which is an extension of the Tibetan plateau into Sikkimese territory, reside the nomadic Dokpa community. The presence of the local governing body, the Dzumsa, is the most unique feature of Lachen and Lachung.

These are the main villages that constitute the Upper Teesta Basin and form the gateway communities to the snow leopard habitat. Both Lachen and Lachung are situated along the banks of two rivers (Lachen Chu and Lachung Chu), which come together at Chungthang to flow down as Teesta River.

Situated lower in elevation to Lachen and Lachung is the Chungthang cluster, which has Chungthang and Theng, which are part of the same Gram Panchayat Unit and Naga Namgor Gram Panchayat Unit further down the highway. Naga Namgor village is also located along the same highway toward Chungthang from Mangan. It lies along the lush and thick forests of the Naga Forest Range, which has contiguity with the Tamze Forests of East Sikkim.

Dzongu in North Sikkim is a special Lepcha Reserve formed and recognized by the State Government as a reserve for protection of the indigenous Lepcha community of Sikkim. The villages are all considered part of the Khangchendzonga Biosphere Reserve, named after the mountain peak that the community has the utmost reverence for. The villages are clustered as Upper and Lower Dzongu depending on their location.

In West Sikkim, the village clusters are around two main protected areas: Khangchendzonga National Park and Barsey Rhododendron Sanctuary. Maneybung Cluster with Maneybung Gram Panchayat Unit, Sopakha Gram Panchayat Unit, and Gitang (also known as Gyaten) ward of the Karmatar Gitang GPU is a large cluster situated close to Barsey Rhododendron Sanctuary. The villages under this cluster are heterogeneous with a mix of many communities. These villages are important in being transboundary in nature with both Nepal and West Bengal.

The Labdang–Karjee Cluster has the villages of Labdang and Karjee. Labdang has three wards under it Upper, Middle, and Lower Labdang, while Karjee has two wards: Upper and Lower Karjee. These villages are located along the same highway axis that begins from Tashiding in West Sikkim and are separated by the Dhupi Narkhola GPU in between, whereas Karjee has a mix of Gurungs, Lepchas, and Subbas. These are the most remote villages in West Sikkim and have remained disconnected from most government schemes and programs.

Nesa Gangyap cluster consists of the villages of Nesa, which is one ward in the Arithang Chongrang GPU and Gangyap (Lower and Upper), which falls in the Tashiding GPU. Both of these are reached from the Tashiding Highway axis and after taking diversions from the main road.

Sindrabong and Topung are clustered together due to their proximity to each other. This cluster is above Nambu, which is about an hour drive from Pelling town. Sindrabong falls is in the Darap Nambu Gram Panchayat Unit and Topung falls in the Rimbi Tingbrum Gram Panchayat Unit, and these GPUs are adjacent to each other. These villages are also closer to the Nepal border and home to the Himal Rakshaks. Yuksam Kyoungtey Cluster has two wards, which are under the same Gram Panchayat Unit of Yuksam. Yuksam is one of the gateways into Khangchendzonga National Park, and owing to that, the village has good tourism potential and is one of the main trekking destinations in the state. Khangchendzonga Conservation Committee has a strong presence in this village, which has led to a number of successful conservation initiatives.

3 Review of Literature

3.1 Climate Change and Impacts in the Himalayan Region

Climate change and its impacts in the Himalayas are existent. Erratic and unpredictable weather, changing rainfall patterns, increasing temperature causing glacier melts, and its impact in the runoff patterns of the rivers are evident.

Some reports and data on the Himalaya by Bhutiyani et al. (2010) classified 1876–1892 period as a warm and 1893–1939 period as a cool episode. It has also been mentioned that many glaciers are receding, and some have expanded over the past 40 years (Archer & Fowler, 2004).

In the Himalaya, climate change also has severe impacts on precipitation, the mountains become vulnerable, and landslides are the most common causes impacting the lives of people; however, exposure to severe events is location-specific, and communities settled in diverse regions are impacted inversely.

3.2 Community Perceptions of Climate Change in Sikkim: Impacts on Livestock, Agriculture, and Springs

Local communities have been facing and reciprocating to climate change for ages; hence, data based on local observations are more as compared to limited scientific data. Local communities from villages of Sikkim have considerable experience of the changing climate to which they have already started to adapt. A lot of visible changes are seen owing to the climate change experienced across the state. Communities mention different levels of climatic variation over the years in the form of erratic rainfall/snowfall, prolonged dry spells, warmer winters, unpredictable monsoon, emergence of new diseases and pests in crops/fodder trees, drying up of water resources, etc. A paper submitted in 2017 at ICAR Research complex mentions that land races of rice, maize, pulses, finger millet, yams, and pumpkin are rapidly disappearing due to climate change impacts. There is a huge decline in the plantation and production of large cardamom, which used to be one of the cash crops for Sikkim. The paper also mentions that the sowing of maize in the subtropical zone has shifted by 15–20 days, while sowing at temperate zones remains the same. Similarly, the harvest of maize remains the same in the subtropical zone, while harvest time has shortened by 15–20 days in the temperate zone. Also, Sikkim mandarin orange has declined both in terms of productivity and plantation area.

The farmers have mentioned a huge change in the agroecology of Sikkim, and they have mentioned about the decrease in productivity of several crops, while there is an increase in crop diversification. Senthil (2012) documented changes in the quality of the milk from cattle. The amount of sour milk being gathered by milk cooperatives has increased, which presumably is being caused by the warm weather. Colonization of invasive species such as *Chromolaena adenophorum*, *Eupatorium odoratum*, and *Bidens biternata* is rapidly inhabiting the farmlands, fallow lands, and croplands, which have spread from subtropical to temperate agroclimatic region causing productivity decline. Impacts of climate change in springs are also mentioned by Tambe et al. (2012), who have mentioned indirect evidence for declines in spring water availability during the dry season. In a report by Kothari et al. (2017), community consultation echoed the cases of receding glaciers and less snow in the winter months, it further mentions the village received 3–4 feet snow earlier but hardly receives 1–2 feet snow now. Further Telwala et al. (2013), observed that the mean temperature in villages like Lachen and Lhonak valleys in 2010, when compared to 1850, increased in the summer and winter months by $0.76 \pm 0.25°$ C and $3.65 \pm 2°$ C, respectively.

However, a need for study regarding the impact of climate change on discharge rates of streams and rivers in Sikkim is also suggested. Scientific research is required to monitor the gradual changes of climate change and adaptation that have taken over in the region.

4 Objectives

Sikkim being a hilly state in the Eastern Himalayas, agricultural practices and adaptations are highly variable in time and space due to varying altitudes and agroclimatic situations. Agriculture is the primary activity of the people of Sikkim. About 15.36% of the total geographical area of the land is devoted to agriculture, but the actual area available for agricultural purpose is declining due to factors related to climate (increase in pest and other diseases), diversion of cultivable land for nonagricultural purposes (establishment of industries, township expansion), and switching to other livelihood options like tourism, which has now become one of major sectors of the state economy. The primary objective of the study is to document gradual changes of life and livelihood with a case study from Sikkim Himalaya. This study would also help in the following:

- Mapping current livelihood practices in the project area
- Developing strategies to enhance income and nutritional security from the existing livelihoods through improved productivity, access to inputs and markets, capacity building, and collective action in marketing to achieve economic scale
- Preparing value chain map for key products/commodities/services reflecting economic return at every stage, product movement from the rural producers to the final consumers
- Identifying major players in each field, constraints and institutional obstacles, social process, and vulnerabilities and risks hampering benefits to the poor along the value chain
- Identifying critical investment needs in the value chain that can accrue better income
- Identifying infrastructure availability, institutional arrangements, and schemes as per the needs
- Exploring partnership possibilities and collaborations along the value chain

5 Data Sources

The study and analysis is under the task assigned by UNDP, SECURE Himalaya Project for Planning and Implementation of Livelihood Activities through Value Chain Analysis, Piloting of livelihood sub-projects, Product Development and Marketing in Sikkim, to WWF-India, Khangchendzonga Landscape program, Sikkim Office from the year 2018–2020.

5.1 About SECURE Himalaya Project

The project aims to support the Government of India and State Governments to effectively promote sustainable land and forest management in alpine pastures and forests in high-range Indian Himalayan ecosystems that secure sustainable livelihoods and ensure conservation of globally significant biodiversity and threatened species.

The project would be implemented over a period of 5 years in the high-altitude trans-Himalayan region and has the following major outcomes:

1. Improved management of high-range Himalayan landscapes for conservation of Snow Leopard and other endangered species and their habitats and sustaining ecosystem services
2. Improved and diversified livelihood strategies and improved capacities of community and government institutions for sustainable natural resource management and conservation

3. Enhanced enforcement, monitoring, and cooperation to reduce wildlife crime and related threats
4. Effective knowledge management and information systems established for promotion of sustainable management practices in the High range Himalayan ecosystems

In Sikkim, the SECURE Himalaya's geographic focus is around the Khangchendzonga Biosphere Reserve covering both North and West District, of which these areas are important habitats for the enigmatic snow leopard, which is at the center of the project. Twenty-seven villages have been identified in the two districts, which are the main areas for project implementation. In this report, a total of 25 villages have been listed as some of the villages were merged in secondary data.

6 Research Methods

6.1 Inception at District Level

Two inception meetings were organized at Geyzing (West Sikkim) and Mangan (North Sikkim), which brought together the key stakeholders from the identified SECURE Himalaya villages of the respective districts. These were Panchayati Raj Institution's elected representatives, Forest Department Field functionaries, and representatives of community-based organizations. The main objectives of the meeting were to apprise the participants about the SECURE Himalaya project and its objectives and to build rapport with the key community stakeholders who would be the nodal persons for taking forward the planning work in their respective villages.

Preliminary understanding about the existing livelihood activities and potential future options was also presented by the community representatives at the meeting. The gaps and challenges of various government schemes and programs were also discussed. These enabled the team to delineate strategy for the village planning process to be organized. It was noted that community members had limited knowledge about the process of village planning, and therefore, a training on microplanning was also felt as necessary to be organized at the village level.

6.2 Community Consultations

Community consultations were organized at ward or GPU level to understand the existing livelihood activities, the skills available in the village, programs being implemented, and their gaps and challenges as well as to understand community perspective for future livelihood options. Various PRA tools were used to understand the existing situation of the village. Exercises like the pairwise ranking enabled the communities to identify and shortlist potential future activities based on the SWOT analysis conducted during the consultation process. Key informants from the villages provided further detailed insights into the history of the villages and enabled identification of historical trends in the livelihood activities of the villages and various drivers of change.

6.3 Review of Existing Village Plans

Microplans prepared at the JFMC and EDC level of the Forest Department and Gram Panchayat Development Plans prepared by the Rural Management and Development Department were reviewed to understand the existing plans and as well as to generate secondary information on the village profile. In West Sikkim, for some villages, Community Development Plans of the North East Rural Livelihood Programme was also referred to.

6.4 Stakeholder Consultations and Departmental Dialogue

Stakeholders consultations involving all the key stakeholders were conducted to assess the feasibility of the livelihoods shortlisted and also to develop a road map for the pilot projects. For Dairy, Cardamom, Apiary, Handloom, and Handicrafts, the team members consulted key persons of the department and relevant stakeholder to get feedback and assess the feasibility of the livelihoods shortlisted by the communities.

6.5 Baseline Survey at Household Level Across the Identified Villages

A baseline survey was also conducted to assess the current livelihood status of the communities in the target villages, and information from the survey was used to strengthen the livelihood strategies further. The baseline survey was conducted in 25 villages of North and West Sikkim. A total of 14 villages in North and 11 in West District were surveyed out of the 27 villages targeted. The survey methodology was a random geographical survey targeting at least 10% of the total village households barring certain exceptions.

Within the 25 surveyed villages, a total of 425 households, 188 households in North and 241 households in West, were covered. The survey encompasses the range of livelihood activities and incomes, which include both farm and nonfarm activities that people are currently engaged.

The findings outline the socioeconomic situation of the identified villages drawing from the village-level community discussions, PRA exercises, and the data from the household-level baseline study that was conducted. The current livelihood options that communities in the village are engaged in also contribute to the result along with the ecological transformation that was observed.

7 Results

7.1 Demographics

North District is the largest in the state though with the least population (Fig. 8.4).

The total households in the identified villages in North Sikkim are 1841 out of which 794 are categorized under poor and 580 households under the very poor category. A total of 123 households are women headed out of the 1841 households in 14 villages (Annex: Tables 8.1 and 8.2). Most of the villages are situated close to the border with China (Tibetan Autonomous Region), which has a large presence of Military and Para Military forces, and the communities living in these high-altitude areas share a close affinity with the army people, depending on them for many of their needs.

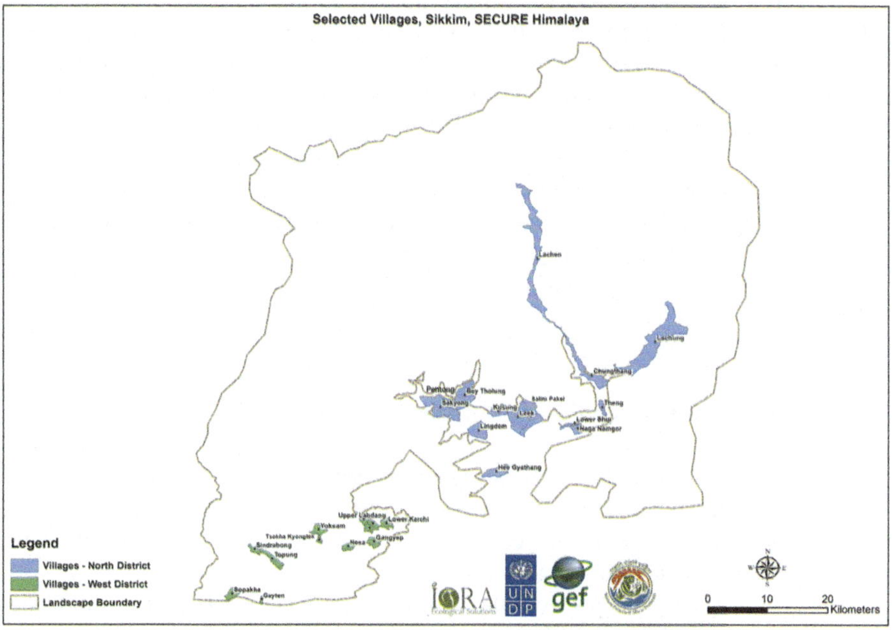

Fig. 8.4 Map of selected villages, Sikkim. (Source: Secure Himalaya Project 2018–2020)

The villages at lower altitudes like Chungthang village had a mixed population with very different dynamics with the settlement of outsiders due to the presence of mega hydropower project, and in villages like Dzongu the Lepchas mostly have been hunter-gatherers and foragers, and their lifestyle has a strong religio-cultural connect to nature, which still remains today.

Furthemore, in West Sikkim, out of 2783 households, 636 households are categorized under poor and 93 households under very poor. Only 61 households are women headed out of the 2783 households in 11 villages (Annex: Tables 8.3 and 8.4). Moreover, for villages in West Sikkim, villages like Labdang are largely dominated by the Gurungs, who were primarily sheepherders in the past, along with a few families from other communities (Fig. 8.5).

7.2 *Dependency on Natural Resources*

Communities have age-old dependence on natural resources across the study area in the SECURE Himalaya landscape. In communication with older members of the villages, the history of being hunter-gatherers and forest foragers was not too long back in the past. Though in every village, the narrative of decreasing dependence on the forest was also very common. Firewood and fodder were the most commonly procured forest resource across all the villages as mentioned during the community consultations organized in all clusters for livelihood planning. This was corroborated further during the baseline survey. Every household had an LPG connection, but refilling of cylinders for remote villages was a costly affair, and firewood stove was the primary cookstove used. Firewood usage in the villages ranged from 3 to 8 kg per day per household. Both fodder and firewood were collected from private as well reserve forests (Fig. 8.6).

The other nontimber forest produce that was harvested from the forests seasonally wa medicinal plants (Annex Table 8.5). Most of which were extracted mainly for self-consumption, along with other edibles that were also sold at nearby markets. High-value plants such as Cordyceps and *Paris polyphylla* are also extracted from the forests area as mentioned during the community consultations (Figs. 8.7, 8.8, 8.9, 8.10, and 8.11).

7.2.1 Dependent Households on Natural Resources

Identification of specific households that were dependent on natural resources as their main livelihood during the community consultations was not feasible. The general understanding was that economically deprived households would have higher dependence on natural resources.

However, a quick analysis from the baseline survey across the villages indicates the following at the cluster village level, and the summary of the ranking of the

Fig. 8.5 Map showing population of Sikkim. (Source ENVIS Centre, MoEF, GOI)

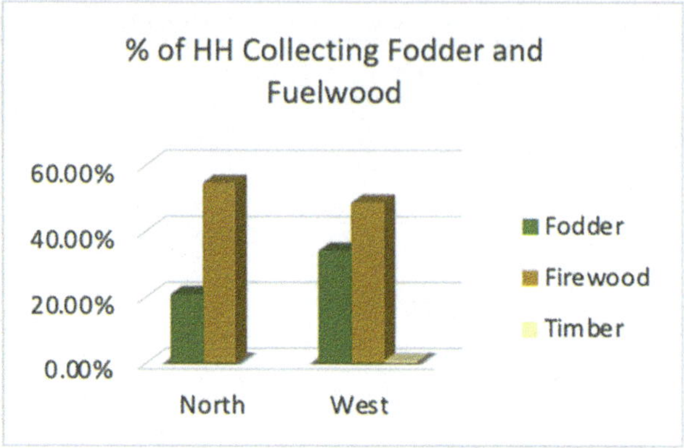

Fig. 8.6 Percent of HH collecting fodder and fuelwood. (Source: SECURE Baseline Survey, 2020)

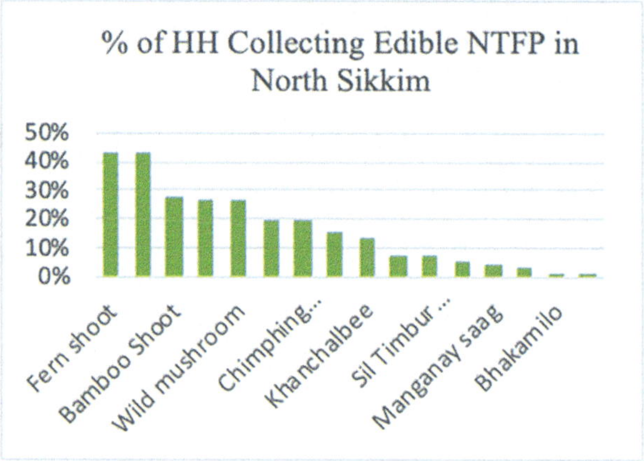

Fig. 8.7 Percent of HH collecting edible NTFP (North). (Source: SECURE Baseline Survey, 2020)

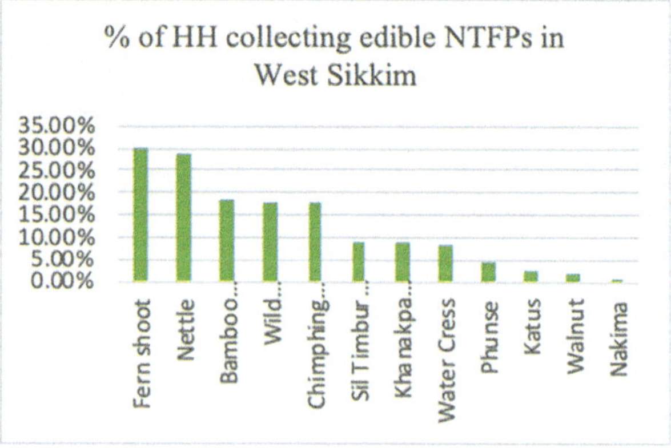

Fig. 8.8 Percent of HH collecting edible NTFP (West). (Source: SECURE Baseline Survey, 2020)

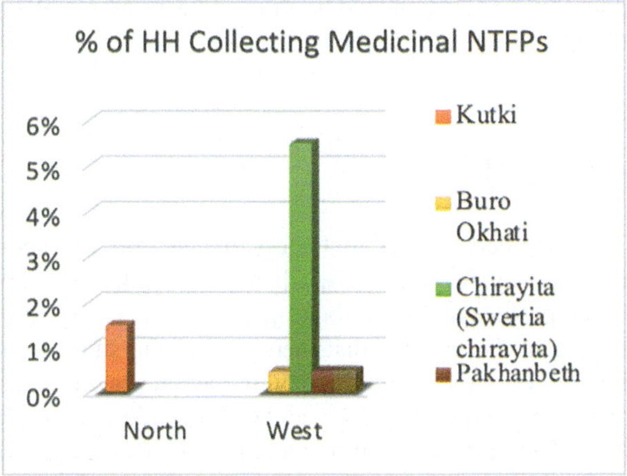

Fig. 8.9 Percent of HH collecting medicinal NTFPs

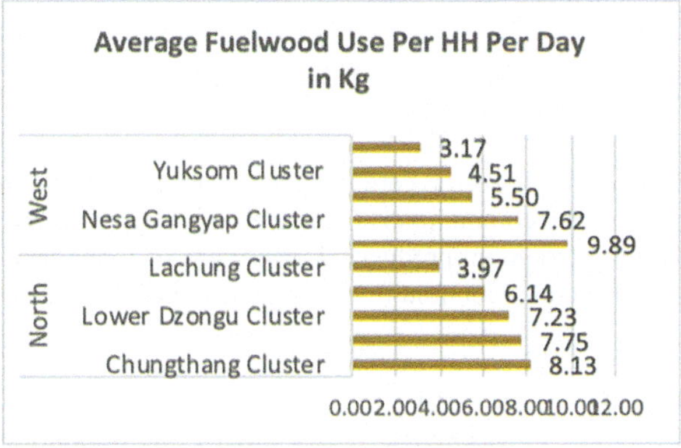

Fig. 8.10 Fuelwood usage. (Source: SECURE Baseline Survey, 2020)

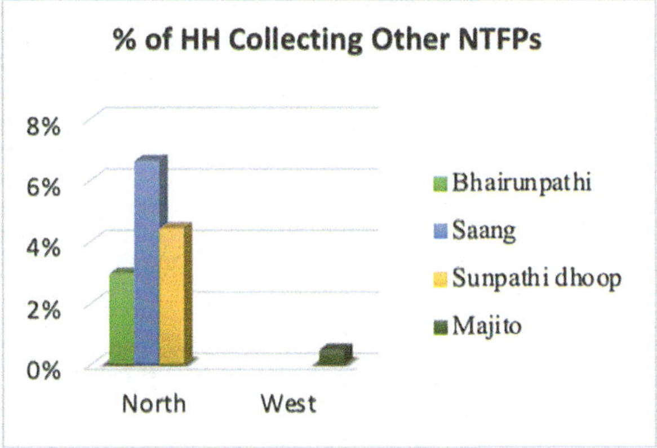

Fig. 8.11 Percent of HH collecting other NTFPs

households based on land holding and earning members for the clusters is presented (Annex: Table 8.6) in the table below.

7.3 *Livelihood*

The identified villages are hugely diverse from one another in terms of their location as well as their ethnicities, which also results in diversity in their current occupational profiles and livelihood opportunities. The existing livelihood activities of the people in the villages that were part of the study in North and West Districts are broadly centered around agriculture and horticulture activities, dairy, livestock, and animal husbandry activities, and in the nonfarm sector, MGNREGA and government service are the two main income providers besides other labor jobs, private sector, and the tourism sector. The recent One Family One Job has also been mentioned as one of the main job providers for many of the communities in the surveyed villages (Figs. 8.12 and 8.13). A brief from both North and West Sikkim villages are presented below.

7.3.1 Agriculture

In the farming sector, cardamom cultivation has been the major source of income for most of the communities. Cultivated in an agroforestry system, Sikkim was world's largest cardamom producer. Large tracts of forest land are covered under this crop that demanded minimum intervention and effort, with labor required only during the

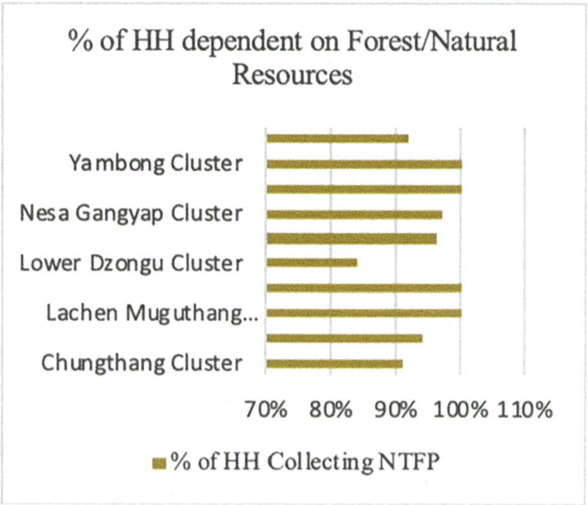

Fig. 8.12 Percent of HH dependent on forest/natural resources

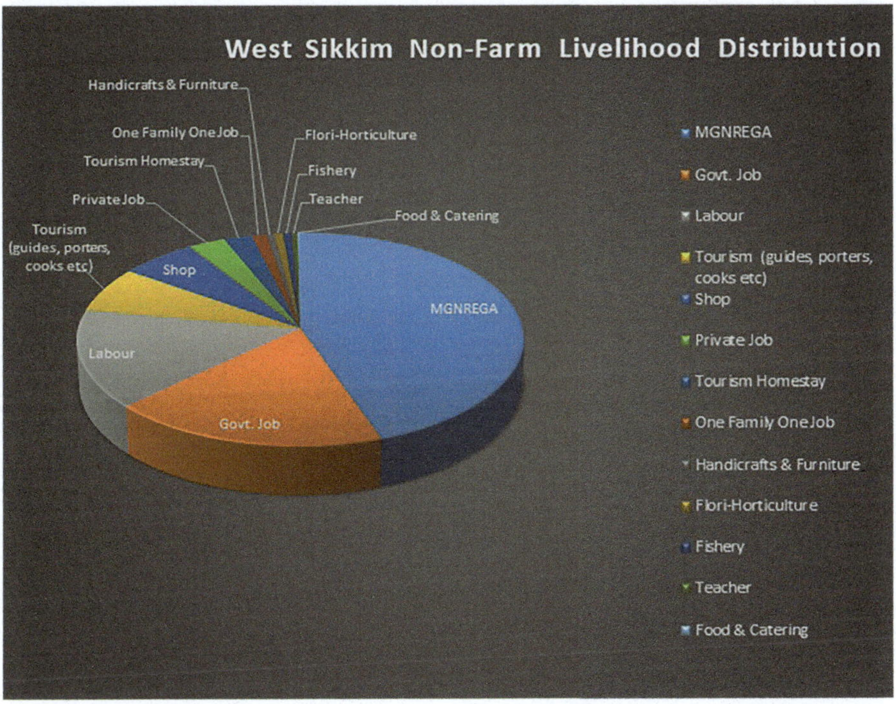

Fig. 8.13 Non Farm Livelihood distribution for West Sikkim. (Source: SECURE Himalaya Baseline Survey, 2020)

weeding and harvesting season once the crop was planted. The current land under cultivation of the cardamom species is around 23,312 hectare, which produces around 5000 tons annually. With changing climate and increased pests and diseases, there has been a massive die-off in cardamom, and this was expressed across all the villages that were part of the study.

Crops that were grown traditionally were maize, buckwheat, millets, paddy, barley, etc. Maize is one of the most important cereal crops of Sikkim. State-wide data shows that it is grown over an area of about 36,000–40,000 ha, which is about 35–40% of total cultivable area (Source: Annual Reports, Food Security & Agriculture Development Department, Government of Sikkim). There has been marked decline in cultivation of traditional food crops in most of the villages as many turned to more cash crop cultivation, mainly cardamom.

Mandarin orange is also an important cash crop though only for specific belts in Lower Dzongu Cluster (Hee Gyathang). Ginger, potatoes, and round chilly are other crops that are grown for cash in many of the villages. Vegetable farming is widely practiced, and it is a mix of both for cash as well as for self-consumption.

Farming has been practiced by the Lachenpas mostly in Thangu Valley, which is their summer place, whereas the village of Lachung has larger farming areas adjacent and within the village limits. In both villages, agriculture practices show a declining

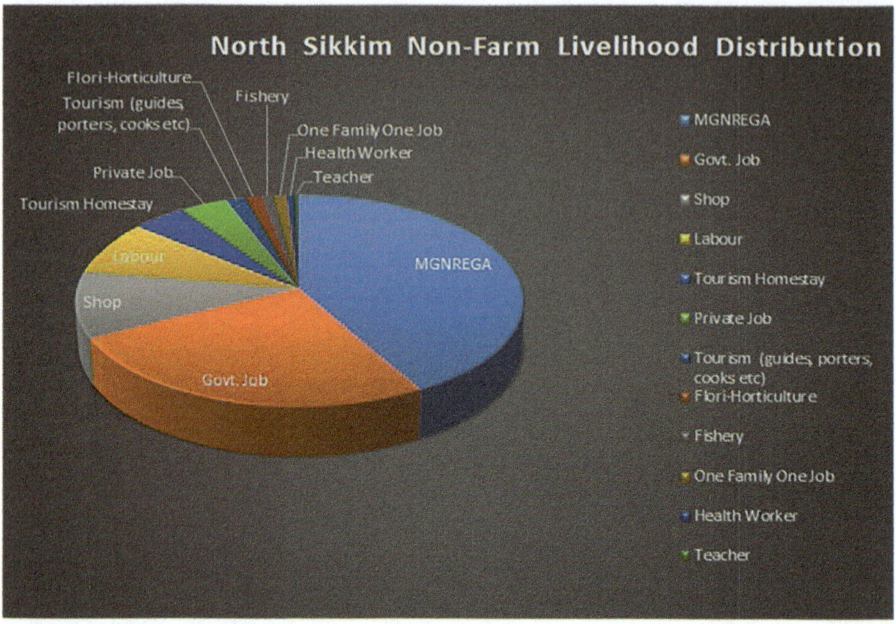

Fig. 8.14 Non Farm Livelihood distribution for North Sikkim. (Source: SECURE Himalaya Baseline Survey, 2020)

trend, which can be attributed to a number of reasons, such as rapid development of other sectors such as tourism and MGNREGA as well as increased instances of human–wildlife conflict. Most of the farming land below Lachen now lie barren and fallow with people now maintaining small patches of vegetable patches behind their houses. Traditional crop cultivation overall has decreased with forests being more inaccessible due to the ban on slash-and-burn practices (Figs. 8.14 and 8.15).

For villages of West Sikkim, larger tracts of land are set aside for cardamom, and farming of other food crops is practiced in smaller patches of land where a variety of crops are grown. Maize cultivation was also practiced widely, but of late, it has decreased. Human–wildlife conflict and access to market emerged as the top challenges communities were facing in the farming sector. Apart from a few villages, which have better road connectivity, most of the crops grown are for self-consumption. Other cash crops such as potato, round chilly, and recently of kiwi were also practiced in smaller pockets, which were replacing traditional crop species (Figs. 8.16 and 8.17).

Above all, cardamom cultivation has remained the most critical of all livelihood options for most of the villages in Sikkim, and despite heavy decline in productivity in the last decade brought about by pests and diseases leading to setbacks for the community, it still remains the main livelihood to fall back on. The main livelihood activities under various sectors and other associated nonmonetary benefits derived from natural resources are presented below (Annex: Tables 8.7 and 8.8) for North and West District (Fig. 8.18).

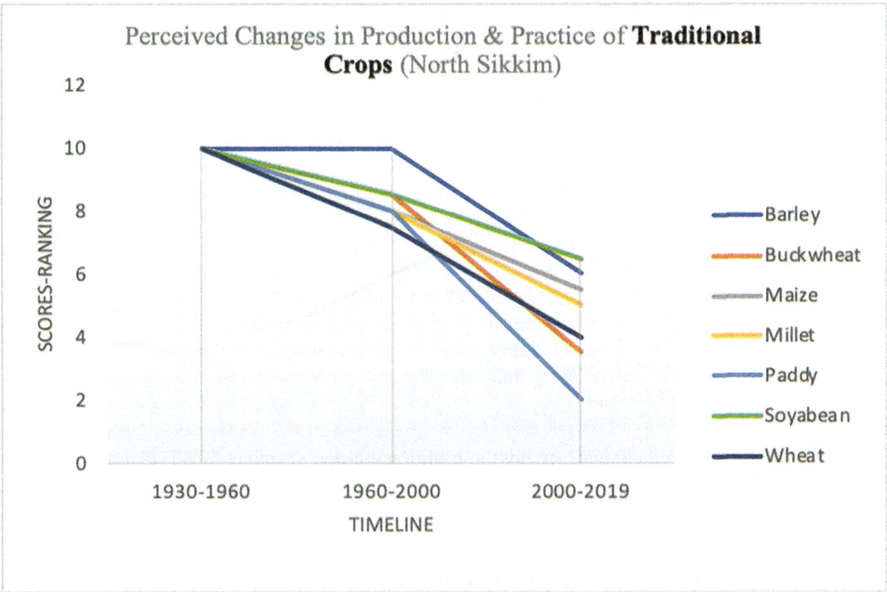

Fig. 8.15 North changes (trends) in traditional crops. (Source: WWF Planning Meeting (Heegyathang and Kusung))

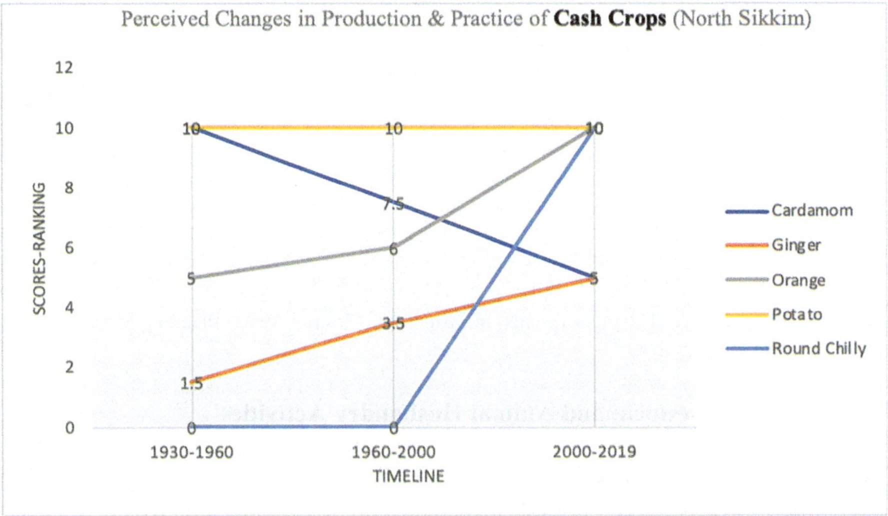

Fig. 8.16 North changes (trends) in cash crops. (Source: WWF Planning Meeting (Heegyathang and Kusung))

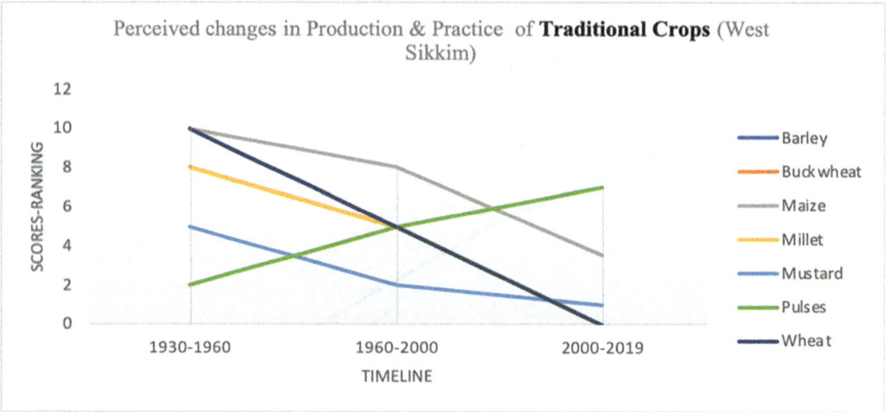

Fig. 8.17 West perceived changes (trends) in traditional crops. (Source: WWF Planning Meeting (Nesa, Maneybong))

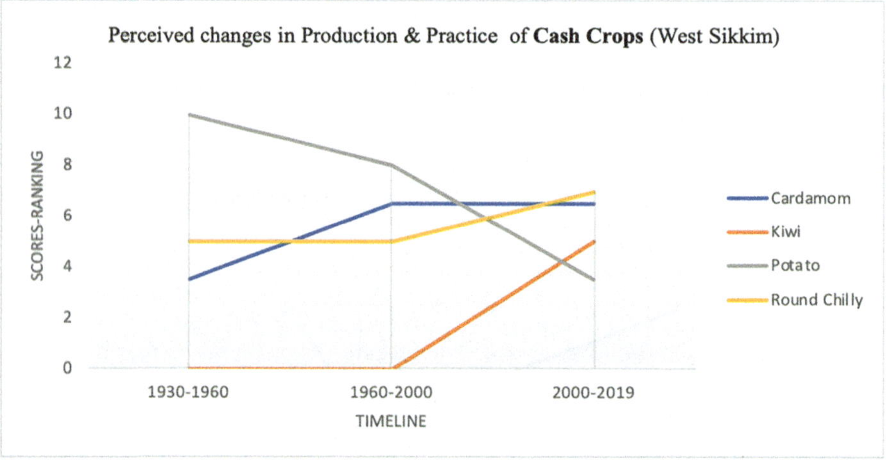

Fig. 8.18 West perceived changes (trends) in cash crops. (Source: WWF Planning Meeting (Nesa, Maneybong))

7.3.2 Dairy, Livestock, and Animal Husbandry Activities

High up the villages of Muguthang, Thangu, Lachen, and Lachung, communities have traditionally practiced agro-pastoralism, and the practice is still continued though it has been seeing a declining trend over the years. Dokpa families still continue their practice of yak herding moving up and down with the animals according to the seasons, though the number of families has dwindled to only about eight to nine who permanently reside in Muguthang. There are a number of reasons for the decline in this tradition. There has been a shrinking of the pasture land available for yaks over the years as climatic

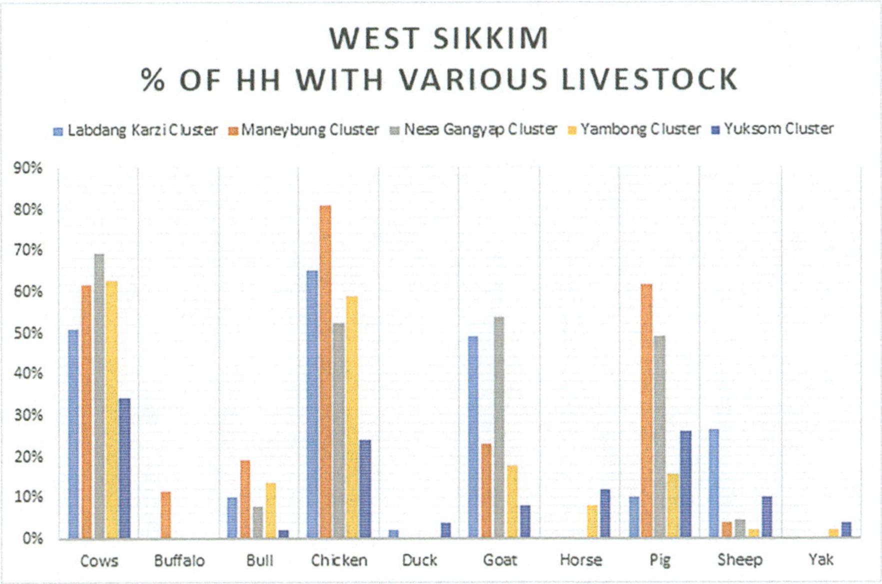

Fig. 8.19 Percent of HH with various Livestock. (Source: SECURE Baseline Survey, 2020)

conditions have changed, while on the other hand, military presence has also expanded. The younger generation of herders does not want to continue with this hard life, which is also affected seriously by wildlife conflict. Free-ranging dogs have also multiplied in large numbers that prey upon younger yaks causing herders to lose out on their income. Sheep herding has seen a bigger decline than yaks with only one sheepherder left in the entire North Sikkim.

Meanwhile, livestock rearing has been a common livelihood activity for West Sikkim communities. Traditionally, different communities have been involved in dairy farming, sheep rearing, goatery, piggery, and poultry. The government policy of ban on grazing in forest areas has witnessed a reduction in cattle rearing in most of the villages in West Sikkim, where implementation of the policy was the strictest. Another direct impact has been on the availability of manure for the farms, which has led to severe shortage of fertilizers for the farmers. A snapshot of various animal husbandry–related activities across the village clusters of the two districts is presented below (Figs. 8.19, 8.20, and 8.21):

Dairy farming is one of the traditional livelihood activities that many households are engaged in the villages surveyed, though the ban on grazing led to a steep decline in the number of cows. Milk is usually sold within the village. Few villages in West Sikkim were connected to the Cheese Plant set up at Dentam by the Sikkim Dairy Producers Pvt. Ltd. But overall, linkage for milk collection to any collectives or cooperatives was limited in most villages. The percentage of dairy households is presented below.

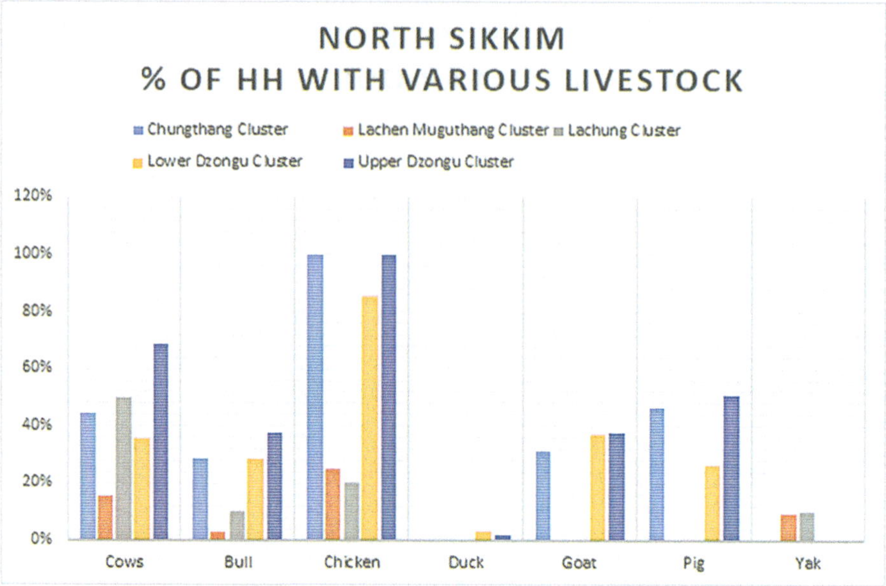

Fig. 8.20 North percent of HH with Various Livestock. (Source: SECURE Baseline Survey, 2020)

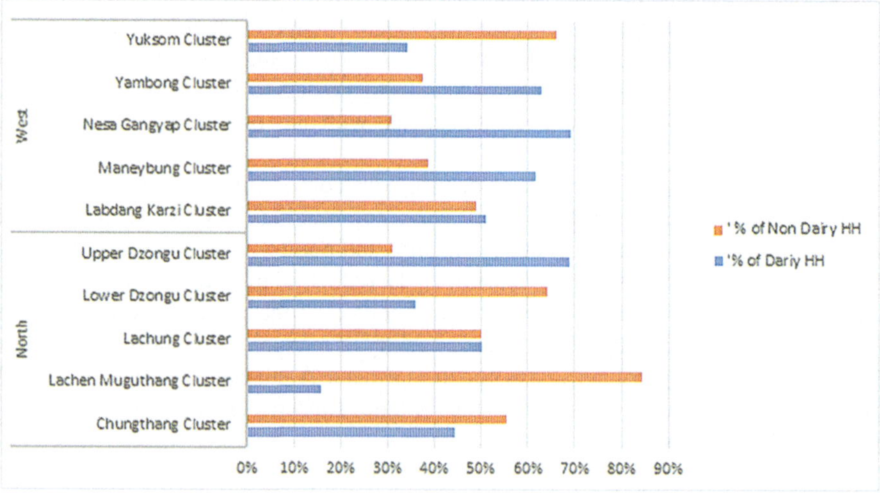

Fig. 8.21 Percent of dairy and nondairy HH. (Source: SECURE Baseline Survey, 2020)

Milk productivity was also very low for most villages. Maximum number of HHs produce 1–2 l of milk (average) in North Sikkim, and in West Sikkim, maximum number of HHs produce 3–5 l of milk as shown in the figure (Fig. 8.22). This made the whole sector highly non-remunerative; however, Sikkim Dairy Producers Private

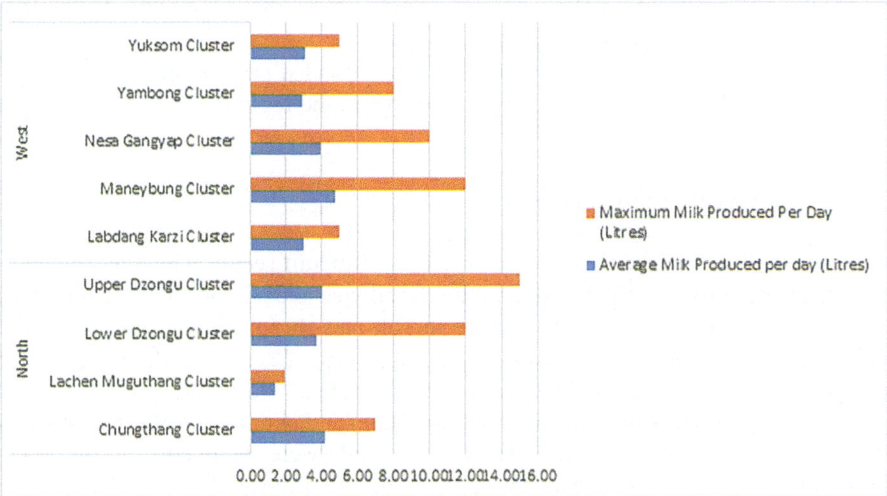

Fig. 8.22 Milk production details. (Source: SECURE Baseline Survey, 2020)

Limited (SDPPL) has served to organize many of the dairy farmers into successfully functioning cooperatives in Sikkim.

7.3.3 Tourism

In the nonfarming, tourism emerged as the priority sector that featured in villages of both North and West Sikkim and one that reflects the aspirations of the communities especially the youth. There is also a strong push from the state government for tourism promotion providing space for convergence. Aspirations of the communities are changing, and there is a strong trend to move from traditional primary occupations such as agriculture and farming to service sector occupations such as tourism.

Communities have also seen the economic benefits that the tourism sector brings in places where tourism is thriving. As per the Department of Tourism and Civil Aviation figures tourist footfalls in Sikkim increased to 14.25 lakhs in 2017, a far cry from 61,000 arrivals in 1990 to 720,500, ten fold more arrivals in 2010.

Rural tourism has shown a marked increase with the construction of homestays by the Government under the 13th Finance Commission project. There are over 1200 homestays today in the fours district of the state, which along with a spurt in construction of resorts and hotels in rural areas has resulted in drawing an increasing number of tourists visitors to the countryside.

In North Sikkim (villages like Lachen and Lachung), the advent of tourism around two decades back has shaped the development of what they are now. These villages are now established as the top tourist destinations within Sikkim with hotels that have mushroomed over the years. Majority of the villagers are now involved in the tourism trade directly or indirectly, whereas, in West Sikkim, tourism promotion was initiated post ban on grazing as an alternative activity for the herding community in places like Uttarey and Sindrabong.

Overall, it was understood that tourism and its allied sectors provided employment and livelihood for a vast number of people in the region.

7.3.4 Handicraft and Handloom

Connected closely to the tourism sector is the handicraft sector, which has income-generating opportunity for the community, but it was understood from the meetings and surveys that along with the decline in yak and sheep herding, the associated skills of weaving and handicrafts is also disappearing. Women in the villages still retained the necessary skills of weaving, but raw material, however, was a challenge, most of which are procured from outside the state, and similarly in West Sikkim, as per community discussions, large numbers of cattle were removed from the forest and either sold to Nepal or Darjeeling. This has also meant the disappearance of allied/associated activities like wool yarn weaving, which used to be major traditional handloom activity in the villages.

7.3.5 Cordycep Collection

In North Sikkim, besides agriculture and tourism, the other livelihood opportunity that opened up around the same time has been through cordyceps or the caterpillar fungus harvest and sale. This high-value species with medicinal properties is only found in the high alpine meadows of North Sikkim and has been one of the main reasons for ushering an economic boom in the village. Most of the villagers are involved in this activity during the harvesting season, which begins around May and lasts till July.

7.3.6 MGNREGA and Other Related Activities

The coverage of the MGNREGA in rural Sikkim has been exemplary. As per the government records in a total of 92,000 rural households in Sikkim, 84,931 households have been provided job cards, while 65,454 (71%) households have been provided employment at an average of 66 person/days per annum.

Since its inception, MGNREGA has generated 326.31 lakh person days of employment, with an average of 40.7 lakh person days of employment per year.

The lifestyle of the rural poor has been enhanced tremendously through this livelihood scheme in the region. During the last 10 years, a total investment of 663.63 crores, with an average of 66.36 crores per year, has been made under MGNREGA, out of which 391.62 crores (59% of total), with an average of 39.16 crores per year have been directly paid as wage to the job card holders' bank accounts.

Because of which it was observed that in most of the villages, agriculture practices showed a declining trend, which can be attributed to a number of reasons, such as rapid development of other sectors such as tourism and MGNREGA as well as increased instances of human–wildlife conflict.

Besides, the strategic location of some villages at the international border with China also increases their significance. Schemes such as the Border Area Development Fund (BADP) have been made easily accessible to the communities of these villages, which has led to infrastructure development in the villages. Execution of these schemes is mainly done by the local youth, and this contractual work has also been an important income source.

7.4 Human Wildlife Conflicts

Sikkim covering just 0.2% of the geographical area of the country has tremendous biodiversity. 46.93% of state's total geographical area comes under the protected area with one national park, and seven wildlife sanctuaries, and 82.31% of the total geographical area of the state is under the administrative control of the State Forest Department. All the villages under the SECURE Himalaya project are the fringe villages to the protected areas, which are home to species such as the elusive red panda and snow leopard, as well as clouded leopard, Himalayan black bear, musk deer, pangolin, blue sheep, and much more.

These forests are also used by local communities for energy, water, and livelihood needs. Human–wildlife conflict is the most pressing concerns for forest fringe dwellers in all areas of Sikkim. In the last 20 years, there has been a considerable increase in conflict cases recorded by the Forest Department, particularly around the eight protected areas and adjoining forests. Communities depend on agriculture, agroforestry, and animal husbandry as their main source of livelihood, which is severely impacted in the case of wildlife attacks. Crop raiding by animals such as wild boar, black bear, barking deer, monkeys, civets, peahens, and more has reached such heights that, in some instances, farmers have abandoned agriculture as a livelihood, leading to significant drop in household income. In FGDs conducted by WWF-India revealed that crop damage ranged from 10% to as high as 64% of the estimated yield across crops such as maize, potatoes, and legumes. Further, damage to property and attacks on livestock as well as humans, particularly by bears, are also on the rise. Rise in conflict and heavy losses suffered due to crop damage have also resulted in people moving away from farming or shifting to cash crops such as cardamom that are least damaged by wild animals. Large-scale conversion of farms that produced food earlier to monoculture of a cash crop prone to diseases raises the risk factor for communities in terms of food security and makes them highly vulnerable, a situation already faced by several villages in Sikkim.

7.5 Vanishing Skills and Changing Landscape

The Himalayan region is a retreat for 16% of the world's human population, but with the resources that are being increasingly exploited and with the visible changes of climate change, the larger population in the villages especially the younger generation is finding it difficult to adjust to the changing scenario, creating a huge knowledge gaps in the traditional systems that were followed and practiced in the community. The Dokpa family in the North Sikkim has decreased in number as the younger generation finds it difficult to continue with the harsh environment and look forward to luring tourism opportunities as future livelihood option. Sheep herding has seen a bigger decline than yaks with only one sheepherder left in the entire North Sikkim. Along with the decline in yaks and sheep herding, the connected abilities of handicrafts are vanishing. Likewise, in the agriculture sector, there has been marked decline in cultivation of traditional food crops in most of the villages as many turned to more cash crop cultivation, mainly cardamom, which would give good returns.

The FGDs and community discussions have also mentioned a good rise on the number of educated youths migrating to towns and cities in search of jobs in private sectors (pharmaceutical companies/hotel industries/drivers. Etc.); moreover, government schemes like One Family One Job (OFOJ), which has been created to give equal job opportunities to all the houses of Sikkim by enrolling at least one family member in Government Job, have played an analytical role for the educated youths to move out of the villages.

The landscape has also seen a lot of ecological transformation with unplanned growth of the villages, which have been largely detrimental to the environment as well as the culture of the villages. Vehicular pollution is a major issue with thousands of vehicles making their way up to these remote sites. Increasing waste, with huge volumes of nonbiodegradable plastic waste, is also a matter of serious concern. Being close to the international borders, the increase in the paramilitary forces in the region has also led to unplanned development of roads and infrastructure, which has impacted the biodiversity and its catchment heavily, of which one visible example is the increase in feral dog population, which is a major threat to the biodiversity.

8 Discussions

Collating all data and observations during the surveys and FGDs conducted in the villages, it can be mentioned that the traditional livelihood activities were on the decline.

The dying pastoralism culture in the mountain was of high concern, with only few families engaged in yak herding practices. Yak herders mentioned getting hardly any snow in lower altitudes, which used to get abundant snowfall 15–20 years before. The herders also mention about the degradation of pasture lands, which they feel could be due to increased temperature and reduced snowfall; however, in the last two decades, the region (north Sikkim) has experienced receding snowfall, and the timing of snowfall has shifted from November to December to January to March; however, between December 2018 and May 2019, 300 yaks starved to death in Muguthang and Yumthang in north Sikkim district. Newspaper reports quoted locals as saying that while snowfall had thinned over the decade, the 2018–2019 season's snowfall was relatively heavy leading to starvation of the animals. In Lachen and Lachung, the Dzumsa decides the migration track and time of the yaks to conserve pastures at high altitudes through a rotation policy, which has caused a shift in the upward migration of yak herders that has advanced by 15–30 days due to a rise in March temperature, and the biggest challenge is to keep the younger generation engaged in yak husbandry because of the hardships encountered in rearing them.

Apart from that, the heavy presence of defense personnel along the borders brings about competition for pasture lands for the animals. Sheep farming has remained only in certain pockets of the state and there are very few elders remaining in the village who possess the age old tradition and skills of sheep wool weaving which is now a declining cultural tradition. In Thangu, North Sikkim, where the farmers had been dependent on sheep manure for their agricultural fields, there has been a direct impact on their farming practices as well since scarcity of manure is one of the pressing concerns for decline in agriculture.

Animal husbandry has been a traditional occupation for the diverse communities that inhabit the mountains, and keeping animals for dairy requirements has been an age-old custom. In view of limited cultivable land holdings, mixed farming systems integrated with livestock farming have emerged as a traditional approach to diversify farmer's income. As per report of economic census, as much as 75.62% of total agricultural enterprises are found in animal husbandry sector, which includes cattle breeding, rearing, and production of milk. Dairy farming was widely practiced in most of the villages though the number of livestock had come down after the grazing ban. Knowledge on proper animal management was however low among the farmers who had been more used to open grazing and had not adapted to the practice of stall feeding. Milk production was very low across the dairy-farming villages. The dairy sector though having great potential to be a major livelihood earner for Sikkim farmers has not had the opportunity to develop fully. Although there are numerous government schemes of NABARD on dairy, the rural population is not aware of subsidized financial support. This is particularly because of the remoteness of the area and limited outreach of these institutions; they too are taking support of NGOs to increase their outreach and information. Penetration of banking services in the rural areas is also quite low for accessing subsidized financial support, and facilitating this linkage with the farmers is critical in implementation of these schemes.

Agroforestry practices has decreased, and overall community reported that agricultural lands were lying fallow or converted into private forests over time, which has led to loss of many traditional varieties of crops and seeds. Many factors contributed to this such as changing aspirations of people, easily available labor work such as MGNREGA and One Family One Job Scheme leading to labor shortage for agriculture work, etc. The increasing instances of wildlife conflict were also one of the factors that was pushing people away from their farming practices. Not just that, the fragile region also had challenges in marketing their produce, which had reasons like remoteness of the villages, which remained isolated due to road blockage caused by unpredicted heavy rainfall and cloud bursts.

Wild harvesting of high-value medicinal plants was also practiced, the main ones being Caterpillar fungus and *Paris polyphylla*. Communities reported that there was marked decline in the availability of these species with every passing year. Besides unsustainable harvesting practices, the process of collection from the wild also led to other secondary impacts in high-altitude areas from waste accumulation, firewood burning, and general disturbance in the fragile areas. Other NTFP such as wild edibles was collected seasonally across the villages for sale at the nearest local markets.

Large cardamom is one of the most important spice crops grown in Sikkim; consequently, cardamom is no longer cultivated as an agroforestry crop but has now moved into open fields. The cardamom crop was favored over other traditional crops, and huge tracts of land were converted into cardamom cultivation. Increase in human–wildlife conflict and escalating instances of crop depredation aggravated the situation even further, and farmers were pushed into converting even more land into cardamom plantations. Now it has replaced both paddy and maize to a large extent, so much so that in many areas, except for tiny backyard vegetable gardens, most of the land holds cardamom. Wildlife damage cardamom crop as well, but in comparison to the losses suffered in other crops, the risks are far less in cardamom.

Today, cardamom is no longer a crop that can be cultivated with minimum interventions but one that demands intensive management, which traditional practices did not warrant. Removal of old plants, large-scale creation of nurseries for new saplings, and replanting of disease-free saplings following a standard package of practices may reduce the outbreak of pest and diseases to some extent if rigorously followed. Change in rainfall patterns, long spells of dry winter months, and warmer temperatures have triggered changes in phenology and increased the risk from pathogens as well. Cardamom crop has been prone to a host of diseases as well as pest attacks. Infection of rhizome rot disease has increased with increase in the number of leaf-eating caterpillar, aphids, stem borer, and thrives that destroy the cardamom crop and reduce productivity. The incidence of Chirkey or Foorkey disease is still a major problem, and several local varieties are susceptible to these diseases.

The other challenges that were visible were of unsustainable drying technologies, increasing human wildlife conflict and absence of price support. Animals such as civets and monkeys destroyed the cardamom crop, forcing the farmers to practice early harvesting, which resulted in lower price in the market. Cardamom drying process still remains a major challenge. The drying process through age-old practice of using traditional bhattis uses heavy logs by the combined effect of heat and smoke generated through the burning of wood. While many agencies have worked on demonstration of technology that reduces firewood usage during the drying process, the acceptability of such pilots was very low in the state. The main reason for farmers unwilling to shift to improved drying process is because there is no price support mechanism for the improved quality of cardamom, thereby providing no incentives to farmers to shift to sustainable practices. There is also no grading or sorting of the cardamom, and farmers usually sell it to the traders in one lot, thus losing out on the premium price that larger-sized cardamom could have fetched in the market. But despite the hardship and dip in production from around 10–15 years ago, cardamom still remains the main crop to fall back on due to its long shelf life and market availability.

Though the communities showed deep interest in tourism to many shifting to tourism livelihood, some places were already well established, while others were just getting started. In places with heavy tourist inflow, the impacts of tourism were also showing through increased waste and pollution and disturbance in natural areas. While there were some examples of good practices of sustainable tourism, these remained few in numbers, and capacities in this sector for interpretation of nature and culture as well as on concepts of community-based tourism remained a challenge. It was observed that many youth were initially inspired to get hold on tourism; however, as has often happened, many abandon it when they secure a government job or other or take up more lucrative business opportunity such as contracts. The recent One Family One Job Scheme of the Government is a case in point. The scheme has absorbed many potential and existing tourism service providers in the villages leaving a vacuum difficult to fill up. It is, therefore, necessary to engage not only youth with leadership qualities but also a judicious mix of senior leaders of the community who are there to stay in the village in the composition of the office bearers of the community-based tourism. Marketing was also seen as the major shortcoming in the promotion of rural tourism across all service providers and particularly with homestays. Some of the sites are also remote and difficult to access and without good internet connectivity. It is therefore necessary to incorporate a detailed marketing component in the training programs and also have a clear marketing strategy with a branding exercise especially for the more remote destinations and for specific service providers as trained guides.

Some of the other issues and challenges for the promotion of sustainable tourism in the project sites were lack of trained manpower, absence of accommodation infrastructure (and funding mechanisms), lack of information (including identified flora and fauna at sites) and proper maps of the destinations, and a lack of properly identified and developed trekking trails, which needed to be addressed as well.

It was also observed that the strong presence of the Panchayati Raj Institutions together with community-based institutions and government stakeholders played significant role in the development pathway of the community. Institutions of the Panchayat and Joint Forest Management are local self-governance institutions that plan, implement, and monitor development interventions along with government; likewise in villages like Lachen and Lachung of North Sikkim, the traditional village level governing body is the Dzumsa, which consists of the elders of every household in the village with special representatives of lamas from the village monastery. Two persons were elected among the members of the Dzumsa and are known as the Pipon I and Pipon II. The Dzumsa performs all the developmental functions that are assigned to the Panchayat in other areas and also have customary judicial powers for trials of cases in their respective villages. The term of office of the Pipon is for

1–2 years The movement of livestock (yak, sheep, horses, and cows) is regulated by traditional local bodies. Every year, the seasonal movement calendar is developed by Dzumsa, based on the Tibetan lunar calendar. The communities abide by the rules and regulations of the Dzumsa. Dates are fixed by Dzumsa for seasonal movement of herds; thus, all herders are asked to move on the same date, thus making them an instrumental body to implement every activity in this landscape.

9 Conclusions

Despite the richness in biodiversity and rich ecology, the region has very limited scientific data, and there have been limited studies to support the climate change scenario. Most of the data remains in the form of people's perceptions and experiences; therefore, the state can enhance capacity and research for studies related to climate change for facilitating research to develop policies for climate change in the Sikkim Himalayan Region.

Integrating the identification of climate-related risks, impacts, and vulnerabilities into all biodiversity conservation and developmental planning activities needs to be promoted at both the state and district levels to ensure local adaptation planning is reflected in state-wide plans.

The change in the livelihood patterns brings in a lot of scope for developing strategies in the region desired for sustainable livelihoods through natural resource management.

Sectors that show high potential for rural development, such as sustainable agroecology, ecotourism, etc., can be highlighted by encouraging skills of the communities and village-level institutions. Traditional livelihood options that communities have practiced over years are more resilient in the face of climate change while also having high cultural significance. Policies for identifying, safeguarding and improving such livelihood options are urgently needed. Integrating the local knowledge-based into science-based policy decisions and implementation process can be thought of though the communities value immediate economic benefits rather than the long-term benefits.

The region needs sustainable development approach on both economic and ecological concerns based on location- and area-specific models.

Acknowledgments This work was carried out by WWF-India, Khangchendzonga Landscape Program, Sikkim Office, for preparing the Livelihood Strategy Document of 27 villages selected under the UNDP SECURE Himalaya Project in Sikkim. We are thankful to all the stakeholders for providing us with necessary inputs during our field visits and planning process. We would also like to thank the communities of the study sites for being a part of all the community discussions and providing us with necessary feedback and recommendation.

Appendix

Table 8.1 Demographic details of villages in North Sikkim

Cluster/village	Total HH	Women headed HH	Female	Male	Total population
North	1841	123	4071	4837	9603
Chungthang Cluster	231	8	452	544	989
Naga Namgor	156	6	299	336	635
Theng-Chungthang	75	2	153	208	354
Lachen Muguthang Cluster	391	0	589	886	1375
Lachen-Thangu	338	0	589	736	1325
Muguthang	14	0	0	0	50
Thangu	39	0	0	150	0
Lachung Cluster	368	5	1309	1423	2732
Lachung	368	5	1309	1423	2732
Lower Dzongu Cluster	486	68	848	1120	2770
Hee-Gyathang	244	40	209	476	1487
Leek	25	1	54	58	112
Salim Pakel	87	12	199	181	380
Shipgyer	130	15	386	405	791
Upper Dzongu Cluster	365	42	873	864	1737
Bay-Pentong	43	12	92	113	205
Kusung-Tingvong	207	27	551	510	1061
Lingdem-Laven	73	0	151	158	309
Sakyong	42	3	79	83	162
Grand total	1841	123	4071	4837	9603

Table 8.2 Socioeconomic details of villages in North Sikkim

Cluster/village	Rich HH	Middle class HH	Poor HH	Very poor HH
North	47	757	794	580
Chungthang Cluster	5	21	49	128
Naga Namgor	0	0	0	128
Theng-Chungthang	5	21	49	0
Lachen Muguthang Cluster	0	0	0	391
Lachen-Thangu	0	0	0	338
Muguthang	0	0	0	14
Thangu	0	0	0	39
Lachung Cluster	2	89	259	18
Lachung	2	89	259	18
Lower Dzongu Cluster	40	568	267	13
Hee-Gyathang	40	410	196	0
Leek	0	5	20	0
Salim Pakel	0	75	12	0

(continued)

Table 8.2 (continued)

Cluster/village	Rich HH	Middle class HH	Poor HH	Very poor HH
Shipgyer	0	78	39	13
Upper Dzongu Cluster	0	79	219	30
Bay-Pentong	0	13	25	5
Kusung-Tingvong	0	35	110	25
Lingdem-Laven	0	1	72	0
Sakyong	0	30	12	0
Grand total	47	757	794	580

Table 8.3 Demographic details of villages in West Sikkim

Cluster/village	Rich HH	Middle class HH	Poor HH	Very poor HH
West	142	708	636	93
Labdang Karzi Cluster	10	91	96	16
Karzi	5	81	30	0
Labdang	5	10	66	16
Maneybung Cluster	5	20	35	5
Gyaten	5	20	35	5
Maneybung GPU	0	0	0	0
Sopakha	0	0	0	0
Nesa Gangyap Cluster	92	467	229	27
Gangyap	52	57	33	27
Nesa	40	410	196	0
Yambong Cluster	3	28	181	20
Sindrabong	3	23	104	17
Topung	0	5	77	3
Yuksom Cluster	32	102	95	25
Tsokha-Kyoungtey	7	18	65	25
Yuksom	25	84	30	0
Grand total	142	708	636	93

Table 8.4 Socioeconomic details of villages in West Sikkim

Cluster/village	Total HH	Women headed HH	Female	Male	Total population
West	2783	61	3980	4243	13,249
Labdang Karzi Cluster	255	4	560	622	1182
Karzi	116	4	253	286	539
Labdang	139	0	307	336	643
Maneybung Cluster	1199	0	770	754	6451
Gyaten	68	0	126	148	274
Maneybung GPU	894	0	0	0	4927
Sopakha	237	0	644	606	1250
Nesa Gangyap Cluster	828	1	1650	1805	3455
Gangyap	182	1	364	401	765
Nesa	646	0	1286	1404	2690

(continued)

Table 8.4 (continued)

Cluster/village	Total HH	Women headed HH	Female	Male	Total population
Yambong Cluster	232	22	402	459	867
Sindrabong	147	7	222	266	488
Topung	85	15	180	193	379
Yuksom Cluster	269	34	598	603	1294
Tsokha-Kyoungtey	130	15	260	290	643
Yuksom	139	19	338	313	651
Grand total	2783	61	3980	4243	13,249

Table 8.5 Different types of NTFPs collected and used

Medicinal plants	Fodder plants	Wild edibles	Other uses
Bhairun path	Gagun	Katus	Juniper
Bhakimlo	Nebhara	Nettle	Rhododendron anthopogon
Buro Okhati	Amliso (Broom Grass)	Manganay saag	Amliso (broom grass)
Chimphing (*Heracleum wallichii*)		Phunse	
Chirayita (*Swertia chirayita*)		Thotnay	
Khanakpa (*Evodia sp.*)		Sil Timbur (*Zanthoxylum*)	
Khanchal bee		Water cress	
Kutki		Nettle	
Majito		Wild orchid (Nakima) Tupistra nutans	
Pakhanbeth		Ferns	
Caterpillar fungus (*Ophiocordyceps sinensis*)		Mushrooms	
Satwa (*Paris polyphylla*)			

Source: SECURE Baseline Survey (2020)

Table 8.6 Wealth ranking

District/cluster	Number of villages in cluster	Rich HH	Middle class HH	Poor HH	Very poor HH
North	14	47	757	794	580
Chungthang Cluster	2	5	21	49	128
Lachen Muguthang Cluster	3	0	0	0	391
Lachung Cluster	1	2	89	259	18
Lower Dzongu Cluster	4	40	568	267	13
Upper Dzongu Cluster	4	0	79	219	30
West	11	142	708	636	93
Labdang Karzi Cluster	2	10	91	96	16
Maneybung Cluster	3	5	20	35	5
Nesa Gangyap Cluster	2	92	467	229	27
Yambong Cluster	2	3	28	181	20
Yuksom Cluster	2	32	102	95	25
Grand total	25	189	1465	1430	673
Rich	Gazetted Govt. employee, land holding of 10 ha, property in city/town, more than one private car				
Middle class	One or two earning member, more than 2 ha. of land holding, 1 no. of local taxi.				
Poor	Two ha of land holding, earning members				
Very poor	Landless, 1 earning member only, kutcha house, Rs. 10,000/ annual income				

Table 8.7 Main livelihood activities of North Sikkim

Village clusters	Livestock/ animals	Agriculture	Main cash Crops	Nonfarm	NTFP	Forest products
Lachen - Muguthang	Yaks Dzo Cow Bull Sheep	Barley Radish Cabbage Spinach Beans	Potato Apple	Tourism (guides, porters, hotels/ homestay, etc.) MGNREGA Govt Sector Handicrafts	Cordyceps *Paris Polyphylla* Junipers *Rhododendrons* Wild mushrooms	Firewood Bamboo Timber Fodder Poles
Lachung	Yaks Dzo Cow Bull Poultry	Barley Radish Cabbage Spinach Beans	Potato Apple	Tourism (guides, porters, hotels/ homestay, etc) MGNREGA Handicrafts	Cordyceps *Paris* *Polyphylla* Junipers *Rhododendrons*	
Chungthang Naga	Piggery Dairy	Beans Greens	Cardamom Round Chillies Potato	Private Sector Govt Sector MGNREGA Shops Apiary	*Paris* *Polyphylla* Wild Mush-rooms Bamboo Shoots Wild avocado Ferns Nettles	
Upper Dzongu	Dairy cows Bull Piggery Goats Poultry	Buckwheat Barley Maize Radish Pulses Millet Greens Paddy Peas	Cardamom Orange Ginger	MGNREGA Govt Sector Apiary Handicrafts	Cordyceps Bamboo Shoot Fern Nettle Walnut Water cress Wild avocados	

(continued)

Table 8.7 (continued)

Village clusters	Livestock/ animals	Agriculture	Main cash Crops	Nonfarm	NTFP	Forest products
Lower Dzongu	Dairy cows Bull Piggery Goats Poultry	Buckwheat Barley Maize Radish Pulses Millet Greens Paddy Peas	Cardamom Orange Ginger	MGNREGA Tourism Govt Sector Handicrafts	Bamboo Shoot Fern Nettle Walnut Water cress Wild avocados	

Table 8.8 Main livelihood activities of West Sikkim

Name of cluster	Livestock/ animals	Agriculture	Main cash crops	Nonfarm	NTFP	Forest products
Labdang Karjee	Sheep Cow Poultry Buffalo	Maize, Millet Soyabean, Beans Buckwheat Green Vegeta-bles Peas, Potato Pulses, Radish Squash, Turmeric	Cardamom	MGNREGA Govt. Job Apiary	Nakima Bamboo shoot Fern Wild mush-rooms Wild orchid Broom grass	Firewood Fodder Bamboo Timber Firewood Fodder
Nesa Gangyap	Cow Poultry Buffalo Bull Poultry Goat Pig	Green vegetables	Round Chillies Cardamom Potato	MGNREGA Govt. Job		
Maneybung	Cow Poultry Buffalo Bull Poultry Goat Pig		Cardamom	MGNREGA Tourism (guides/porters/ homestays)		
Sindrabong-Topung	Cow Poultry Buffalo Bull		Cardamom	MGNREGA Tourism (guides/porters/home-stays) Himal Rakshaks		

(continued)

Table 8.8 (continued)

Name of cluster	Livestock/ animals	Agriculture	Main cash crops	Nonfarm	NTFP	Forest products
	Poultry Goat Pig					
Yuksam Kyoungtey	Cow Poultry Buffalo Bull Poultry Goat Pig		Cardamom	MGNREGA Govt Job Tourism (guides/porters/cooks) Himal Rakshaks		

References

Archer, D. R., & Hayley J. F. (2004). Spatial and temporal variations in precipitation in the Upper Indus Basin, global tele connections and hydrological implications. *Hydrology and Earth System Sciences 8*(1), 47–61.

Basnett, S., Kulkarni, A. V., & Bolch, T. (2013). The influence of debris cover and glacial lakes on the recession of glaciers in Sikkim Himalaya, India. *Journal of Glaciology, 59*(218), 1035–1046.

Bawa, K. S., & Ingty, T. (2012). Climate change in Sikkim: A synthesis. In M. L. Arrawatia & S. Tambe (Eds.), *Climate change in Sikkim-patterns, impacts and initiatives* (pp. 413–424). Information and Public Relations Department, Government of Sikkim.

Bhutiyani, M. R. (2010). Climate change in the northwestern Himalayas-A myth or reality? *Climate Change*, 1387–1403.

Das, S. K., Avasthe, R. R. K., Sharma, P., & Sharma, K. (2017, December). Rainfall characteristics pattern and distribution analysis at Tadong East Sikkim. *Indian Journal of Hill Farming, 30*(2), pp. 326–330.

Department of Science and Technology and Climate Change. (2016). *The study of South Lhonak Glacial Lake in terms of Glacial Lake Outburst Flood (GLOF)*. Department of Science and Technology, Ministry of Science and Technology, Government of India.

G-SHE. (2009). Governance for Sustaining Himalayan Ecosystem: Guidelines and Best Practices. New Delhi: Ministry of Environment and Forest (MoEF), Government of India.

Ganguli-Lachungpa, U., & Rahmani, A. R. (2003). Development of Conservation Strategy for the Alpine Grasslands of Sikkim. Annual Report 2003. Bombay Natural History Society, Mumbai, Unpublished. Pp. 106.

Hajra, P. K., & Verma, D. M. (1996). Flora of Sikkim. Botanical Survey of India, Calcutta.

India State of Forest Report. (2015). https://fsi.nic.in/isfr-2015/isfr-2015-forest-and-tree-resources-in-states-and-union-territories.pdf

Kothari, R., Upadhya, D., Theengh, L. T., Ghose, P. S., & Shrestha, P. (2017). *Climate change vulnerability assessment in snow leopard habitat.* https://www.worldwildlife.org/publications/climate-change-vulnerability-assessment-in-snow-leopard-habitat-gateway-communities-in-north-sikkim

Sharma, E., Chettri, N., Tse-ring, K., Shrestha, A. B., Jing, F., Mool, P., & Eriksson, M. (2009). *Climate change impacts and vulnerability in the Eastern Himalayas.* ICIMOD.

Tambe, S. (2007). Alpine vegetation ecology and livestock grazing in Khangchendzonga National Park, Sikkim. Ph.D. Thesis, FRI University, Dehradun. pp. 232.

Tambe, S., Arrawatia, M. L., Bhutia, N. T., & Swaroop, B. (2012). Rapid, cost effective and high resolution assessment of climate-related vulnerability of rural communities of Sikkim Himalaya, India.

Telwala, Y., Brook, B. W., Manish, K., & Pandit, M. K. (2013). Climate-induced elevational range shifts and increase in plant species richness in a Himalayan biodiversity epicentre. *PLoS One, 8*(2), e57103. https://doi.org/10.1371/journal.pone.0057103.

Chapter 9
Socioeconomic Aspect of Disaster Risk in Kashmir: Contextualizing Village Vulnerability in Sindh Basin

Mohd Sharjeel Sofi, Sami Ullah Bhat, Irfan Rashid, and Jagdish Chandra Kuniyal

Abstract Social vulnerability is one of the main elements that determine the coping capacity to climate variability and natural disasters, and identification of responsive populations and factors that weigh in to magnify the vulnerability is crucial for effective disaster risk reduction. Here, we investigated the relationship between socioeconomic status and the vulnerability of households and communities to natural disasters in the Sindh Basin of Hindu Kush Himalayas (HKH). In this study, 13 main villages of the Sindh basin were randomly selected for the survey. The survey was conducted between November 2017 and May 2019, and 650 household residents (~30% of households in each village) were interviewed. Our research found that the families living below poverty line experience frequent disasters with losses higher than the households living above poverty line. The results show that there is a link between economic status, disaster damage, and vulnerability. The findings have the potential to assist disaster risk managers and policymakers in the formulation of more efficacious and terrain-targeted disaster management policies and ensure the sustainable human development of the region.

Keywords Social vulnerability · Natural disasters · Disaster management · Sustainable development

---

M. S. Sofi · S. U. Bhat (✉)
Department of Environmental Science, University of Kashmir, Srinagar, India

I. Rashid
Department of Botany, University of Kashmir, Srinagar, India

J. C. Kuniyal
G.B. Pant National Institute of Himalayan Environment (NIHE), Almora, Uttarakhand, India

S. Sharma et al. (eds.), *Climate Change Adaptation, Risk Management and Sustainable Practices in the Himalaya*,
https://doi.org/10.1007/978-3-031-24659-3_9

1 Introduction

The Hindu Kush Himalaya (HKH), an important geo-ecological asset, is the source of 10 major river basins and includes an area of over 4.2 million km^2 (Bajracharya et al., 2015). This HKH region and Tien Shan mountains include four global biodiversity hotspots, 330 key bird habitats, and hundreds of high peaks (Chettri et al., 2008). The area is known for various cultures, languages, religious beliefs, and traditional knowledge (Sharma et al., 2019). The greater Kashmir region, a vast area at the Western side of the HKH region, is home to a multi-ethnolinguistic community consisting of Kashmiris, Paharis, Ladakhis, and Dogras. The Sind Basin of the HKH region is one of the highly developed side valleys of the Kashmir region that spreads between 34° 11′ 17″ N to 34° 46′ 30″ North and 74° 57′ 10″ East to 75° 63′ 4″ East (Sofi et al., 2022a, b). The region is one of the most hazard-prone areas and is susceptible to disasters like floods, landslides, snow avalanches, and earthquakes.

Most of the terrestrial surface of the earth has been altered by anthropogenic actions (Vitousek et al., 1997), and "it is clear that we live on a human-dominated planet." Acknowledging anthropological supremacy, researchers opine that we have moved into the Anthropocene (Jamieson, 2017). The Anthropocene era is one in which a single species—*Homo sapiens*— has arisen as a global force, influencing the morphology and the functioning of the planet. Under these circumstances, vulnerability science revolves around the logic that human populations mediate changes in environmental, economic, and social spheres. Therefore, the bottom-up approach appears to be a viable alternative for making key policies and other practical applications with emphasis on the nature-society relationship. Assessing vulnerability over time and space using various indicators provides some nifty means of classifying the processes that subsidize vulnerability, prioritize approaches for plummeting vulnerability, and assess the efficacy of these approaches in diverse social and ecological environments (Adger, 1999).

The "vulnerability" assessment argues in favor of a bottom-up form of assistance that is expected to allow adaptive capacities and flexible governance structures to flourish. Having said that, one of the most critical tasks in generating vulnerability indicators is their unit of the assay. At this stage, local elucidations and estimations must take into consideration the socioeconomic, political, and institutional factors as proposed by "contextual vulnerability" to determine the level of vulnerability. Traditional science adopts a scenario-driven approach and makes projections based on past analogy. This top-down approach does not take into account the existing disparity in the vulnerabilities of human populations to environmental risks and therefore may be inadequate for policies. This sturdily argues that the likely impacts of disasters may be adjudged best through bottom-up-approach such as social vulnerability. Social vulnerability is partly the corollary of factors that characterize the vulnerability of communities to disasters and those that affect their capability to respond. The socioeconomic inequities in the societies are widening and urgently merit a much deeper, rational, and cohesive approach to reduce disaster risk and formulate recovery plans and policies.

2 Introducing Environmental Humanities in the Field of Hazards, Disasters, and Vulnerability

Since the middle of the twentieth century, research on risks and disasters has been classified into two major categories: the "hazards" and the "alternatives" (Dominey-Howes, 2018). The "hazards paradigm" consists of both physical and engineering sciences, whereas the "alternative hazards paradigm" studies developmental, behavioral, and complexity sciences (Smith & Petley, 2009). Physical and engineering disciplines made their best efforts in disaster management, yet people continued to die, and damages kept increasing across the earth. This led to combining political economics and political ecology with physical system principles (Dominey-Howes, 2018). The increase of environment-related problems and disasters can't be managed by the old-style cooperation of scientists and elected policymakers. But a hybrid of scientific and political negotiation away from prevailing institutional arrangements such as environmental humanities is required. Swedish environmental historian Sverker Sörlin has stated this shift by highlighting that:

> [o]ur belief that science alone could deliver us from the planetary quagmire is long dead. For some time, hopes were high for economics and incentive-driven new public management solutions. . . . It seems this time that our hopes are tied to the humanities . . . in a world where cultural values, political and religious ideas, and deep-seated human behaviours still rule the way people lead their lives, produce, and consume, the idea of environmentally relevant knowledge must change. We cannot dream of sustainability unless we start to pay more attention to the human agents of the planetary pressure that environmental experts are masters at measuring but that they seem unable to prevent. (Sörlin, 2012)

The environmental humanities visualize environmental predicaments primarily as problems of socioeconomic disparity, cultural variance, and different histories, ideals, and moral contexts. Environmental humanity is a term used for the collection of multidimensional scholarly approaches that recognize environmental problems as inseparable from sociocultural and anthropogenic influences. The scientific perceptions and technical methods for solving the problems are influenced by such factors and benefit by positioning themselves in these historical and sociocultural backgrounds (Heise, 2017). Astrida Niemanis, Cecilia Åsberg, and Johan Hedrén provide a noteworthy summary of environmental humanities, "acknowledging the differences and diffractions in worldviews, histories, subjectivities, relations and practices that various communities (both human and non-human) engage in, concerning their environment" and "cultivating environmental humanities that is well-placed for research and analysis of these differences" (Garrard, 2017). The growth of the environmental humanities is a consequence of an increased desire to include environmental issues within the humanities and social sciences (Rose et al., 2012).

3 Redefining "Natural Disasters" in the Context of Present Anthropocence

Crutzen opined that humans "should shift mission from crusade to management, so we can steer nature's course symbiotically instead of enslaving the formerly natural world" (Crutzen & Schwagerl, 2011). Steffen et al. (2011) and other workers declared that "the concept of the Anthropocene sharpens the focus on an overarching long-term goal for humanity keeping the Earth's environment in a state conducive for further human development." Anthropocene presents humanity with two choices, whether we have to love it or leave it. For instance, if we want to accept or want to leave the Anthropocene, we have to develop methods that will permit the human race to thrive in this period (Jamieson, 2017). Deliberations concerning the connotations and influence of the environment have played a vital role in the development of the political ecology of hazards. Throughout the twentieth century, literature hypothesized that disasters were caused by extreme natural events that could be fixed by modernist scientific methods like engineering and centralized planning (Burton et al., 1978). Many critiques of this concept from various disciplines have challenged this concept, which they call as the "hazards paradigm" (Blaikie et al., 1994; Steinberg, 2006). The basic premise on which their challenge is based in the new analysis, which is expressed often as the assertion that "natural disasters are not natural." For the last many years, numerous political ecologists have maintained that natural disasters don't exist. They claim that the concept of a "natural" disaster obscures the social and political practices that make people more or less susceptible to natural hazards (Gaillard & Kelman, 2012). While this intervention has initiated a debate for redefining disaster for risk reduction, this criticism of the concept of "natural" can further be expanded to reach conclusions in more useful ways. From Marxian dialectics to social construction to realist policy methods, people have agreed for a long time that the word "natural" in disaster literature makes it hard to see the causative factors of disaster risk and, by extension, the policies that are needed to lower that risk (Gould et al., 2016). Eventually, the critique of the "nature" of the "hazard paradigm" produced a new method for investigating and handling disasters that paid attention to the political, societal, and financial aspects that produce social vulnerability (Collins, 2008, 2009). For instance, the critiques of the naturalness of natural disasters are not repudiating the natural hazards as a whole; on the contrary, they argue that it is social, political, and economic dimensions that determine the vulnerability of the people to hazards. For instance, the same losses or physical effects in two different countries with different political and economic systems could have very different impacts. Due to the difference in the absorption capacity of each of the social systems involved, an incident that would go relatively unnoticed in a large nation might be a catastrophe in a small one. For instance, if the damage due to a particular hazard is same in both rich and poor countries, the effects on society are harsher in poor countries, where the poor and disadvantaged are usually affected the most (Wijkman & Timberlake, 1984). Hence vulnerability "is the degree to which the different social classes are differentially at risk" (Susman et al., 1984). This defines vulnerability in the context of political, social, and economic circumstances, which is also evident from this case study that households under the poverty ceiling suffer greater relative losses from disasters, even though they own far less than APL families. Leaving 85% of the households vulnerable to disasters and 91% of the household in a state where they are not prepared to handle a disaster of high magnitude in the Sindh basin opens larger debate for analyzing the multidimensional aspects that produce social vulnerability. Thus, it is suggested that social discrimination, expropriation, exploitation, political oppression, socioeconomic conditions, and governance issues be taken into consideration while defining "natural" disasters.

4 Socioeconomic Vulnerability and Disaster Risk: Level and Trends in HKH Region

Livelihood opportunities in the mountainous regions are extremely vulnerable due to the effects of climate change. The vulnerability is rooted in daily power undercurrents; thus, its exposure and intensity vary according to political economy, social capital (Turner, 2016), gender (Morchain et al., 2015), and ethnicity (Bolin, 2007). The lack of institutional and market structures, in the form of infrastructure, insurance, social security, and a wide range of livelihood alternatives, contributes to social vulnerability (Adger & Kelly, 1999). Rich societies with equities in all domains of social spheres are normally less vulnerable as compared to poor societies. According to this principle, disasters impact wealthy and poor communities differently because economically stronger communities may afford to invest in robust infrastructure and superior emergency services. Thus, to diminish the vulnerability, it is important to take account of those social systems which produce vulnerability, besides decreasing the hazard exposure (Blaikie et al., 1994). The United Nation's Human Index (HDI) is commonly used for assessing the quality of life, and Hindu Kush Himalayan states (except China) have low HDI values when compared to the world average. The state of Jammu and Kashmir has HDI values of 0.684 for the year 2017, which shows that citizenry of these nations has a higher social vulnerability (Radhakrishnan & Sen, 2019) (Fig. 9.1 and Table 9.1).

Individual-level poverty assessed using Multidimensional Poverty Index (MPI) incorporates deficiencies that each person experiences in terms of education, health, and livelihood in addition to the standard income-based poverty metrics (Alkire & Robles, 2017). The values for the MPI in the HKH nations indicate that all the countries except China are highly vulnerable (Garschagen et al., 2016) (Table 9.2). The index uses 28 individual indicators, associated to exposure, and 23 linked to elements that are representative of vulnerability (susceptibility, coping capacity, and adaptive capacity). About 17% of the population continue to live below the poverty line in the rural Indian Himalayas (Excluding West Bengal and Uttarakhand), and 15% of the population of the state of Jammu and Kashmir continue to live in

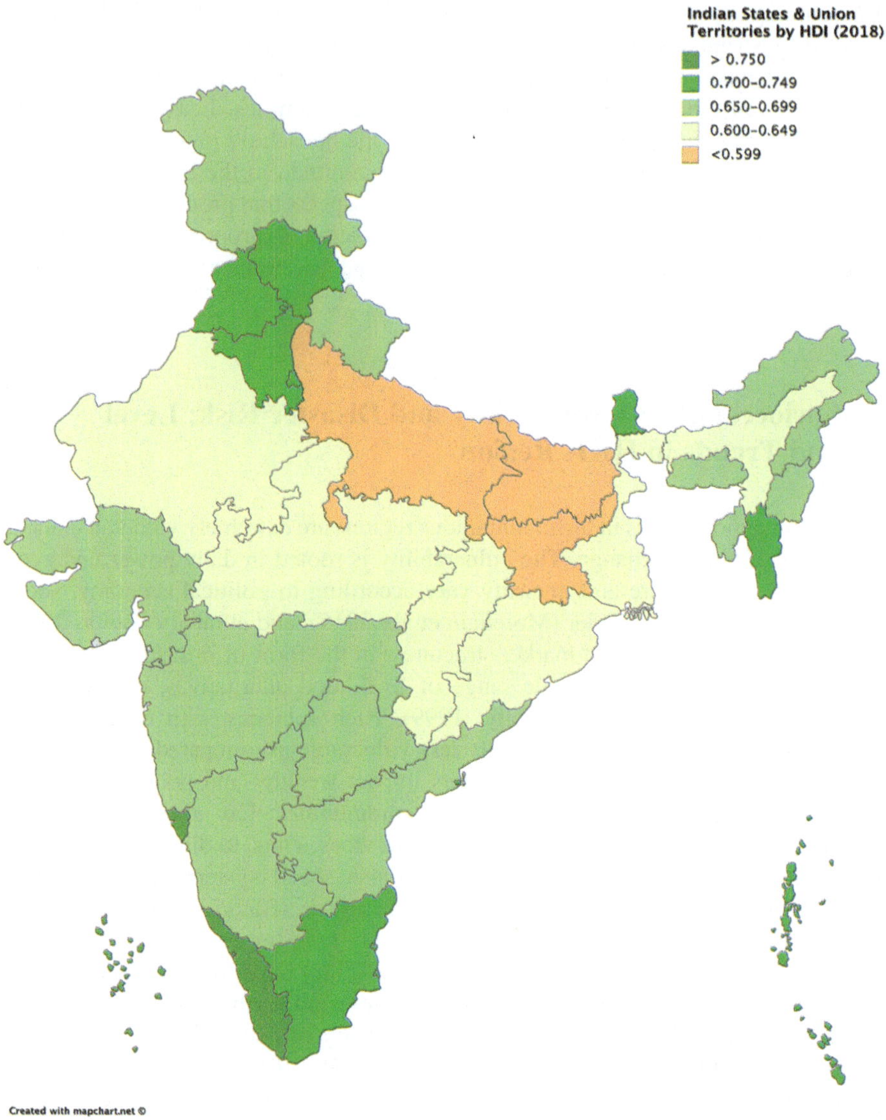

Fig. 9.1 Indian states and Union Territories by HDI 2018. (Map adapted from https://en.wikipedia.org/wiki/List_of_Indian_states_and_union_territories_by_Human_Development_Index#/media/File:Indian_States_and_Union_Territories_by_HDI_(2018).png)

multidimensional poverty. The results obtained in the Sindh Basin indicate that poverty incidence (headcount ratio) and its depth (poverty gap) are high for households that rely on one income source and are dependent on natural resources. Though the significance level isn't very high, poorer households are still at higher risk to frequent disaster.

Table 9.1 Human Development Index of the countries of HKH region

Country	Human Development Index (HDI)
China	0.758
India	0.645
Bhutan	0.617
Bangladesh	0.614
Nepal	0.579
Myanmar	0.584
Pakistan	0.560
Afghanistan	0.496

Source: http://hdr.undp.org/en/content/2019-human-development-index-ranking

Table 9.2 Value of Multidimensional Poverty (MPI) Index of the countries of HKH region (UNDP, 2019)

Country	Year and survey	MPI
China	2014 N[i]	0.016[j, k]
India	2015/2016 D	0.123
Bhutan	2010 M	0.175[g]
Bangladesh	2014 D	0.198
Nepal	2016 D	0.148
Myanmar	2015/2016 D	0.176
Pakistan	2014 D	0.198
Afghanistan	2015/2016 D	0.272[d]

Notes

D indicates data from Demographic and Health Surveys, M indicates data from Multiple Indicator Cluster Surveys, and N indicates data from national surveys (see http://hdr.undp.org/en/faq-page/multidimensional-poverty-index-mpi for the list of national surveys)

d-Missing indicator on nutrition

j-Missing indicator on housing. k-Child mortality was constructed based on deaths that occurred between surveys—that is, between 2012 and 2014. Child deaths reported by an adult man in the household were taken into account because the date of death was reported

i-Missing indicator on cooking fuel

g-Considers child deaths that occurred at any time because the survey did not collect the date of child deaths

5 Evidence of the Intersection of Poverty, Vulnerability, and Disaster Preparedness from the Sindh Basin of HKH Region

The main aim of this study was to study the convoluted relationship between community-level preparedness for natural disasters and socioeconomic vulnerability. The study focuses on individual households as basic units of a community and tries to answer two basic questions: First, how do natural disasters impact households belonging to various socioeconomic groups? Secondly, we tried to examine the extent of community disaster preparedness among the villages of the Upper Sindh basin of Kashmir Himalayas. Community disaster preparedness in HKH river basins is vital to protect human lives from adverse health and social and economic effects arising from unpredicted natural disasters. Various studies have investigated the vulnerability of diverse economic stratas to specific disasters and their responses. Some studies have examined the liaison between poverty and natural disasters by cogitating upon the effect of a single kind of disaster. Taking into account the disparities in economic situations, regions, and scale, most of the studies endorse that natural disasters have a substantial impact on poverty and human development. But only a few studies have been carried out to find the impact of natural disasters on poverty (Foltz et al., 2013), and no such study has been carried out in the Kashmir valley. Our study is a pioneering work that tries to analyze the direct impact of natural disasters on various economic groups and assesses the extent of community-level preparedness for the disasters in the region.

A standard questionnaire was used during the study, and the questions were framed to reduce sociodemographic profile and disaster preparedness among the residents of the Sindh Basin. In this study, 650 rural household residents selected randomly were interviewed. Each face-to-face interview, conducted in the local language (Kashmiri), took about 15–20 min. Interviewees, including both men and women of various ages, were asked about their occupation, socioeconomic conditions, exposure to natural disasters and their preparedness toward them, and how disasters have impacted them in the past. In this regard, the respondents were asked to assess their vulnerability toward natural disasters. During the survey, the respondents considered inaccessibility and low access to markets and basic infrastructure, access to natural resources, and political and socioeconomic marginalization as the basic elements of their poverty. We categorized the whole sample based on the type of ration card Above Poverty Line (APL) and Below Poverty Line (BPL) issued to the households by the competent authority of the local government. The sample includes 60% of poor households among a total of 650 households. The case study threw up some important findings. For example, almost 85% of households were found to be vulnerable to natural disasters of one or the other type. Damage costs per household imply that the average household loses about 35–45% average annual household income when struck by the disaster. Financial loss incurred is mainly due to the damage to agricultural products followed by property damage. The results show that a greater percentage of BPL households can suffer more damages from natural disasters than APL households. The disaster loss to annual income ratio, also known as relative loss, is more for BPL households than for APL households. The disaster loss was found to be highest for households whose livelihood was solely dependent on agriculture. Natural disasters have a greater impact on families with lower per-capita incomes. Hence, financial potential remains an important factor influencing disaster preparedness, and this necessitates the provision of providing the necessary financial help for the households lying below the poverty level. One of the most striking findings of the present case study is related to those poor households who are solely dependent on agriculture, and it was found that frequent exposure to natural disasters can be ravaging for the existence of such households.

An important observation made during this study is that if members of such households were able to find some source of income beyond agriculture, the impact of disasters tends to decrease. Although individuals may take precautionary procedures to lessen the consequences of future disasters, poor households do not have this choice. Thus, BPL households ceiling face higher relative losses from disasters, even though they own substantially less property as compared to APL households (De Silva & Kawasaki, 2018). The argument that "environmental justice and political ecology has demonstrated, that burden of environmental risks, whether climate change-related or not, falls unevenly on different social groups, mediated by class, race, gender, and ethnicity" (Chakrabarty, 2016).

6 Is Rural Community in the Sindh Basin Ill-Prepared for Natural Disasters?

There is almost an oblivious retort to disaster preparedness, sometimes fringing the realms of fatalism. This can be understood from the data obtained during the study as the majority of households (91%) believed that they are not prepared to handle a disaster of high magnitude at all, whereas 3.3% of the households reposed faith that their households are well prepared. The level of concern among the rural community studied was highest for earthquakes followed by floods among the disasters. When the respondents were asked about their knowledge of the Family Disaster Management Plan, majority of them were found to have no idea about the disaster management plan at all. A large majority of respondents from both economic groups were found to have no knowledge about the presence of any search and rescue team in their locality. Ninety-eight percent answered "No" when questioned about whether the concerned agency is conducting the mock drills.

When asked about the reasons for not being prepared for a disaster, the top reason mentioned by respondents was "lack of knowledge about disasters." Awareness about the toll-free emergency number for disaster response was very low among the respondents. The respondents believed that at the community level they don't have sufficient preparedness to face any eventuality in the form of any disaster. One of the most important factors that have contributed to the low community preparedness for disasters is poor knowledge, which determines the attitude of the community toward these hazards. The level of unawareness can be understood by the fact that respondents appeared to have no faith in the effectiveness of the preparedness at the household level. These findings indicate the potential value of educating communities about disaster preparedness. A general response to why people do not have a Family Disaster Plan is that people hang around until something happens. Most people perceive disasters as an act of "God." This may be due to the prevalence of religious dogmatism in society. "Trusting God" during disasters is an argument reported by people in many previous studies, similar trends were observed in the Sindh Basin. A similar trend in the results was reported by the Council for

Excellence in Government reported that 32% of US citizens took no concrete actions to prepare for potential disasters, with lack of awareness, beliefs, and resources being cited as the key reasons. Most of the respondents felt that the motivation to overcome suffering should come from themselves and highlighted the importance and significance of their faith and religion and sought to compensate their losses through religious beliefs.

7 Religious Conceptions and Coping Mechanism of Disasters in Local Villages

The relationship between religion and natural disaster leads individuals to fatalistic belief systems. This idea alludes to the view that individuals have no control over natural disasters since they are God's will. Every time a natural calamity occurs, the question of whether or not it is God's wrath is asked.

A similar thought line was observed in every village during the survey. People in the region being Muslims strongly believed the idea of disasters as "God's punishment." These ideas are strongly supported by religious scriptures, for example, in the Hadith (record of the traditions or sayings of the Prophet Muhammad^(PBUH)) as explained by At-Thabrani, Al-Hakim, and Al-Baihaqi (Kosim, 2012);

> Immoral deeds will result in God's Wrath

The Holy Quran (religious scripture) encourages people to undertake preparation for disasters, as mentioned in (Chapter 6, verse 131 and Chapter 59, verse 18), which can be interpreted as "it is necessary for people living in the vulnerable regions to prepare for disasters." However, we found people oblivious to the religious aspect of disaster preparedness.

8 Way Forward

Disasters should be seen as unresolved developmental crises as they are not natural occurrences only, but rather the result of the interaction between natural and society's organizational structures. It is thus essential to emphasize that disaster is not synonymous with any natural event. As the foundation for the idea of vulnerability, it is essential to take into account community's capacity for adaptation or adjustment when confronted with natural hazards. As it is generally agreed that disaster occurs only when the damages surpass the community's capacity to sustain or tolerate them, or when the repercussions impede rapid recovery. Thus, to assess vulnerability, it is important to take into consideration the physical and social aspects, like family fragility, economic condition, lack of basic social services, absence of access to property and credit, the existence of ethnic and political prejudice, unhygienic air and water resources, high illiteracy rate, and the lack of educational opportunities (Lavell, 1996; Cardona, 2004).

In some behavioral theories, information and attitudes are undeniably seen as the primary determinants of human behavior. Effective education programs and exercises may assist community inhabitants in recognizing and identifying threats, improving their awareness of how to react during and after a specific disaster. Risk awareness should also be included as a basic part in disaster-related education programs. Risk awareness plays a key role in the transformation from information attainment to behavioral changes. The role of women is central to the idea of community emergency preparedness because women often realize that if they are dislocated from their homes, they become more vulnerable, and therefore, they understand the issue in a more instinctive and deeper sense than men do. Despite the fact that women play an important role in decision-making, their presence is not obvious in most disaster management strategies. To improve women's roles in governance, particularly disaster risk reduction (DRR), the first phase would be to identify the obstacles that prohibit them from participating in decision-making, governance, and DRR activities.

9 Policy Implication

In terms of policy, there is an urgent need to address the income disparity gap in mountain communities via more effective poverty-reduction initiatives, including increased expenditures in education and development. Overall, it is obvious that residents in the Sindh basin area are more vulnerable to natural catastrophes in the era of climate change. To reduce the community's susceptibility, structural changes must be implemented. Furthermore, women must be considered the primary audience for all schemes and programs since they are in charge of housing arrangements and are expected to be more prepared than men. Furthermore, it is critical that the academia, governments, and relevant stakeholders address this problem at a time when a consistent message about the causes of disasters has never been more important. It would also assist to enhance management effectiveness by identifying and prioritizing practical and efficient risk-reduction actions by authorities and communities, who are unquestionably the primary stakeholders in the management of disasters.

10 Conclusion

Disaster events are too powerful, too monstrous, and too perilous. The disaster events are magnified by cumulative human action and poor management practices. These events have an intimate connection with the humans as we all have contributed them in some measure, to their making. This making of our own hands is now returning to haunt us in shapes that are too grotesque and unthinkable. The risk factors added to the natural disaster by anthropogenic activities and official mismanagement are annihilating and startlingly real. However, we believe that disaster management is not rocket science. It is just about our daily lives. In light of the Sendai Disaster Risk Reduction Framework, socioeconomic vulnerability and disaster risk assessments should consider a variety of factors, including income inequality, gender inequality, governance, religious aspects, cultural considerations, and national progress for disaster risk reduction. All the contributing factors of vulnerability, disaster risk, and emergency preparedness must be understood, linked up, and elucidated simply and convincingly. A reliable and clear approach to risk management that takes into account numerous aspects could help disaster managers make decisions in a certain area. The disaster management approach should be based on a complex theoretical foundation that considers not only geological and structural elements but also economic, social, political, and cultural factors. This type of approach evaluates the nonlinear relationships between contextual characteristics and the complexity and dynamics of social systems in a more consistent way.

Acknowledgments The research work was conducted as part of a research project titled "Anthropogenic Impacts and Their Management Options in Different Ecosystems of the Indian Himalayan Region," sponsored and funded by National Mission on Himalayan Studies (NMHS) under the aegis of Ministry of Environment, Forest, and Climate Change (MoEFandCC), Govt. of India, New Delhi, is thankfully acknowledged. The authors also thank Director Govind Ballabh Pant National Institute of Himalayan Environment (NIHE), Kosi-Katarmal, Almora-263 643, Uttarakhand, India as well as the Department of Environmental Science, University of Kashmir for providing facilities, which together could make the present study possible.

Author Contributions

Mohd Sharjeel Sofi: Conceptualization, Writing - original draft, Visualisation.

Sami Ullah Bhat: Writing, Review and editing, Supervision, Funding acquisition.

Irfan Rashid: Writing, Review and editing, Supervision, Funding acquisition.

Jagdish Chandra Kuniyal: Review and editing, Funding acquisition.

Data Availability Statement The data presented in this study are available within the manuscript.

Conflicts of Interest The authors declare no conflict of interest.

Funding This study was funded by National Mission on Himalayan Studies Implemented by the Ministry of Environment, Forest and Climate Change (MoEF and CC) Govt. of India, under Grant number NMHS/SG/-2017/260.

References

Adger, W. N. (1999). Social vulnerability to climate change and extremes in coastal Vietnam. *World Development, 27*(2), 249–269. (Lars Otto Naess). https://doi.org/10.1016/S0305-750X(98)00136-3

Adger, W. N., & Kelly, P. M. (1999). Social vulnerability to climate change and the architecture of entitlements. *Mitigation and Adaptation Strategies for Global Change, 4*(3–4), 253–266.

Alkire, S., & Robles, G. (2017). *Multidimensional poverty index summer 2017: Brief methodological note and results* (OPHI methodological note 44). Oxford University of Oxford.

Bajracharya, S. R., Maharjan, S. B., Shrestha, F., Guo, W., Liu, S., Immerzeel, W., & Shrestha, B. (2015). The glaciers of the Hindu Kush Himalayas: Current status and observed changes from the 1980s to 2010. *International Journal of Water Resources Development, 31*(2), 161–173. https://doi.org/10.1080/07900627.2015.1005731

Blaikie, P., Cannon, T., Davis, I., & Wisner, B. (1994). *At risk: Natural hazards, people's vulnerability and disasters*. Routledge.

Bolin, B. (2007). Race, class, ethnicity, and disaster vulnerability. In *Handbook of disaster research* (Handbooks of sociology & social research, pp. 113–129). Springer. https://doi.org/10.1007/978-0-387-32353-4_7

Burton, I., Kates, R. W., & White, G. F. (1978). *The environment as hazard*. Oxford University Press.

Cardona, O. D. (2004). The need for rethinking the concepts of vulnerability and risk from a holistic perspective. In G. Bankoff et al. (Eds.), *Mapping vulnerability*. Earthscan Publications.

Chakrabarty, D. (2016). Whose anthropocene? A response. In R. Emmett & T. Lekan, *RCC perspectives: Transformations in environment and society, whose anthropocene? Revisiting Dipesh Chakrabarty's 'four theses,', 2* (pp. 103–113). Rachel Carson Center.

Chettri, N., Shakya, B., Thapa, R., & Sharma, E. (2008). Status of a protected area system in the Hindu Kush-Himalayas: An analysis of PA coverage. *International Journal of Biodiversity Science & Management, 4*(3), 164–178. https://doi.org/10.3843/Biodiv.4.3:4

Collins, T. W. (2008). The political ecology of hazard vulnerability: Marginalization, facilitation and the production of differential risk to urban wildfires in Arizona's White mountains. *Journal of Political Ecology, 15*(1), 21–43. https://doi.org/10.2458/v15i1.21686

Collins, T. W. (2009). The production of unequal risk in hazardscapes: An explanatory frame applied to disaster at the US–Mexico border. *Geoforum, 40*(4), 589–601. https://doi.org/10.1016/j.geoforum.2009.04.009

Crutzen, P., & Schwagerl, C. (2011). Living in the Anthropocene: Towards a new global ethos. *Yale Environment, 360*. https://e360.yale.edu/features/living_in_the_anthropocene_toward_a_new_global_ethos. Assessed on 12 Nov 2020.

De Silva, M. M. G. T., & Kawasaki, A. (2018). Socioeconomic vulnerability to disaster risk: A case study of flood and drought impact in a rural Sri Lankan community. *Ecological Economics, 152*, 131–140. https://doi.org/10.1016/j.ecolecon.2018.05.010

Dominey-Howes, D. (2018). Hazards and disasters in the Anthropocene: Some critical reflections for the future. *Geoscience Letters, 5*(1), 1–15. https://doi.org/10.1186/s40562-018-0107-x

Foltz, J., Gars, J., Özdoğan, M., Simane, B., & Zaitchik, B. (2013). Weather and welfare in Ethiopia. In *Agricultural and applied economics association's 2013, AAEA and CAES joint annual meeting*, Washington, DC.

Gaillard, J. C., & Kelman, I. (2012). Framing disaster: Theories and stories seeking to understand hazards, vulnerability and risk. In B. Wisner, J. C. Gaillard, & I. Kelman (Eds.), *The Routledge handbook of hazards and disaster risk reduction* (pp. 18–33). Routledge.

Garrard, G. (2017). Environmental humanities: notes towards a summary for policymakers. In U. Heise, J. Christensen, & M. Niemann (Eds.), *The Routledge companion to the environmental humanities* (1st ed.). https://doi.org/10.4324/9781315766355

Garschagen, M., Hagenlocher, M., Comes, M., Dubbert, M., Sabelfeld, R., Lee, Y. J., et al. (2016). World risk report 2016. World Risk Report. *Bündnis Entwicklung Hilft & UNU-EHS*. http://weltrisikobericht.de/wp-content/uploads/2016/08/WorldRiskReport2016.pdf

Gould, K. A., Magdalena Garcia, M., & Remes, J. A. C. (2016). Beyond "natural-disasters-are-not-natural": The work of state and nature after the 2010 earthquake in Chile. *Journal of Political Ecology, 23*(1), 93–114. https://doi.org/10.2458/v23i1.20181

Heise, U. K. (2017). Introduction: planet, species, justice—and the stories we tell about them. In *The Routledge companion to the environmental humanities* (pp. 17–26). Routledge.

Jamieson, D. (2017). The anthropocene: Love it or leave it. In U. K. Heise, J. Christensen, & M. Niemann (Eds.), *The Routledge companion to the environmental humanities* (pp. 29–36). Routledge.

Kosim, M. (2012). Bencana moral Picu bencana alam. *Berita Haluan*. http://www.harianhaluan.com/index.php?option=com_content&view=article&id=1062:bencana-moral-picu-bencana-alam-&catid=11:opini&Itemid=83

Lavell, A. (1996). Degradacion ambiental, riesgo y desastre urbano: Problemas y conceptos. *La Red*. In M. A. Fernandez (Ed.), *Ciudades en Riesgo*. United States Agency for International Development.

Morchain, D., Prati, G., Kelsey, F., & Ravon, L. (2015). What if gender became an essential, standard element of vulnerability assessments? *Gender and Development, 23*(3), 481–496. https://doi.org/10.1080/13552074.2015.1096620

Radhakrishnan, V., & Sen, S. (2019). *Is Jammu and Kashmir underdeveloped as stated by Amit Shah?* https://www.thehindu.com/data/where-does-jammu-and-kashmir-stand-in-comparison-to-other-states-in-key-indicators-of-growth-and.*Development*/article28855512.ece. Assessed on 31 Oct 2020.

Rose, D. B., van Dooren, T., Chrulew, M., Cooke, S., Kearnes, M., & O'Gorman, E. (2012). Thinking through the environment, unsettling the humanities. *Environmental Humanities, 1*(1), 1–5. https://doi.org/10.1215/22011919-3609940

Sharma, E., Molden, D., Rahman, A., Khatiwada, Y. R., Zhang, L., Singh, S. P., et al. (2019). Introduction to the Hindu Kush Himalaya assessment. In *The Hindu Kush Himalaya assessment* (pp. 1–16). Springer.

Smith, K., & Petley, D. (2009). *Environmental hazards: Assessing risk and reducing disaster* (4th ed.). Routledge.

Sofi, M. S., Hamid, A., Bhat, S. U. et al. (2022a). Impact evaluation of the run-of-river hydropower projects on the water quality dynamics of the Sindh River in the Northwestern Himalayas. *Environmental Monitoring and Assessment, 194*, 626. https://doi.org/10.1007/s10661-022-10303-5

Sofi, M. S., Hamid, A., Bhat, S. U., Rashid, I., & Kuniyal, J. C. (2022b). Biotic alteration of benthic macroinvertebrate communities based on multispatial-scale environmental variables in a regulated river system of Kashmir Himalaya. *Ecological Engineering, 177*, 106560. https://doi.org/10.1016/j.ecoleng.2022.106560

Sörlin, S. (2012). Environmental humanities: Why should biologists interested in the environment take the humanities seriously? *BioScience, 62*(9), 788–789. https://doi.org/10.1525/bio.2012.62.9.2

Steffen, W., Persson, A., Deutsch, L., Zalasiewicz, J., Williams, M., Richardson, K., Crumley, C., Crutzen, P., Folke, C., Gordon, L., Molina, M., Ramanathan, V., Rockström, J., Scheffer, M., Schellnhuber, H. J., & Svedin, U. (2011). The Anthropocene: From global change to planetary stewardship. *Ambio, 40*(7), 739–761. https://doi.org/10.1007/s13280-011-0185-x. PubMed: 22338713

Steinberg, T. (2006). *Acts of God: The unnatural history of natural disasters in America* (2nd ed.). Oxford University Press.

Susman, P., O'Keefe, P., & Wisner, B. (1984). Global disasters: A radical interpretation. In K. Hewitt (Ed.), *Interpretations of calamity from the viewpoint of human ecology* (pp. 264–283). Allen Press & Unwin.

Turner, M. D. (2016). Climate vulnerability as a relational concept. *Geoforum, 68*, 29–38. https://doi.org/10.1016/j.geoforum.2015.11.006

United Nations Development Programme (UNDP). (2019). *Human Development Report 2019: Beyond income, beyond averages, beyond today: Inequalities in human development in the 21st century*. New York.

Vitousek, P. M., Mooney, H. A., Lubchenco, J., & Melillo, J. M. (1997). Human domination of Earth's ecosystems. *Science, 277*(5325), 494–499. https://doi.org/10.1126/science.277.5325.494

Wijkman, A., & Timberlake, L. (1984). *Natural disasters acts of god or acts of man?* Earthscan Publications.

Chapter 10
Indigenous Knowledge System and Livelihood Option of Natives of Lahaul and Spiti District, Himachal Pradesh

Sumati Rathore and Sarla Shashni

Abstract Indigenous knowledge system (IKS) is a knowledge system developed and adapted by any particular communities in a changing environment condition and generally passed from one generation to next. The current study aims to catalogue the IKS of the tribal people dwelling in the Lahaul and Spiti district of Himachal Pradesh. The study was conducted from July, 2015, to December, 2019, in the region. Aspects such as use of bioresources and its bioprospecting, land and soil management, water conservation and management were assessed for the indigenous knowledge system. Documentation of around 98 medicinal plants, 14 species bioprospecting, 20 traditional food and beverages, unique traditional architecture, resource utilization pattern, agricultural practices, and their livelihood were done from the region. The region remained cut off from the outer world for around 5–6 months during winter. Due to the remoteness of the region, people residing in the area also had a limited options for their livelihood. But, recent opening of the Atal Tunnel connecting outer world to the region throughout the year has brought some unprecedented changes to the culture, economy, and tourism sector. Therefore, documentation of such valuable IKS is essential to strengthening the residing communities for sustainable development in the future.

Keywords Indigenous · Communities · Natives · Lahaul and Spiti · Himachal Pradesh

1 Introduction

The Indian Himalayan Region (IHR) holds a special place in the world's ecosystems with its rugged beauty, snow covered mountains, and unique cultural and biological diversity (Schild, 2008). IHR covers a northern boundary of the country representing 16.2% (537 Lakh km^2) of the total area (Samal et al., 2000) inhabited by 3.86%

S. Rathore · S. Shashni (✉)
G. B. Pant National Institute of Himalayan Environment, Himachal Regional Center, Mohal-Kullu, Himachal Pradesh, India

S. Sharma et al. (eds.), *Climate Change Adaptation, Risk Management and Sustainable Practices in the Himalaya*,
https://doi.org/10.1007/978-3-031-24659-3_10

(3,96,28,311) of total population. It includes 11 states. Himachal Pradesh and Uttarakhand connect to the Jammu and Kashmir and Ladakh union territories, as well as the seven sister's north-eastern states Sikkim, Arunachal Pradesh, Meghalaya, Nagaland, Manipur, Mizoram, Tripura, and Sikkim. The area is home to a remarkable array of communities, flora, and animals, as well as cultural, racial, and biological diversity (Nandy et al., 2006; Singh, 2006). Communities in Himalayan regions have historically adapted their way of life to rely on nature for their daily needs for ages (Das & Mishra, 2022). For their day-to-day need and livelihood pattern, they have developed a unique knowledge system which sustained through time (Ellan & Harris, 1997). This knowledge system of local people for sustaining their environment, ecology, etc. is also called peoples science (Tharakan, 2017) and indigenous knowledge system (IKS). IKS is defined as a body of accumulated wisdom, which has evolved in years through the experiences by group of people to meet the challenges they face in their environment (Green, 1996), and it confines to particular area, culture, and geography. This knowledge system generally passes orally or practical application from one generation to another (Mazzocchi, 2006; Savitri & Bhalla, 2012; Kumar et al., 2013). In last 2 or 3 decades, due to unscientific exploitation, unsustainable developmental practices and increase in population, modernization, etc. has imposing a significant risk to the distinctive knowledge system of the area. Lack of interest for various indegenous methods among younger generations is also the reason due to which the traditional knowledge system depleting fast. Therefore, there is an immediate need to document these traditional community-based knowledge systems from IHR for overall sustainable development of the region in future.

2 Study Area

Himachal Pradesh, one of the major states in IHR, has an area of 55,673 km^2, and is located between latitudes of 30°22′40″ and 33°12′40″ north and longitudes 75°45′55″ and 79°40′20″ east with an altitudinal ranges of 300–7000 m amsl (above mean sea level). Communities living in the region belongs to various ethnic societies and tribal communities having diverse knowledge on traditional handicrafts, food, beverages, traditional farming and ethnomedicinal uses of wild plants for curing various ailments and diseases (Vidyarthi et al., 2013). Census data from 2011 shows that 68.64 lakh people live in 59 towns, 20,690 villages, and other places around the state, with a density of 123 people per square kilometre (Census, 2011). Villages in the state is scattered in small hamlets in the remote area. Major scheduled tribes (ST)' population in the district is concentrated in tribal districts of Kinnaur, Kullu, Lahaul and Spiti, and Pangi valley of Chamba district. The current study was carried out in the Lahaul and Spiti district of Himachal Pradesh. Study is concentrated to the indigenous knowledge system of the tribal community of the district in their day-to-day life, such as ethnomedicinal uses of natural resources, agricultural practices, food and beverages made by the communities, land and soil management and water conservation practices, bioprospecting, etc.

3 Lahaul and Spiti District

In the Alpine arid region of Himachal Pradesh, Lahaul and Spiti constitute a crucial component of the Indian cold desert region in the north-western Himalayas. The district which spans an area of 13,835 km^2, is located between latitudes of 31°44′57″ and 32°59′57″ north and longitudes of 76°46′29″ and 78°41′34″ east. It is divided into two subdivisions: Lahaul, which has its headquarters in Keylong, and Spiti, which has its headquarters in Kaza. District is mainly known for its snow-peaked mountains, glaciers, ice caves, and snow fields (Kumar et al., 2018). District is thinly populated having a population density of just two people per km^2, just 31,564 people with sex ratio of 903 female every 1000 male, and more than 80% literacy rate (Census, 2011). At an elevation of 4270 m amsl, the village of Gete in this district is the highest village with electricity and road in the world. The *Lahaulis* are of Tibetan and Indo-Aryan decent, while the Bhotia tribes of Spiti are similar to the Tibetans, owing their proximity to the Tibet (Chand et al., 2019). The region has a significant presence of Buddhism, which is frequently infused with elements of Hinduism. The tribal communities existing in the district are *Swangla*, *Todpa*, *Gari*, *Tinnanba*, *Pitishag*, and *Bhot*. The polyandry system was prevailing in the valley in the past which is steadily declining in recent times. The law of primogeniture prevails in the district, where ancestral property is inherited by the eldest son in the family and younger generation aspiring their career in various government and private jobs. Economy of the district is mainly agrarian, where traditionally people used to grow buckwheat, barley, and kuth which later are replaced by the cash crops, like potato, pea, hops, and vegetables like lettuce, cabbage, cauliflower, iceberg, broccoli, etc. The good quality crop of peas and potato from the region is always in demand not only in India but in foreign country also. Among the schedule tribe, 77% of the population is communities like *Pitishag*, *Tinnanba*, *Bhot*, *Gari*, *Todpa*, and *Pitishag*, while 24% population is *Swangla* with marginal section of scheduled caste as *Chan/ Shipi*. Three major dialects are spoken in the valley as *Bunan*, *Tinnan*, and *Manchat* in Lahaul and Pattan valley, whereas in Spiti valley, Tibetan dialects are in use.

4 Materials and Methods

The representative communities of the district were surveyed from years 2015 to 2019 several times a year. All these communities were residing in remote villages of the district in different altitudes. For the collection of primary data or to unfold people's attitude, behaviour, and their opinion, the tool used was questionnaire (Valentine, 2005). The information was also collected through personal interview and group discussion. The informants were consisting of local knowledgeable persons, *Amchis,* village heads, elderly women, and community herb practitioners. Hindi language was used during the field survey to communicate with locals. Information about the plants' local name and parts that are used to treat the illness and its mode of utilization was gathered through structured questionnaire. Information about plants of herbal value was recorded as per the local names and was further confirmed by their field photographs. The species were identified with the aid of local and regional floras (Khullar, 1994, 2000, etc.). The secondary source of information was used to validate in each case to ascertain the facts about the local use of each plant species (Samant, 1998). The Plant List database, 2011, was used to verify the botanical name of the plants as well as their families.

TKS of the tribal communities in the region has been documented in aspects, such as the use of bioresources, bioprospecting, foods and beverages, architecture, agriculture, land and soil management practices, night-soil composting and water conservation practices, etc.

5 Results and Discussion

(a) *Bioresources:* A total of 98 medicinal plants from the Lahaul and Spiti district is mainly used by the five dominant communities *Swangla*, *Todpa*, *Gari*, *Tinnanba*, and *Pitishag* (Table 10.1). The district has a very sparse population as well as vegetation. The medicinal plants, along with each one's scientific name, common name, family, part used, and ailments/diseases, for which it is used, are also given in Table 10.1. In the time of modernization, these resource base and IKS about its utilization pattern are depleting fast from the region. There are a number of studies done by various researchers regarding the population assessment of the plant species, which are rare, endemic, high value, vulnerable, and threatened and need conservation measures. Most of the species of the region are commercially viable also due to which it has been over exploited in its natural habitat. The study also gives us an insight about its conservation aspect, reason, and threat identification for sustainable development of the region in future. Documentation of these resource base is very important for its conservation as well as the associated traditional knowledge of the communities living in the region. Other anthropogenic activities like increase in built-up area, road construction, unsustainable harvesting, grazing, etc. imposed a great threat on the resource (Fig. 10.1).

(b) *Bio-prospecting:* The region is also rich in value-added products of the bioresources such as oil, dye, gums, wax, resin, fibres, etc. These bioresources are essential for boosting the local tribal populations' economies and giving them a means of subsistence. These resources are used as various value-added products and marketed by locals in local or regional market. Total 14 plant resources are used by the local communities for its value addition and livelihood generation (Table 10.2). Throughout the area, there is a plant species called sea buckthorn (*Hippophae rhamnoides*) that is used for making tea (from its leaves and berries), juice, jam, and as raw material for medicines, cosmetics, etc. due to its various nutritional components. It is also called the golden shrub of cold desert. Bioprospecting is a process of examination, extraction, and screening of

Table 10.1 List of medicinal plants utilized by the locals in Lahaul and Spiti district

S. no.	Taxa	Family	Vernacular name	LF	Part used	Used for disease/ailment	Community
1.	*Berberis aristata*	Berberidaceae	Kerpameto	Sh	Rt	Eye disease	Swangla
2.	*Chenopodium album*	Chenopodiaceae	Bathu/Aem	H	WP	Constipation	Swangla
3.	*Cynoglossum glochidiatum*	Boraginaceae	Gava	H	Lf	Used for boil treatment	Swangla
4.	*Dactylorhiza hatagirea*	Orchidaceae	Wang-luck	H	Rt	Cut and wounds	Swangla
5.	*Equisetum arvense*	Equisetaceae	Rugosika	P	AP	Kidney stone	Swangla
6.	*Gentiana kurroo*	Gentianaceae	Pangyin	H	Rt	Cough	Swangla
7.	*Inula royleana*	Asteraceae	Mannu	H	Rt	Dermatitis	Swangla
8.	*Hyoscyamus niger*	Solanaceae	Datura	H	Sd	Tooth ache	Swangla
9.	*Juniperus communis*	Cupressaceae	Hauber	T	Tw	Joints pain	Swangla
10.	*Meconopsis aculeata*	Papaveraceae	Kahiharu	H	Lf	Swelling	Swangla
11.	*Pedicularis oederi*	Scrophulariaceae	–	H	Lf	Kidney pain	Swangla
12.	*Polygonum barbatum*	Polygonaceae	Neyoloe	H	Lf	Constipation	Swangla
13.	*Saussurea costus*	Asteraceae	Kuth	H	Rt	Dysentery	Swangla
14.	*Verbascum thapsus*	Scrophulariaceae	Gidar tambaku	H	Lf	Wounds and sores	Swangla
15.	*Hippophae rhamnoides*	Elaeagnaceae	Chharma	Sh	Fr	Headache	Swangla
16.	*Anemone rivularis*	Ranunculaceae	Talpal	Sh	AP	Asthma	Swangla
17.	*Convolvulus arvensis*	Solanaceae	Dhenchi	H	Fl	Kidney problem	Swangla
18.	*Epilobium latifolium*	Onagraceae	–	Sh	Lf.	Joint pain	Swangla
19.	*Galium* spp.	Rubiaceae	–	H	AP	Fever	Swangla
20.	*Juniperus* spp.	Cupressaceae	–	T	Br	Joint pain	Swangla
21.	*Artemisia maritima*	Asteraceae	Shoma/atong	H	Rt	Boils	Gari
22.	*Arnebia euchroma*	Boraginaceae	Khamed/ratanjot	H	Rt	Skin infections	Gari
23.	*Dactylorhiza hatagirea*	Orchidaceae	Wangpo-lakpa	H	Rh	Cut, wounds, and pyorrhoea	Gari
24.	*Galium aparine*	Rubiaceae	Zangchi	H	AP	Wound healing	Gari

(continued)

Table 10.1 (continued)

S. no.	Taxa	Family	Vernacular name	LF	Part used	Used for disease/ailment	Community
25.	*Hyssopus officinalis*	Lamiaceae	Tiyanku	H	Fl	Cough, bronchitis, and cold	Gari
26.	*Inula racemosa*	Asteraceae	Manu	H	Rt	Skin eruptions and blisters	Gari
27.	*Lindelofia longiflora*	Boraginaceae	Showara	H	Lf	Cut and wounds	Gari
28.	*Medicago falcata*	Fabaceae	Kathoama	H	Lf	Cut and wounds	Gari
29.	*Origanum vulgare*	Lamiaceae	Ban-tulsi	H	Lf	Indigestion	Gari
30.	*Rheum emodi*	Polygonaceae	Tukshu, Lichu	H	Rt	Wound healing, boils	Gari
31.	*Saussurea costus*	Asteraceae	Kuth	H	Rt	Pimples and leprosy	Gari
32.	*Taraxacum officinale*	Asteraceae	Sirsim	H	Fl	Boils, blisters	Gari
33.	*Verbascum thapsus*	Scrophulariaceae	Pungchunpuru	H	Lf	Boils	Gari
34.	*Juniperus communis*	Cupressaceae	Bethar	Sh	Br	Blisters	Gari
35.	*Ephedra gerardiana*	Ephedraceae	–	Sh	AP	Toothache	Gari
36.	*Taraxacum officinale*	Asteraceae	Poplar	T	Rt	Backache	Gari
37.	*Rosa webbiana*	Rosaceae	Shyabala	Sh	Fl	Headache	Gari
38.	*Cotoneaster* spp.	Rosaceae	–	Sh	Tw	Boils	Gari
39.	*Aquilegia fragrance*	Ranunculaceae	Ludud- Dorje	H	St	Joint pains	Pitishag
40.	*Arnebia euchroma*	Boraginaceae	Damok	H	Rt	Dandruff, hairfall	Pitishag
41.	*Artemisia brevifolia*	Asteraceae	Atong-Carpo	Sh	Lf	Cold and cough	Pitishag
42.	*Bistorta affinis*	Polygonaceae	Re-Tharam	H	Fl	Dysentery, vomiting	Pitishag
43.	*Capparis spinosa*	Capparaceae	Rutokpa	Sh	Fr	Anaemia	Pitishag
44.	*Carum carvi*	Apiaceae	Mao	H	Sd	Cold	Pitishag
45.	*Codonopsis clematidea*	Campanulaceae	Ludud-dorjw-nakpo	H	AP	Joint pain	Pitishag
46.	*Corydalis govaniana*	Papaveraceae	Tongsil	H	Fl	Joint pain	Pitishag
47.	*Cousinia thomsonii*	Asteraceae	Tultang	H	St	Body aches, muscular pains	Pitishag
48.	*Cuscuta reflexa*	Convolvulaceae	Dita-san-gin	H	St	Nose bleeding	Pitishag
49.	*Ephedra gerardiana*	Ephedraceae	Chhedum	Sh	AP	Liver problems	Pitishag
50.	*Gentianella paludosa*	Gentianaceae	Tikta	H	AP	Jaundice, digestive problems	Pitishag

51.	*Gynura angulosa*	Asteraceae	Aachak	H	AP	Blood circulation	Pitishag
52.	*Hippophae rhamnoides*	Elaeagnaceae	Charmma/Tarkuk	Sh	Fr	Constipation, purify blood	Pitishag
53.	*Melilotus officinalis*	Leguminosae	Pusukhang	H	Lf	Fever, boils, chest infection	Pitishag
54.	*Oxyria digyna*	Polygonaceae	Chumcha	H	AP	Gastritis, digestive disorder	Pitishag
55.	*Plantago major*	Plantaginaceae	Lung-ke-naram	H	AP	Diarrhoea/dysentery	Pitishag
56.	*Rheum austral*	Polygonaceae	Thu-la-lachu	H	Rt	Joint pain	Pitishag
57.	*Rhodiola wallichiana*	Crassulaceae	Solo-Marpo	H	Rt	Asthma	Pitishag
58.	*Rosa webbiana*	Rosaceae	Saveemantok	Sh	Fl	Headache, earache	Pitishag
59.	*Taraxacum officinale*	Asteraceae	Khur-Khang	H	AP	Stomach ache	Pitishag
60.	*Mentha longifolia*	Lamiaceae	Taling	H	AP	Medicinal antiseptic, carminative, digestive	Pitishag
61.	*Fagopyrum dibotrys*	Polygonaceae	–	H	AP	Stomach ache	Pitishag
62.	*Chenopodium* spp.	Chenopodiaceae	–	H	AP	Stomach ache	Pitishag
63.	*Geranium* spp.	Geraniaceae	–	H	AP	Diarrhoea	Pitishag
64.	*Thymus linearis*	Lamiaceae	–	H	WP	Indigestion	Pitishag
65.	*Trigonella* spp.	Papilionaceae	–	H	AP	Fever	Pitishag
66.	*Rumex* spp.	Polygonaceae	–	H	AP	Gastric problem	Pitishag
67.	*Aster* spp.	Asteraceae	Jachipa	H	AP	Cold	Pitishag
68.	*Hyoscyamus niger*	Solanaceae	–	H	Sd	Toothache	Pitishag
69.	*Codonopsis* spp.	Campanulaceae	–	H	Rt	Joint pain	Pitishag
70.	*Corydalis* spp.	Fumariaceae	–	Sh	Fl	Blood purification	Pitishag
71.	Nepeta spp.	Lamiaceae	Sheeng	Sh	Lf	Fever	Pitishag
72.	*Primula reptans*	Primulaceae	–	H	Fl	Cough	Pitishag
73.	*Bergenia* spp.	Saxifragaceae	–	H	Rt	Body pain	Pitishag
74.	*Geranium pratense*	Geraniaceae	–	H	Lf	Liver problem	Pitishag
75.	*Cedrus deodara*	Pinaceae	Devdar	T	Wd	Skin irritation	Tinnanba
76.	*Dactylorhiza hatagirea*	Orchidaceae	Salam-Panja, Thebrang	H	Rt	Cut and wounds	Tinnanba

(continued)

Table 10.1 (continued)

S. no.	Taxa	Family	Vernacular name	LF	Part used	Used for disease/ailment	Community
77.	*Equisetum arvense*	Equisetaceae	–	P	AP	Stones of the kidney	Tinnanba
78.	*Gentiana kurroo*	Gentianaceae	–	H	Rt	Cough	Tinnanba
79.	*Inula royleana*	Asteraceae	Mannu-kuth	H	Rt	Dermatitis	Tinnanba
80.	*Hyoscyamus niger*	Solanaceae	–	H	Lf	Cuts and wounds	Tinnanba
81.	*Juniperus communis*	Cupressaceae	Dhoop	T	Tw	Joints pain; especially knee pain	Tinnanba
82.	*Saussurea costus*	Asteraceae	Kuth	H	Rt	Leprosy, locally known as *Kod*	Tinnanba
83.	*Morchella esculenta*	Helvellaceae	Chunchuru	H	AP	Fever and cold	Tinnanba
84.	*Tagetes minuta*	Asteraceae	–	H	Fl	Cut and wounds	Tinnanba
85.	*Capparis spinosa*	Capparaceae	Safeda	Sh	Fr	Anaemia	Tinnanba
86.	*Plantago major*	Plantaginaceae	Lung-ke-naram	H	AP	Dysentery	Tinnanba
87.	*Melilotus officinalis*	Leguminous	–	H	Lf	Boils	Tinnanba
88.	*Juniperus* spp.	Cupressaceae	–	T	Tw	Body ache	Tinnanba
89.	*Origanum vulgare*	Lamiaceae	Ban-tulsi	H	Lf	Indigestion	Todpa
90.	*Plantago major*	Plantaginaceae	–	H	Lf	Feet inflammation	Todpa
91.	*Rheum emodi*	Polygonaceae	Lichu-ga	H	Rt	Inflammation	Todpa
92.	*R.*spp.	Polygonaceae	–	H	Rt	Cuts and wounds	Todpa
93.	*Taraxacum officinale*	Asteraceae	Koadi	H	Fl	Boils	Todpa
94.	*Verbascum thapsus*	Scrophulariaceae	–	H	Lf	Joint pain	Todpa
95.	*Gentiana* spp.	Gentianaceae	Beli	H	Rt	Cough	Todpa
96.	*Artemisia* spp.	Asteraceae	–	T	Lf	Cough	Todpa
97.	*Thymus linearis*	Lamiaceae	–	H	AP	Indigestion	Todpa
98.	*Jurinea* spp.	Asteraceae	–	H	Rt	Skin eruption	Todpa

Abbreviations used: *LF* life form, *H* herb, *Sh* shrub, *T* tree, *F* fungus, *P* pteridophyte, *Rh* rhizome, *Lf* leaf, *St* stem, *Rt* root, *Br* bark, *Sd* seed, *Lt* latex, *Tw* twigs, *LF* life form, *WP* whole plant, *AP* aerial part, *FR* frond, *Fb* fruiting body

Fig. 10.1 (**a**) Medicinal plants of (L & S) *Dactylorhiza hatagirea,* (**b**) *Equisetum arvense,* and (**c**) Hyoscyamus niger

the natural resources of social and commercial value (Sharma & Singh, 2012). Moreover, harnessing these bioresources and using the concept of value addition makes them into high-cost products, which lead to both short- and long-term subsistence of value chain of marketing and good economic profits (Maikhuri et al., 1998). It plays a significant role in the conservation of biodiversity, which increases the bio-prospecting value of resources (Rausser & Small, 2000). According to Kumar (2004), bio-prospecting of the plant resources can be helpful for the conservation of the resources while providing livelihood to local communities. Traditional knowledge systems and biological resources of any particular region have been used by the agencies for the development of different products in pharmaceutical, cosmetic, food, beverages, and other industries. This value addition processed from raw material to an end product offers revenue generation to all the stakeholders involved in the process with the overall economic development of the country (Natarajan & Govind, 2006). Increase in demand of these resources as a raw material has also increased the unsustainable harvesting practices in its natural habitat, which has an adverse impact on the environment. In order to maintain a sustainable development of the region, collector groups must be aware on the sustainable collection of these resources from its natural habitat.

(c) *Traditional foods and beverages:* Traditional food and beverages play a vital role in the health security and well-being of the indigenous people of IHR. At the same time, it provides livelihood to the local people. Table 10.3 shows around 20 number of foods and beverages prepared and consumed by the tribal people of the district. The table also describes its raw material used, inoculum used, and communities who make it and its mode of preparation. Various researchers also identified and analysed the microbial characteristics of these traditional foods and beverages (Thakur et al., 2004; Savitri & Bhalla, 2012). These traditional foods from the district are largely linked with the availability of raw material,

Table 10.2 List of plants used for bio-processing (commercially exploited for small-scale industries) in the study areas

S. no.	Botanical name	Local name	Family	Location	Part used	Use
1.	*Juniperus* spp.	Bethar	Cupressaceae	L & S	Tw	For making dhoop
2.	*Arnebia euchroma*	Lal-jadhi, Ratanjot	Boraginaceae	L & S	Rt	Used as colouring agents in pickle formation
3.	*Humulus lupulus*	Hops	Cannabaceae	L & S	Fl	Used for making beverages
4.	*Carum carvi*	Kala-jeera	Apiaceae	L & S	Sd	Used as condiment
5.	*Artemisia* spp.	Jhau	Asteraceae	L & S	AP	Essential oil is extracted from aerial parts
6.	*Hippophae rhamnoides*	Sea buckthorn	Elaeagnaceae	L & S	Fr	Fruit berries are used to prepare jam and juice leaves are used for green tea
7.	*Cedrus deodara*	Dyar	Pinaceae	L & S	Wd	Essential oil is extracted from wood
8.	*Rhododendron anthopogon*	–	Ericaceae	L & S	Lf	Used for making herbal tea
9.	*Malus pumila*	Apple	Rosaceae	L & S	Fr	Used to prepare jam and juice
10.	*Saussurea costus*	Kuth	Asteraceae	L & S	Rt	Roots are used to generate aroma and also used for cold and cough
11.	*Inula racemosa*	Manu	Asteraceae	L & S	Rt	Dried roots are used to cure stomach troubles
12.	*Thymus linearis*	Ban-ajwain	Lamiaceae	L & S	AP	As a flavouring agent
13.	*Origanum vulgare*	Ban-tulsi	Lamiaceae	L & S	AP	As a spice and also for making chutney
14.	*Dactylorhiza hatagirea*	Salam-panja	Orchidaceae	L & S	Tb	Tubers are edible

Abbreviation used: *Tw* twigs, *Rt* root, *Lf* leaf, *Sd* seed, *AP* aerial part, *Fl* flower, *Fr* fruit, *Wd* wood, *Rh* rhizome, *Tb* tuber, *FR* frond, *FB* fruiting body

environmental conditions, and different ethnic or tribal groups. This also shows how each community has its own culture of making traditional dishes in different season and ceremony (Fig. 10.2).

(d) *Traditional architecture:* The architecture of traditional house in the cold desert area is influenced by the harsh climatic condition of the region and as per the raw material available for construction. These traditional houses are made of stone, wood, and mud consisting mainly of two or three stories with livestock at ground

Table 10.3 List of traditional food and beverages of Lahaul and Spiti district

S. no.	Product	Raw material used	Source of inoculums	Community	Methodology
Fermented food					
1.	*Lwad*	Buckwheat flour and wheat flour	*Malera/ khameer*	Gari, Todpa, Tinnanba, Pitishag, Swangla	A semi-solid paste is created by combining wheat flour and buckwheat flour (4:1). In order to ferment, khameer is added. The mixture is fermented for 8–10 h at 25–30 °C. with the presence of gas, its quantity doubles. Flattened cakes are created from it
2.	*Jhanchang*	Barley grains	*Phab*	Gari, Todpa, Tinnanba, Pitishag, Swangla	Rinsed grains of barley are mashed into a paste and then cooked in an uncovered pot for 2 h. The mixture is chilled before *phab* is added. Mixed and covered woollen cloth. For 2–3 days, it is fermented at 25–30 °C temperature. It is also given shape by squeezing them in the hands
3.	*Bhatooru*	Wheat flour	*Malera/ khameer*	Gari, Todpa, Tinnanba, Pitishag, Swangla	Inoculums are combined with wheat flour and water to create dough and then fermented overnight or for 7–8 h at 25–30 °C. Small chapattis are made, which is around 1 centimetre thick, and then deep-fry in oil
4.	*Babru*	Wheat flour	*Malera*	Swangla, Tinnanba, Pitishag	Rice flour and water are combined in a 2:1 ratio to create a semi-solid paste. For flavour, salt or spices are used and kept for 3–4 h at 25–30 °C for fermentation. Flat pancakes fried in oil
5.	*Marchu*	Wheat flour	*Malera*	Gari, Todpa, Tinnanba, Pitishag, Swangla	Prepared with wheat flour by fermentation using *Malera*. It is flattened in a wood plate having specific design and deep-fried in oil. Specially prepared during religious and marriage ceremonies

(continued)

Table 10.3 (continued)

S. no.	Product	Raw material used	Source of inoculums	Community	Methodology
Fermented beverages					
6.	*Chhang*	Wheat	*Phab* or *dhaeli*	Gari, Todpa	After being cleaned, wheat grains are steeped in water for 8–10 h. Cooked grains are dusted with cooled, powdered dhaeli after being cooked for 1–2 h in an open pot (for about 2–3 kg wheat, 1–2 granules or cakes are added). Unsorted and stacked on bags kept for 2–3 weeks at 25–35 °C for fermentation. To maintain its temperature, sacks are wrapped in blankets and the finished product is known as *Chhang*
7.	*Lugri*	Rice	*Phab*	Gari, Todpa	After being thoroughly cleaned, rice is then soaked for 20–24 h in a drum of water (1: 2). Grain and 'Phab' (traditional inoculums) are combined. The mixture is maintained and heated (15–30 days, 25–37 °C) for fermentation. A cloth is used to squeeze the liquid portion from the solid portion. It is offered to guests at traditional wedding ceremonies
8.	*Sara/Arak*	Wheat	*Phab*	Gari, Todpa, Tinnanba, Pitishag, Swangla	The grains of wheat are cleaned before being steeped in water (1:1) for 8–10 h. Grains are cooked in an open vessel for 1–2 h. After cooling, powdered *dhaeli* is dusted on grains (for about 2–3 kg wheat, 1–2 granules or cakes are added). Mixed and piled on bags and kept for 2–3 days at 25–35 °C for fermentation. Fermented grains are further distilled through a traditional vessel to get the end product arak or sara

(continued)

Table 10.3 (continued)

S. no.	Product	Raw material used	Source of inoculums	Community	Methodology
Traditional food/beverages					
9.	*Aktori*	Buckwheat	–	Gari, Todpa, Tinnanba, Pitishag	Buckwheat flour is used to prepare thick rotis and is one of the staple foods. Not practiced by the communities nowadays
10.	*Chhangpa*	Roasted barley flour, lassi, ghee	–	Tinnanba, Pitishag, Swangla	Solid dough specially made for wedding ceremonies by combining sattu and lassi
11.	*Churpy* (dry cheese)	Dried buttermilk	–	Gari, Todpa, Tinnanba, Pitishag, Swangla	Buttermilk is boiled, water is removed, and the solids are dried to create *churpa*. It is used in thukpa. *Churpy* enhances the taste and flavour of their ethnic food and therefore added to all types of vegetables as well as meat preparations
12.	*Dosha*	Wheat flour	–	Gari, Todpa, Tinnanba, Pitishag, Swangla	*Dosha* is prepared by deep-frying the spirals made by wheat flour dough; end product is used as snack food
13.	*Marpinni*	Sattu, ghee	–	Gari, Todpa, Tinnanba, Pitishag, Swangla	A sweet dish called *Marpinni* is produced by combining roasted barley flour, ghee, and sugar
14.	*Nudu*	Milk, wheat flour	–	Tinnanba, Pitishag, Swangla	*Nudu* is prepared with the help of wheat flour cooked in milk and is prepared by slow cooking and mixing and served with ghee
15.	*Geri*	Mutton	–	Gari, Todpa, Tinnanba, Pitishag, Swangla	Geri is prepared by boiling large pieces of mutton and is one of the foods eaten in winters
16.	*Tsati*	Mutton pieces	–	Gari, Todpa, Tinnanba, Pitishag, Swangla	Mutton pieces are boiled in water adding small amount of sattu and salt
17.	*Juma*	Sheep intestine	–	Gari, Todpa, Tinnanba, Pitishag, Swangla	Sheep intestine is filled with the wheat flour mixed with spices and steamed

(continued)

Table 10.3 (continued)

S. no.	Product	Raw material used	Source of inoculums	Community	Methodology
18.	*Tsaku cha*	*Taxus* bark	–	Gari, Todpa, Tinnanba, Pitishag, Swangla	The bark of the *Taxus* is boiled in water until it turns pink in colour. A little bit of cow milk, butter, and salt are added and mixed in a traditional container called dongmo for better flavour
19.	*Manna*	Rice flour	*Malera*	Swangla, Tinnanba	Prepared with rice flour by adding water and then fermented with help of *Malera*, and finally rotis are prepared, eaten with ghee

Fig. 10.2 Traditional food and beverages prepared in L & S district (**a**, **b**). Preparation of mannu/manna (**c**), nudu (**d**, **e**), butter tea (*tsakucha*), (**f**) dried cheese (*churpy*), (**g**) bhaturu (**h**) tsati, (**i**) marchu and dosha

Fig. 10.3 Internal view of mud house showing ceiling centre of the room

floor and upper or first floor for humans. This type of houses is also useful for keeping the warmth during harsh winters. Windows and doors are usually small sized made of wood. These types of buildings are typically found in both sloped and hilly terrains. The winter for 4–5 months is harsh for the locals due to heavy snowfall with restricted movement to their houses. Nowadays with the increase in economic activities and mixing with outer world, the people living the region have shifted towards the construction of cemented modern houses. Research conducted by Aggarwal (1982), suggested the suitability of the traditional mud houses in the region than cemented ones (Fig. 10.3).

(e) *Agricultural practices:* Tribal communities living in the region are mainly dependent on agriculture for their day-to-day food security and at the same time livelihood opportunities (Maikhuri et al., 1998). Currently, various developmental activities have adversely affected the traditional economy and social and cultural practices of communities around the world (Shukla et al., 2018), for example, modern agriculture led to soil erosion, loss of soil fertility, chemical contamination of soil, etc. To reduce invulnerability of these mountain communities, indigenous knowledge system is recognised as a key component to enhance ecological sustainability. This also implies to cultivation of traditional crops (Immanuel et al., 2010). The tribal people of the district pursue agriculture as their predominant occupation, and they cultivate traditional agro crops, which are well adapted in the variable climatic conditions. Recently, traditional crops, like barley and black pea, have been increasingly replaced by green pea, vegetables, and fruit crops. The reason behind this replacement is mainly good economic returns. Similar results have been authenticated by Sharma and Chauhan (2013), who found that initially cultivation of green pea has promoted economic viability of the region. Traditional food grain crop has almost lost its relevance. The farmers shifted from traditional farming to cash crop cultivation. Cultivation of exotic vegetables, fruits, and flowers (oil yielding and ornamental flowers), lettuce, Chinese cabbage, European carrot, hops, apple, grapes, lilium, etc. are some of the exotic plant species which are grown by the farmers of the district (Rana et al., 2008; Rana & Samant, 2011).

(f) *Land and Soil Management Practices:* Land is the most fundamental natural resource on which agricultural production depends (Sigunga & Wandahwa, 2011), and the soil is the ultimate source of mineral nutrients to plants (White et al., 2012). Presently, there is an exponential growth of population in India, which has brought higher demands on agriculture to produce more food (Kundu, 2015). But, due to unsustainable agricultural practices such as the cultivation of hybrid crops, injudicious use of chemical fertilizers and pesticides has led to harm agriculture land, increased soil erosion, and decreased soil fertility and productivity (Komatsuzaki & Ohta, 2007). Generally, there is a degradation of land and soil resources. To achieve future food security, the management of land and soil in a sustainable manner is need of the hour, which can only be done by the traditional knowledge system of the communities living in mountain ecosystem, since the tribal communities of the region have been using sustainable land and soil management practices, which include proper nutrient management and appropriate soil conservation for a long time. Therefore, documentation of practices, such as terrace farming, vegetative barriers, biopesticides, traditional fertilizers, green manuring, mulching, crop rotation, and the intercropping system, is needed. It is the primitive styles of farming that involves the intensive use of indigenous knowledge of the farmers. This sustains the environment due to its eco-friendly nature. This helps in reducing soil erosion and safeguarding the hilly region from landslides as well as sustains ecosystem services provided by the soil ecosystem. In Lahaul and Spiti district, tribal communities practice soil conservation and nutrient management strategies by adopting terrace farming, vegetative barriers, traditional fertilizers, etc. Construction of stone barriers is also an age-old practice for conserving both soil and water. For increasing soil fertility in the agriculture field, local people mostly use cattle manure, night-soil compost, and leaf litters; besides these, they also practice burning of crop residues, mulching, mixed cropping, etc. For example, the cultivation of potato (*Solanum tuberosum*) in the terrace farming system in Lahaul and Spiti district is very popular.

(g) *Night-Soil Composting:* Night-soil or human waste has been one of the major agricultural resources of the cold desert area of Lahaul and Spiti since many decades. Low soil fertility, landslide and erosions, and high nutrient leaching are some of the major hurdles for sustainable agriculture in Lahaul and Spiti district. Also, due to high snowfall and low-temperature conditions, it is not possible to produce an adequate amount of farmyard manure. A new impetus for the restoration of this ancient practice has come from the rising demand for organic manure in recent years. Usually, laypeople do not anticipate the serious ecological effects of chemical fertilizers, which to some extent disrupt the environment. For most of the small communities that make up the district, agriculture is their main source of income.

The harsh and adverse climatic conditions of this area make this a challenging prospect (Singh, 1997; Kuniyal et al., 2004; Oinam et al., 2005). With these current situations, local communities have been relied on preparing night-soil compost at the same time of its management. In the freezing desert environment, where snow deposits and avalanches constantly have negative effects on the topsoil's and its nutrients, it may be extremely useful to promote recycling human excrement as fertilizer in the fields. The high nutrient needs of the region's cash crops, including the potato (*Solanum tuberosum*), pea (*Pisum sativum*), and hop (*Humulus lupulus*), are also met by this bio-compost.

The demand of manure in the area is achieved by recycling human waste into night-soil or finally used as farmyard manure (FYM). For this reason, people constructed conventional toilets on the first floor of the house attached to their restrooms. The night-soil automatically descends to the ground floor through a rectangular hole ($12'' \times 6''$) in the toilet floor, where the composting is allowed to take place. The use of water in these toilets is absolutely prohibited to avoid extra moisture content during composting. The use of water not only slows down the process of decomposition but also produces an unpleasant odour. The materials locally known as *fot* are used by the people to hide the faces after defecating. Two basics purposes: first off, it enriches the compost rich in nutrients, and secondly, it prevents bad odour and keeps the flies away. This is mainly a mixture of ash, leave litter, and cow dung. This composting process takes 6–8 months to complete. NSC from compost chamber is typically emptied/removed twice a year, first in March/April and second in October/November. However, the removal of the compost is highly dependent on the appropriate decomposition of the materials and the workforce's availability at the time. For the purpose of removing the compost, a particular door is provided in the composting room. In the past, women in the family were in charge of managing the night-soil composting process. The wealthy homes started hiring outside labour to perform the composting process, though, as the village economy improved as a result of the development of cash crops and a few medicinal herbs. The NSC is carried to the fields and put in a number of heaps after being empty from the composting room. It is dispersed around the fields soon after the snow melts and before the start of the growing season. This task was typically carried out at night because of the social stigma, especially during a full moon. The night-soil compost is delivered to the fields using a *kilta* (a carrier). In order to ensure that the manure is distributed evenly when it is dispersed across the fields, the *kiltas* are unloaded in the fields while maintaining a consistent gap of 4–5 metres between each heap. However, because each farmer has a distinct capability for producing night-soil compost based on their landholdings, the per unit area application of NSC in the fields varies for each farmer. The amount of NSC used, the number of people working on the project, and the distance between the fields and the composting chambers all affect how long the procedure will take in total. The NSC heaps stay in the fields for around 4 or 5 months. The farmers appropriately distribute the NSC in each field once the snow has melted. For the majority of their agricultural operations, the communities often share their available labour force. However, when it comes to the management of composted night soil, this is never the case.

Villagers prepare NSC by mixing excreta with ash, grass, leaves, and animal waste from sheep, cattle, and goats, among other animals. Its population gradually began to decline over the past 20 years, mostly as a result of the traditional toilets unclean conditions, a labour shortage, and advancements in education, social standing, and the accessibility of chemical fertilizers with financial subsidies. But in some parts, this old method is still used in the present period for a variety of reasons. These reasons include its applicability as an agricultural resource or also ineffectiveness of modern toilets during harsh winter months. The use of this night-soil as a composting technique and fertilizer for maintaining soil health as well as an increase in yield is also documented by Oinam et al. (2008) (Fig. 10.4).

Fig. 10.4 (**a**) Outside view of traditional toilet in Lahaul and Spiti district. (**b**) Inner view. (**c**) Night-soil heaps in agricultural fields

Table 10.4 Traditional water bodies in Lahaul & Spiti

S. no.	Local name	Water right holder	Nature	Catchment recharge source	Artificial recharge structures	Current status
For drinking purpose						
1.	Spring	Native community	Perennial	Snowfall recharge	Storage tanks	Well managed
For livestock, irrigation, and other domestic use						
2.	Naa (Nallah)	Native community	Seasonal	Snowfall recharge	Storage ponds	Well managed
3.	Phil	Native community	Seasonal	Snowfall recharge	Storage ponds	Well managed
4.	Shrik	Native community	Seasonal	Snowfall recharge	Storage ponds	Well managed
5.	Phizur	Native community	Seasonal	Snowfall recharge	Storage ponds	Well managed

(h) *Water Conservation Practices:* The local inhabitants of the Indian Himalayan Region are surrounded by several natural water sources, which help them to fulfil daily works, such as cooking, washing, agriculture, forestry, livestock, etc. (Negi et al., 2017). In addition, as population has grown, it has become more dependent on these limited water resources, which has led in the depletion and deterioration of such resources (Seth et al., 2016). Recently, it has been observed that traditional water sources are deteriorated, poorly maintained, and disused due to modern facilities, such as borewells, hand pumps, and electric pumps (Singh et al., 2010). This has led the transformation of many perennial water sources to become seasonal, and some of them even have dried up. But the local communities have been very much strengthened with traditional knowledge systems related to traditional water sources. Numerous traditional water sources, such as *khuls*, *baories*, *magru*, *kua (well)*, *khads*, *nahllas*, etc., by the local communities has played a significant role in supporting daily life activities, conserves soil by increasing groundwater table, increasing biodiversity and overall sustaining various ecosystem services (Sharma & Kanwar, 2009). Based on the interpretation of information given by the local communities, a list of water storage practices, their usage, and conservation approaches by the local communities in the district has been prepared as shown in Table 10.4. The Lahaul and Spiti district is the cold desert region of Himachal Pradesh, receives scanty rainfall and abundant snowfall, and is under cold conditions for more than 6 months in a year (Bajpai, 1987). The primary supply of water for drinking, irrigation, and other household uses is freshwater, which is temporarily stored at higher altitudes by the perennial ice forms known as glaciers (Mark et al., 2015). Based on the interaction with the local people of the district, it was revealed that they have created customary methods of water management, which involve channelling and transferring water in order to supply freshwater for drinking and irrigating agriculture fields. This has been authenticated by Padigala (2013) who found that traditional 'Farmers Managed Irrigation System' (FIMS) is being

Fig. 10.5 Water conservation and management practices in Lahaul & Spiti district

used by the local people of the district in form of open channels called *Kuhls*. These *Kuhls* are ancient traditional irrigation channels that collect and deliver glacier snowmelt water to agriculture fields. Also, it has been documented that the command area of individual *Kuhl* for the irrigation of landholding of the farmers, which includes agriculture, horticulture, forest land, and cultivated grasslands, is estimated. Based on this management system, water sharing is done and every field receives timely irrigation, often from early in the mornings. During the night, water is gathered in storage structures for the use of the next day. Hence, there is no disagreement or dispute regarding the operation, usage, and upkeep of *Kuhls* and irrigation channels. Moreover, Demul village in Spiti valley is a great example of community-based water storage and distribution system. They have developed a local water tank to store glacier melt water and then channelizes the water into small streams, i.e. *Kuhls* to their crop fields.

The perusal of data presented in Table 10.4 shows the local names of traditional water storage structures, holder rights, nature of the flow, catchment recharge, name of the artificial recharge structures, and the current status. The water structures of the region recharge from different sources, like glaciers, snowfall, rainfall, rivers, etc. Water sources of the L & S district are both perennial and seasonal. They act as a major source of drinking, irrigation, as well as other domestic uses (Fig. 10.5).

6 Conclusion

The tribal district of Lahaul & Spiti has a unique traditional knowledge system for their culture, traditional, food, bioresource use, land and soil management, water resource conservation, etc. This knowledge system is depleting fast from the region mainly due to modernization and lack of interest among younger generation. Recent opening of Atal AtuTunnel has provided the all-weather connectivity to the valley, which remained cut off for nearly 6 months in a year earlier. This has led the tourist to visit the valley throughout the year with no planned tourist activity in the region. As a result, the unique diversity, culture, traditional, and ecology of the region is under threat. Therefore, there is a dire need of documentation of these knowledge systems, so that it can be used with its scientific validation by the future generation for regional sustainability in its whole.

Acknowledgement Authors are thankful to the Director, G. B. Pant National Institute of Himalayan Environment, Kosi-Katarmal, Almora, Uttarakhand, India, for providing necessary facilities at Himachal Regional Centre. We are really grateful to the Department of Science and Technology, New Delhi, for providing financial support for the NMSHE TF 5 project. Authors are thanked to native community and elderly people for sharing their knowledge.

References

Aggarwal, A. (1982). Research: Mud as traditional building material. In B. B. Taylor (Ed.), *The changing rural habitat; Volume I: Case studies*. Concept Media/The Aga Khan Award for Architecture.

Bajpai, S. C. (1987). *Lahaul-Spiti: A forbidden land in the Himalayas* (p. 186). Indus Publishing Company.

Census. (2011). *Primary census abstracts, registrar general of India*. Ministry of Home Affairs, Government of India. Available at http://www.censusindia.gov.in

Chand, K., Shashni, S., Rathore, S., Sood, S., Lata, R., & Sundriyal, R. C. (2019). Promotion of cultural and natural heritage of the tribal district of Lahaul & Spiti as a potential tourism destination in Northwest Himalaya. In A. Chauhan, C. K. Bharti, & P. K. Bharti (Eds.), *Rural development and techno innovation* (pp. 8–27). Discovery Publishing House.

Das, S., & Mishra, A. J. (2022). Dynamics of indigenous community's food and culture in the time of climate change in the Himalayan Regions. *Journal of Ethnic Foods, 9*(1). https://doi.org/10.1186/s42779-022-00118-7

Ellan, R. F., & Harris, H. (1997). *Indigenous environmental knowledge in scientific and developmental literature: A critical assessment*. University of Kent at Canterbury.

Green, E. C. (1996). Indigenous knowledge systems and health promotion in Mozambique. In H. Normann, I. Snyma, & M. Cohen (Eds.), *Indigenous knowledge and its use*. SRP, Exeter.

Immanuel, R. R., Imayavaramban, V., Elizabeth, L. L., Kannan, T., & Murugan, G. (2010). Traditional farming knowledge on agroecosystem conservation in northeast coastal Tamil Nadu. *Indian Journal of Traditional Knowledge, 9*(2), 366–374.

Khullar, S. P. (1994). *An illustrated fern flora of the West Himalaya* (Vol. I). International Book Distributors.

Khullar, S. P. (2000). *An illustrated fern flora of the West Himalaya* (Vol. II, p. 538). International Book Distributors.

Komatsuzaki, M., & Ohta, H. (2007). Soil management practices for sustainable agro-ecosystems. *Sustainable Science, 2*, 103–120.
Kumar, P. (2004). Valuation of medicinal plants for pharmaceutical uses. *Current Science, 86*(7), 930–937.
Kumar, S., Chand, G., & Sankhyan, P. (2013). Herbal folk remedies for curing various ailments in Lug Valley of Kullu district, Himachal Pradesh (N. W. Himalaya). *International Journal of Ayurvedic and Herbal Medicine, 3*(5), 1308–1314.
Kumar, P., Singh, V., & Singh, A. (2018). Seabuckthorn (Hippophae spp.) conserve plant diversity in the fragile mountain ecosystem of cold desert Himalaya. *Journal of Biodiversity, 9*, 53–68.
Kundu, P. (2015). Population explosions as a threat to food security in India. *International Journal of Novel Research in Humanity and Social Sciences, 2*(4), 55–62.
Kuniyal, J. C., Vishvakarma, S. C., & Singh, G. S. (2004). Changing crop biodiversity and resource use efficiency of traditional versus introduced crops in the cold desert of the north-western Indian Himalaya: A case of Lahaul valley. *Biodiversity and Conservation, 13*(7), 1271–1304.
Maikhuri, R. K., Nautiyal, S., Rao, K. S., & Saxena, K. G. (1998). Role of medicinal plants in traditional health care system: A case study form Nanda Devi Biosphere Reserve. *Current Science, 75*(2), 152–157.
Mark, B. G., Baraer, M., Fernandez, A., Immerzeel, W., More, R., & Weingartner, R. (2015). Glaciers as water resources. In C. Huggel, M. Carey, M. Clague, & A. Kaab (Eds.), *The high-mountain cryosphere environmental changes and human risks* (pp. 184–203). Cambridge University Press.
Mazzocchi, F. (2006). Western science and traditional knowledge. Despite their variations, different forms of knowledge can learn from each other. *EMBO Reports, 7*(5), 463–466. https://doi.org/10.1038/sj.embor.7400693
Nandy, S. N., Dhyani, P. P., & Samal, P. K. (2006). *Resource information database of the Indian Himalaya* (Envis Monograph 3, p. 123). ISSN 0972 – 1819.
Natarajan, & Govind, S. (2006). Indigenous agricultural practices among tribal women. *Indian Journal of Traditional Knowledge, 5*, 118–126.
Negi, V. S., Maikhuri, R. K., Pharswan, D., Thakur, S., & Dhyani, P. P. (2017). Climate change impact in the Western Himalayas: People's perception and adaptive strategies. *Journal of Mountain Science, 14*, 403–416.
Oinam, S. S., Rawat, Y. S., Khoiyangbam, R. S., Gajanand, K., Kuniyal, J. C., & Vishvakarma, S. C. (2005). Land use and land changes in Jahlma watershed of the Lahaul valley, cold desert region of the north-western Himalaya. *Indian Journal of Mountains Science, 2*, 129–136.
Oinam, S. S., Rawat, Y. S., Kuniyal, J. C., Vishvakarma, S. C., & Pandey, D. C. (2008). Thermal supplementing soil nutrients through bio composting of night soil in the north western Indian Himalaya. *Waste Management, 28*, 1008–1019.
Padigala, B. (2013). Traditional water management by mountain communities of Lahaul and Spiti, Himachal Pradesh. *Spandrel, 6*, 106–117.
Rana, M. S., & Samant, S. S. (2011). Diversity, Indigenous uses and conservation status of medicinal plants in Manali Wildlife Sanctuary, Northwestern Himalaya. *Indian Journal of Traditional Knowledge, 10*(3), 439–459.
Rana, M. S., Lal, M., Sharma, A., & Samant, S. S. (2008). Ecological evaluation of orchid diversity of Kullu District, Himachal Pradesh. *Journal of Orchid Society of India, 22*(1–2), 77–84.
Rausser, G. C., & Small, A. A. (2000). Valuing research leads: Bioprospecting and the conservation of genetic resources. *Journal of Political Economy, 108*(1). Available at SSRN https://ssrn.com/abstract=201051
Samal, P. K., Fernando, R., & Rawat, D. S. (2000). Influences of economy and culture in development among mountain tribes of Indian central Himalaya. *International Journal of Sustainable Development and World Ecology, 7*, 41–49.
Samant, S. S. (1998). Diversity, distribution and conservation of fodder resource of west Himalaya, India. In B. Misri (Ed.), *Proceedings of the Third Temperate Pasture and Fodder Network (TAPAFON)*, Pokhra, Nepal, 9–13 March, 1998, sponsored by F.A.O, Rome, pp. 109–128.

Savitri, & Bhalla, S. T. (2012). Characterization of *bhatooru*, a traditional fermented food of Himachal Pradesh: Microbiology and biochemical aspects. *3 Biotech, 3*, 247–254.

Schild, A. (2008). ICIMOD's position on climate change and mountain systems. *Mountain Research and Development, 28*, 328–331.

Seth, R., Mohan, M., Singh, P., Singh, R., Dobhal, R., Singh, K. P., & Gupta, S. (2016). Water quality evaluation of Himalayan Rivers of Kumaun region, Uttarakhand, India. *Applied Water Science, 6*, 137–147.

Sharma, H. R., & Chauhan, S. K. (2013). Agricultural transformation in Trans Himalayan Region of Himachal Pradesh: Cropping pattern, technology adoption and emerging challenges. *Agricultural Economics Research Review, 26*, 173–179.

Sharma, N., & Kanwar, P. (2009). Indigenous water conservation systems – A rich tradition of rural Himachal Pradesh. *Indian Journal of Traditional Knowledge, 8*(4), 510–513.

Sharma, N., & Singh, A. (2012). An insight into traditional foods of Northwestern areas of Himachal Pradesh. *Indian Journal of Traditional Knowledge, 11*(1), 58–65.

Shukla, R., Chakraborty, A., Sachdeva, K., & Joshi, P. K. (2018). Agriculture in the western Himalayas – An asset turning into a liability. *Development in Practice, 28*, 318–324.

Sigunga, D. O., & Wandahwa, P. J. (2011). Land and soil resources and their management for sustainable agriculture production in Kenya: Current position and future challenges. *Egerton Journal of Science and Technology, 11*, 66–86.

Singh, G. S. (1997). Socio-cultural evaluation of sacred groves for biodiversity conservation in north western Himalaya. *Journal of Hill Research, 10*(1), 43–50.

Singh, J. S. (2006). Sustainable development of the Indian Himalayan region: Linking ecological and economic concerns. *Current Science, 90*, 784–788.

Singh, H. P., Sharma, M. R., Hassan, Q., & Ahsan, N. (2010). Sustainability of traditional drinking water sources in Himachal Pradesh. *Nature, Environment and Pollution Technology, 9*(3), 587–592.

Thakur, N., Savitri, & Bhalla, T. C. (2004). Characteristics of some traditional fermented foods and beverages of Himachal Pradesh. *Indian Journal of Traditional Knowledge, 3*(3), 325–335.

Tharakan, J. (2017). Indigenous knowledge system for appropriate technology development. In P. Venkatesan (Ed.), *Indigenous people*. IntechOpen. https://doi.org/10.5772/intechopen.69889

The Plant List, Version 1.1. Published on the internet. http://www.theplantlist.org/. Accessed 17 Dec 2021.

Valentine, G. (2005). Tell me about using interviews as a research methodology. In R. A. Flowerdew & D. Martin (Eds.), *Methods in human geography guide for students doing research projects*. Harlow.

Vidyarthi, S., Samant, S. S., & Sharma, P. (2013). Traditional and indigenous uses of medicinal plants by local residents in Himachal Pradesh, North Western Himalaya, India. *International Journal of Biodiversity Science, Ecosystems Services and Management, 9*(3), 185–200.

White, P. J., Crawford, J. W., Alvarez, M. C. & Moreno, R. G. (2012). Soil management for sustainable agriculture. *Applied and Environmental Soil Science*. 2012: Article ID 850739, 3p.

Chapter 11
Impacts of Climate Change on Plants with Special Reference to the Himalayan Region

Zishan Ahmad Wani, Taskeen Fatima, Omkar Verma, Virbala Sharma, and Shreekar Pant

Abstract The Himalayan region is home to one of the world's most diverse and multifaceted mountain systems and is particularly prone to climate change. Climate change is a serious worry in the Himalaya because of its probable impacts on the economy, ecology, and environment, as well as downstream areas. Climate change has the potential to irrevocably affect distinct forest ecosystems and biodiversity, leading to the extinction of several species. Keeping the fact in mind that climate change affects various ecological as well as physiological parameters of the plant species, the present study was aimed to review the impacts of climate change on floral diversity in the Himalayan region. Some of the effects of climate change on the plants of Himalayan region include phenological changes, timberline shift, spread of invasive species, pests and diseases, habitat loss, and rise in the frequency and intensity of forest fires. Thus, climate change poses a serious threat to plant survival, which in turn poses a threat to the entire biosphere and human life. Urgent action plan is required to address this alarming situation. To accomplish this, a consolidation of greenhouse gases concentrations in the atmosphere, brisk global investments, and stationing of mitigation technology, along with research into new-fangled energy sources, is requisite. Appraisal of the biological invasions by exotic species, organization of decisive landscape linkages for flagship species, scrutiny of

Z. A. Wani · T. Fatima
Conservation Ecology Lab, Department of Botany, Baba Ghulam Shah Badshah University, Rajouri, Jammu and Kashmir, India

O. Verma
Geology Discipline Group, School of Sciences, Indira Gandhi National Open University (IGNOU), New Delhi, India

V. Sharma
Central University of Himachal Pradesh, Kangra, Himachal Pradesh, India

S. Pant (✉)
Coordinator, Centre for Biodiversity Studies, Baba Ghulam Shah Badshah University, Rajouri, Jammu and Kashmir, India

S. Sharma et al. (eds.), *Climate Change Adaptation, Risk Management and Sustainable Practices in the Himalaya*,
https://doi.org/10.1007/978-3-031-24659-3_11

population trends of flagship and vulnerable species, intensification of the efficiency and extent of protected area coverage, and fire management are all tactics and policies that should be implemented as soon as possible at micro-level. Further, regional cooperation is critical for informed decision-making, risk and vulnerability mapping, efficient biodiversity and conservation management, and a comprehensive method to designing climate change adaption tactics, not only in the Himalayan region but also throughout the globe.

Keywords Himalayan region · Climate change · Global warming · Phenological changes · Invasive species · Management strategies

1 Introduction

Climate change is a term used to describe long-term changes (~30 years) in the earth's global and regional climates. It encompasses both anthropogenic greenhouse gas emission-induced global warming and the resulting large-scale weather pattern shifts. The global temperature has been continuously rising for the past 130 years, having important implications for a range of climate-related problems (Verma, 2021). During 1900, the Earth's global surface temperature has risen by around 0.8 °C and by about 0.5 °C since the 1970s. By 2100, global temperature is expected to rise by up to 4 °C, causing changes in climatic patterns (Thuiller, 2007). Considerable increase in atmospheric concentrations of certain greenhouse gases, like carbon dioxide (CO_2) and methane (CH_4), has led to rise in atmospheric temperatures. CO_2 and CH_4 are being pumped into the atmosphere at an unparalleled rate as a result of the industrial revolution. There are chemical spills and gas flare-ups all over the place. The accumulation of greenhouse gases has the potential to alter the Earth's climate zones together with putting human health, safety, welfare, and ecosystems at risk. Organisms and ecosystems can alter and adapt in response to their surroundings. Organisms and ecosystems will evolve together and adapt to new habitats as climate change progresses. When climate change occurs abruptly, many species fail to adapt in a timely manner and therefore face serious impacts of climate change. Climate change has a several impacts on ecosystems, and figuring out what that means for biodiversity and how to minimize it is a challenging task in ecology (Thuiller, 2007). Climate change causes a shift in the Earth's temperature equilibrium, which has far-reaching implications for plants and ecosystems. Global warming changes the energy balances, due to increase in temperature of the Earth as a result of higher concentrations of greenhouse gases, and has direct impacts on climatically sensitive plants and ecosystems. Droughts, storms, heat waves, rising sea levels, melting glaciers, and warming oceans are all becoming more often and severe, wreaking havoc on vegetation. Climate change may exacerbate soil erosion, loss of organic matter, salinization, biodiversity loss, landslides, desertification, and flooding. As a result of climate change, changes in ambient carbon dioxide concentrations, higher temperature, and altered precipitation patterns may all have an impact on soil carbon storage. Despite the widespread effects of climate change on biodiversity and ecosystems, the mountainous regions appear to be the most vulnerable and sensitive habitats to changed climate settings and have undergone more than the average global warming over the preceding century (Beniston, 2003).

The Himalaya mountain range comprises of about 50 mountain peaks that are higher than 7200 m above mean sea level, including 10 of the 14 summits that are more than 8000 m above mean sea level. The hills rise sharply in the Himalayan area, resulting in complex ecology that includes everything from alluvial grasslands to subtropical broad leaf forests to alpine meadows above the tree line. Biogeographically, the Himalayan mountain range connects the Palearctic and Indo-Malayan worlds (Packert et al., 2012). The Himalaya, which stretches for over 2400 kms in length from northwest to southeast, encompasses a huge range of climatic zones ranging from subtropical to boreal (Rana et al., 2021). Out of the estimated 10,000 plant species, the Himalayan hotspot is home to 3160 indigenous species and 71 genera (Singh & Pusalkar, 2020). The plant families, Tetracentraceae, Hamamelidaceae, Circaesteraceae, Butomaceae, and Stachyuraceae, are endemic to Himalayan region (Singh & Pusalkar, 2020). Orchidaceae is the most prominent flowering plant family in the hotspot, with 750 species, and a large number of orchids have recently been identified from the hotspot, many of which are relatively young endemic species, implying that further exploration would most likely reveal a much higher degree of endemic plants. The Eastern Himalaya is home to several widely widespread plant *taxa*, including *Rhododendron*, *Primula*, and *Pedicularis*. In addition, the Himalayan region is home to 980 birds, 300 mammals, 175 reptiles, 105 amphibian species, and 270 fish species. Among the many large birds and mammals that call the hotspot home are vultures, tigers, elephants, rhinos, and wild water buffalo (www.cepf.net.). Nearly 52.7 million people live in the Himalayan region including the people from five countries: Bhutan, China, India, Nepal, and Pakistan (Apollo, 2017). The region is especially vulnerable to climate change and is currently undergoing fast changes (Negi et al., 2016). The Himalaya has been reported to have a three-fold faster rate of warming than the global average (Shrestha et al., 2012). This rapid warming in this mountain habitat has resulted in changes in the richness and distribution of plant communities in this mountain habitat (Shrestha et al., 2012). Climate change is seen as a threat to people's livelihoods, particularly in areas where people rely on natural resources that are particularly vulnerable to climatic changes. In the face of already diminishing natural resource availability, declining natural resource availability and the uncertainty created by climate variability are seen as posing a danger to agriculture and related sectors' viability (Rautela & Karki, 2015). However, despite rising curiosity in revealing the overall impacts of climate change in biodiversity hotspots, scientific data on the Himalaya is lacking (Manish et al., 2016). In terms of observations of climate change consequences on ecosystems and biodiversity, the Himalayan region is woefully lacking in data (IPCC, 2007; Tewari et al., 2017). Furthermore, systematic studies and actual data about the impacts of climate change on plant species in the Himalaya are severely lacking (Gautam et al., 2013). The current research is an attempt to review the impacts of changing climatic scenarios on plants species and communities in the Himalayan region.

2 Impacts of Climate Change on Himalayan Vegetation

Mountains are very diverse and important hotspots for biological diversity all over the world. The Himalayas have an equally imperative role in delivering ecosystem services that have allowed big populations and high levels of biodiversity to thrive for so long. Anthropogenic activities, like timber extraction, intense grazing, and agricultural expansion, coupled with climate change, are prompting loss of biodiversity and degradation of habitats (Dar & Khuroo, 2013). The Himalayan region is home to one of the world's most diverse and complex mountain systems, and it is especially vulnerable to climate change (Bandyopadhyay & Gyawali, 1994). According to some studies, it has been predicted that temperatures in the Eastern Himalaya will continue to climb, and alterations in precipitation patterns will grow more complex (Tse-ring et al., 2010). The extent of climate change in the Eastern Himalayan region is likely to be greater than what the IPCCs prediction for Asia. People in the area, however, have limited adapting potential due to economic, sociopolitical, and technological deficiencies; thus, the impacts felt will be more severe. The International Centre for Integrated Mountain Development (ICIMOD) has designed landscape and transect approach to bridge the medium- and long-term timescale gaps, since there is a requirement for dependable data generation (Tse-ring et al., 2010). Climate change affects forest ecosystem features and functions (Stenseth et al., 2002; Woodall et al., 2009). Thus, in order to adjust to the changing climatic scenarios, forest ecosystems require a considerable response period (Kumar & Chopra, 2009). Some of the major effects of climate change and global warming on plants and plant communities are depicted and described in Fig. 11.1.

3 Phenological Changes

Alterations of phenological events are the most significant indicators of climate change (Parmesan & Yohe, 2003; Root et al., 2003). The phenological shifts are due to rising temperatures, shorter and milder winters, and the start of summer earlier than usual (Rautela & Karki, 2015). Climate change leads to variations in pollinator and seed dispersion agents, either alone or in combination, thus affecting the timing of life cycle or phonological events in plants (Primack et al., 2004; Lieberman, 1982). Failure to find partners (Visser et al., 1998), or for a pollinator to find pollen and nectar, or for a flower to be pollinated due to change in temperature, can all be caused due to the alteration in flowering and fruiting, in advance or after the required season (Singh & Negi, 2016). Early flowering has been reported in many plant species *Arnebia benthamii* (10 days earlier than expected), *Meconopsis aculeata* (22–31 days), *Sinopodophyllum hexandrum* (17–21 days), *Delphinium denudatum* (40–48 days), *Dioscorea deltoidea* (19–21 days), and *Swertia cordata* (8–15 days) (Gaira & Dhar, 2020). Over the last 100 years, *Rhododendron arboreum* has also recorded 88–97 days of early flowering (Gaira & Dhar, 2020). Flowering time is

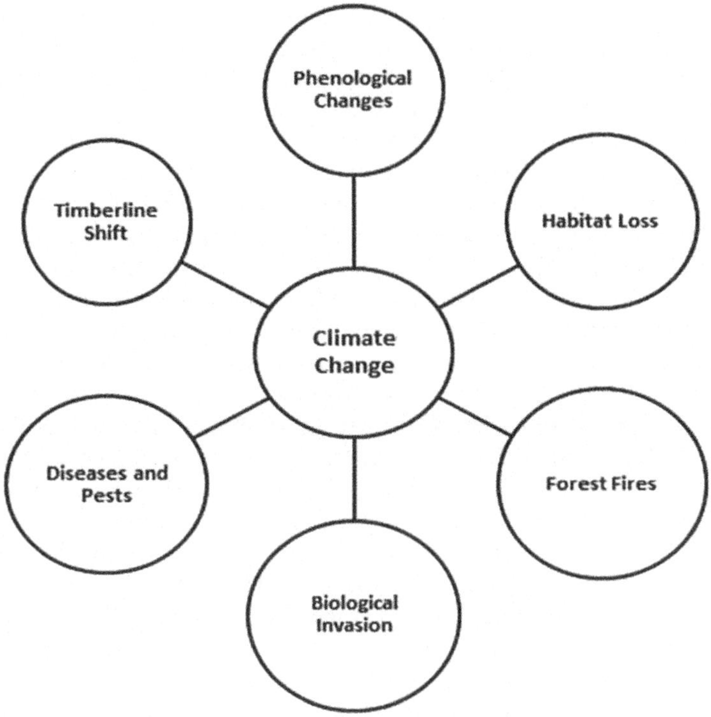

Fig. 11.1 Impacts of climate change on plants in Himalayan region

generally agreed to be postponed at higher elevations. However, at higher elevations, earlier flowering has also been noticed. This may be a result of winter warming, as warmer winters cause snow to melt earlier in the alpine zone, allowing for more favorable conditions for plant development. *Aconitum heterophyllum*, initiating growth in May and flowering in August (Vashistha et al., 2009) with a 20-day peak flowering time, has progressed (17–25 days) in the alpine and subalpine parts of the Indian Himalayan Region (IHR) (Nautiyal et al., 2009). Its flowering pattern is thought to be vulnerable to rising winter temperatures, which favors early blossoming (Gaira et al., 2011).

4 Timberline Shift

The "tree line" marks the ecotone between the forest and alpine zone. Tree line species are those that shape the tree line in an area, and tree line elevation (TLE) refers to the elevation of the tree line at a certain site (Singh et al., 2019). Individuals of at least 2 m height are distributed across the tree line (Singh & Rawal, 2017). It is expected that due to climate change, mountains regions will lose the cooler climatic zones at mountain peaks, and there will be upward shifting of tree line (Gottfried et al., 2012; Engler et al., 2011). In future, as the world warms, lower-altitude forests will migrate to higher-altitude woods (Verma, 2021). Some plant species will be able to adapt and shift, while others will be unable to do so and will go extinct. The Himalayan region has the highest tree line and the broadest tree line elevational range in the Northern Hemisphere, meaning that most sites are below the climatic tree line. Owing to the biotic and abiotic strains over a period of time, tree lines are severely miserable, particularly in the western Himalaya. Some of the most important tree line species include *Juniperus squamata*, *Betula utilis*, *Abies pindrow*, *Rhododendron anthopogon*, and *Rhododendron campanulatum*. Majority of these genera besides habitat degradation are now feeling the adverse effects of the rising temperatures. The junipers, over the ages, has suffered greatly by the vagaries of nature, such as snow and rockslides, avalanches, high wind velocity, and the extremes of temperature (Jishtu & Rawat, 2014). *Betula utilis* defines the natural tree line in Himalayas and is a potential indicator species of changing climatic scenarios (Maroof et al., 2019). In a warming environment, steep slopes, recently deglaciated surfaces (Donato, 2013; Macais-Fauria & Johnson, 2013), pastoral operations, firewood extraction, and combustion, minor forest product collection can all obstruct the upward advance of tree lines. TLE falls with increasing latitude (global trend, decreasing temperature) but increases from the NW to the SW of the Himalayan arc (regional pattern, increased precipitation), with evergreen species dominating broad-leaved trees and deciduous species declining (Singh et al., 2019). According to scientists, the oak woodlands have been overrun by pine trees (between 1000 and 1600 m) in the Mandakini Valley of northern India, particularly on south facing slopes. Many water sources, such as springs, have dried up as a result of losing oak trees and invading pines (FAO, 2012). However, Schickhoff et al. (2015) expected the possible niche of *Betula utilis* to migrate from lower to elevated altitudes and to extend into new habitats north of the Himalayan range due to considerable variations in bioclimatic variables, like mean diurnal range (monthly temperature, minimum temperature of coldest month, slope and annual precipitation, and mean temperature of wettest month). In the Sikkim Himalaya, Manish and Pandit (2019) found that under the current climate scenario (2000–4500 m), the majority of plant species prefer mid- and higher elevations, and almost all of these plant species show a northward shift from their possible habitats toward higher elevations in future climates (2050s, 2070s) under all pollution scenarios.

Climate change has been linked to shifts in species distribution (Permesan, 2006) and carbon accumulation in plants and soils (Shaver et al., 2000). The survival and management of NTFPs, including the important medicinal plants, needs an understanding of their future distribution in response to climate change. According to Wang et al. (2014), temperature played the most important role in deciding the best habitat for *Fritillaria cirrhosa* and *Lilium nepalense*. Wang et al. (2014) also discovered that isothermality and average annual precipitation influence the distribution of *Fritillaria cirrhosa* in the Chinese highlands, while temperature is more significant in the Nepal Himalaya. It is discovered that both species need a constant temperature of at least −10 °C. Precipitation defines the suitability of both the species; for example, the likelihood of *F. cirrhosa* increases as the wettest quarter precipitation increases from 250 to 500 mm, while the probability of *L. nepalense* decreases as the driest quarter precipitation increases from 30 mm. Although over-population, encroachment, and forest degradation have immediate effects on population of both the species due to their therapeutic potentials, there would be a climate-related reduction in habitat suitability, and by 2050, both these species would have shifted to the northwest, losing possible habitat in hilly and lower mountainous regions (Kumar et al., 2017).

5 Biological Invasions

Biological invasion has evolved as one of the most significant reasons of environmental degradation in most countries throughout the globe (Pimentel et al., 2001; Bellard et al., 2012), and its consequences are expected to aggravate under future climatic conditions (Early & Sax, 2014). Mountain habitats are one of the few that are unaffected by plant invasions. Climate change, on the other hand, is likely to enhance the threat of invasion (McDougall et al., 2011). Invasion of exotic species has been triggered by changing climatic scenarios, resulting in expansions in climatically favorable environments (Bellard et al., 2013). According to Thapa et al. (2018), bioclimatic variables, such as the minimum temperature of the coldest month, the mean temperature of the driest quarter, and the mean diurnal range, play a significant role in the invasion of 11 invasive species in Nepal, and these exotic species may invade natural ecosystems. Invasive species such as *Ageratum houstonianum* and *Mesosphaerum suaveolens*, which are less suited to the current environment, are projected to develop fast as a result of climate change (Shrestha et al., 2018). As a result, future climate change is anticipated to produce more appropriate regions for invasive alien species, facilitating their spread, which has already had detrimental implications for livelihoods, the economy, and biodiversity (Shrestha et al., 2018). Similarly, various alien invasive species have been found infiltrating forest, grassland, and wetland ecosystems across the Kashmir Himalaya (Masoodi et al., 2013). One of the key elements contributing to the growth of exotic species in the area has been recognized as climate change (Khuroo et al., 2007). Climate change would cause considerable alterations in the distribution and type of vegetation in the region, reducing the ecosystem services, which is currently supporting the livelihoods of the local populations. The vegetation in the alpine zones, which includes meadows and shrub areas and offers habitat for many rare and therapeutic plant species, is very vulnerable to even modest temperature changes. The region's grasslands and tropical deciduous forests will be severely injured, while the land currently covered in harsh desert, rock, and ice would be colonized by scrub, temperate evergreen broad-leaved forest, boreal evergreen forest, and mixed forest forms. The invasion of *Leucanthemum vulgare* and *Sambucus wightiana* in forest areas, subalpine and alpine meadows, and other natural environments is posing a major danger to these ecosystems (McDougall et al., 2011). A vast area of land currently covered in permanent snow and ice is predicted to vanish by the end of the century, posing a threat to agricultural productivity and biodiversity in the region (Rashid et al., 2015).

6 Habitat Loss

The potential distribution modelling of *Dactylorhiza hatagirea*, *Paris polyphylla*, and *Taxus* species in Nepal Himalaya indicates habitat degradation as a result of climate change. Furthermore, the suitable forest region would be at risk of habitat loss as a result of global climate change (Gajural et al., 2014; Ranjitkar et al., 2016). Potential habitats of *Hippophae salicifolia* are quickly declining; by 2050, the species will have lost 87.2% of its potential habitats, with a marginal gain of 30 km square by 2070 compared to 2050 (Dhyani et al., 2018). Possible distribution and habitat appropriateness of *H. salicifolia* in the central Himalayas have been discovered to have reached a critical point, putting it at risk in its natural habitats (Dhyani et al., 2018).

7 Diseases and Pests

Food must be inexpensive and accessible in the quantity and forms required. This is contingent on the processing, distribution, and trade infrastructure and mechanisms in place. All of these parameters may be impacted by changing climatic scenarios, and some are impacted both directly and indirectly as a result of pest and pathogen-mediated changes brought on by climate change. Agriculture and horticulture, as well as the centuries-old traditional food crops, have become disease-prone (Eriksson, 2006). The weather is the most essential influential element in the incidence, severity, and relative relevance of wheat *Fusarium* head blight and crown rot in cereals (Chakraborty et al., 2006). Plant pests and diseases may diminish the agricultural yield up to 82% of the achievable yield in cotton and more than 50% in other major crops (Oerke, 2006), and when combined with post-harvest spoilage and quality degradation, these losses become critical, especially in resource-deficient regions. Rice losses averaged 37.4% between 2001 and 2003, with 15.1% attributable to pests, 10.8% due to illnesses, and 1.4% due to viruses, with the remaining 10.2% owing to weeds (Oerke, 2006). EIDs (emerging infectious diseases) pose a threat to human health and the environment. Crop diseases including late blight of the potato, caused by the oomycete *Phytophthora infestans*, and Karnal bunt of wheat, durum, rye, and triticale, caused by *Tilletia indica*, have been linked to climate change (Singh, 1989). Furthermore, as a result of global climate change, the distribution and quantity of arthropod vectors, as well as the frequency of unexpected weather occurrences, which is one of the key drivers of plant emerging infectious diseases, are predicted to vary (Anderson et al., 2004). In case of high elevation, trees, in stressful habitats, have reports of greater incidence of parasitic plants. *Juniperus polycarpos* in Lahaul and Spiti is susceptible to attack of a partial parasite, *Arceuthobium* (dwarf mistletoe), which causes considerable damage to the tree (Jishtu & Rawat, 2014).

8 Extinction and Vulnerability

Climate change has a direct or indirect impact on species survival, and expected extinction rates are heavily impacted by climate change mitigation efforts (Jantz et al., 2015). Arctic, alpine, and coastal forest species, as well as island endemics, can decrease or become extinct if they can't spread or move quickly enough or have nowhere to go (Verma, 2021). Variations in weather patterns, followed by changes in river flow, will disturb the successional stage of alluvial grasslands, jeopardizing the viability of species like *Sus salvanius* (Chettri et al., 2010). Similarly, Myanmar's Irrawaddy River's middle plain contains a dense damp deciduous forest that shelters a small number of high-level endemic species that are vulnerable to changes in river patterns caused by climate change (Chettri et al., 2010). *Taxus wallichiana*, commonly known as the Himalayan Yew, is an evergreen gymnosperm found throughout the Northern Hemisphere's temperate zone. *Taxus wallichiana* is facing severe threats of extinction owing to its small population size and climate change (Samant, 1999; Pant & Samant, 2008). The long-term consequences of harsh climatic circumstances are affecting the morphology, physiology, and behavioral individualities of *Taxus*, with considerable disruptions due to solar radiation, temperature extremes, and cold winds that are linked to both population and individual stands of yew (Gegechkori, 2018).

9 Forest Fire

A rise in the number of forest fires is another significant result of climate change. The drying of organic materials in forests has been hastened by increasing temperature, resulting in a twofold increase in forest fires. The frequency, scale, severity, seasonality, and form of fire are all influenced by the weather and climate, as well as forest structure and composition (Bhatta, 2007). Long periods of summer drought can turn regions that are already prone to fire into locations, where fire will always be a threat. Fire could spread to places that are presently comparatively undisturbed as dire climatic, environmental, and biological thresholds for fire outbreaks are exceeded (Johnson, 1992). Wildfires can cause a shift in species by exposing a segment of the Himalayan flora to their impacts. *Pinus roxburghii* had the highest shifts in distribution range, whereas *Rhododendron arboretum* had the least shifts (Chitale & Behara, 2019). In Himachal Pradesh, fire destroyed 8195 ha of forest in 2005–2006 (Bhatta, 2007). One of Uttarakhand's most damaging forest fires, which occurred on May 27, 1995, severely burned a total of 2115 km^2 in four districts (Almora, Chamoli, Tehri, and Pauri Garhwal) (Semwal & Mehta, 1996). A comparison of air temperatures in Binsar Wildlife Sanctuary, Kumaun Himalaya, during the fire season in 1994 (no fire event) and 1995 (major fire events) revealed that air temperatures in 1995 were warmer (Sharma & Rikhari, 1997). The direct effects of climate change on species distribution, migration, substitution, and extinction may be more important than the indirect effects of climate change on forest fires when it comes to the influence on vegetation (Weber & Flannigan, 1997). Fire has the capacity to operate as a change agent, hastening the transformation of the plant landscape into a new equilibrium and so hastening vegetation changes (Flannigan et al., 2000).

10 Conclusion and Future Policy and Recommendations

Today, climate change is, without a doubt, one of the grave threats to biodiversity and must be addressed immediately. Climate change is causing noticeable harm to the entire biodiversity in one way or another. Climate change's effects are now obvious all over the world, and the entire ecosystem is suffering in one way or another way. Climate change poses a threat to plant survival, which in turn poses a threat to the entire biosphere including human life. As a result, urgent action is required to address this alarming situation. It seems that we are gifting the deadliest ecosystem to future generations if effective and clear measures are not taken now. Intolerable temperatures are severely impacting the important ecosystem ecosystems, making Earth an unlivable planet in distant future. Additionally, it is posing new threats to livelihoods, particularly for those who rely on the ecosystem services provided by mountainous ecosystems (Singh et al., 2010; Houet et al., 2010; Janpeter et al., 2012). We must address this issue as soon as possible to protect our own interests as well as the interests of the entire biosphere. Prior attention, smart research, advanced technology, sufficient financing, and unbiased plans must all be on the table at the same time at global level. Data deficiency due to a lack of environmental and historical bioclimatic data and systematic monitoring, as well as political insensitivities within the region, is critical for formulating scientifically comprehensive transboundary climate change adaptation policies. As a result, regional cooperation is critical for decision-making, vulnerability mapping, pertinent biodiversity and conservation management, and a comprehensive attitude to designing adaption strategies to cope with the changing climatic scenarios, not only in the Himalayan region but also internationally. Global greenhouse gases emission will continue to rise in the coming decades and beyond without new mitigation initiatives. In order to achieve a stabilization of greenhouse gases concentration in the atmosphere, ample global investments and deployment of mitigation technology, as well as research into new energy sources, will be desirable without any delay. The extreme weather events and coastal erosion are some of the effects of climate change that can be mitigated through improved physical infrastructure. Changes in temperature and water availability can be mitigated in agriculture by switching to new crops, seeds, or agricultural practices. Improved weather and flood predictions, as well as improved communications, can help with evacuation, relief, and recovery. It is vital that we begin making decisions that will reduce greenhouse gases emissions, and the best way to do so is to educate ourselves and future generations through the educational systems and other public education and training opportunities (Nwankwoala, 2015). Even if it is hard to foretell or forecast the effects of climate change, one thing is certain: the environment we are accustomed to is no longer a good indicator of what to expect in the future. Policies based on rigorous scientific analysis must be strengthened and tailored to meet the challenges posed by climate change and other drivers of change to biodiversity. Climate change research, ecosystem vulnerability, and the degree of biodiversity consequences based on collaborative modelling are insufficient. Many protected areas in the Himalaya lack comprehensive inventories of species and their populations, and their coverage is limited. Appraisal of the biological invasions and population trends of flagship and vulnerable species, extension of protected area (PA) networks with more reasonable protection, and fire management are all strategies and policies that should be implemented as soon as possible.

References

Anderson, P. K., Cunningham, A. A., Patel, N. G., Morales, F. J., Epstein, P. R., & Daszak, P. (2004). Emerging infectious diseases of plants: Pathogen pollution, climate change and agrotechnology drivers. *Trends in Ecology and Evolution, 19*, 10.

Apollo, M. (2017). The population of Himalayan regions – By the numbers: Past, present and future. In R. In Efe & M. Oeturk (Eds.), *Contemporary studies in environment and tourism* (pp. 143–154). Cambridge Scholars Publishing.

Bandyopadhyay, J., & Gyawali, D. (1994). Himalayan water resources: Ecological and political aspects of management. *Mountain Research and Development, 14*, 1–24.

Bellard, C., Bertelsmeier, C., Leadley, P., Thuiller, W., & Courchamp, F. (2012). Impacts of climate change on the future of biodiversity. *Ecology Letters, 15*(4), 365–377.

Bellard, C., Thuiller, W., Leroy, B., Genovesi, P., Bakkenes, M., & Courchamp, F. (2013). Will climate change promote future invasions? *Global Change Biology, 19*(12), 3740–3748.

Beniston, M. (2003). Climate change in mountain regions; a review of possible impacts. *Climate Change, 59*, 5–31.

Bhatta, A. (2007). Himachal villagers resist pine monoculture reclaim forests for fodder. *Down to Earth, 16*, 6.

Chakraborty, S., Liu, C. J., & Mitter, V. (2006). Pathogen population structure and epidemiology are keys to wheat crown rot and *Fusarium* head blight management. *Australasian Plant Pathology, 35*, 643–655.

Chettri, N., Sharma, E., Shakya, B., Thapa, R., Bajracharya, B., Vddin, K., Oli, K. P., & Choudhary, D. (2010). *ICIMOD, technical report 2, Kathmandu, Nepal.*

Chitale, V., & Behara, M. D. (2019). How will forest fires impact the distribution of endemic plants in the Himalayan biodiversity hotspot. *Biodiversity and Conservation, 28*. https://doi.org/10.1007/S10531-019-01733-8

Dar, G. H., & Khuroo, A. A. (2013). Floristic diversity in the Kashmir Himalaya: Problems, progress and prospectus. *Sains Malaysiana, 42*(10), 1377–1387.

Dhyani, S., Kadaverugu, R., Dhyani, D., Verma, P., & Pujari, P. (2018). Predicting impacts of climate variability on habitats of *Hippophae salicifolia* (Sea buck thorn) in Central Himalayas: Future challenges. *Ecological Informatics*. https://doi.org/10.1016/j.ecoinf.2018.09.003

Donato, D. C. (2013). Limits to upward movement of subalpine forests in a warming climate. *Proceedings of the National Academy of Sciences of the United States of America, 110*(20), 7971–7972.

Early, R., & Sax, D. F. (2014). Climatic niche shifts between species' native and naturalized ranges raise concern for ecological forecasts during invasions and climate change. *Global Ecology and Biogeography, 23*(12), 1356–1365.

Engler, R., Randin, C. F., Thuiller, W., Dullinger, S., Zimmermann, N. E., Araujo, M. B., Pearman, P. B., et al. (2011). 21st century climate change threatens mountain flora unequally across Europe. *Global Change Biology, 17*, 2330. https://doi.org/10.1111/j.1365-2486.2010.02393.x

Eriksson, J. G. (2006). Patterns of growth; relevance to developmental origins of health and diseases. In P. Gluckman & M. Hanson (Eds.), *Developmental origins of health and diseases* (pp. 223–232). Cambridge University Press.

FAO (2012). Wildlife in a changing climate. FAO Forestry Paper 176. Eds (Edgar Kaeslin, Ian Redmond, Nigel Dudley). Rome, p. 108.

Flannigan, M. D., Stocks, B. J., & Wotton, B. M. (2000). Climate change and forest fires. *Science of the Total Environment, 262*, 221–229.

Gaira, K., & Dhar, U. (2020). Phenological change modelling for selected Himalayan medicinal herbs using herbarium records. *Ecological Informatics, 60*. https://doi.org/10.1016/j.ecoinf.2020.101177

Gaira, K. S., Dhar, V., & Belwal, O. K. (2011). *Potential of herbarium records to sequence phenological pattern: A case study of Aconitum heterophyllum in the Himalaya* (Vol. 20, pp. 2201–2210). Springer Science, Business Media Biodiversity Conservation.

Gajural, J. P., Werth, S., Shrestha, K. K., & Scheidegger, C. (2014). Species distribution modelling of Taxus wallichiana (Himalayan yew) in Nepal Himalaya. *Asian Journal of Conservation Biology, 3*(2), 127–134.

Gautam, M. R., Tamilsina, G. R., & Acharya, K. (2013). *Climate change in the Himalayas current state of knowledge* (Policy Research Working Paper WP-6516) (p. 47). The World Bank Development Research Group Environment and Energy Team.

Gegechkori, A. (2018). Patterns of distribution and survival of European Yew (*Taxus baccata L.*) in an alpine tree line ecotone in the greater Caucasus (Georgia). *Annals of Agrarian Science*. https://doi.org/10.1016/j.aasci.2018.02.006

Gottfried, M., Pauli, H., & Grabherr, G. (2012). Continent wide response of mountain vegetation to climate change. *Nature Climate Change, 2*, 111–115.

Houet, T., Loveland, T. R., Hubert-Moy, L., et al. (2010). Exploring subtle land use and land cover changes: A framework for future landscape studies. *Landscape Ecology, 25*, 249–266.

IPCC. (2007). *Contribution of working Group II to the fourth assessment report of the Intergovernmental Panel on Climate Change*. Cambridge University Press.

Janpeter, S., Freier Korbinian, P., Hertige, E., & Jurgen, S. (2012). Climate change, vulnerability and adaptation in North Africa with focus on Morocco. *Agriculture, Ecosystems and Environment, 156*, 12–26.

Jantz, S. M., Barker, B., Brooks, T. M., Chini, L. P., Huang, Q., Moore, R. M., Noel, J., & Hurtt, G. C. (2015). Future habitat loss and extinction driven by land use change in biodiversity hotspots under four scenarios of climate change mitigation. *Conservation Biology, 29*(4), 1122–1131.

Jishtu, V., & Rawat, R. S. (2014). *Juniperus polycarpos* C. Koch Forests and its conservation status in Cold Deserts of Himachal Pradesh, North West Himalayas. *Indian Forester, 140*(4), 378–383.

Johnson, E. A. (1992). *Fire and vegetation dynamics: Studies from the North American boreal forest* (p. 129). Cambridge University Press.

Khuroo, A. A., Rashid, I., & Reshi, Z. (2007). The alien flora of Kashmir Himalaya. *Biological Invasions, 9*(3), 269–292.

Kumar, V., & Chopra, A. K. (2009). Impact of climate change on biodiversity of India with special reference to Himalayan region-an overview. *Journal of Applied and Natural Science, 1*(1), 117–122.

Kumar, R. S., Kala, R. H., Kumar, G. S., Kumar, S. K., & Sailesh, R. (2017). Predicting the impact of climate change on the distribution of two threatened Himalayan medicinal plants of Liliaceae in Nepal. *Journal of Mountain Science, 14*(3), 558–570.

Lieberman, D. (1982). Seasonality and phenology in a dry tropical forest in Ghana. *Journal of Ecology, 70*, 791–806.

Macais- Fauria, M., & Johnson, E. A. (2013). Warming induced upslope advance of subalpine forest is severely limited by geomorphic processes. *Proceedings of the National Academy of Sciences of the United States of America, 110*(20), 8117–8122.

Manish, K., & Pandit, M. K. (2019). Identifying conservation priorities for plant species in the Himalaya in current and future climates: A case study from Sikkim Himalaya, India. *Biological Conservation, 233*, 176–184.

Manish, K., Telwala, Y., Nautiyal, D. C., & Pandit, M. K. (2016). Modelling the impacts of future climate change on plant communities in the Himalaya: A case study from Eastern Himalaya, India. *Modelling Earth Systems and Environment, 2*, 92.

Maroof, H., Khuroo, A. A., Charles, B., Ahmad, R., Singh, C. P., & Aravind, N. A. (2019). Impact of climate change on distribution range and niche dynamics of Himalayan birch, a typical tree line species in Himalayas. *Biodiversity and Conservation, 28*, 2345–2370.

Masoodi, A., Sengupta, A., Khan, F. A., & Sharma, G. P. (2013). Predicting the spread of alligator weed (*Alternanthera philoxeroides*) in Wular lake, India: A mathematical approach. *Ecological Modelling, 263*, 119–125.

McDougall, K. L., Khuroo, A. A., Loope, L. L., Parks, C. G., Pauchard, A., Reshi, Z. A., Rushworth, I., & Kueffer, C. (2011). Plant invasions in mountains: Global lessons for better management. *Mountain Research and Development, 31*(4), 380–387.

Nautiyal, B. P., Nautiyal, B. C., Khanduri, V. P., & Rawat, N. (2009). Floral biology of Aconitum heterophyllum Wall: A critically endangered alpine medicinal plant of Himalaya, India. *Turkish Journal of Botany, 33*, 13–20.

Negi, V. S., Maikhuri, R. K., Pharswan, D., Thakur, S., & Dhyani, P. P. (2016). Climate change impact in the western Himalaya; people's perception and adaptive strategies. *Journal of Mountain Science, 14*(2), 403–416.

Nwankwoala, H. N. L. (2015). Causes of climate and environmental- friendly education policy in Nigeria. *Journal of Education and Practice, 6*, 30.

Oerke, E. C. (2006). Crop losses to pests. *Journal of Agricultural Science, 144*, 31–43.

Packert, M., Martens, J., Sun, Y. H., Severinghaus, L. L., Nazarenko, A. A., Ting, J., Topfer, T., & Tietze, D. T. (2012). Horizontal and elevational phylogeographic patterns of Himalayan and Southeast Asian forest passerines (Aves: Passeriformes). *Journal of Biogeography, 39*, 556–573.

Pant, S. S., & Samant, S. (2008). Population ecology of the endangered Himalayan Yew in Khokhan Wildlife Sanctuary of North Western Himalaya for conservation management. *Journal of Mountain Science, 5*(3), 257–264.

Parmesan, C., & Yohe, G. (2003). A globally coherent fingerprint of climate change impacts across natural systems. *Nature, 421*, 37–42.

Permesan, C. (2006). Ecological and evolutionary responses to recent climate change. *Annual Review of Ecology, Evolution and Systematics, 37*, 637–669.

Pimentel, D., McNair, S., Janecka, J., Wightman, J., Simmonds, C., O'Connell, C., Wong, E., Russel, L., Zern, J., Aquino, T., & Tsomondo, T. (2001). Economical and environmental threats of alien plant, animal and microbe invasions. *Agriculture, Ecosystems and Environment, 84*(1), 1–20.

Primack, D., Imbres, C., Primack, R. B., & Rushing, A. J. M. (2004). Herbarium specimens demonstrate earlier flowering times in response to warming in Boston. *American Journal of Botany, 91*, 1260–1264.

Rana, S. K., Rawal, R. S., Dangwal, B., Bhatt, I. D., & Price, T. D. (2021). 200 years of research on Himalayan biodiversity: Trends, gaps, and policy implications. *Frontiers in Ecology and Evolution*. https://doi.org/10.3389/fevo.2020.603422

Ranjitkar, S., Sajakhu, N. M., Lu, Y., Wang, Q., Wnag, M., He, J., Mortimer, P. E., Xu, J., Kindt, R., & Zomer, R. J. (2016). Climate modelling for agroforestry species selection in Yunnan Province, China. *Environmental Modelling and Software, 75*, 263–272.

Rashid, I., Romshoo, S. K., Chaturvedi, R. K., Ravindranath, N. H., Sukumar, R., Jayaraman, M., Lakshmi, T. V., & Sharma, J. (2015). Projected climate change impacts on vegetation distribution over Kashmir Himalaya. *Climate Change, 132*, 601–613.

Rautela, P., & Karki, B. (2015). Impact of climate change on life and livelihood of indigenous people of Higher Himalaya in Uttarakhand, India. *American Journal of Environmental Protection, 3*(4), 112–124.

Root, T. L., Price, J. T., Hall, K. R., Schneider, S. H., Rosenzweig, E., & Pounds, J. A. (2003). Fingerprints of global warming on wild animals and plants. *Nature, 421*, 57–60.

Samant, S. S. (1999). Diversity, nativity and endemism of vascular plants in a part of Nanda Devi Biosphere Reserve in West Himalaya. *Biannual Bulletin, 1*, 1–28.

Schickhoff, U., Bobrowski, M., Bohner, J., Burzle, B., Chaudhary, R. P., Gerlitz, L., Heykess, H., Lange, I., Muller, M., Scholten, T., Schwab, N., & Wedegartner, R. (2015). Do Himalayan treelines respond to recent climate change? An evaluation of sensitivity indicators. *Earth System Dynamics, 6*, 245–265.

Semwal, R. L., & Mehta, J. P. (1996). Ecology of forest fires in Chir pine (*Pinus roxburghii*), forests of Garhwal Himalaya. *Current Science, 70*, 426–427.

Sharma, S. & Rikhari, H. C. (1997). Forest fire in the central Himalaya: climate and recovery of trees. *International Journal of Biometeorology 40*, 63–70.

Shaver, G. R., Canadell, J., Chapin, F. S., Gurevitch, J., Harte, J., Henry, G., Ineson, P., Jonasson, S., Melillo, J., Pitelka, L., & Rustad, L. (2000). Global warming and terrestrial ecosystems: A conceptual framework for analysis: Ecosystem responses to global warming will be complex and varied. Ecosystem warming experiments hold great potential for providing insights on ways terrestrial ecosystems will respond to upcoming decades of climate change. Documentation of initial conditions provides the context for understanding and predicting ecosystem responses. *Bioscience, 50*(10), 871–882.

Shrestha, U. B., Gautam, S., & Bawa, K. S. (2012). Widespread climate change in the Himalayas and associated changes in local ecosystems. *PloS one, 7*(5), e36741.

Shrestha, U. B., Sharma, K. P., Devkota, A., Siwakoti, M., & Shrestha, B. B. (2018). Potential impact of climate change on the distribution of six invasive alien plants in Nepal. *Ecological Indicators*. https://doi.org/10.1016/j.ecolind.2018.07.009

Singh, D. V. (1989). First report of *Neovossia indica* on wheat in Nepal. *Plant Diseases, 73*, 277.

Singh, P., & Negi, G. C. S. (2016). *Impact of climate change on phenological responses of major forest trees of Kumaun Himalaya* (ENVIS bulletin Himalayan ecology. 24: B). Pant National Institute of Himalayan Environment and Sustainable Development.

Singh, D. K., & Pusalkar, P. K. (2020). Floristic diversity of the Indian Himalaya. In Biodiversity of the Himalaya: Jammu and Kashmir State (pp. 93–126). Springer, Singapore.

Singh, S. P., & Rawal, R. S. (2017). *Manual of field methods* (p. 99). Central Himalayas Environmental Association.

Singh, S. P., Singh, V., & Skutsch, M. (2010). Rapid warming in the Himalayas: Ecosystem response and developmental options. *Climate and Development, 2*, 1–13.

Singh, S. P., Sharma, S., & Dhyani, P. P. (2019). Himalayan arc and treeline distribution, climate change responses and ecosystem properties. *Biodiversity and Conservation, 28*, 1997–2016.

Stenseth, N. C., Mysterud, A., Ottersen, G., Hurrell, J. W., Chan, K. S., & Lima, M. (2002). Ecological effects of climate fluctuations. *Science, 297*(5585), 1292–1296.

Tewari, V. P., Verma, R. K., & von Gadow, K. (2017). Climate change effects in the Western Himalayan ecosystems of India; evidence and strategies. *Forest Ecosystems, 4*, 13.

Thapa, S., Chitale, V., Rijal, R. J., Bisht, N., & Shrestha, B. B. (2018). Understanding the dynamics in distribution of invasive alien plant species under predicted climate change in Western Himalaya. *Plos One, 13*(4).

Thuiller, W. (2007). Climate change and the ecologist. *Nature, 448*, 550–552.

Tse-ring, K., Sharma, E., Chettri, N., & Shrestha, A. B. (2010). Climate change vulnerability of mountain ecosystems in the Eastern Himalayas. International centre for integrated mountain development (ICIMOD).

Vashistha, R. K., Rawat, N., Chaturvedi, A. K., Nautiyal, B. P., Prasad, P., & Nautiyal, M. P. (2009). An exploration on the phenology of different growth forms of an alpine expanse of North-West Himalaya, India. *New York Science Journal, 2*(6), 29–41.

Verma, O. (2021). Climate change and its impacts with special reference to India. In A. K. Taloor, B. S. Kotlia, & K. Kumar (Eds.), *Water, cryosphere and climate change in Himalaya: A geospatial approach*. Springer.

Weber, M. G., & Flannigan, M. D. (1997). Canadian boreal forest ecosystem structure and function in a changing climate: impact on fire regimes. *Environmental Reviews, 5*(3–4), 145–166.

Wang, J. J., Cao, B., & Bai, C. K. (2014). Potential distribution prediction and suitability evaluation of *Fritillaria cirrhosa* D. Don based on MaxEnt modelling and GIS. *Bulletin of Botanical Research, 34*(5), 642–649.

Woodall, C. W., Oswalt, C. M., Westfall, J. A., Perry, C. H., & Nelson, M. D. (2009). Tree migration detection through comparisons of historic and current forest inventories. In *USDA Forest Service Proceedings*, pp. 56.

www.Cepf.net. Assessed on 27 Mar 2021.

Chapter 12
Promotion, Utilization, and Commercial Cultivation of Local Spices with Special Reference to Eryngo (*Eryngium foetidum* L.) as a Measure for Livelihood Improvement Towards Achieving the Goal of Sustainable Development in the Indo-Burma Biodiversity Hotspot: A Case Study From Manipur, North-East India

Pangeijam Bijaya Devi, Panna Deb, and Huidrom Birkumar Singh

Abstract Local spices play an important role in the life of traditional communities and are immensely useful in the preparation of traditional medicines in various countries. Some species are used exclusively in food cuisine, while others were found to be used as medicine. The spices are found to be useful in management of cough, loss of appetite, stomach-ache, pain, and fever. A total of 38 plant species was used as spices by the different ethnic groups of Manipur, which falls in the Indo-Burma centre of biodiversity 'hotspot' of global significance. Eryngo (*Eryngium foetidum* L.), which is a tropical rosette herbaceous plant belonging to Apiaceae, is an important high-value and underutilized spice crop of north-eastern region of India that is consumed mostly locally, but the consumption rate has increased rapidly in the recent times. Eryngo had remained either wild or semi-domesticated and is traded and consumed locally. An average leaf yield of 11,497.04 kg/ha/year could be obtained with C:B ratio of 1: 5.34. Eryngo sold in markets of Manipur is either collected from the wild or maintained in kitchen gardens. The indigenous spices

---

P. B. Devi
Krishi Vigyan Kendra, Utlou, Manipur, India

P. Deb
Centre for Biodiversity and Natural Resource Conservation, Assam University, Silchar, Assam, India

Department of Ecology & Environmental Science, Assam University, Silchar, Assam, India

H. B. Singh (✉)
CSIR-North East Institute of Science & Technology, Branch Laboratory, Imphal, Manipur, India

S. Sharma et al. (eds.), *Climate Change Adaptation, Risk Management and Sustainable Practices in the Himalaya*,
https://doi.org/10.1007/978-3-031-24659-3_12

have a major role in income generation and in ensuring the livelihood security of the poor people, and Eryngo is one such important spice crop of the region. Thus, it is essential to assess the usage, conservation issues, and cultivation of Eryngo with the application of readily available organic inputs with a target to promote sustainable agriculture and use the spice in traditional healthcare and also promote self-employment, thereby enhancing the income for farmers and helping in balancing the social, economic, and environmental sustainability of the region.

Keywords Eryngo · Spice · Cultivation · Sustainable development · Livelihood · Manipur · Indo-Burma hotspot

1 Introduction

India is the abode of spices, exhibiting diverse spice plants like bay leaf, cardamom, garlic, chili, pepper, cinnamon, and different varieties of tree and seed spices. Spices have an important place in the food habit of Indians (Singh & Sundriyal, 2003). The agro-climatic condition of the country provides an ideal situation for the natural growth of varieties of aromatic and medicinal plants used in the spice and pharmaceutical industry. The consumption of spices in India is very high as compared to other countries. The demand for spices is very high both in the developed and the developing nations. A large number of underutilized aromatic herbs are used as spices by the ethnic people of the north-east region of India. Generally, they maintain some spice crops in their home gardens. The lesser amount of expenditure on spices in north-east India might be due to their dependence on spices mostly from wild habitats and kitchen gardens and not from the market; otherwise, spice consumption in the north-east region of the country is comparatively very high in terms of both variety and quantity. Consumption of fresh or raw spices in the form of traditional salad is also very prevalent in the north-eastern part of the country. Several high potential plants from the wild are still collected by indigenous communities for diverse purposes but comparatively less known to outside the world (Singh & Sundriyal, 2003). More than 34 aromatic plants are used as local spices in one form or the other in the state of Manipur alone: out of which 13 species are already under cultivation, while 8 species were collected from the wild habitat, and the remaining 17 species were either cultivated or collected from the wild habitat (Singh & Sundriyal, 2003).

In Manipur, Eryngo (*Eryngium foetidum* L.), which is known as *Awa-phadigom* or occasionally called *Sha-maroi* (meaning meat spice), is a tropical annual or biennial rosette herb from the Apiaceae family. Eryngo is one of the most preferred spices for the different ethnic communities of Manipur and the entire north-east India (Devi et al., 2021). It is reported to be native to Mexico and South America. It is perceived to be native to the Caribbean area of the West Indies (Singh & Devi, 2014). It is generally cultivated as a spice plant in India, more prominently in north-east and eastern Himalayan regions. It has different names, such as Eryngo, sea holly, spirit weed, Burma dhania, Nepali dhania, Bhutia dhania, Bilati dhonia, and

Naga dhania (Singh & Devi, 2014). In Manipur, Eryngo has been found growing in the wild since ages and is cultivated in the kitchen gardens in the recent times for home consumption and also for selling in the local markets for cash needs. This spice is used in various cuisines, especially in meat and fish cooking (Devi, 2021; Devi et al., 2021), and also for treatment of various diseases and ailments, like epilepsy, high blood pressure, cold and cough, paralysis, etc. (Singh & Devi, 2014). Eryngo is grown especially for its leaves, which is used as vegetables, medicines, and others (Ignacimuthu et al., 1999). Similar to Manipur, Eryngo is also regarded as an important crop in the hilly region of Bangladesh that helps in the revenue generation (Moniruzzaman et al., 2007). Thus, this underutilized crop has a huge potential for the socio-economic development of traditional farmers through its commercial cultivation. It is accepted that enhancement of socio-economic status of farmers will help to reduce their dependence on the surrounding forest, leading to conservation of biodiversity and sustainability in the long run.

2 Methodology

A survey was conducted in the state of Manipur, north-east India, for the documentation of important spice plants, which are used by various ethnic communities of the state during 2015–2019, through community participatory appraisal, interaction with resource persons (*maiba* for male and *maibi* for female in Manipuri vernacular), and market survey in important markets, namely, Khwairamband or Ima Keithel, Bishnupur Keithel, Lamlong Keithel, Kakching Keithel, Thoubal Keithel, and Nambol Keithel of the valley districts of Manipur through a questionnaire survey of vegetable vendors. The plant species were identified by consulting regional flora and available literatures on the subject (BSI, 2000; Singh et al., 2003; Sinha, 1996; Singh & Devi, 2014; Deb 1961; Singh & Sundriyal, 2003). Information pertaining to plant habit, parts used, mode of uses, source of collection, and mode of propagation of the species were recorded. Traditional uses of Eryngo in food/cuisine and medicine were also documented. The experimentation on development of package of practices for Eryngo was carried out in the experimental farm of the Krishi Vigyan Kendra, Utlou, Bishnupur district, and Manipur, India, during 2015–2018 (Fig. 12.1). The nutrient (proximate) analysis of the aerial parts of Eryngo was carried out in CSIR-CFTRI, Mysuru.

3 Results and Discussion

A total of 38 plants used as local spices by various ethnic communities of the state of Manipur are documented (Table 12.1). In case of parts used, a maximum number of nine species were used as a whole plant, followed by fruits (eight species), shoot (five species), rhizome (three species), bark (three species), leaf and fruits together

Fig. 12.1 Experimentation of the Eryngo crop for development of agro-practice

(two species), leaf and inflorescence together (two species), inflorescence (two species), leaf (single species), shoot and leaf together (single species), runners and leaf together (single species), and leaf and root (single species) (Table 12.1). Out of the 38 spices recorded, 18 species were grown in kitchen garden, while six species were collected from the wild habitat, and the remaining 14 species were found to be both cultivated and collected from the wild habitat (Table 12.1).

Among the 38 spices recorded, the highest number of 23 species were generally propagated through seeds, followed by rhizome (eight species), seed (two species), seed and stolon (two species), seed and bulb (single species), clove (single species), and seed and vegetative parts (single species) (Table 12.1). Some spices were used daily along with vegetables, while few others were used occasionally. The species of *Allium* are the most commonly and frequently used one by the ethnic communities. The maximum species were represented by the families Lamiaceae and Zingiberaceae (18.42% each).

Eryngo is an important ingredient in many dishes and soups in many American, African, and south-east Asian Countries. It is also used as a seasoning material in culinary items and is used in vegetable, meat dishes, sauces, and snacks. The spicy aroma of the herb renders a unique flavour and makes it a much sought-after seasoning material in many oriental dishes and is very popular among the ethnic populace. It has a huge market across the world and specially in United Kingdom, United States, and other Latin American nations. In the north-eastern states of India, Eryngo is cultivated as an important spice plant. The plant is famous for its medicinal properties, mainly prepared from its leaves and root decoctions. It has been in use for treating gastrointestinal problems, pain, and flu-like ailments. The plant formulations have been reported to be used widely in herbal medicines (Singh et al., 2003; Sinha, 1996). Eryngo is regarded as a healthy food as it contains good proportion of minerals, vitamins, and proteins (Paul et al., 2011). Out of the 38 important spice plants of Manipur including Eryngo, 23 species were used to treat different ailments in the indigenous healing system of Manipur (Singh & Sundriyal, 2003). Eryngo has remained either wild or semi-domesticated and is one of the most underutilized spices of the north-eastern region of India that is mostly consumed locally (Devi

Table 12.1 Common spices used by ethnic communities of Manipur valley, NE India

Sl. no.	Species	Family	Habit	Source	Parts used	Method of propagation
1.	*Allium cepa* L.	Alliaceae	Herb	Cultivated	Whole plant	Seed and bulb
2.	*Allium hookeri* Thw.	Alliaceae	Herb	Cultivated	Whole plant	Rhizome
3	*Allium porrum* L.	Alliaceae	Herb	Cultivated	Whole plant	Seed
4	*Allium sativum* L.	Alliaceae	Herb	Cultivated	Whole plant	Cloves
5.	*Allium tuberosum* L.	Alliaceae	Herb	Cultivated	Whole plant	Rhizome
6.	*Anisochilus carnosus* Wall.	Lamiaceae	Herb	Cultivated	Leaf and inflorescence	Seeds
7.	*Artabotrys hexapetalus* (L.f.) Bhandari	Annonaceae	Shrub	Cultivated and wild	Bark	Seed and cloning
8.	*Alpinia allughas* Rosc.	Zingiberaceae	Herb	Wild	Rhizome	Rhizome
9.	*Capsicum annum* L.	Solanaceae	Herb	Cultivated	Fruits	Seed
10.	*Capsicum sinensis* L.	Solanaceae	Herb	Cultivated	Fruits	Seed
11.	*Capsicum maxi-mum* L.	Solanaceae	Shrub	Cultivated and wild	Fruits	Seed
12.	*Capsicum minimum* Roxb.	Solanaceae	Shrub	Cultivated	Fruits	Seed
13.	*Cinnamomum tamala* Nees.	Lauraceae	Tree	Cultivated and wild	Leaf	Seed
14.	*Cinnamomum camphora* Nees.	Lauraceae	Tree	Cultivated and wild	Bark	Seed
15.	*Cardamine hirsuta* L.	Brassicaceae	Herb	Cultivated	Whole plant	Seed
16.	*Cinnamomum zeylanicum* Breyn.	Lauraceae	Tree	Cultivated and wild	Bark	Seed and cloning
17.	*Coriandrum sativum* L.	Apiaceae	Herb	Cultivated	Whole plant	Seed
18.	*Curcuma zedoaria* Rosc.	Zingiberaceae	Herb	Wild	Inflorescence	Seed
19.	*Curcuma amada* Roxb.	Zingiberaceae	Herb	Wild	Inflorescence	Rhizome
20.	*Curcuma caesia* Roxb.	Zingiberaceae	Herb	Cultivated and wild	Rhizome	Rhizome
21.	*Citrus macroptera* Montrouz	Rutaceae	Tree	Cultivated	Fruits	Seed and vegetative
22.	*Curcuma longa* L.	Zingiberaceae	Herb	Cultivated	Whole plant	Rhizome

(continued)

Table 12.1 (continued)

Sl. no.	Species	Family	Habit	Source	Parts used	Method of propagation
23.	*Citrus latipes* Swingle	Rutaceae	Shrub	Cultivated	Whole plant	Rhizome
24.	*Elsholtzia blanda* Benth.	Lamiaceae	Herb	Cultivated	Leaves and inflorescence	Seed
25.	*Eryngium foetidum* L.	Apiaceae	Herb	Cultivated and wild	Leaf and roots	Seed
26.	*Foeniculum vulgare* Gaertn.	Apiaceae	Herb	Cultivated	Fruits	Seed
27.	*Ferula asafoetida* Boiss.	Apiaceae	Herb	Cultivated	Fruits	Seed
28.	*Houttuynia cordata* Thunb.	Saururaceae	Herb	Cultivated and wild	Runners and leaves	Seed and stolon
29.	*Kaempferia galanga* L.	Zingiberaceae	Herb	Cultivated and wild	Rhizome	Rhizome
30.	*Knoxia sumatrensis* DC.	Lamiaceae	Shrub	Wild	Shoot	Seed
31.	*Leucas aspera* Spreng	Lamiaceae	Shrub	Wild	Shoot	Seed
32.	*Litsea citrata* BI.	Lauraceae	Herb	Wild	Shoot	Seed
33.	*Mentha arvensis* L.	Lamiaceae	Herb	Cultivated and wild	Shoot and leaves	Seed and Stolon
34.	*Meriandra bengalensis* Benth.	Lamiaceae	Shrub	Cultivated and wild	Shoot	Seed
35.	*Ocimum canum* Sims.	Lamiaceae	Shrub	Cultivated and wild	Shoot	Seed
36.	*Zingiber officinale* Rosc.	Zingiberaceae	Herb	Cultivated	Leaves and fruits	Seed
37.	*Piper nigrum* L.	Piperaceae	Climber	Cultivated and wild	Fruits	Seed
38.	*Zanthoxylum acanthopodium* DC.	Rutaceae	Tree	Cultivated and wild	Leaves and fruits	Seed

et al., 2021). Several works related to medicinal and aromatic plants conservation have been attempted in different north-east regions, including Manipur, harbouring huge diversity of plant species (Mao & Hynniewta, 2000). Conservation of indigenous plants should be done by introducing the priority species in the kitchen garden and traditional home gardens.

Domestication and cultivation of traditional plants can be done by using scientifically developed package of practices as a conservation initiative, which will be helpful both for income generation and conservation of the germplasm. For the betterment of human community, ex situ and in situ conservations of selected indigenous plants are necessary to go along with preservation of traditional cultural diversity. Taking clue from traditional knowledge, high-value indigenous plants need to be explored for mass cultivation in order to meet the local requirements as well as for export, which will help in generating income for the rural populace and also for the conservation of biodiversity. Manipur located in the north-eastern region is considered as a biodiversity 'hotspot', and many interesting plant species, such as *Polygonum* sp., *Plumbago zeylanica*, *Cycas pectinata*, *Chenopodium album*, *Meriandra bengalensis*, *Stellaria media*, *Curcuma angustifolia*, *Hedychium coronarium*, *Alpinia nigra*, *Zizania latifolia*, *Oenanthe javanica*, *Ipomea aquatica*, *Antidesma diandrum*, *Houttuynia cordata*, *Enhydra fluctuans*, etc., are abundantly found in the wild habitat of the region. Due to vast area of its land mass under wetlands or water bodies, many edible, aromatic, and medicinal aquatic plant species are also available for the use of its inhabitants in the state of Manipur (BSI, 2000). Among the local spice plants of Manipur, Eryngo is found having a promising potential for the enhancement of farmer's income due to its wider use in food/ cuisines and in the traditional healing system (Table 12.2).

All the parts of Eryngo plant, except the roots, are used as a spice in preparation of fish and meat curry items. Yuhlung and Bhattacharyya (2015) observed that the Chothe tribe of Manipur used the leaves of Eryngo often in the preparation of beef curry and meat chutney to enhance its aroma and taste. The use of Eryngo plant in indigenous healing therapy for the treatment of various diseases, like body ache, cough, fever, cut and wounds, epilepsy, hypertension, paralysis, ulcer, vomiting, diarrhoea, is reported by several workers (Shavandi et al., 2012; Singh & Sundriyal, 2003; Singh et al., 2003; Singh & Devi, 2014; Devi et al., 2016, 2021). Eryngo has been considered very good for health, since it has a good amount of minerals, vitamins, and proteins and other health beneficial elements (Paul et al., 2011). The high content of 46% crude fibre and 23.2% protein in the aerial parts of Eryngo (Table 12.3) reflects its health benefits.

4 Cultivation of Eryngo (*Eryngium foetidum* L.)

Agriculture is the main occupation for rural populace of any region. Balanced use of fertilizer in crop production is indispensable for the maximization of yield. Application of nitrogen is required throughout the growing period. Application of artificial fertilizers offers short-term nutrient supply to crops. Application of organic manures or composts is required for the present-day trend that involves the use of different technologies to get a good yield. Organic agriculture has an important part to play in the present-day scenario to ensure food security. Organic manures help in improving the soil properties and also conserve soil moisture to enhance crop production (Premsekhar & Rajashree, 2009). In crop production, mulching technology implies spreading of the mulches on the ground around plants, which helps to prevent excessive evaporation and soil erosion, inhibits weed growth and conserves soil moisture. It is a suitable technology that can make the farming system a profitable proposition. Mulching prevents moisture loss and also reduces soil erosion and

Table 12.2 Usage of Eryngo plants by different communities inhabiting in Manipur

Use type	Uses and mode of preparation/application
Food/ cuisine	The tender shoots and leaf of Eryngo are extensively used as spice in preparation of curry, especially in fish and meat cooking. Chopped shoots or leaf are semi-fried in edible oil or can garnish meat cuisines. Slices of leaf are put in preparation of omelette. The whole plants and leaf are macerated with gram flour and fried in edible oil and taken as snacks as well as along with major meals. The fresh whole plants or leaves are crushed into paste and taken as chutney. It can be smashed with chili and fermented fish and taken along with major meals.
	The plant shoots or leaves are used as an important ingredient in preparation of 'pakoura' or 'bora'. In Manipuri traditional cuisine called 'paknam', the chopped plants are added as an important ingredient. Chopped plants are mixed with gram flour, preferably green chili, leaf of onion, and common salt, sometimes added with fresh prawn, and wrapped with the leaf of turmeric 2–3 layers and baked. This food can be taken as snacks or along with major meals
Medicine	*Epilepsy*: The fresh plant shoots or leaves of Eryngo are washed thoroughly with water and slightly crushed, and the mass is applied to mouth while a patient is suffering from epileptic attack. Also, put a lump just on the nose to inhale it. Regular consumption of it, along with vegetables, is generally practiced to control epilepsy
	Hypertension: Local villagers consume the plant regularly as a vegetable to control hypertension. The fresh plant paste is also taken along with little honey
	Vomiting and diarrhoea: The plant is used to treat vomiting and diarrhoea. Fresh plant paste is given to patient along with little common salt. This preparation is also effective against dyspepsia
	Body ache: The fresh plant paste is applied against body ache. The paste is also applied against mumps
	Cough and fever: The plants are boiled in water, and the soup, after adding with little common salt, is given a glass daily for a week period to treat prolong cough and fever. In fever, the plant paste is applied on forehead. Children are applied with less dose
	Cut and wounds: In fresh cut and wounds, fresh plant paste is applied as a bandage, which immediately stops bleeding and heals quickly. It also enhances for early suppuration and helps to remove foreign bodies, like spine, pins, etc.
	Ulcer: The fresh root paste along with little honey is applied against sore throat. The tincture made from fresh shoot or leaf is gurgling twice a day to treat throat ulcer

Table 12.3 Nutritional contents of Eryngo aerial parts

Sl. no.	Parameters	Value
1.	Moisture (%)	6.20
2.	Total ash (%)	11.50
3.	Fat (%)	2.70
4.	Crude fibre (%)	46.00
5.	Protein (%)	23.20
6.	Carbohydrates (by difference) (%)	10.40
7.	Energy value (Kcals/100 g)	158.00
8.	Total sugar (%)	7.10

reduces irrigation costs. Mulching immediately after germination should be done at a prescribed rate to achieve the maximum benefits of mulching (Sarolia & Bharadwaj, 2012). The cultivation could be done throughout the year, but for cultivation in open field, rainy season is the best-suited period for rapid establishment of the seedlings. The seed of Eryngo can be sown during February in the well-prepared raised nursery beds. Local cultivar at the rate of 300 g/ha is generally prescribed.

4.1 Raising of Seedlings

Nursery bed of 1 m in width with 15–20 cm high from the ground level and 1.2 m length as per need and convenience should be prepared in east and west direction. The soil type of the nursery beds should be clay-loam, rich in organic manure content and well-drained. All the clods, stones, and weeds from the nursery beds should be removed, and proper levelling should be done. Farm yard manure (FYM) at 4 kg/m^2 should be applied to the nursery beds. Line sowing is found to be the best method of sowing in the nursery beds. Lines should mark parallel to the width at the distance of 10–12 cm from the line, and the seeds should be sown on the marked lines. The finely sieved vermicompost or FYM should be used to cover the seeds just after sowing. Then, the nursery beds should be covered by gunny bags for better and quick germination. The nursery beds should be irrigated optimally till seed germination. When the seeds started germination, the mulch should be removed with caution to prevent any harm to the plant. The mulches are generally put away, when the direct sunlight is not there to avoid the impact of sun rays on the tender plants. Watering should be done once in a day with the help of rose cane. The watering should be continued till the seedlings gets ready for transplanting. Hardening of seedlings should be done by holding watering of the plants for 4–5 days before transplanting to enable them to withstand unfavourable weather condition like hot day and high temperature. The seedlings are ready to transplant at 50 days after the sowing at the well-prepared transplanting raised beds (Devi et al., 2021).

4.2 Land Preparation

The experimental field should be ploughed by power tiller to a depth of about 20–25 cm during the month of March. All the weeds and stumbles should be prepared to a fine tilt and levelled properly. The field should be prepared in raised bed methods. Vermicompost at 2.5 tons per hectare can be added during final land preparation.

4.3 Transplanting

The seedlings are ready for transplanting after 50 days of sowing. Transplanting should be done in the evening time at a spacing of 20 cm × 10 cm. The seedlings at three leaves stage should be transplanted in the evening hours in well-prepared raised beds at prescribed spacings. The irrigation should be given immediately after transplanting for proper establishment of seedlings.

4.4 Mulching

The cropped area should be mulched with rice straw and rice husk at 2.5 tons per hectare; first mulching should be carried out at 7 DAT (days after transplanting) and after every harvesting. The mulches should not be applied in incomplete doses at a single point of time, because there is always a chance of blowing away of the mulches by high-speed wind during the cropping period.

4.5 Irrigation

The first irrigation should be given immediately after the transplantation. Subsequent irrigations should be given, depending upon the moisture content of the cropped area. After each harvest, irrigation should be done.

4.6 Inter Cultural Operation

Care should be taken to maintain the cropped area, which should be free from weeds and other unwanted materials.

4.7 Plant Protection

There should be no incidence of insect pests and diseases during the cropping period.

4.8 Harvesting and Leaf Yield

The inflorescence may appear at an early stage, but pinching should be practiced to produce more leaves. The leaf of the plants should be harvested at marketable or edible stage. Approximately, an average leaf yield of 11,497.04 kg/ha could be obtained in a year.

4.9 Economy

The analysis in terms of economic impact revealed that growing of Eryngo can earn up to Rs 526,420.56 per hectare per annum with a C:B ratio of 1:5.34. In local markets of Manipur, the Eryngo is sold at Rs 10 per bundle, which is made of 3–5 plants based on the size and stage of growth and season (Fig. 12.2). The use of naturally available mulches is advocated for farmers as the cost of the mulching is economical, which could help in the reduction of cost of production. Therefore, application of vermicompost, followed by mulching with rice straw in the prescribed quantity, and application time are critically important inputs for the improved yield of Eryngo (Devi et al., 2021). It has been observed that Eryngo has been sold in most of the markets of Manipur, both in rural and urban localities almost throughout the year. Once, the plant was solely collected from the wild habitat, but now it is hardly seen in the wild, which might be due to its rampant collection for dietary requirements and medicinal uses. Due to its high demand and cost, now people are showing good interest for its cultivation as a remunerative crop. Due to the traditional mode of cultivation, the yield of the crop is not obtained up to the desired level. The scientific cultivation of the crop will definitely increase the income of the farmers to a great extent. With increased appreciation about its dietary and therapeutic values, the

Fig. 12.2 Eryngo is sold at Rs 10 per bundle in the local markets of Manipur

demand for this herb is gradually increasing. Now it has attained the status of one of the major cash crops with a flourishing scope for income generation for the ethnic communities of Manipur. A thorough market survey and assessment of the market demand can initiate the commercial-scale growing and expansion of the targeted species by the rural farmers. Market surveys can help to have better understanding of the indigenous spices found in the region. Eryngo being a high potential crop, there are hardly any instance where it has not been left unsold, resulting in a left-over stock in any of the local markets. Given the popularity and the market potential of Eryngo, one can devise a sound strategy for scientific cultivation and production throughout the year.

Food security has been a major challenge in several developing nations. Nations are still finding ways to make use of available resources to combat hunger (Alemu, 2017). Drought, land degradation, marketing constraints, improper post-harvest handling, and anthropogenic factors are aggravating the problems related to the production and productivity of agricultural land resources. In this scenario, the use of an improved technologies and adaptable and high-yielding cash crops can help combat the issues of food insecurity.

5 Conclusion

Eryngo (*Eryngium foetidum* L.) is an important local leafy spice plant used by the different ethnic communities of Manipur. The Eryngo plant is generally maintained in homestead gardens for self-use and also for income generation by selling it in the local markets. Earlier, it abundantly grew in the wild habitats but is now gradually getting lost from its natural habitat due to the unscientific and over exploitation, and it also might be due to the effect of weedicides used by the farmers in their paddy fields, leading to loss of its associated species. The demand of this plant in the local markets has increased substantially in the recent times. Now, the time has come to conserve this high-value local underutilized spice species as well as to cultivate it in private lands using the proper scientific technique. The commercial cultivation of Eryngo, by adopting the scientific method of cultivation, can help in ensuring the livelihood security of the marginal and small-scale farmers of Manipur. Every local market of Manipur sold Eryngo and farmers got good returns. It is also known that the cultivation of Eryngo with a judicious application of organic manure (vermicompost), along with organic mulches in open-field condition, can be beneficial for small and marginal farmers. The recommended dose of application of vermicompost at 2.5 tons per hectare with 5 tons of rice straw can give the best yield, which in turn can provide high-cost benefit ratio. Thus, by adoption of this technology, farmers can be benefited with higher returns while inflicting no harm to the environment. Due to exceptionally high content of nutritional values, especially the crude fibre and protein, Eryngo might help in nutrition and the healthcare sector, besides its utility in treating various diseases and ailments in the traditional healing system. Therefore, Eryngo plant is one of the most prized crops of Manipur, which needs to be promoted for big-scale commercial cultivation, production, and promotion. This spice plant can be a very important cash crop for farmers that might help in increasing the farmers' income, thus ensuring the livelihood security of the communities. Efforts must be made to formulate Eryngo plant-based value-added spice or food products, which might be in paste or dried forms for use during off-season and for profitable marketing. Though results are inconclusive, yet agricultural diversification and commercialization continue to remain as an effective solution to reduce poverty, improve household food and nutrition security, and promote growth in rural areas. As this crop is a low volume but high cost evident through its high market demand, scientific and intensive farming will significantly enhance and ensure the farmers' income, leading to balanced social, economic, and environmental sustainability.

References

Alemu, M. M. (2017). Banana as a cash crop and it's food security and socioeconomic contribution: The case of Southern Ethiopia, Arba Minch. *Journal of Environmental Protection, 8*, 319–329.

BSI. (2000). *Flora of Manipur* (Vol. 1). Botanical Survey of India.

Deb, D. B. (1961). Angiospermic Plants (monocot+ dicot) of Manipur Territory, *Bulletin, Botanical Survey of India, 3*(2), 115–138 & *3*(3&4), 253–350.

Devi, P. B. (2021). Ecological studies on Eryngo (*Eryngium foetidum* L.) in Manipur valley and development of its Agro-techniques. Unpublished PhD thesis, Assam University, Silchar.

Devi, P. B., Deb, P., & Singh, H. B. (2016). Economic uses of Eryngo/Culantro (*Eryngium foetidum* L.): A review. *Pleione, 10*(2), 356–361.

Devi, P. B., Deb, P., & Singh, H. B. (2021). Ethno-medicinal uses of Eryngo (*Eryngium foetidum* L.) by Meitei community of Manipur, Northeast India. *Indian Journal of Traditional Knowledge, 20*(3), 767–774.

Ignacimuthu, S., Arockiasam, S., & Ravichandran, P. (1999). Plant regeneration through somatic embryogenesis from mature leaf explants of *Eryngium foetidum*, a condiment. *Journal Plant Cell, Tissue & Organ Culture, 56*, 131–137.

Mao, A. A., & Hynniewta, T. M. (2000). Floristic diversity of North East India. *Journal Assam Science Society, 41*(4), 255–266.

Moniruzzaman, M., Islam, S., Hossain, M. M., Hossain, T., & Miah, M. G. (2007). Growth and yield response of Bilati Dhonia to light intensity and nitrogen levels. *Bangladesh Journal of Agricultural Research, 32*(1), 151–162.

Paul, J. H., Seaforth, C. E., & Tikasingh, T. (2011). *Eryngium foetidum* L.: A review. *Fitoterapia, 82*(3), 302–308.

Premsekhar, M., & Rajashree, V. (2009). Influence of organic manures on growth, yield and quality of Okra. *American-Eurasian Journal of Sustainable Agriculture, 3*(1), 6–8.

Sarolia, D. K., & Bharadwaj, R. L. (2012). Effect of mulching on crop production under rained condition: A review. *International Journal of Research in Chemistry and Environment, 2*(2), 8–20.

Shavandi, M. A., Haddadian, Z., & Ismail, H. M. H. (2012). *Eryngium foetidum* L., *Coriandrum sativum* L. and *Persicaria odorata* L.: A review. *Journal of Asian Scientific Research, 2*(8), 410–426.

Singh, H. B., & Devi, P. (2014). *Hidak Kanaraba Pambishing: A treatise of medicinal and aromatic plants (Manipuri medium)*. Padma Printers.

Singh, H. B., & Sundriyal, R. C. (2003). Common spices and their use in traditional system of ethnic groups of Manipur state, North eastern India. *Indian Journal of Traditional Knowledge, 2*(2), 148–158.

Singh, H. B., Singh, R. S., & Sandhu, J. S. (2003). *Herbal medicine of Manipur*. Daya Publishing House.

Sinha, S. C. (1996). *Medicinal plants of Manipur*. Mass Publication.

Yuhlung, C. C., & Bhattacharyya, M. (2015). Ethno-medicine practice among the Chothe tribe of Manipur, North-east India. *Tribal Health Bulletin, 22*(1 & 2), 40–58.

Chapter 13
Seasonal Variation of Ecosystem Fluxes of a Himalayan Banj-Oak-Dominated Vegetation

Priyanka Lohani, Sandipan Mukherjee, and K. Chandra Sekar

Abstract The *Quercus leucotrichophora*, commonly known as the Banj-oak in Uttarakhand, Indian Himalaya, provides a large number of ecosystem services to the people of Himalaya. However, due to the unavailability of direct flux observations, sub-daily to seasonal-scale variability of the Banj-oak ecosystem biosphere-atmosphere carbon exchanges have not been studied in detail. To address this lacuna, this study presents the first-time eddy covariance-based observations of carbon exchanges of a Banj-oak-dominated vegetation of western Himalaya. The carbon exchanges and atmospheric parameters were measured from 1 March, 2020, to 30 November, 2020, using a 10 m eddy covariance tower at Gangolihat, Pithoragarh, Uttarakhand (80.02°E, 29.39°N). The high-frequency carbon exchange data were quality checked and converted to daily net ecosystem exchange (NEE), gross primary productivity (GPP), and ecosystem respiration (RE). The variability of NEE, GPP, and RE was evaluated for the pre-monsoon (March–May), monsoon (June–September), and post-monsoon (October–November) seasons. Additionally, relationships between NEE and meteorological parameters (air temperature, PPFD, and VPD), the major controlling factors of plant photosynthesis, were analysed. Results of our analysis indicated that the Banj-oak-dominated vegetation were a net sink of CO_2 having a carbon uptake of 610.1 gCm^{-2} during the 275 days of 2020. The highest amount of NEE was noted for the pre-monsoon season (-2.72 $gC.m^{-2}.day^{-1}$), followed by monsoon (-1.96 $gC.m^{-2}.day^{-1}$) and post-monsoon seasons (-1.95 $gC.m^{-2}.day^{-1}$). The relationships between meteorological parameters and NEE

P. Lohani
G.B. Pant National Institute of Himalayan Environment, Almora, Uttarakhand, India

S. Mukherjee (✉)
Ladakh Regional Centre, G. B. Pant National Institute of Himalayan Environment, Leh, India
e-mail: sandipan@gbpihed.nic.in

K. C. Sekar
Garhwal Regional Centre, G. B. Pant National Institute of Himalayan Environment, Srinagar-Garhwal, India

S. Sharma et al. (eds.), *Climate Change Adaptation, Risk Management and Sustainable Practices in the Himalaya*,
https://doi.org/10.1007/978-3-031-24659-3_13

indicated a higher NEE coupling of post-monsoon season due to lesser rainfall activities, resulting to limited changes in the relationships between air temperature, radiation, and VPD.

Keywords Carbon exchange · Ecosystem services · Ecosystem flux · Gross primary prodictivity · Meteorological parameters · Net ecosystem exchange · Western Himalaya

1 Introduction

The broad-leaved forest ecosystems, particularly the oak (*Quercus* spp.) forests, have significantly influenced the livelihood of the people of Uttarakhand Himalayas as well as the people of adjacent plains, by providing a number of ecosystem services (Naudiyal & Schmerbeck, 2016). Oaks (*Quercus* spp.) are the zenith species of the Uttarakhand Himalayas and extensively distributed between 1000 to 3500 m elevations. The oaks of Uttarakhand are mostly evergreen, maintaining a dense canopy through strong leaf flushing before the shedding of old leaves. Four oak species are found naturally in Uttarakhand Himalayas, namely, *Q. glauca* known as Phaliyant, *Q. leucotrichophora* known as Banj, *Q. floribunda* known as Tilonj or Moru, and *Q. semecarpifolia* known as Kharsu (Upreti et al., 1985). Out of the four oak species, three species, *Q. leucotrichophora*, *Q. floribunda*, and *Q. semecarpifolia*, have relatively more importance in terms agro-ecosystems and as a life support system to the inhabitants of the hills (Rathore et al., 1985). The Banj-oak (Q. *leucotrichophora*)-dominated forests of the western Himalaya are known for their rich biodiversity with complex branching pattern. Furthermore, the Banj-oak system normally has a significant number of logs and snags on the floor sheltering mosses, lichens, and epiphytes (Singh et al., 2014), as the bark of *Quercus* spp. provides a suitable environment for growth of these lower plant species. According to the recent study by Dhyani et al. (2020), more than 5% (around 1284 km^2) of the total forest area of Uttarakhand is occupied by the Banj-oak forests within the elevation ranges of 1500 m to 2300 m above mean sea level. Out of 1284 km^2 area under Banj-oak, around 840 km^2 falls within the Garhwal region, whereas nearly 445 km^2 area falls within the Kumaon region of Uttarakhand. The Banj-oak vegetation, which is the most common amongst the different oak species, can be found in Uttarakhand, where temperature ranges between −6.7 °C and 35 °C and the annual rainfall varies between 1200 mm and 1600 mm, out of which approx. 75% of the total rainfall occurs in the monsoon season (June to September) (Singh & Singh, 1986, 1987). The soil organic matter and water holding capacity of Banj-oak vegetation result in better recharge of the aquifers and subsurface streams (Nautiyal et al., 2020). However, deforestation, enhanced forest fires, and fluctuating rainfall climatology of the last few decades have impacted variability of the air temperature and vapour pressure deficit (VPD), further affecting the ecosystem responses and regeneration pattern of forests (Naudiyal & Schmerbeck, 2016). Furthermore, the relationships between Banj-oak ecosystem responses and changing meteorological parameters are seldom evaluated. In fact, the daily- to seasonal-scale variation in the net ecosystem exchanges (NEE) of Banj-oak has never been reported. Therefore, this study presents the first-time assessment of seasonal variation of NEE with gross primary productivity (GPP) and ecosystem respiration (RE) of a Banj-oak-dominated forest of Uttarakhand Himalaya. Additionally, the functional relationships between NEE-air temperature, PPFD, and VPD are analysed, as these are considered as the major drivers of ecosystem functioning of terrestrial ecosystems (Cowan, 1977; Farquhar & Sharkey, 1982; Grossiord et al., 2020).

2 Site Description, Instrumentation, and Methodology

The ecosystem fluxes and meteorological parameters were measured at a field station situated in the Gangolihat block, Pithoragarh, Uttarakhand (80.02°E, 29.39°N), at an elevation of 1650 m using a 10 m eddy covariance flux tower (Fig. 13.1). The eddy covariance (eddy covariance measurements, commonly known as EC measurements, are one of the most direct and appropriate ways to measure gas fluxes) flux tower was installed under the 'National Mission on Sustaining the Himalayan

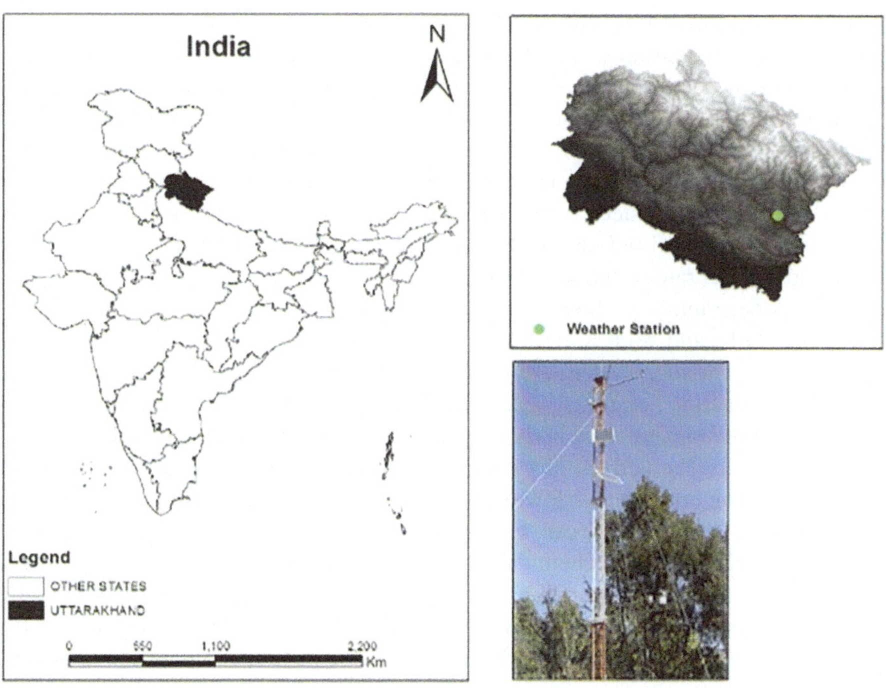

Fig. 13.1 Location of the measurement mast within *Quercus leucotrichophora* patch at the Gangolihat, Pithoragarh of Uttarakhand, India

Ecosystem (NMSHE): Task Force-3 (Phase-1) and is maintained by NMSHE (Phase-2)’. The site falls under Koppen climate classification of C_{wb} with an average monsoon season rainfall of around 704 mm and with average heavy rainy days of 2.48 per season (Mukherjee, 2021; Mukherjee et al., 2015). The site has dense and undisturbed forest patch, and it falls under Hat-Kalika watershed. The land-cover assessment of the watershed shows that out of the 37 km^2 of total watershed area, 54.5% is under forest, of which 38.4% area had dominant *Q. leucotrichophora* trees. The flux tower was surrounded by *Q. leucotrichophora*, maintaining a dense canopy throughout the year. The associated dominant tree species within the forest cover include *Pyrus pashia*, *Cedrus deodara*, and *Juglans regia* (Joshi et al., 2020). The average height of the canopy was around 8 m. The eddy covariance tower had a 3D sonic anemometer (CSAT3, Campbell Scientific Inc., Utah, USA), along with an infrared CO_2/H_2O gas analyser (EC150, Campbell Scientific, Utah, USA), installed at 10 m height having a working frequency of 10 Hz, measuring 3D wind vectors, CO_2/H_2O flux, and sonic temperature. Furthermore, net radiation sensor (NR Lite, Kipp and Zonen, Netherlands), rain gauge (TE525, Campbell Scientific, Utah, USA), and relative humidity-air temperature sensors were also installed. The raw data files were produced at a 30 min interval.

Data in this study were analysed for the period of 1 March, 2020, to 30 November, 2020, with special emphasis to pre-monsoon (March–May), monsoon (June–September), and post-monsoon (October–November) seasons. The rationale for considering pre-monsoon, monsoon, and post-monsoon was the different surface meteorological conditions, which changes with respect to seasons. As surface meteorological conditions modulate ecosystem responses, particularly the stomata opening and closure in the plants and subsequent change in the plant physiological properties, impacts of different surface meteorological conditions of three different seasons were quantified. Special emphasis to seasonal changes in the ecosystem fluxes of Banj-oak system could also be corroborated to the leaf flushing and flowering of the Banj-oak-dominated ecosystem during April–October, signifying enhanced growth rate of the system.

A brief description of how ecosystem fluxes are measured is provided below following Burba and Anderson (2009). The ecosystem flux exchange measurements using eddy covariance system are generally carried out under the assumption that airflow is primarily horizontal with each eddy parcel moving in the upward and downward direction and each eddy having 3D component of temperature, chemical constituents, etc. The vertical flux could be computed at a point once the number of particles moving upward and downward within a known volume of air is measured. Subsequently, vertical flux can be mathematically represented as covariance of vertical velocity and the chemical constituent. In atmospheric turbulent flows, vertical flux can be represented as $F = \overline{\rho_a w_s}$, where $s = \rho_c / \rho_a$indicting a mixing ratio of a chemical constituent c in the air. After Reynold’s decomposition of the covariance term to mean and perturbation, the flux equation could be elaborated as:

$$F = \bar{\rho}_a \bar{w} \bar{s} + \bar{\rho}_a \overline{w' s'} + \bar{w} \overline{\rho_a' s'} + \bar{s} \overline{\rho_a' w'} + \overline{\rho_a' s' w'} \quad (13.1)$$

Subsequently, two assumptions are made – (i) density fluctuations are negligible and (ii) mean vertical flow is negligible – and the final flux of a chemical constituent is calculated as follows:

$$F \approx \bar{\rho}_a \overline{W'S'} \tag{13.2}$$

In case of NEE, CO_2 molecules are transported by eddies from the ecosystem surface to the sensor at a measurement height, subsequently by using eddy covariance method; the net exchange rate of CO_2 between the atmosphere and the ecosystem is calculated.

The primary eddy covariance data filtering was carried out by using the EddyPro software (v6.0, LiCor-Bioscience, USA). Here, the displacement height ($d = 5.33$ m) was estimated using the canopy height (8 m) and parameterization of Foken and Nappo (2008). Moreover, sonic temperature correction for the air temperature was done using water vapor density and formulations of Schotanus et al. (1983), followed by the WPL correction (Webb et al., 1980). A planar fit correction with 30° wind sectors is used for tilt correction in the streamlines following the traditional method of Wilczak et al. (2001). After EddyPro processing, the eddy covariance data were used for flux partitioning, and the 30 min values of NEE, GPP, and RE values in $\mu mol.m^{-2}.s^{-1}$ were produced using the methods of Lasslop et al. (2010) of REddyProc (v1.1.3). The ambient water vapour mass densities were used to compute 30 min VPD values (kPa). The global radiation was used to compute 30 min photosynthetic photon flux density (PPFD) values ($\mu mol.m^{-2}.s^{-1}$). Furthermore, a spike detection method of Papale et al. (2006) was used for all the variables to improve data quality. However, the additional amount of CO_2, which does not take part in exchange processes, assumed to be stored in canopy, left unmeasured by eddy covariance system. At last, the daily averages of NEE, GPP, RE, VPD, and PPFD were computed along with the daily averages of wind speed (ms^{-1}), air temperature (°C), relative humidity (%), and daily total rainfall (mm). Here, a negative NEE implies carbon assimilation by the ecosystem. The flux variability assessments were carried out for the pre-monsoon (March–May), monsoon (June–September), and post-monsoon (October–November) seasons. The NEE-VPD relationships were quantified for each season using the power law relationship NEE = aVPD^b, and a and b were optimised.

3 Results and Discussions

3.1 Surface Meteorological Conditions

This section presents analyses of basic meteorological parameters of pre-monsoon, monsoon, and post-monsoon seasons with the rationale that better comprehension of the seasonal variations in the surface meteorological conditions would provide linkages to ecosystem response and changing environment. Variability of the surface

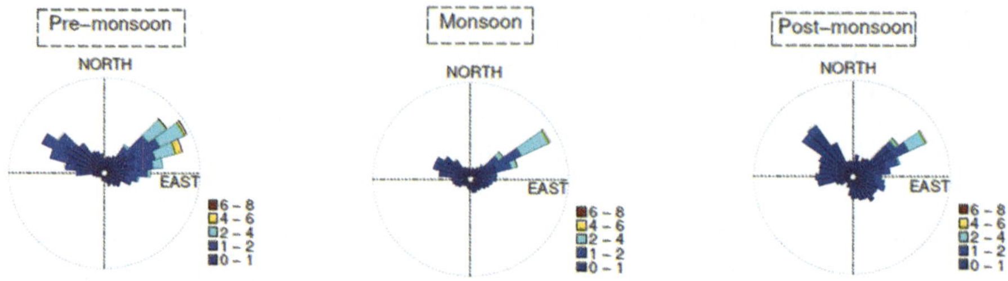

Fig. 13.2 Wind distribution observed at the site in different seasons

meteorological parameters during 00:00 Hrs of 1 March 2020 to 23:30 Hrs 30 November 2020 was assessed by producing wind roses and contour plots of air temperature, relative humidity, PPFD, and VPD. The reason for considering only these four environmental parameters was the direct linkages of all these parameters to physiological properties of oak trees, such as the opening and closure of the stomata that are primarily controlled by the availability of light and vapour pressure deficit; similarly, stand-level transpiration and water use efficiencies of trees are also modulated by air temperature and PPFD. Prior to more detailed analysis, the wind roses for pre-monsoon, monsoon, and post-monsoon seasons are provided in Fig. 13.2. The average wind speed for the entire observation period was 0.82 ms^{-1}. The pre-monsoon months were relatively windy (average = 1.01 ms^{-1}), where the maximum wind speed reaches up to 7.38 ms^{-1} during 11:00 to 13:00 Hrs IST. On the contrary, post-monsoon months had the lowest wind variability (average = 0.67 ms^{-1}) with a minimum wind speed of 0.008 ms^{-1}. Irrespective of seasons, the wind direction was dominantly north-westerly and north-easterly (Fig. 13.2). The lower wind speed in the monsoon season indicates lower advection of moisture and heat in the ecosystem, hence comparatively lesser responses of the ecosystem to environmental conditions.

The season-wise diurnal variations of air temperature, relative humidity, VPD, and PPFD are provided in Fig. 13.3. The diurnal variation in the air temperature clearly indicates that at the beginning of pre-monsoon season, the temperature of the site was mild, and the maximum air temperature during noon was around 15 °C; however, by the end of the pre-monsoon season, the maximum air temperature during noon was around 30.8 °C (Fig. 13.3a). Very clear diurnal variation in the air temperature was noted during monsoon with a seasonal maximum of 28.1 °C, which became 26.3 °C in post-monsoon. Again, by the end of the post-monsoon season, the maximum air temperature during noon was around 16 °C (Fig. 13.3c). The average air temperatures during pre-monsoon, monsoon, and post-monsoon periods were 17.0, 22.0, and 17.5 °C, respectively. The relative humidity of the site ranged from 7.1 to 100% during the observation period with an average of 58.5% (Fig. 13.3b, f, j). As expected, the monsoon season had the highest average relative humidity (75.2%), and the driest season was post-monsoon (42.6%). Since relative humidity modulates VPD, the lowest seasonal average VPD was noted during monsoon (0.67 kPa), and the highest VPD was noted during post-monsoon (1.16 kPa) (Fig. 13.3c, g, k). The diurnal variation of PPFD is provided in

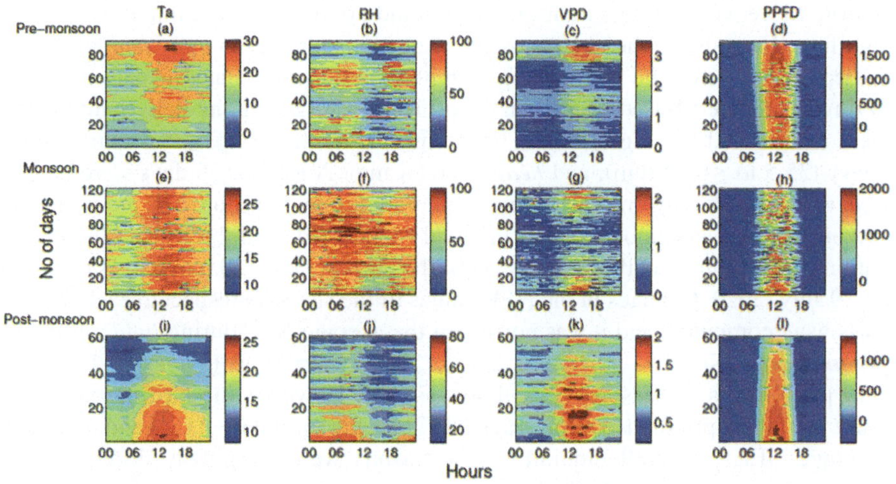

Fig. 13.3 Diurnal variations of (**a**, **e**, **f**) air temperature (°C), (**b**, **f**, **j**) relative humidity (%), (**c**, **g**, **k**) VPD (kPa), and (**d**, **h**, **l**) PPFD (μ mol.m $^{-2}$ .s $^{-1}$) are presented, respectively

Table 13.1 Distribution of rainfall during observation periods of 2020 is presented

	Light		Moderate		Rather Heavy		Heavy	
Seasons	No. days	Avg. rain (mm)	No. days	Avg. rain (mm)	No. days	Avg. rain (mm)	No. days	Avg. rain (mm)
Pre-monsoon	27	3.81	06	18.6	01	51.5	–	–
Monsoon	34	2.98	19	18.9	03	45.8	02	94.6
Post-monsoon	–	–	–	–	–	–	–	–

The rainfall classifications are as per Pattanaik and Rajeevan (2010)

Fig. 13.3d, h, l. As expected, significant variations in the PPFD values were noted during monsoon season due to enhanced cloud activity; however, the seasonal average values during pre-monsoon, monsoon, and post-monsoon seasons were 302.7, 294.5, and 173.2 μmol.m^{-2}.s^{-1}, respectively.

In general, rainfall has a strong effect on the vegetation growth over the central Himalayan region (Singh & Singh, 1986); however, detailed analyses of rainfall spell and Banj-oak ecosystem responses are not yet carried out. Prior to a more detailed assessment of Banj-oak ecosystem response to rainfall spell, the rainfall statistics at the site during the observation period was computed. Irrespective of seasons, the site received a total of 1075 mm rainfall (Table 13.1). A total of 286.1 mm and 788.9 mm rainfall was observed during the pre-monsoon and monsoon seasons, respectively, and the monsoon season rainfall was almost similar to the long-term average of 704 mm as reported by Mukherjee et al. (2015). The highest cumulative rainfall was noted at the site during August 2020 (314.1 mm), and approximately 73% of the total observed rainfall was recorded in monsoon season. Moreover, the post-monsoon season did not receive any rainfall, which can significantly modulate the plant activities. The monsoon season daily rainfall started to decrease towards the end of August with a few more isolated events observed during September. Based on Pattanaik and Rajeevan (2010), the daily rainfall data were categorized into *light* (0 to ≤10.0 mm), *moderate* (10 to ≤35.5 mm), *rather heavy* (35.5 to ≤64.4 mm), and *heavy* > 64.4 mm. A total of 56 days were having *light* rainfall with only one case of *heavy* rainfall during monsoon. The pre-monsoon season rainfall was almost 26% of the total rainfall received throughout the observation period. Out of all the rainfall events, the highest rain (120.6 mm) was recorded on 29 July, 2020. Changes in the NEE values of the site with respect to each rainfall class were computed, and it was noted that the average NEE during *light* rainfall of pre-monsoon (NEE = −2.19 μmol.m^{-2}.s^{-1}) and monsoon (average NEE = −2.19 μmol.m^{-2}.s^{-1}) seasons were almost similar. However, the NEE during *moderate* rainfall of pre-monsoon (average NEE = −2.81 μmol.m^{-2}.s^{-1}) and monsoon (average NEE = −2.39 μmol.m^{-2}.s^{-1}) seasons were marginally different. The difference in the average NEE during *rather heavy* rainfall of pre-monsoon (NEE = −3.56 μmol.m^{-2}.s^{-1}) and monsoon (average NEE = −2.50 μmol.m^{-2}.s^{-1}) seasons was significantly high. Since pre-monsoon season is generally dry at the measurement location, increasing rainfall spell was noted to result in higher Banj ecosystem responses.

3.2 *Daily and Seasonal Variation of Ecosystem Fluxes*

In order to evaluate the sub-daily-scale variation of carbon uptake by the vegetation under consideration, diurnal changes of NEE during the three seasons are plotted (Fig. 13.4). During all the seasons, the NEE has a similar pattern, and irrespective of

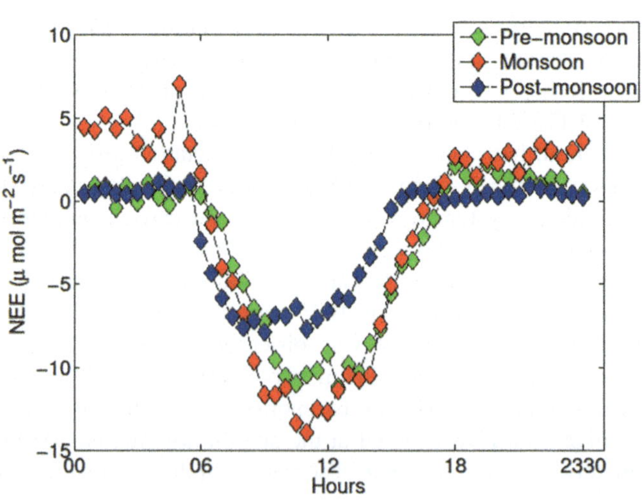

Fig. 13.4 Diurnal variations of NEE in different seasons are represented

the seasons, the lowest NEE due to enhanced carbon assimilation was noted around midday (11:00 to 13:00 IST) time. The diurnal variation of NEE further indicates that the maximum carbon sequestration was observed around 11:00 IST in pre- and post-monsoon seasons, while for the monsoon season, it was around 12:00 IST. As the day progressed, the carbon sequestration rate decreases gradually. It can further be noted from the diurnal variation of NEE that higher carbon released by vegetation had occurred during the morning hours of 00:00–06:00 IST monsoon season than the pre- and post-monsoon months. In the pre-monsoon season, nocturnal fluxes ranged from 0.2 to 1.09 μmol.m^{-2}.s^{-1} during 00:00–06:00 IST. As the day progresses during the pre-monsoon season, enhanced carbon uptake was noted around 11:00 to 13:00 IST. Although, this trend changed gradually with time and around 17:00 IST, the ecosystem started releasing carbon (0.4 to 2.24 μmol.m^{-2}.s^{-1}). In the monsoon season, nocturnal fluxes significantly changed during 00:00 to 06:00 IST (1.65 to 7.04 μmol.m^{-2}.s^{-1}) and 17:00 IST (0.2 to 3.66 μmol.m^{-2}.s^{-1}) due to higher respiration. However, in post-monsoon season, the carbon uptake started at around 06:30 IST, and after 15:00 IST, the ecosystem started releasing the carbon in the atmosphere. Changes in the diurnal patterns of NEE in the monsoon months can be attributed to changes in the PPFD and VPD parameters due to frequent changes in the cloud and moisture content of the atmospheric surface layer. On a sub-daily scale, the highest (lowest) carbon assimilation was noted during monsoon (post-monsoon) season, i.e. −13.9 (−7.68) μmol.m^{-2}.s^{-1}; however, the highest (lowest) carbon release was also noted during monsoon (post-monsoon) season, i.e. 7.04 (1.12) μmol.m^{-2}.s^{-1}. Therefore, the sub-daily-scale variation in the NEE values indicates that the Banj-oak system was most active during monsoon period with enhanced carbon assimilation and release.

The above-ground net biomass carbon of the Banj-oak forest is approximately 9.68 t.C.h^{-1}.$year^{-1}$ (Naudiyal & Schmerbeck, 2016); however, forest disturbances may lead to reduction in the amount of carbon stock of Banj-oak vegetation. The novelty of this study is the measurement of sub-daily- to almost annual-scale carbon uptake and release by an undisturbed Banj-oak vegetation of the Himalayas. The daily averages of NEE, GPP, and RE of the Banj-oak vegetations of this study are presented in Fig. 13.5. Irrespective of seasons, the daily average NEE was found to

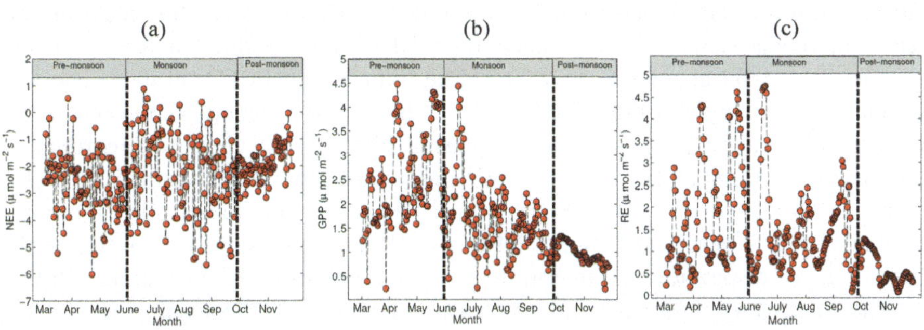

Fig. 13.5 Subplots are showing daily average variations of ecosystem fluxes (**a**) NEE, (**b**) GPP, and (**c**) RE during pre-monsoon, monsoon, and post-monsoon seasons, respectively

be -2.21 gC.m^{-2}.day^{-1}. Consequently, the study site was a net sink of CO_2 having a total carbon uptake of 610.1 gCm^{-2} during the 275 days of 2020. Similarly, the daily averages of GPP and RE were found to be 2.09 gC.m^{-2}.day^{-1} and 1.26 gC.m^{-2}.day^{-1}, respectively, which are relatively smaller than a pine (*Pinus roxburghii*)-dominated mixed vegetation of Uttarakhand (Deb Burman et al., 2021; Mukherjee et al., 2018). It can be further noted from the Fig. 13.5 that pre-monsoon months were having higher daily average NEE (-2.72 gC.m^{-2}.day^{-1}) than monsoon (-1.96 gC.m^{-2}.day^{-1}) and post-monsoon (-1.95 gC.m^{-2}.day^{-1}) seasons. This lower value of carbon assimilation in the monsoon season by the ecosystem corroborates well with Fig. 13.4, as it was earlier indicated that the Banj-oak system also released higher carbon in the atmosphere during monsoon associated to higher respiration. The higher NEE of pre-monsoon could be due to higher air temperature and PPFD providing suitable environment for enhanced leaf flushing and flowering (Singh & Singh, 1986).

3.3 Functional Relationship Between NEE and Surface Meteorological Parameters

Many studies have highlighted that precipitation, air temperature, vapour pressure deficit, and radiation are the major influencing factors of the ecosystem growth at regional and global scale (Grossiord et al., 2020; Hao et al., 2020; Law et al., 2002; Renaud et al., 2011; Renaud & Rebetez, 2009). Therefore, attempts are made in this study to quantify seasonal-scale relationships between these meteorological parameters and NEE of the Banj-oak system (Fig. 13.6). It is already mentioned in Sect. 3.1 that air temperature increases gradually in the pre-monsoon season at the site till the end of May; however, when this air temperature trend was evaluated with respect to the NEE of the Banj-oak system, no statistically significant correlation was observed (Fig. 13.6a), although it can be clearly noticed from the figure that carbon uptake rate of the vegetation decreases under high temperature of pre-monsoon season. Similarly, in the monsoon season, no statistically significant relationship was observed between NEE and air temperature. However, in the post-monsoon season, due to less surface heating, the air temperature was noted to decline, and this declining air temperature was statistically correlated to the NEE values of the system (correlation coefficient = 0.52 at *p-value* < 0.001). The influence of PPFD on net ecosystem exchange is presented in Fig. 13.6d–f. The carbon uptake capacity of the ecosystem under consideration increases with increasing PPFD in pre-monsoon (correlation coefficient = 0.38 at *p-value* < 0.001) and post-monsoon (correlation coefficient = 0.50 at *p-value* < 0.001) seasons. However, no statistically significant relationship was observed between NEE and PPFD in monsoon season due to the high effect of respiration and intermittent cloud cover resulting fluctuating PPFD.

The plants are highly responsive to variation in the VPD leading to stomata closure, which further reduces the carbon uptake (Cowan, 1977; Farquhar & Sharkey, 1982). In actuality, higher VPD values can reduce the carbon assimilation

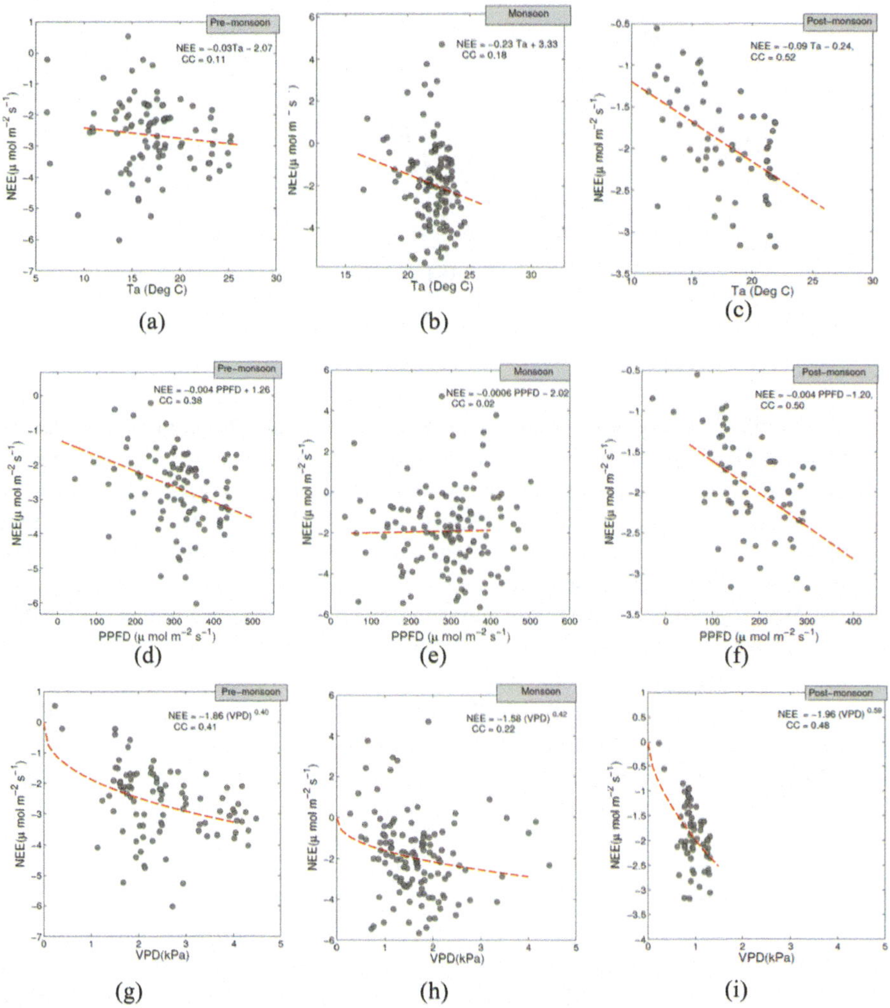

Fig. 13.6 The relationships between NEE-temperature (**a–c**), NEE-PPFD (**d–f**), and NEE-VPD (**g–i**) are presented for pre-monsoon, monsoon, and post-monsoon periods. Red line shows fitted power law/linear equation

and an inverse relation can be observed between NEE and VPD (Zhang et al., 2016; Zhou et al., 2014). Stomatal conductance is observed to decline under high VPD condition, which further increases the transpiration in most plants up until a given VPD threshold; subsequently, photosynthesis and growth are reduced. In order to quantify the impact of VPD on NEE, a power law relationship between NEE and VPD was evaluated by plotting daily average values of VPD against NEE. It can be noted from Fig. 13.6g–i that the NEE-VPD relationship of Banj-oak system, irrespective of seasons, was moderately strong. The power law relationship NEE

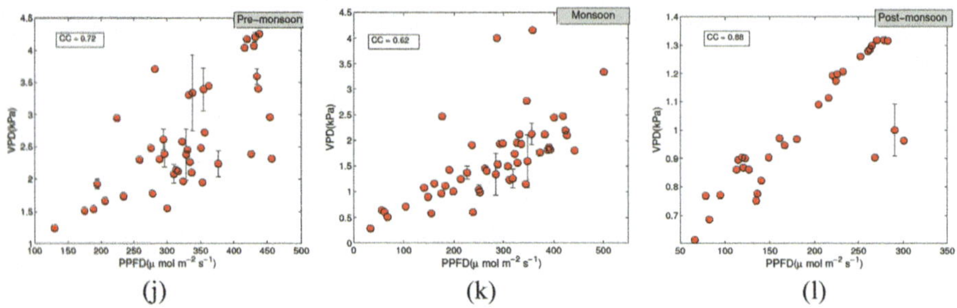

Fig. 13.7 The relationships between PPFD and VPD are presented for (**a**) pre-monsoon, (**b**) monsoon, and (**c**) post-monsoon periods

$= -1.96\text{VPD}^{0.59}$ of post-monsoon season was noted to have the highest correlation coefficient of 0.48 (*p-value* < 0.001), followed by the pre-monsoon season (NEE $=$ $-1.86\text{VPD}^{0.40}$) correlation coefficient of 0.41 (*p-value* < 0.001). The monsoon season relationship was noted to be NEE $= -1.58\text{VPD}^{0.42}$ having correlation coefficient of 0.23 with no statistical significance. The higher NEE-VPD coupling of post-monsoon and pre-monsoon seasons could be linked to lesser rainfall activities resulting in limited changes in the relationships between radiation and VPD, i.e. correlation coefficients between VPD and PPFD were 0.88 and 0.72 (*p-value* < 0.001), respectively, whereas the rainfall events of monsoon season had significantly modulated the VPD fluctuations through changes in PPFD, i.e. correlation coefficient between VPD and PPFD was 0.62 at a *p-value* < 0.001 (Fig. 13.7a–c).

4 Conclusion

Sub-daily to seasonal-scale variation of ecosystem fluxes of a Banj-oak forest (*Quercus leucotrichophora*) of Uttarakhand is studied using the ground observations in 2020. The ecosystem fluxes were measured using an eddy covariance system during 1 March 2020 to 30 November 2020 at 10 m height in Gangolihat, Pithoragarh, Uttarakhand. The high-frequency data are quality checked and used to quantify the pre-monsoon, monsoon, and post-monsoon season patterns of NEE, GPP, and RE. Additionally, relationships between NEE and air temperature, PPFD, and VPD are analysed for identifying seasonal variability. The Banj-oak vegetation is found to be a net sink of CO_2 having a total carbon uptake of -610.1 gCm^{-2} during 275 days of observation. The highest amount of carbon sequestration is noted during pre-monsoon season due to higher temperature and PPFD providing suitable environment for enhanced photosynthesis along with leaf flushing and flowering. The lowest seasonal carbon assimilation is noted for post-monsoon season, often associated to leaf fall. The meteorological parameters that affect the plant physiology are noted to have higher coupling with NEE during the post-monsoon and pre-monsoon seasons due to lesser rainfall activities. This study provides some of the very initial measurements and seasonality assessments of ecosystem fluxes between Banj-oak vegetation and atmosphere in Uttarakhand. Subsequently, it is expected that this ecosystem flux measurement would continue in future, and the flux data would be used for addressing many pertinent research questions related to biosphere-atmosphere interactions of Oak systems of Himalaya.

Acknowledgement The flux tower of Gangolihat, Pithoragarh, India, was established through a research grant from NMSHE-TF3(Phase-1) and is maintained through NMSHE-TF3(Phase-2), Govt. of India. P. Lohani, S. Mukherjee, and K. Chandra Sekar are grateful to NMHS-MoEFCC, Govt. of India, for providing a research grant for analysing the oak system data (GBPI/NMHS-2019-20/MG62/04, 2019) for this study.

References

Burba, & Anderson. (2009). *A brief practical guide to eddy covariance flux measurements: Principles and workflow examples for scientific and industrial applications, LI-COR biosciences* (Version 1.0.1).

Cowan, I. (1977). Stomatal behaviour and environment. *Advances in Botanical Research, 4*, 117–228.

Deb Burman, P. K., Launiainen, S., Mukherjee, S., Chakraborty, S., Gogoi, N., Murkute, C., Lohani, P., Sarma, D., & Kumar, K. (2021). Ecosystem-atmosphere carbon and water exchanges of subtropical evergreen and deciduous forests in India. *Forest Ecology and Management, 495*, 119371.

Dhyani, S., Kadaverugu, R., & Pujari, P. (2020). Predicting impacts of climate variability on Banj oak (Quercus leucotrichophora A. Camus) forests: Understanding future implications for Central Himalayas. *Regional Environmental Change, 20*, 113.

Farquhar, G., & Sharkey, T. (1982). Stomatal conductance and photosynthesis. *Annales Review of Plant Physiology, 33*, 317–345.

Foken, T., & Nappo, C. (2008). *Micrometeorology*. Springer.

Grossiord, C., Buckley, T. N., Cernusak, L. A., Novick, K., Poulter, B., Seigwolf, R., Sperry, J., & McDowell, N. (2020). Plant responses to rising vapour pressure deficit. *New Phytologist, 226*(6), 1550–1566.

Hao, G., Hu, Z., Di, K., & Li, S. (2020). Rainfall pulse response of carbon exchange to the timing of natura intra-annual rainfall in a temperate grass ecosystem. *Ecological Indicators, 118*, 106730.

Joshi, B., Rawal, R. S., Chandra Sekar, K., & Tewari, A. (2020). Assessment of fuelwood resource preference in representative watershed of west Himalaya, India: Conservation and management implications. *Environment, Development and Sustainability, 22*, 1617–1632.

Lasslop, G., Reichstein, M., Papale, D., Richardson, A., Arneth, A., Barr, A., Stoy, P., & Wohlfahrt, G. (2010). Separation of net ecosystem exchange into assimilation and respiration using a light response curve approach: Critical issues and global evaluation. *Global Change Biology, 16*(1), 187–208.

Law, B., Falge, E., Gu, L., Baldocchi, D., Bakwin, P., Berbigier, P., Davis, K., Dolman, A.-J., Falk, M., Fuentes, J., Goldstein, A., Grelle, A., Hollinger, D., Janssens, I.-A., Jarvis, P., Jenssen, N., Katul, G., Mahli, K., Matteucci, G., Meyers, T., Monson, R.-K., Munger, W., Oechel, W., Olson, R., Pilegaard, K., Paw, U.-T., Thorgeirsson, K., Valentini, H., Verma, R., Vesala, T., Wilson, J., & Wofsy, S. (2002). Environmental controls over carbon dioxide and water vapour exchange of terrestrial vegetation. *Agricultural and Forest Meteorology, 113*, 97–120.

Mukherjee, S., Joshi, R., Prasad, R., Vishvakarma, S., & Kumar, K. (2015). Summer monsoon rainfall trends in the Indian Himalayan region. *Theoretical and Applied Climatology, 121*(3–4), 789–802.

Mukherjee, S., Sekar, K. C., Lohani, P., Kumar, K., Patra, P., & Ishijima, K. (2018). Investigation of scale interaction between rainfall and ecosystem carbon exchange of Western Himalayan Pine dominated vegetation. *Biogeosciences Discussions, 2018*, 1–23.

Mukherjee, S. (2021). Summer monsoon rainfall trends in the Indian Himalayan region. *Science of the Total Environment, 789*, 147754.

Naudiyal, N., & Schmerbeck, J. (2016). The changing Himalayan landscape: Pine-Oak forest dynamics and the supply of ecosystem services. *Journal of Forestry Research, 28*(3), 431–443.

Nautiyal, H., Mathur, V., Sinha, A., & Huffman, M.-A. (2020). The Banj oak *Quercus leucotrichophora* as a potential mitigating factor for human-langur interactions in the Garhwal Himalayas, India: People's perceptions and ecological importance. *Global Ecology and Conservation, 22*.

Papale, D., Reichstein, M., Aubinet, M., Canfora, E., Bernhofer, C., Kutsch, W., Longdoz, B., Rambal, S., Valentine, R., Vesala, T., & Yakir, D. (2006). Towards a standardized processing of Net Ecosystem Exchange measured with eddy covariance technique: Algorithms and uncertainty estimation. *Biogeosciences, 3*, 571–583.

Pattanaik, D., & Rajeevan, M. (2010). Variability of extreme rainfall events over India during southwest monsoon season. *Meteorological Applications, 17*, 88–104.

Rathore, P., Roy, A., & Karnatak, H. (1985). Assessing the vulnerability of Oak (Quercus) forest ecosystems under projected climate and land use land cover changes in Western Himalaya. *Biodiversity and Conservation, 28*, 2275–2294.

Renaud, V., Innes, J., Dobbertin, M., & Rebetez, M. (2011). Comparison between open-site and below-canopy climatic conditions in {Switzerland} for different types of forests over 10 years (1998-2007). *Theoretical and Applied Climatology, 105*(1–2), 119–127.

Renaud, V., & Rebetez, M. (2009). Comparison between open-site and below-canopy climatic conditions in Switzerland during the exceptionally hot summer of 2003. *Agricultural and Forest Meteorology, 149*(5), 873–880.

Schotanus, P., Nieuwstadt, F., & DeBruin, H. (1983). Temperature measurement with a sonic anemometer and its application to heat and moisture fluctuations. *Boundary-Layer Meteorology, 26*, 81–93.

Singh, J. S., & Singh, S. P. (1986). Structure and function of the Central Himalayan oak forests. *Proceedings of the Indian Academy of Science (Ptant Sciences), 96*(3), 159–189.

Singh, J. S., & Singh, S. P. (1987). Forest vegetation of the Himalaya. *The Botanical Review, 53*(1), 80–192.

Singh, V., Thadani, R., Tiwari, A., & Ram, J. (2014). Human Influence on Banj Oak (*Quercus leucotrichophora*, A. Camus) Forests of Central Himalaya. *Journal of Sustainable Forestry, 33*, 373–386.

Upreti, N., Tiwari, J. C., & Singh, S. P. (1985). The oak forests of the Kumaun Himalaya (India): Composition, diversity, and regeneration. *Mountain Research and Development, 5*(2), 163–174.

Webb, E., Pearmann, G., & Leuning, R. (1980). Correction of flux measurements for density effects due to heat and water-vapour transfer. *Quarterly Journal of the Royal Meteorological Society, 106*, 85–100.

Wilczak, J., Oncley, S., & Stage, S. (2001). Sonic anemometer tilt correction algorithms. *Boundary-Layer Meteorology, 99*, 127–150.

Zhang, Z., Zhao, P., McCarthy, H., Zhao, X., Niu, J., Zhu, L., Ni, G., Ouyang, L., & Huang, Y. (2016). Influence of the decoupling degree on the estimation of canopy stomatal conductance for two broadleaf tree species. *Agricultural Forest Meteorology, 221*, 230–241.

Zhou, S., Yu, B., Huang, Y., & Wang, G. (2014). The effect of vapour pressure deficit on water use efficiency at the subdaily time scale. *Geophysical Research Letters, 41*(14), 5005–5013.

Chapter 14
Human-Wildlife Conflict in the Western Himalaya: A Systematic Review of Research and Conservation Interventions Implemented Over Three Decades

Christi Sylvia and Rishi Kumar Sharma

Abstract Human-wildlife conflict (HWC) is a daunting challenge for conservation worldwide. Multiple social and ecological factors influence HWC at different spatial scales, making it challenging to determine effective management strategies. Coupled with issues of climate breakdown and its overarching influence on the livelihood security of mountain people, HWC is becoming a prominent issue in the Indian Himalayas. The rising HWC threatens to diminish local communities' historical tolerance towards wildlife. A growing number of scientific studies indicate the urgency of finding a solution to this problem. Yet, a synthesis of the existing knowledge and solutions that work is lacking. Through a systematic review of 88 publications from 1991 to 2020, we examine the current state of information on HWC in the Western Himalayas and identify the strengths and limitations of the existing knowledge to mitigate conflict.

Our analysis revealed (i) a lack of longitudinal studies on conflict, with 42% ($n = 37$) of studies being one-time snapshots of HWC; (ii) even though the majority of the study region is essentially a socio-ecological system, 64% ($n = 56$) of the studies focused only on either ecological or social aspects of conflict; (iii) a considerable proportion of studies (45%, $n = 40$) provided generic conflict mitigation measures; (iv) lack of a standardised approach to quantify monetary losses rendered results of the majority of studies incomparable even within sites; and (v) 22% ($n = 19$) studies evaluated effectiveness of interventions, of which $n = 9$ were external evaluations. Most evaluations were narrative accounts of the efficacy of conservation interventions.

Snow leopards, black bears, and Himalayan wolves were the most studied conflict-prone species, but there were considerable spatial gaps in HWC studies across the Western Himalayas.

C. Sylvia · R. K. Sharma (✉)
Wildlife and Habitats Program WWF India, New Delhi, India
e-mail: rksharma@wwfindia.net

S. Sharma et al. (eds.), *Climate Change Adaptation, Risk Management and Sustainable Practices in the Himalaya*,
https://doi.org/10.1007/978-3-031-24659-3_14

Generating systematic long-term data to understand the underlying spatial and temporal drivers of HWC is urgently needed. A carefully implemented multi-stakeholder approach seemed to be the most successful in managing HWC in the Western Himalayas.

Keywords Human-wildlife interactions · Western Himalaya · Evidence-based conservation · Knowledge gaps · Conservation interventions · Drivers of human-wildlife conflicts

1 Introduction

Human-wildlife interactions (HWI) have always been an intricate part of human evolution, and for many people, these interactions form an intricate part of their life (Nyhus, 2016). Such interactions seem inherent in socio-ecological systems across high-altitude arid rangelands and grasslands globally. The interactions can be positive, negative, or neutral. Human-wildlife conflict (HWC) refers to struggles that arise when the presence or behaviour of wildlife poses actual or perceived direct, recurring threats to human interests or needs, often leading to disagreements between groups of people and negative impacts on people or wildlife (IUCN, 2020). Human-wildlife conflict is a chronic and growing problem globally. Regional assessments of conflicts suggest an increase in HWC in Asia (Seoraj-Pillai & Pillay, 2017) and that HWC afflicts nearly 90% of India (Anand & Radhakrishna, 2017).

Wildlife management involves 'the guidance of decision-making processes and implementing practices to purposefully influence interactions amongst and between people, wildlife, and habitats to achieve impacts valued by stakeholders' (Riley et al., 2002). Effective management of HWC and natural resources needs reliable information through on-ground situation assessments informed by historical assessments, involvement of all stakeholders in decision-making processes, and overall strategic planning, implementation, and evaluation of management measures (Allen & Garmestani, 2015; Marchini et al., 2019). The results of HWC assessment studies should also be inferred against recent developmental and socio-economic changes in any region. Successful conflict management requires a multi-stakeholder approach to recognise common goals to solve problems, awareness of trade-offs, and a transparent evidence base (Redpath et al., 2013). Additionally, conservationists have reiterated the dire need to evaluate conservation interventions (CI) over time (Miteva et al., 2012; Bull et al., 2014; Eklund et al., 2017).

The Indian Himalayan region (IHR) and Hindu-Kush Himalayan region (HKH) are geographic extensions to the Western Himalayas, and most conflict-related studies in these regions highlight the expansion of protected areas, habitat loss, and over-exploitation of natural resources as regional drivers of HWC (Bhatnagar et al., 2001; Chettri & Sharma, 2016; Khan & Baig, 2020; Sharma et al., 2021). Some studies have described the lack of focus on the monetary cost associated with HWC and limitations in selecting species and geographic locations while addressing

HWC (Manral et al., 2016; Anand & Radhakrishna, 2017). And a wealth of information also comes from group or genus-specific (Inskip & Zimmermann, 2009; Khan et al., 2014; Holland et al., 2018; Torres et al., 2018; Krafte, 2019; Ijaz et al., 2020) and species-specific studies (Sanwal & Lone, 2012; Can et al., 2014; Jackson, 2015; Valentová, 2017; Bombieri et al., 2019; Rashid et al., 2020; Sharma & Singh, 2020) that discuss trends and patterns of HWC.

While there is a global consensus on the need for evidence-based conservation (Sutherland et al., 2004; Svancara et al., 2005; van Eeden et al., 2018), reports on rigorous evaluations of conservation interventions are limited or lacking (Ferraro & Pattanayak, 2006; Miteva et al., 2012; Bull et al., 2014; Eklund et al., 2017). Overall, the existing studies provide rich information on conflict-prone species and recommendations on conservation interventions to alleviate HWC in the Western Himalayas. But a comprehensive understanding of site-specific issues and effective interventions is lacking.

Preventive measures aim to increase the safety of people, their farms, and livestock and include measures such as predator-proof corrals, electric fences, and walls. On the other hand, compensation and insurance schemes aim to mitigate the financial losses caused by wildlife. Local communities often implement preventive measures at a small scale due to resource constraints even when they face conflict repeatedly. Where conflict mitigation mandates are in place, other stakeholders, such as conservationists and regional administrative authorities, implement preventive and mitigative conflict management measures. The mitigative measures are usually implemented after the conflict incidents have occurred, after receiving various claims of livestock loss or crop depredation, ideally after an evaluation of the conflict scenario, conservation planning for the species involved, and assessment of socio-economic aspects of the communities.

However, we still do not understand whether (i) commonly suggested and used HWC management measures are designed to address the proximate as well as ultimate drivers of conflict, (ii) which of the HWC measures have been most effective and under what scenario, and (iii) whether HWC studies focus on an empirical evaluation of the impact and effectiveness of the interventions.

We reviewed studies and publications on HWC from 1991 to 2020 in the Western Himalayas, focusing on the Trans-Himalayas. We aimed to examine the patterns, drivers, and impacts of HWC and identify gaps in HWC-related research and conservation interventions implemented in the region.

2 Objectives of the Review

Our objectives were to (a) synthesise the current state of knowledge on HWC in Western Himalayas and identify gaps in knowledge and practice, (b) review whether conflict-assessment studies have led to the identification and implementation of effective interventions, and (c) identify preventive and mitigative measures used

Table 14.1 Objectives addressed, sub-questions answered, and information sought through this review chapter

Objectives of the review chapter	Sub-questions of the review chapter	Information sought
1. Synthesise the current state of knowledge on HWC in Western Himalayas and identify gaps in knowledge and practice	(a) Do HWC studies nuance drivers of conflict following a multidisciplinary approach?	Information related to spatial spread of studies, regions of research focus. Duration of studies and type of data generated through these studies
2. Review whether conflict-assessment studies have led to the identification and implementation of interventions	(b) Do conservation interventions suggested and/or implemented address the underlying drivers of conflict? (c) Is economic loss higher due to livestock depredation, crop depredation, or a combination of factors? (d) What is the relative importance of conflict-related loss in people's lives?	Did conflict-assessment studies and studies with a focus on conservation interventions identify the underlying drivers of conflict? Did they suggest and/or implement interventions to address the drivers of conflict? Were impacts of HWC quantified appropriately?
3. Identify preventive and mitigative measures used and recommended and conservation interventions implemented and evaluated (if any)	(e) How were beneficiaries who receive the benefits of conservation interventions identified? (f) Have there been reports of conservation interventions not working or having an unintended effect?	Types of preventive and mitigative measures used and recommended for HWC. Were interventions evaluated. Identification of those conservation interventions that were effective

and recommended and conservation interventions implemented and evaluated (if any). To develop an in-depth understanding of the current state of HWC in the Western Himalayas, we sought to answer the following questions (Table 14.1):

3 Study Area

Western Himalayas spread across India and Pakistan (Quinye & Du, 2004). The rain-shadow region of the Himalayas in Central and South Asia is called the Trans-Himalayas (Mani, 1974; Mishra et al., 2009). Around 280 mammals have been reported from this region (Chandra et al., 2018), and the Trans-Himalayas supports a faunal diversity that exhibits high endemism. Trans-Himalayan rangelands are natural grasslands, steppes, and wetland areas dominated by grasses and grass-like plants predominantly used for livestock grazing. They serve as traditional food sources for pastoral production systems and provide other vital ecosystem services, such as atmospheric carbon storage, critical habitats for wildlife, and watershed functions (Kumar et al., 2015).

The Trans-Himalayan rangelands of Western Himalayas are contiguous to a larger complex of Alpine regions and rangelands of Central Asia and are limited in their above-ground biomass productivity (Mishra et al., 2001). Nonetheless, the Trans-Himalayan socio-ecological system supports a diversity of local communities and wildlife species, which intersperse, leading to complex human-wildlife interactions. Local economies converging into global markets (Namgail et al., 2010), rapid urbanisation, and concurrent socio-economic changes (Sandhu & Sandhu, 2015; Tiwari et al., 2018; Singh et al., 2020) exasperate the problem of human-wildlife relationships.

4 Methods

4.1 Literature Search and Collation

We used a list of keywords focused on 'human-wildlife conflict' to search on Google Scholar (Gusenbauer & Haddaway, 2020) (Appendix 1). All articles/studies that provided information on preventive or mitigative measures to deal with HWC based on their title and abstract for the specified geographic region (Western Trans-Himalayas) were included in this review. Total results per keyword search were documented, along with the number of articles per keyword. For ten results per page per keyword search, the relevancy of search results in the page was then used to discard the results (end the search) for that keyword from pages beyond 35 pages (350 results per keyword search). A flow chart of the data screening process is provided in Fig. 14.1.

4.2 Data Selection and Eligibility Criteria

Studies that published conflict-related information between 1991 and 2020, including grey literature (reports, thesis, case studies), articles published in peer-reviewed journals in the English language, and those that are indexed so that they are part of the results in a keyword-based Google Search, were included in the initial dataset.

We categorised the articles in the review dataset as 'direct references' or 'indirect references' based on their relevance to 'conflict' and 'intervention' keyword-based Google Search (Inskip & Zimmermann, 2009). Indirect references are those articles that focus on the status and distribution of large mammals in the Trans-Himalayas. We included these too in the review, since they provided information regarding HWC and threats to the species in the Western Himalayas, even when it was not the stated objective of such studies. We also used a snowballing technique to cross-check the references in the articles we reviewed, and additional studies were included based on the relevance of the available information. This resulted in the inclusion of additional 'indirect' references. In total, 88 publications dating from

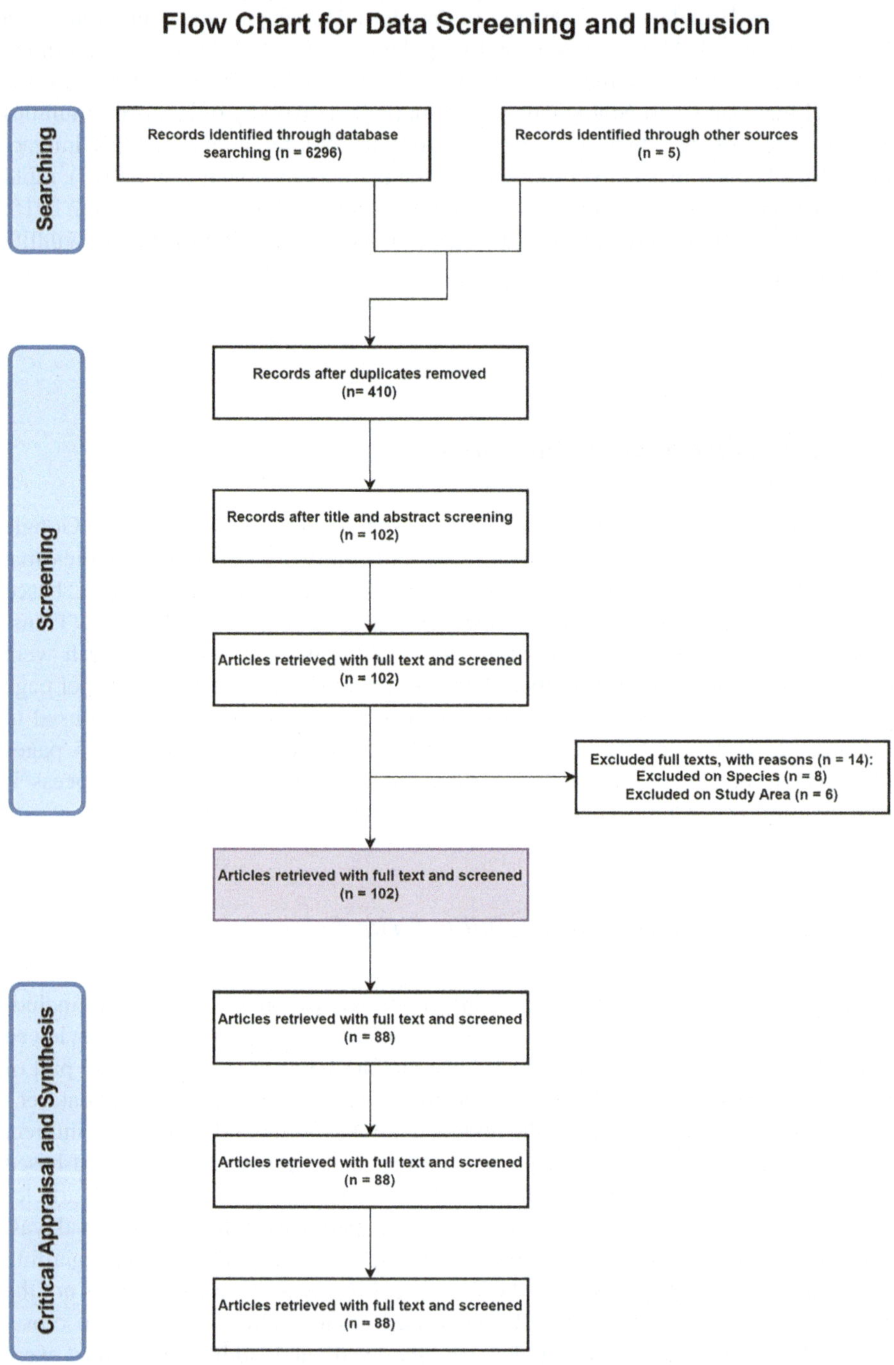

Fig. 14.1 A flow chart of the data screening process followed in the current review chapter

1997 (the earliest article published) to 27 December 2020 (the final search date) were included in the review. To access the entire database of articles reviewed, please see Appendix 2.

4.3 Data Screening, Variable Identification, and Coding

Articles downloaded based on the keyword-based Google Search were scanned and filtered by the first author primarily based on the title and abstract. We created a data extraction form with 80 parameters (Appendix 3) based on predetermined review criteria. Data were collated, managed, and referenced according to these parameters using Microsoft Excel and Mendeley Reference Manager (Version 1.19.8., https://www.mendeley.com/).

The authors independently classified a sub-sample of 25 articles and identified and discussed discrepancies to arrive at a consensus for classifying the articles for consistency and repeatability. We primarily looked at the abstract, the study's objectives, results, and discussion sections to code it per the data extraction form's categories. We observed an 80% match between our independent classification of the articles. We restricted the review to field studies that gathered primary data from field sites and did not include meta-analysis or review studies in the dataset. We treated all articles in the current review as independent studies. Following this, the first author synthesised the information for all 88 studies. We coded articles for the seven sections and 80 parameters (Appendix 3) mentioned in Table 14.2 following Holland et al. (2018). The minimum criteria for studies to be included in data synthesis were to have a relevant title and abstract and be accessible through the search resources available to the review team.

Table 14.2 Sections and parameters used to characterise the human-wildlife conflict studies as per the data extraction form

Sections of the data extraction form	Number of parameters per section
Section 1 – Basic information regarding the studies	13
Section 2 – Purpose of the studies as defined in the articles	7
Section 3 – Type of data generated/methodology utilised in the studies	4
Section 4 – Potential drivers of human-wildlife conflict identified through these studies	11
Section 5 – Impact of conflicts as documented by these studies	12
Section 6 – Types of interventions that can be utilised to manage conflict	24
Section 7 – Regarding interventions being implemented and evaluated	9
Total parameters used for data synthesis	*80*

See Appendix 3 for details

Information for the 80 parameters was sorted as per the sections mentioned in Table 14.2, and content was extracted as part of the data synthesis based on the data extraction form (Appendix 3).

The GPS coordinates of 88 studies from the review dataset were extracted where possible, and point locations were plotted using ArcMap (Version 10.7.1) (Fig. 14.3). We used point locations to illustrate the geographic spread of the studies, since polygons were unavailable for most studies. If a study focused on one protected area (PA) and three villages around the PA, all four-point locations (GPS coordinates) were generated and used to represent the study wherever possible. GPS coordinates were included from various datasets (UNEP-WCMC & IUCN 2021; Meiyappan et al., 2018), and when unavailable, locations from Google Earth were used.

4.4 Data Synthesis and Analysis

Most of the studies in our review dataset provided narrative accounts of the conservation interventions implemented and the evaluations carried out. Hence, content analysis (Krippendorff, 2004; Hsieh & Shannon, 2005) and thematic coding (Gibbs, 2012) following a deductive approach was used to analyse this information. We also used descriptive statistics to summarise the dataset.

5 Limitations of the Current Review

While the keyword-based search for conflict in Western Himalayas resulted in many species, we chose to focus on large mammals in the Trans-Himalayan zone of Western Himalayas (above 3000 m altitude). We also limited ourselves from including any primate-related research and research related to conflict with the common leopard (*Panthera pardus*). In contrast, we have included conflict research about the black bear (*Ursus thibetanus*).

We faced several challenges during this review's coding and content analysis stages. Several studies lacked clarity in defining their purpose, while some stated their objectives minimally yet concluded with extensive observations beyond the stated objectives. We have analysed the information put forth by these studies based on their conceptual and narrative clarity as conveyed by the authors directly. Overall, we feel the analytical details, and the process outlined above allowed us to analyse the dataset objectively.

6 Results

We obtained a total of 6296 records from our preliminary literature search and screening, of which 88 studies were included based on relevance for final data synthesis and coding. Studies that began in 1991 and were published before December 27, 2020, were included.

7 Temporal Patterns of HWC Research in Western Himalayas

The number of conflict-related publications varied considerably over the years but generally showed an increasing trend (Fig. 14.2). While globally and regionally in Asia publications related to HWC show an overall increasing trend (Anand & Radhakrishna, 2017; Seoraj-Pillai & Pillay, 2017; Rashid et al., 2020; Sharma & Singh, 2020; Sharma et al., 2021), our results show a fluctuating trend in conflict-related publications from the Western Himalayas.

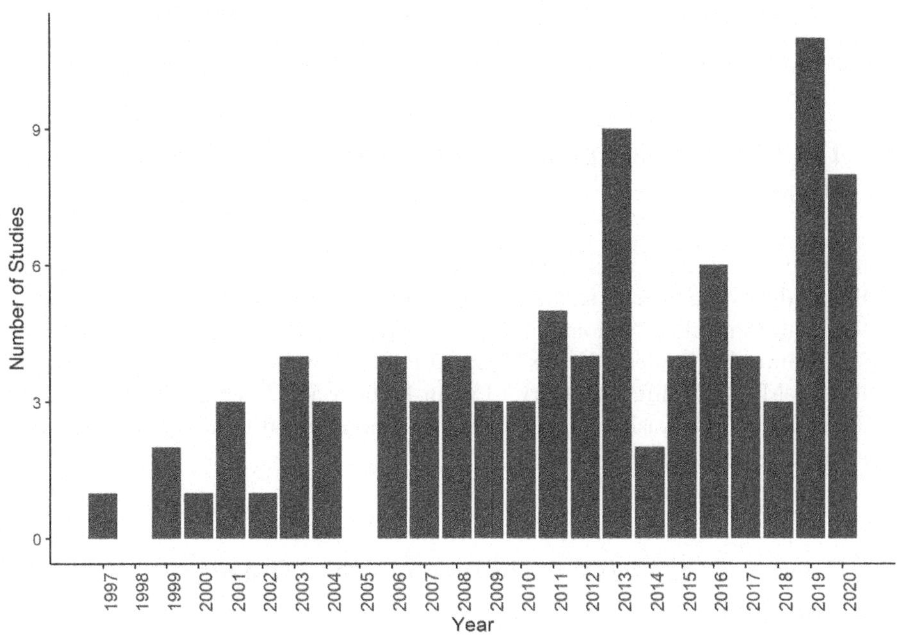

Fig. 14.2 Number of conflict-related publications in Western Himalayas over three decades

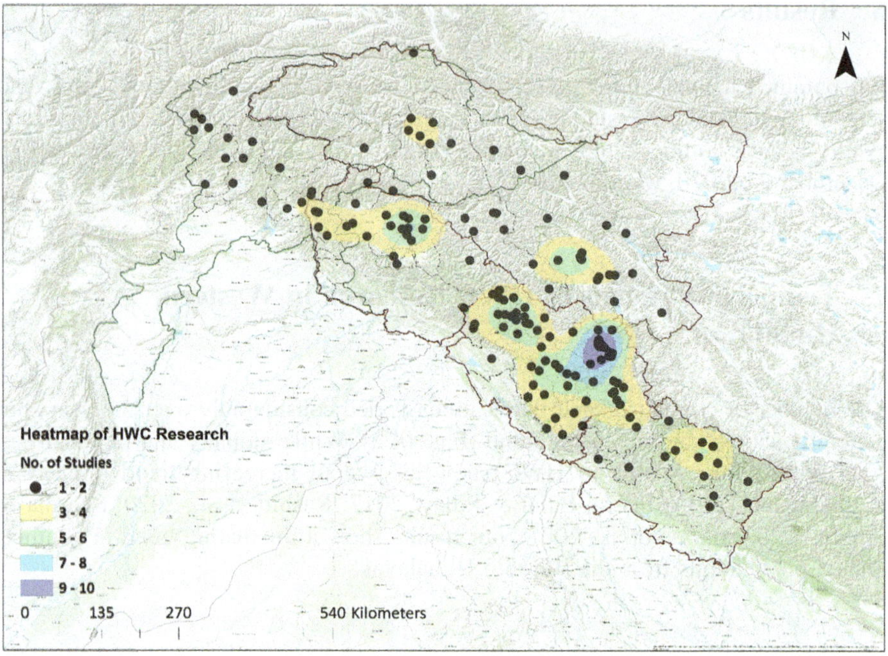

Fig. 14.3 Geographic spread of HWC-related research in Western Himalayas. (If the study focused on one protected area (PA) and three villages around the PA, all four-point locations (GPS coordinates) were generated and used to represent the study wherever possible)

8 The Geographic Spread of HWC Research in Western Himalayas

Our review showed an uneven distribution of studies across Western Himalayas, with the highest number of studies in the Spiti region (36%), in geographic continuation with Pin Valley National Park (7%), and in Great Himalayan National Park (6%) in Himachal Pradesh. Hemis National Park (11%) in geographic continuation with Gya-Miru Wildlife Sanctuary (3%) in Ladakh and Dachigam National Park (6%) in Jammu & Kashmir, and other one-time or short-term studies that were conducted as depicted in Fig. 14.3.

9 Focal Species of HWC Research in the Western Himalayas

The snow leopard (*Panthera uncia*) (41%), followed by the black bear (*Ursus thibetanus*) (38%) and the grey wolf (*Canis lupus*) (28%), emerged as the focal species studied across Western Himalayas. The snow leopard and grey wolf emerged as conflict-prone species, with studies reporting impacts such as livestock loss and loss of household income. In contrast, the black bear emerged as a conflict-prone species, with studies reporting impacts such as livestock loss, crop depredation, and human injuries. Other carnivores commonly studied across Western Himalayas were the Himalayan brown bear (*Ursus arctos isabellinus*) (18%) and Eurasian lynx (*Lynx lynx*) (8%). Amongst herbivores, the blue sheep (*Pseudois nayaur*) (10%) emerged as the most conflict-prone species causing crop depredation across Western Himalayas.

10 The Focus of Conflict Research in the Western Himalayas

A summary of information on HWC-research in Western Himalayas concerning their focus and objectives (purpose defined), drivers and impacts of conflict identified, and implementation and evaluation of conservation interventions is presented in Table 14.3. From $N = 88$ studies included in this review chapter, 47 studies are primary references, i.e. studies which identified that their primary objective was 'to assess the extent of conflict', whereas 41 studies are secondary references that varied considerably in their primary objective and purpose but did touch upon human-wildlife conflicts. Almost half of the studies focused on assessing and quantifying the extent of HWC and examining primary drivers of conflicts. One fourth of the studies focused on conflict mitigation interventions and almost an equal number carried out some form of evaluation. Two studies have identified tolerance towards HWC in certain areas within Western Himalayas (Dar et al., 2009; Ali et al., 2018), and three studies pointed out a lack of government involvement in addressing HWC in the region (Wahid et al., 2017; Ali et al., 2018; Kazmi et al., 2019). Studies that explored people's perceptions noted concerns as well as willingness to share space with wildlife. Still, the inability to address conflict was the prominent reason for the passive tolerance of HWC. Despite owning arms and ammunition, no retaliation was indicated in certain regions, generally showing high tolerance towards wildlife.

11 Gaps in HWC-Related Research in Western Himalayas

An overview of gaps identified in HWC-related research across Western Himalayas through this review is listed below:

1. Of $N = 88$ studies, $n = 37$ (42%) were rapid conflict assessment surveys or one-time studies (e.g. master's theses). The short-term studies represent information from a snapshot of time.
2. Even though most of the study region is essentially a socio-ecological system, of $N = 88$ studies, $n = 56$ (64%) focused solely on either ecological or social aspects

Table 14.3 Summary of HWC-related studies as per their focus (objectives and purpose)

Objectives and purpose as defined in publications	Percentage of studies/ publications	Comments
Purpose – to assess conflict	47 studies (53%)	
Purpose – defined as 'other' – studies with an ecological focus on the status, threats, and distribution of species	62 studies (70%)	HWC was not defined as a key objective, but these studies did tangentially address HWC
Conflict assessment studies that identified drivers and impacts of HWC	All 47 studies that defined their purpose as to assess conflict did identify drivers and impacts (53%)	The quality of studies varied widely. Most studies lacked a rigorous quantitative assessment of drivers of HWC
Studies related to conservation interventions	21 studies (24%)	See Table 14.4 for details
Studies that defined implementation of conservation interventions as their purpose	Five studies (5.6%)	
Studies that implemented conservation interventions	14 studies (16%)	See Table 14.4 for details
Studies that defined evaluation of conservation interventions as their purpose	16 studies (18%)	
Studies that evaluated conservation interventions	19 studies (22%) (9 studies were evaluated by external resources – by members who were not part of the original research team)	Resources engaged by the funding agencies carried out some external evaluations, and some evaluations were carried out as independent studies in the same region and hence were possibly unintentional external evaluations (Table 14.4)
Additional purpose defined to quantify the impact on animals	Six studies (7%)	
Impacts or threats reported for wildlife	35 studies (40%)	

of HWC. We categorised those studies that assessed conflict through such singular focus as 'limited in their approach' to HWC. We emphasise that HWC is multifaceted (Pooley et al., 2017) and should be inspected through multiple lenses to understand the actual drivers of conflict in any region.

3. Of $N = 88$ studies, $n = 40$ (45%) provided generic conflict mitigation recommendations that had little to do with the actual data and results of the investigation. Monetary compensation was the most recommended mitigation measure in such studies.

4. The lack of standard measurement scales hampers a better understanding of human-wildlife conflict in the Western Himalayas, making it difficult to compare across regions. In the present review, we found that crop loss and livestock predation information was not comparable across studies.
5. Very few studies conducted some form of evaluation of the effectiveness of conservation interventions. Of $N = 88$ studies, only $n = 21$ studies (24%) were explicitly about conservation interventions (HWC mitigation measures), and a subset of those $n = 19$ (22%) evaluated conservation interventions, of which only $n = 9$ had conducted external evaluations.
6. Identifying the drivers of conflict in a region is a necessary precursor to identifying, implementing, and evaluating conflict mitigation measures. Most studies (87%, $N = 88$) identified the drivers of HWC, albeit the rigour between the studies varied considerably.
7. The social-economic and sociopolitical disparity has not been addressed as a driver in Western Himalayan studies. However, it has been acknowledged/identified as a driver of conflict (human-human conflict) in some research studies in the Upper Spiti Landscape of Himachal Pradesh, India.

12 Drivers and Impacts of HWC in Western Himalayas

Ease of access to livestock followed by dysfunctional herding practices, developmental pressures, and expanding human settlements emerged as the top four most common drivers of conflict involving livestock loss, whereas ease of access to crops, expanding human settlements, and forest fragmentation or habitat loss were identified as the top three and most common drivers of conflict for crop depredation (Fig. 14.4).

The significant impacts of HWC reported across Western Himalayas included livestock loss (72% of studies, $n = 64/N = 88$), impact on livelihoods (42%), crop depredation and hunting/death of wildlife (36% each), negative attitudes towards wildlife (33%), and human injuries/death caused due to wildlife (19%) (Fig. 14.5).

Other factors such as proximity to protected areas, increased prey availability (of livestock and wild prey) in certain areas, increase in carnivore density owing to conservation efforts, ecological factors such as pasture ruggedness for wolves, lack of den sites or vegetation providing shelter for bears, authorities lacking capacity to deal with HWC, and climate change influencing habitat loss were mentioned by 30% ($n = 27/N = 88$) studies as drivers of HWC in Western Himalayas.

Due to the absence of a standardised approach to quantify conflict, we could not discern any patterns or compare the extent of conflict across different sites in the Western Himalayas. Less than 6% of studies, $n = 5$ (of $N = 88$), quantified and reported the loss caused due to HWC in their respective study areas. Crop or livestock depredation was reported either exclusively in monetary terms or percentage of crop or livestock loss and sometimes even in terms of the percentage of respondents who reported crop or livestock loss. The top five repercussions of

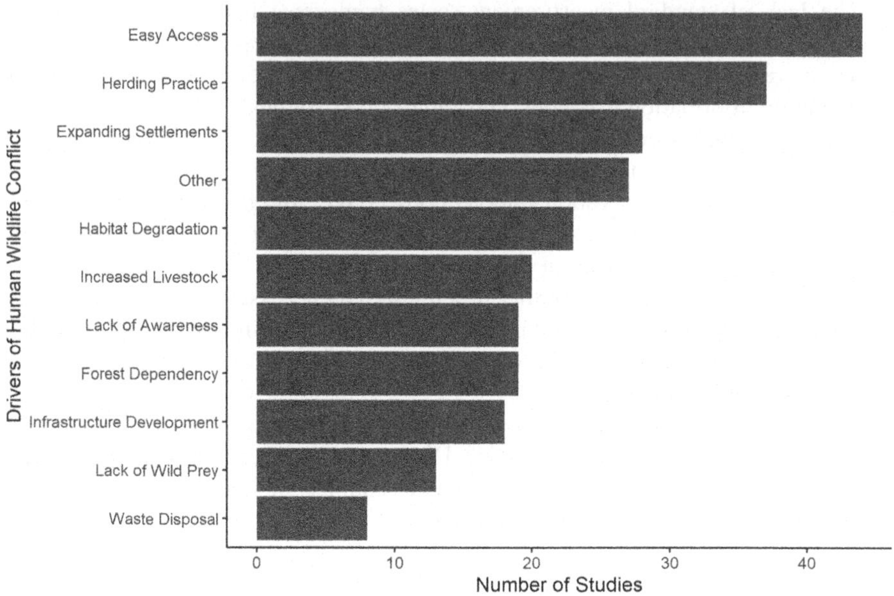

Fig. 14.4 Drivers of HWC across Western Himalayas. (Easy access-to crop and livestock; herding practice-inefficient herding practices; expanding settlements-refers to expanding human settlements; habitat degradation-includes forest fragmentation and habitat loss; increased livestock-implies an overall increase in livestock numbers as well as livestock stocking densities; forest dependency- dependency on forest resources; infrastructure development- includes linear infrastructure and tourism; waste disposal- improper waste disposal)

human-wildlife conflicts included livestock loss, impacts on livelihoods and economic security, hunting/poaching/retaliatory killing of wildlife, crop depredation, and negative attitudes towards wildlife.

13 Conflict Management in the Western Himalayas

13.1 Preventive and Mitigative Measures for HWC

The data extraction form (Appendix 3) created for the current review chapter was based on identifying predetermined criteria involving various parameters used for content analysis and information synthesis. Twenty-one parameters (Table 14.2) about the conservation interventions implemented and evaluated across Western Himalayas were categorised broadly into preventive and mitigative measures (Fig. 14.6).

Preventive measures qualify as direct interventions (Distefano, 2005; Treves et al., 2009) and consist of deterrents and barriers (like corrals or fences) or work towards removing attractants like food or garbage (through crop rotation, crop

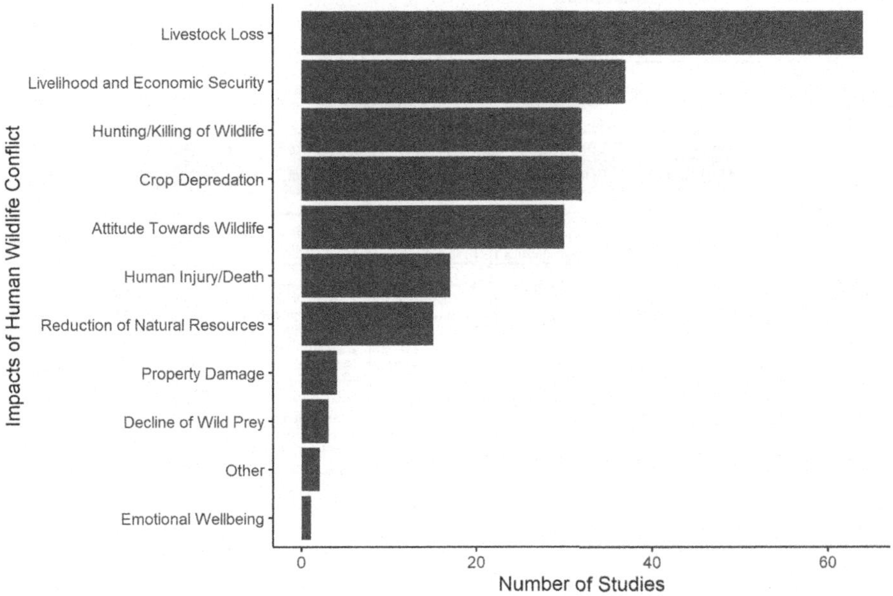

Fig. 14.5 Impacts of HWC across Western Himalayas. (Livestock loss- to wild carnivores; livelihood and economic security- impact on livelihood and economic security; hunting/killing of wildlife- as a preventive or retributive measure; crop depredation; attitude towards wildlife-negative attitudes towards wildlife; emotional well-being- where peoples emotional well-being has been negatively affected by HWC)

change, waste removal). Mitigative measures qualify as indirect strategies or interventions that may take considerable resources and time to implement. Mitigative measures may also be implemented after thoroughly assessing the on-ground conflict. We identified five preventive and 16 mitigative measures currently under use in the Western Himalayas. Although interlinked, the mitigation measures can be further classified into five broad approaches to mitigation (Fig. 14.6), and they often involve multiple stakeholders.

Here, we provide a summary of specifics from the suite of preventive and mitigative measures identified across the Western Himalayas (Fig. 14.6):

- *Of the five types of preventive measures* (direct interventions) identified above, improving corrals/livestock stocking pens was the most recommended intervention across Western Himalayas. Around 25% of studies ($n = 22/N = 88$) recommended improving corrals in their respective study areas, and $n = 4$ studies have implemented and/or evaluated corrals, recognising that corrals were effective against snow leopards and wolves.
- Secondly, the use of guard dogs, increased vigilance, skilled shepherds, or chasing away wild animals was documented and recommended by about 32% ($n = 28/N = 88$) of studies across Western Himalayas. Most studies suggested

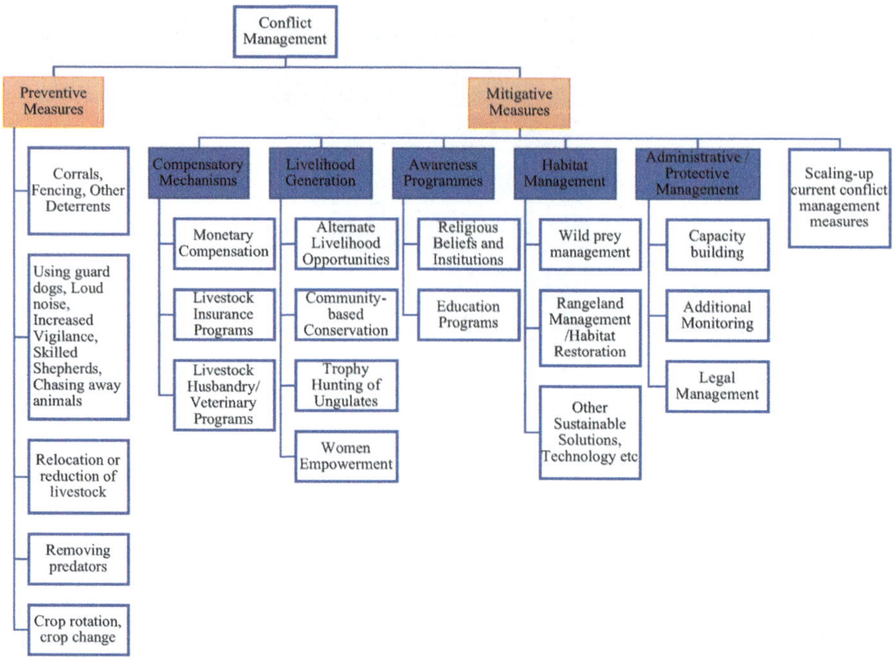

Fig. 14.6 Common conflict management measures classified into preventive and mitigative measures used in the Western Himalayas

both corrals and the above-stated suite of preventive measures, recognising that these measures were effective against all carnivores and herbivores.

- *From the suite of mitigative measures*: Monetary compensation for livestock loss was recommended by 34% ($n = 30/N = 88$) of studies across Western Himalayas. Livestock insurance programs were recommended by $n = 10$ studies (11%) and evaluated by $n = 2$ studies across Western Himalayas. The requirement of husbandry and veterinary programmes was identified by $n = 14$ (16%) studies, while $n = 2$ studies across Western Himalayas evaluated it.
- Livelihood generation was strongly recommended across the Western Himalayas, and homestays were the most common mode of alternate livelihood interventions in India. In contrast, trophy hunting along with incentive schemes and community-based conservation approaches was commonly implemented in Pakistan. Education and awareness programmes were recommended across the Western Himalayas. Rangeland management and monitoring of rangelands have been predominantly recommended across the Western Himalayas.

Although human-wildlife conflict is a complex problem to address, conservation interventions focused on alleviating HWC in the Western Himalayas seem to have worked to a large extent. The conservation interventions implemented and evaluated in Western Himalayas are described in Table 14.4.

Table 14.4 Summary of studies that implemented and evaluated conservation interventions to alleviate HWC in the Western Himalayas

No.	References	Were interventions implemented	Type of interventions implemented or discussed	Intervened by	Were interventions evaluated	Comments on effectiveness of interventions (as described in the original study)	Excerpts of comments (as described in the original study)	Type of evaluation and observations
1	Jackson and Wangchuk (2001)	Yes	Predator-proof corrals, homestays	Snow Leopard Conservancy (SLC)	Yes	Predator-proof corrals were effective	NA	Internal evaluation – CI have been implemented over the long-term and hence seem effective
2	Mishra et al. (2003)	Yes	Grazing-free areas, other incentives	Nature Conservation Foundation (NCF) and International Snow Leopard Trust (ISLT)	Yes	Interventions were effective	NA	Internal evaluation – CI have been implemented over the long-term and hence seem effective
3	Jackson and Wangchuk (2004)	Yes	Corrals, homestays training, pasture resting, monitoring, etc.	ISLT	Yes	Interventions were effective	Appreciative participatory planning and action has worked	Internal evaluation – CI have been implemented over the long-term and hence seem effective

(continued)

Table 14.4 (continued)

No.	References	Were interventions implemented	Type of interventions implemented or discussed	Intervened by	Were interventions evaluated	Comments on effectiveness of interventions (as described in the original study)	Excerpts of comments (as described in the original study)	Type of evaluation and observations
4	Namgail et al. (2007)	Yes	Conservation linked-livelihood, insurance program	ISLT	NA	Interventions had been initiated at the end of this study	NA	NA
5	Satterfield (2009)	NA	Homestays, conflict control, education	Snow Leopard Conservancy India Trust (SLC-IT) and Department of Wildlife Protection-Jammu and Kashmir (DWP-J & K)	Yes	Interventions were effective at the time	Results show neutral or positive reactions to human-wildlife conflicts	External evaluation
6	Peaty (2009)	NA	Homestays	SLC-IT	Yes	Interventions were effective	Homestays provide significant monetary and non-monetary benefits to the community	External evaluation
7	Maheshwari et al. (2012)	Yes	Predator proof corrals	Locals, DWP-J & K, World Wide Fund for Nature (WWF)	NA	Interventions had been initiated at the end of this study	NA	NA

8	Huyett (2013)	NA	Corrals, home-stays, insurance programs, handi-craft livelihoods, environment education	SLC-IT	Yes	Interventions were effective to a large extent	Research showed the importance of material incen-tives in gaining initial participa-tion from vil-lagers, however they were not often mentioned as benefits experi-enced from participation	External evaluation
9	Mishra et al. (2016)	Yes	Grazing-free vil-lage reserves	NCF	Yes	Interventions were effective	Prey base has improved	Internal evaluation – CI have been implemented over the long-term and hence seem effective
10	Jamwal et al. (2019)	Yes	Predator proof corrals, homestays	SLC	Yes	Interventions have been effec-tive over time	NA	External evaluation – authors assume no other relevant variables have changed over the years; they investigate the effect of inter-ventions in silos (some ecological variables used); changes in SE profile have not been assessed either

(continued)

Table 14.4 (continued)

No.	References	Were interventions implemented	Type of interventions implemented or discussed	Intervened by	Were interventions evaluated	Comments on effectiveness of interventions (as described in the original study)	Excerpts of comments (as described in the original study)	Type of evaluation and observations
11	Bagchi et al. (2019)	NA	Grazing-free areas, other incentives	NCT and ISLT	Yes	Interventions were effective	Wild prey increased to a certain extent in intervention site, and livestock loss reduced as compared to control site	Internal evaluation, CI have been implemented over the long-term and hence seem effective
12	Maheshwari and Sathyakumar (2019)	NA	Tourism income	Various	Yes	Interventions have been effective over time	Income from ecotourism was able to offset economic loss due to livestock predation and improve perceptions of SL	External evaluation – authors have not directly compared the financial loss due to HWC to the financial gain through ecotourism. Only descriptive trends and comments have been mentioned

13	Vannelli et al. (2019)	NA	Homestays	SLC-IT	Yes	Interventions were effective	Wildlife has a commodified value, and this should be changed. Interventions were effective in influencing people's perceptions towards SL	External evaluation
14	Maheshwari and Sathyakumar (2020)	Yes	Predator proof corrals and awareness programs	NCF and ISLT earlier, followed by WWF	Yes	Interventions were effective	NA	Internal evaluation
15	Nawaz et al. (2008)	Yes	Community participation, incentivisation, reduced grazing pressure, strict patrolling, etc.	Himalayan Wildlife Foundation (HWF)	Yes	Interventions were effective	Brown bear population increased, poaching managed	Internal evaluation
16	Mir (2006)	Yes	Trophy hunting of ungulates	Local communities, various	Yes	Interventions were effective to a certain extent	Transparency within stakeholders was lacking. Price for the trophies and marketing opportunities was an issue	External evaluation

(continued)

Table 14.4 (continued)

No.	References	Were interventions implemented	Type of interventions implemented or discussed	Intervened by	Were interventions evaluated	Comments on effectiveness of interventions (as described in the original study)	Excerpts of comments (as described in the original study)	Type of evaluation and observations
17	Ali (2008)	NA	Trophy hunting, CBC, incentivisation	Local communities, various	Yes	Interventions were effective to a certain extent	NA	External evaluation – multiple complications within stakeholders affect the overall success of CI implemented
18	Rosen et al. (2012)	Yes	Incentive scheme through PSL, predator proof corrals, etc.	Local communities, various	Yes	Interventions were effective	Skoyo village backed out of PSL since Markhor was involved. Some gaps identified in the multi-pronged strategy of interventions	Internal evaluation
19	Ahmad (2016)	Yes	Vaccination trainings, village conservation committee, awareness workshops	Local communities, various	Yes	Interventions were effective	NA	Internal evaluation – vaccination drive was very effective

20	Hussain (2000)	Yes	Livestock insurance, income through ecotourism, incentive scheme through PSL	Local communities, Project Snow Leopard (PSL), SLC, Full Moon Night Trekking company (FMNT)	Yes	Interventions were effective	NA	Internal evaluation – CI have been implemented over the long-term and hence seem effective
21	Shackleton (2001)	Yes	Trophy hunting of ungulates	Local communities, various	Yes	Interventions were effective to a certain extent	The government needs to improve their engagement, areas where the government was not involved performed better	External evaluation

13.2 *Implementation and Evaluation of Conservation Interventions to Alleviate HWC*

The following is a summary of studies that implemented and evaluated the conservation interventions to alleviate HWC in the Western Himalayas (Table 14.4).

1. While 84% ($n = 74/N = 88$) of conflict-related studies identified the drivers of HWC in the study area, conservation interventions were implemented by only 16% of studies and evaluated by 22% of studies. External evaluations of conservation interventions are generally considered more rigorous (Eklund et al., 2017) and highlight the problems with the implementation of CI. A similar trend was observed from our review dataset. But only 10% of studies ($n = 9/N = 88$) had conducted external evaluations of the CI they had implemented. Some of our observations regarding the evaluations carried out for studies included in this review are stated in Table 14.4. For additional reference, we have also mentioned excerpts of comments specifically shared by authors regarding these evaluations.
2. Notably, only those studies ($n = 3$) that monitored, collated, and published information on conservation interventions for a longer duration (around 5–7 years) could establish whether interventions worked and had the intended effect.
3. Studies that examined or included multiple stakeholders identified lack of transparency, funding, and lack of capacity as primary reasons for the ineffectiveness of conservation interventions. Additionally, $n = 2$ studies (Hussain, 2000, 2003) indicated that trophy hunting, a commonly prescribed conservation intervention for generating livelihood, may backfire. People would kill carnivores to protect the 'trophy' wild herbivores in the region, while illegal hunting persisted (Shackleton, 2001).
4. Two significant limitations in the existing knowledge of HWC in the Western Himalayas are (i) the generic recommendation of conservation interventions without context and linkage to the drivers of HWC and (ii) conservation interventions recommended/implemented devoid of local socio-ecological context. These limitations are partly because most of the studies on HWC were short-term, one-time studies.
5. Multi-stakeholder adaptive management emerged as a potential approach to successfully address HWC and promote coexistence in Western Himalayas. We have discussed these aspects at length below.

14 Discussion

Below, we discuss the main implications of our findings from the perspective of existing knowledge on HWC research, gaps in the state of knowledge, and the state of conservation interventions and their effectiveness. We provide a set of recommendations that we believe can improve the science and management of HWC in the Himalayas.

15 Trends of HWC and Species Involved

Despite HWC emerging as an increasingly intractable problem in the Western Himalayas, the number of studies focusing on HWC in the regions shows a fluctuating trend in recent years. This calls for a greater collaborative effort amongst researchers, institutions, policymakers, and affected communities and a higher level of funding support towards understanding and mitigating HWC. Supporting long-term, longitudinal, and multidisciplinary studies would be critical to examining the HWC and devise effective and scalable solutions. Historically, the remoteness of the Himalayan region, the rugged terrain, and logistic challenges have posed significant challenges for research and continue to be a limitation to some extent.

Our review highlights the species that have been most studied in the context of HWC in the Western Himalayas. We focused on large mammals as these species directly hamper people's livelihood, security, and well-being by causing livestock and crop depredation and human injuries. Identifying the carnivore species responsible for livestock depredation is challenging in the Trans-Himalayas, since livestock depredation often occurs in pastures far away from human settlements, especially when livestock is not accompanied by skilled herders (Pahuja & Sharma, 2021). In such cases, higher negative perceptions towards one species, for example, against wolves in Western Himalayas, can result in wolves being reported more often. Conservation of apex predators like snow leopards generates spillover benefits for other wildlife (Alexander et al., 2015). Hence, such species may receive increased research and conservation attention and higher public support. Therefore, the species most studied in the HWC literature could in part represent these biases. Although pervasive conflict is reported due to other species such as blue sheep or kiang in some areas of Western Himalayas, it receives limited attention compared to charismatic species such as the snow leopard.

16 Critical Gaps in HWC Knowledge

While useful as baselines, short-term studies and rapid surveys are snapshots in time, and inference based on them is limited. Organisations and research institutions should invest in long-term studies to rigorously examine the HWC and find solutions that work. The lack of focused long-term research undermines efforts to alleviate HWC in the Western Himalayas. Long-term studies and datasets that examine the underlying drivers and trends of HWC can successfully identify and implement effective HWC measures (e.g. see – Mishra, 1997; Hussain, 2000, 2003; Jackson, 2015; Mishra et al., 2016). Most studies take a narrow view of HWC, focusing solely on ecological aspects of conflict, while in complex socio-ecological systems, interdisciplinary research can be far more effective.

We contend that studies on HWC should pay attention to the underlying drivers of HWC. The drivers can be manifold and can vary locally and regionally. The ecology, economy, politics, developmental changes and aspirations of people, and impacts of government and non-governmental institutions can all impact human-wildlife conflicts. For example, a predominant fortress conservation model that involves establishing protected areas in a region often influences people's perceptions about increased carnivore numbers and conflicts. Such assumptions also lead people to perceive higher conflicts with carnivores in a region (Mishra, 1997; Jackson, 2015; Bhatnagar et al., 1999). The underlying factors here are likely to be conflicts between local people and government agencies on how an area should be managed, the rights of people, denial of access to resources, and local disenchantment with the protected area management. Ecological studies can reveal the difference between perceptions and realities (Suryawanshi et al., 2014), yet people's perceptions are an essential part of their lived experience (Hill et al., 2017; Talbert et al., 2020), and the conflict in a region can accentuate when people's negative perceptions are not addressed.

Conservation research has emphasised how human-human conflict (Maikhuri et al., 2001; Redpath et al., 2013, 2015; Dhee et al., 2019; Marchini et al., 2019) and sociopolitical conflict (Treves et al., 2006) are the underlying and, at times, the ultimate drivers of conflict in a region. Once addressed, the negative effects of HWC can be reduced to a significant extent. As mentioned earlier, HWC is multifaceted (Pooley et al., 2017) and should be assessed accordingly. Multidisciplinary, longitudinal studies can help achieve that for priority sites and help avoid the trap of general recommendations. A multifaceted understanding of conflicts remains a critical knowledge gap that needs to be addressed urgently.

Conflict-assessment studies should provide recommendations based on the data and information inferred from the research and not extrapolate or generalise such recommendations (Holland et al., 2018; Pahuja & Sharma, 2021). Conservation organisations often cannot operate at a large geographical scale, which exasperates human-human conflict (Huyett, 2013). Given the overarching impact of HWC on people and wildlife, it is crucial that governments at national and provincial levels treat HWC as a critical challenge, formulate policies that can alleviate conflicts, and set aside resources for testing and replicating interventions at scale.

Adaptive management through a multi-stakeholder approach emerged as one of the best solutions to deal with HWC across Western Himalayas. We identified how multi-stakeholder approach that involved local communities and enabled them to deal with conflict through preventive measures or offsetting the cost of conflict had worked well in the Western Himalayas. Relevant examples here would be long-term conservation programs around Spiti valley (Mishra et al., 2003, 2017), Hemis National Park (Jackson, 2015), Baltistan (Hussain, 2000, 2003), and Suleiman mountain regions of Pakistan (Ahmed et al., 2003; Woodford et al., 2004).

Since equity is unequally distributed in many regions that face HWC, the relative importance of HWC in people's lives may vary. This should be ascertained by including the social drivers of conflict in HWC studies. Conservation interventions that focus on mitigating HWC can also help alleviate poverty (Dickman et al., 2011) and should be part of HWC mitigation strategies.

The geographic coverage of studies indicates a clear bias towards studies in and around protected areas. This is expected as there is no system to monitor HWC at scale in the Western Himalayas. We urge that annual monitoring of HWC should become an essential part of National Programs, such as the Project Snow Leopard of the Government of India. Conservationists also need to look beyond formally protected areas as more than 90% of the Trans-Himalaya is outside of the protected area. Addressing HWC in such multiple-use landscapes is essential for the continued coexistence of people and wildlife. We also strongly urge the conservation community to develop robust monitoring and evaluation systems in their HWC mitigation programs, so that we know what is working and what is not and under what conditions and which interventions hold a promise of being effective at scale.

17 Use of Ad Hoc Measures to Address Conflict

While HWC management is typically an iterative process, delays in implementing practical/actionable measures that alleviate conflicts – especially in conflict hotspots – contribute to a lack of tolerance amongst communities suffering from human-wildlife conflicts.

Many studies from the review dataset and elsewhere (Dickman et al., 2011) have noted how monetary compensation works as a 'Band-Aid' to alleviate the effects of HWC. Still, they continue to be recommended, perhaps due to an expected delay in finding and implementing appropriate conservation interventions at scale. Monetary compensation sometimes aggravates human-human conflict, especially when wildlife is viewed as a 'property' of the authorities/government agencies operating in any region. Compensation and insurance schemes may commodify wildlife (Vannelli et al., 2019) and fail to motivate people to deliver conservation (Dickman et al., 2011). Although human-wildlife conflict incidences cannot be eliminated, the constant pressure and need to address HWC has led to the use and replication of ad hoc measures to manage HWC. Moreover, a rigorous evaluation of the interventions that have been implemented is limited or lacking entirely (Eklund et al., 2017). This results in a lack of understanding of what solution fits in a particular scenario and leads to stakeholders having to reexamine mitigation measures that effectively address HWC.

18 Conclusion

Human-wildlife conflict is being increasingly explored through a social-ecological perspective globally (Lozano et al., 2019). However, the current research on the HWC in the Western Himalayas is limited in several aspects and hampers finding effective and scalable solutions. We urge conservationists and managers to invest in designing and implementing rigorous interdisciplinary studies that address multiple aspects of conflicts and provide solutions that have the highest potential to address HWC. In addition, an unhindered sharing of data and information is crucial for all stakeholders. We also urge conservationists to conduct evaluations at multiple project stages (Woodhouse et al., 2015) and alter the course of investigation and resource investments if necessary.

Formalising HWC monitoring as part of National Programs, such as the Project Snow Leopard, and using modern tools, such as public datasets and citizen science, can enable the rapid generation and sharing of information, especially when local communities are involved (Karanth & Vanamamalai, 2020). Integrating local and scientific knowledge through a continuous engagement model and approach to multi-stakeholder engagements has also worked well in dynamic socio-ecological settings of East African savannas (Reid et al., 2016).

Partnering with local communities right from the initial designing to implementation and monitoring of conservation interventions and maintaining transparency and trust in a multi-stakeholder approach has worked well in dealing with human-wildlife conflict in the Western Himalayas (Mishra, 1997; Hussain, 2000, 2003; Jackson, 2015; Mishra et al., 2016), and we urge these approaches be widely adopted.

Another significant problem of HWC in Trans-Himalayas is the limited understanding of rangeland dynamics and extensive gaps in research (Sharma & Singh, 2020). Little information on the stocking density of livestock in the Trans-Himalayan rangelands, the economics of Pashmina production, and the importance of and tolerance towards HWC (Kansky et al., 2021) limits our understanding of the multiple facets of human-wildlife relationships. Understanding the dynamics between livestock, wild herbivores, carnivores, and people who depend on the Trans-Himalayas for survival requires insights into socio-ecological and sociopolitical aspects that can be the ultimate drivers of HWC in the Western Himalayas (Sharma & Singh, 2020).

We urge conservation programs to invest in partnering with local communities in reducing the cost of living with wildlife, empowering local communities, poverty alleviation, and conservation-oriented development. These approaches resonate with the principles of inclusive conservation, community empowerment, and community-based conservation and are likely to succeed in the long run.

Appendixes

Appendix 1: Summary of Keyword Search on Google Scholar

Google search keyword	Status of download with date	Total number of results returned on Google	Number of Google pages reviewed as per relevance of articles	Total results through database search	Search URL	Comments (why the search was terminated)
"human wild-life" conflict "western himalayas"	Scan & download completed – 26 May	Around 125 results (0.05 s)	12 pages/ 125 results scanned	125	https://scholar.google.com/scholar?hl=en&as_sdt=0%2C5&as_ylo=1990&as_yhi=2020&q=%22human+wildlife%22+conflict+%22western+himalayas%22&btnG=	Search ended
"human wild-life" conflict western himalayas	Scan & download completed – 8 July	Around 2250 results (0.10 s)	35 pages/ 350 results scanned	350	https://scholar.google.com/scholar?hl=en&as_sdt=0%2C5&as_ylo=1990&as_yhi=2020&q=%22human+wildlife%22+conflict+western+himalayas&btnG=	Search turned into generic himalayas results after this
human wild-life conflict "western himalayas"	Scan & download completed – 8 July	Around 2890 results (0.09 s)	10 pages/ 100 results scanned	100	https://scholar.google.com/scholar?hl=en&as_sdt=0%2C5&q=human+wildlife+conflict+%22western+himalayas%22&btnG=	Search turned into generic botanical and other kind of results for the geography
"human wild-life conflict" "western himalayas"	Scan & download completed – 8 July	Around 98 results (0.08 s)	10 pages/ 98 results scanned	98	https://scholar.google.com/scholar?hl=en&as_sdt=0%2C5&q=%22human+wildlife+conflict%22+%22western+himalayas%22&btnG=	Search ended

(continued)

Google search keyword	Status of download with date	Total number of results returned on Google	Number of Google pages reviewed as per relevance of articles	Total results through database search	Search URL	Comments (why the search was terminated)
"livestock" depredation or loss "western himalayas"	Scan & download completed – 31 July	Around 135 results (0.07 s)	13 pages/ 130 results scanned	130	https://scholar.google.com/scholar?start=0&q=%22livestock%22+depredation+or+loss+%22western+himalayas%22&hl=en&as_sdt=0,5	Search ended
"livestock depredation" "western himalayas"	Scan & download completed – 31 July	Around 71 results (0.06 s)	8 pages/ 80 results scanned	80	https://scholar.google.com/scholar?start=0&q=%22livestock+depredation%22++%22western+himalayas%22&hl=en&as_sdt=0,5	Replicates of previous search
"livestock loss" "western himalayas"	Scan & download completed – 31 July	Around 25 results (0.05 s)	3 pages/ 30 results scanned	30	https://scholar.google.com/scholar?start=0&q=%22livestock+loss%22+%22western+himalayas%22&hl=en&as_sdt=0,5	Replicates of previous search
"human wildlife" conflicts or interactions "jammu and kashmir"	Scan & download completed – 31 July	Around 205 results (0.05 s)	20 pages/ 200 results scanned	200	https://scholar.google.com/scholar?start=5&q=%22human+wildlife%22+conflicts+or+interactions+%22jammu+and+kashmir%22&hl=en&as_sdt=0,5	Replicates of previous search
"human wildlife" conflicts or interactions "uttarakhand"	Scan & download completed – 27 Aug	Around 598 results (0.04 s)	60 pages/ 600 results scanned	600	https://scholar.google.com/scholar?hl=en&as_sdt=0%2C5&q=%22human+wildlife%22+conflicts+or+interactions+%22uttarakhand%22&btnG=	Search was irrelevant for the most part

(continued)

Google search keyword	Status of download with date	Total number of results returned on Google	Number of Google pages reviewed as per relevance of articles	Total results through database search	Search URL	Comments (why the search was terminated)
"human wild-life" conflicts or interactions "himachal pradesh"	Scan & down-load completed – 28 Sep	Around 398 results (0.04 s)	40 pages/ 400 results scanned	400	https://scholar.google.com/scholar?hl=en&as_sdt=0%2C5&q=%22human+wildlife%22+conflicts+or+interactions+%22himachal+pradesh%22&btnG=	Search was irrelevant and led to repetitive results for the most part
"human wild-life" conflicts or interactions "ladakh"	Scan & down-load completed – 29 Sep	Around 244 results (0.05 s)	23 pages/ 230 results scanned	230	https://scholar.google.com/scholar?start=20&q=%22human+wildlife%22+conflicts+or+interactions+%22ladakh%22&hl=en&as_sdt=0,5	Search was irrelevant and led to repetitive results for the most part
"human wild-life" conflicts or interactions "jammu and kashmir"	Scan & down-load completed – 19 Oct	Around 217 results (0.08 s)	21 pages/ 210 results scanned	210	https://scholar.google.com/scholar?hl=en&as_sdt=0%2C5&q=%22human+wildlife%22+conflicts+or+interactions+%22jammu+and+kashmir%22&btnG=	Search was irrelevant for the most part
"livestock depredation" "himachal pradesh"	Scan & down-load completed – 19 Oct	Around 285 results (0.04 s)	28 pages/ 280 results scanned	280	https://scholar.google.com/scholar?hl=en&as_sdt=0%2C5&q=%22livestock+depredation%22+%22himachal+pradesh%22&btnG=	Search was irrelevant for the most part

(continued)

(continued)

Google search keyword	Status of download with date	Total number of results returned on Google	Number of Google pages reviewed as per relevance of articles	Total results through database search	Search URL	Comments (why the search was terminated)
“livestock loss” “himachal pradesh”	Scan & download completed – 2 Nov	Around 121 results (0.04 s)	11 pages/ 110 results scanned	110	https://scholar.google.com/scholar?start=0&q=%22livestock+loss%22+%22himachal+pradesh%22&hl=en&as_sdt=0,5	Search was limited and repetitive for the most part
“livestock depredation” “uttarakhand”	Scan & download completed – 2 Nov	Around 344 results (0.05 s)	33 pages/ 330 results scanned	330	https://scholar.google.com/scholar?start=0&q=%22livestock+depredation%22+%22uttarakhand%22&hl=en&as_sdt=0,5	Search was limited and not relevant for the most part
“livestock loss” “uttarakhand”	Scan & download completed – 2 Nov	Around 144 results (0.05 s)	14 pages/ 140 results scanned	140	https://scholar.google.com/scholar?hl=en&as_sdt=0%2C5&q=%22livestock+loss%22+%22uttarakhand%22&btnG=	Search results were mostly irrelevant
“livestock depredation” “ladakh”	Scan & download completed – 2 Nov	Around 295 results (0.11 s)	29 pages/ 290 results scanned	290	https://scholar.google.com/scholar?start=280&q=%22livestock+depredation%22+%22ladakh%22&hl=en&as_sdt=0,5	Search results were mostly irrelevant
“livestock loss” “ladakh”	Scan & download completed – 4 Nov	Around 125 results (0.03 s)	13 pages/ 125 results scanned	125	https://scholar.google.com/scholar?start=0&q=%22livestock+loss%22+%22ladakh%22&hl=en&as_sdt=0,5	Search was limited and repetitive for the most part

(continued)

Google search keyword	Status of download with date	Total number of results returned on Google	Number of Google pages reviewed as per relevance of articles	Total results through database search	Search URL	Comments (why the search was terminated)
"livestock depredation" "jammu and kashmir"	Scan & download completed – 4 Nov	Around 209 results (0.04 s)	19 pages/ 190 results scanned	190	https://scholar.google.com/scholar?start=0&q=%22livestock+depredation%22+%22jammu+and+kashmir%22&hl=en&as_sdt=0,5	Search was decent
"livestock loss" "jammu and kashmir"	Scan & download completed – 4 Nov	Around 104 results (0.10 s)	11 pages/ 104 results scanned	104	https://scholar.google.com/scholar?hl=en&as_sdt=0%2C5&q=%22livestock+loss%22+%22jammu+and+kashmir%22&btnG=	Search was limited
"livestock mortality" "western himalayas"	Scan & download completed – 4 Nov	Around 22 results (0.06 s)	3 pages/ 30 results scanned	30	https://scholar.google.com/scholar?hl=en&as_sdt=0%2C5&q=%22livestock+mortality%22+%22western+himalayas%22&btnG=	Search was limited
"livestock mortality" "jammu and kashmir"	Scan & download completed – 4 Nov	Around 68 results (0.09 s)	7 pages/ 70 results scanned	70	https://scholar.google.com/scholar?hl=en&as_sdt=0%2C5&q=%22livestock+mortality%22+%22jammu+and+kashmir%22&btnG=	Search was limited
"livestock mortality" "himachal pradesh"	Scan & download completed – 4 Nov	Around 88 results (0.07 s)	9 pages/ 90 results scanned	90	https://scholar.google.com/scholar?start=0&q=%22livestock+mortality%22+%22himachal+pradesh%22&hl=en&as_sdt=0,5	Search was limited

(continued)

Google search keyword	Status of download with date	Total number of results returned on Google	Number of Google pages reviewed as per relevance of articles	Total results through database search	Search URL	Comments (why the search was terminated)
"livestock mortality" "uttarakhand"	Scan & download completed – 4 Nov	Around 48 results (0.05 s)	5 pages/ 50 results scanned	50	https://scholar.google.com/scholar?start=0&q=%22livestock+mortality%22+%22uttarakhand%22&hl=en&as_sdt=0,5	Search was limited
"livestock mortality" "ladakh"	Scan & download completed – 4 Nov	Around 91 results (0.09 s)	10 pages/ 100 results scanned	100	https://scholar.google.com/scholar?hl=en&as_sdt=0%2C5&q=%22livestock+mortality%22+%22ladakh%22&btnG=	Search was limited
"economic loss by wildlife" "western himalayas"	Scan & download completed – 4 Nov	Around 73 results (0.09 s)	8 pages/ 80 results scanned	80	https://scholar.google.com/scholar?hl=en&as_sdt=0%2C5&q=%22economic+loss%22+by+%22wildlife%22+%22western+himalayas%22&btnG=	Search was limited
"economic loss by wildlife" "jammu and kashmir"	Scan & download completed – 4 Nov	Around 214 results (0.04 s)	21 pages/ 210 results scanned	210	https://scholar.google.com/scholar?start=0&q=%22economic+loss%22+by+%22wildlife%22+%22jammu+and+kashmir%22&hl=en&as_sdt=0,5	Search was not relevant for the most part
"economic loss by wildlife" "himachal pradesh"	Scan&download completed – 4 Nov	Around 361 results (0.05 s)	31 pages/ 310 results scanned	310	https://scholar.google.com/scholar?start=0&q=%22economic+loss%22+by+%22wildlife%22+%22himachal+pradesh%22&hl=en&as_sdt=0,5	Search was not relevant for the most part

(continued)

Google search keyword	Status of download with date	Total number of results returned on Google	Number of Google pages reviewed as per relevance of articles	Total results through database search	Search URL	Comments (why the search was terminated)
"economic loss by wildlife" "uttarakhand"	Scan & download completed – 4 Nov	Around 316 results (0.07 s)	35 pages/ 350 results scanned	350	https://scholar.google.com/scholar?hl=en&as_sdt=0%2C5&q=%22economic+loss%22+by+%22wildlife%22+%22uttarakhand%22&btnG=	Search was not relevant for the most part
"economic loss by wildlife" "ladakh"	Scan & download completed – 5 Nov	Around 117 results (0.04 s)	11 pages/ 110 results scanned	110	https://scholar.google.com/scholar?start=0&q=%22economic+loss%22+by+%22wildlife%22+%22ladakh%22&hl=en&as_sdt=0,5	Search was limited
"conflict mitigation" "western himalayas"	Scan & download completed – 5 Nov	Around 12 results (0.03 s)	2 pages/ 20 results scanned	20	https://scholar.google.com/scholar?start=0&q=%22conflict+mitigation%22+%22western+himalayas%22&hl=en&as_sdt=0,5	Search was limited
"conflict mitigation" "jammu and kashmir"	Scan & download completed – 5 Nov	Around 112 results (0.06 s)	11 pages/ 110 results scanned	110	https://scholar.google.com/scholar?hl=en&as_sdt=0%2C5&q=%22conflict+mitigation%22+%22jammu+and+kashmir%22&btnG=	Search was limited
"conflict mitigation" "himachal pradesh"	Scan&download completed – 5 Nov	Around 129 results (0.18 s)	10 pages/ 100 results scanned	100	https://scholar.google.com/scholar?hl=en&as_sdt=0%2C5&q=%22conflict+mitigation%22+%22himachal+pradesh%22&btnG=	Search was limited

(continued)

Google search keyword	Status of download with date	Total number of results returned on Google	Number of Google pages reviewed as per relevance of articles	Total results through database search	Search URL	Comments (why the search was terminated)
"conflict mitigation" "uttarakhand"	Scan&download completed – 5 Nov	Around 236 results (0.04 s)	23 pages/ 230 results scanned	230	https://scholar.google.com/scholar?hl=en&as_sdt=0%2C5&q=%22conflict+mitigation%22+%22uttarakhand%22&btnG=	Search was limited
"conflict mitigation" "ladakh"	Scan&download completed – 5 Nov	Around 74 results (0.09 s)	8 pages/ 80 results scanned	80	https://scholar.google.com/scholar?hl=en&as_sdt=0%2C5&q=%22conflict+mitigation%22+%22ladakh%22&btnG=	Search was limited
"patterns" of "human wildlife conflict" "western himalayas"	Scan & download completed – 5 Nov	Around 87 results (0.07 s)	9 pages/ 90 results scanned	90	https://scholar.google.com/scholar?start=0&q=%22patterns%22+of+%22human+wildlife+conflict%22+%22western+himalayas%22&hl=en&as_sdt=0,5	Search was limited
"conservation interventions" "western himalayas"	Scan & download completed – 5 Nov	Around 46 results (0.05 s)	5 pages/ 50 results scanned	50	https://scholar.google.com/scholar?hl=en&as_sdt=0%2C5&q=%22conservation+interventions%22+%22western+himalayas%22&btnG=	Search was limited
Random keywords from the ones listed above	Scan & download completed – 4 Dec	Around 48 results (0.05 s)	5 pages/ 48 results scanned	48		Search was limited
Random keywords from the ones listed above	Scan&download completed – s27 Dec	Around 46 results (0.05 s)	5 pages/ 46 results scanned	46		Search was limited

6296 records were scanned from the Google Search pages – based on search results – and a total of 405 files were downloaded from these records

Appendix 2: Database of 88 Articles Reviewed

Sr. no.	Year published and author(s) names used as file names	Year published	Journal name/ organisation or institution name	Title of the studies/ publications/ reports/theses/ short papers	Country
1	1997_Mishra_HP_IND	1997	Environmental Conservation	Livestock Depredation by Large Carnivores in the Indian Trans-Himalaya: Conflict Perceptions and Conservation Prospects	IND
2	1999_Bhatnagar&_HemisNP_LAD_IND	1999	International Snow Leopard Trust	A Survey of Depredation and Related Wildlife-Human Conflicts in Hemis National Park, Ladakh (India)	IND
3	1999_Chauhan_IND	1999	Wildlife Institute of India	Evaluation of Crop Damage in the Eco-development Project Area to Suggest Mitigation Measures	IND
4	2001_Sathyakumar_PAs_IND	2001	Ursus	Status and Management of Asiatic Black Bear and Himalayan Brown Bear in India	IND
5	2001_Jackson&Wangchuk_LAD_IND	2001	Endangered Species Update	Linking Snow Leopard Conservation and People-Wildlife Conflict Resolution: Grassroots Measures to Protect the Endangered Snow Leopard from Herder Retribution	IND
6	2002_Spearing_J&K_IND_Pg175	2002	Snow Leopard Survival Strategy Summit 2002	The Snow Leopard in Zanskar, Jammu & Kashmir, NW India	IND
7	2003_Mishra & _HP_IND	2003	Conservation Biology	The role of incentive programs in conserving snow leopards	IND
8	2003_Jackson & _HemisNP_LAD_IND	2003	Snow Leopard Conservancy Field Document Series	Local People's Attitudes toward Wildlife Conservation in the Hemis National Park with Special Reference to the Conservation of Large Predators	IND

(continued)

Sr. no.	Year published and author(s) names used as file names	Year published	Journal name/ organisation or institution name	Title of the studies/ publications/ reports/theses/ short papers	Country
9	2003_Chauhan & _HP_IND	2003	Ursus	Human Casualties and Livestock Depredation by Black and Brown Bears in the Indian Himalaya, 1989–1998	IND
10	2004_Jackson & Wangchuk_LAD_IND	2004	Human Dimensions of Wildlife	A Community-Based Approach to Mitigating Livestock Depredation by Snow Leopards	IND
11	2004_Bagchi & _HP_IND	2004	Animal Conservation	Conflicts between Traditional Pastoralism and Conservation of Himalayan ibex (Capra sibirica) in the Trans-Himalayan Mountains	IND
12	2004_Namgail & _GMWS_IND	2004	International Snow Leopard Trust	Interactions between Argali and Livestock, Gya-Miru Wildlife Sanctuary, Ladakh, India	IND
13	2006_Bagchi & Mishra_IND	2006	Journal of Zoology	Living with Large Carnivores: Mitigating Large Carnivore-Human Conflicts in Kargil, Ladakh, India	IND
14	2006_Bhatnagar & _Ladakh	2006	Environmental Management	Perceived Conflicts Between Pastoralism and Conservation of the Kiang Equus kiang in the Ladakh Trans-Himalaya, India	IND
15	2006_Sathyakumar_IND	2006	Indian Forester	Status and Distribution of Himalayan Brown Bear in India: An Assessment of Change Over 10 Years	IND
16	2007_Namgail & _IND	2007	Environmental Management	Carnivore Caused Livestock Mortality in Trans-Himalaya	IND

(continued)

Sr. no.	Year published and author(s) names used as file names	Year published	Journal name/ organisation or institution name	Title of the studies/ publications/ reports/theses/ short papers	Country
17	2007_Sathyakumar & Choudhury_Pas_IND	2007	Journal of Bombay Natural History Society	Distribution and Status of Black Bear in India	IND
18	2008_RathoreBC_PhD_Thesis	2008	Saurashtra University	Ecology of Brown Bear (Ursus arctos) with Special Reference to Assessment of Human-Brown Bear Conflicts in Kugti Wildlife Sanctuary, Himachal Pradesh and Mitigation Strategies	IND
19	2008_Choudhury & _J & K_IND	2008	Predator Alert-Conservation Action Series	Attacks on Humans by Leopards and Asiatic Black Bear in the Kashmir valley – Analysis of Case Studies and Spatial Patterns of Elevated Conflict	IND
20	2009_Namgail & _J & K_IND	2009	Oryx	Status and distribution of the Near Threatened Tibetan argali (Ovis ammon hodgsoni) in Ladakh, India: Effect of a Hunting Ban	IND
21	2009_Lauren Satterfield_Ladakh	2009	Mount Holyoke College	Trailing the Snow Leopard: Sustainable Wildlife Conservation in Ladakh (India)	IND
22	2010_Maheshwari & _J & K_IND	2010	World Wide Fund for Nature – India	An investigation of Carnivore Human Conflicts in Kargil	IND
23	2010_Peaty_IND	2010	Journal of Ritsumeikan Social Science and Humanities	Community Based Tourism in the Indian Himalayas: Homestays and Lodges	IND
24	2011_Bhatta & Satya_IND	2011	Mountain Research and Development	Natural Resource Use by Humans and Response of Wild Ungulates	IND

(continued)

Sr. no.	Year published and author(s) names used as file names	Year published	Journal name/ organisation or institution name	Title of the studies/ publications/ reports/theses/ short papers	Country
25	2011_Charoo & _J & K_IND	2011	Ursus	Asiatic Black Bear–Human Interactions around Dachigam National Park, Kashmir, India	IND
26	2011_Sabic_UK_IND	2011	University of Michigan	Human-Wildlife Conflict in Nanda Devi Biosphere Reserve, Uttarakhand, India	IND
27	2012_ChandolaShivani_PhD_Thesis	2012	Saurashtra University	An Assessment of Human-Wildlife Interaction in the Indus Valley, Ladakh, Trans-Himalaya	IND
28	2012_Sanwal & Lone_KupwaraDist_J & K	2012	Indian Forester	An Assessment of the-Asiatic Black Bear-Human Conflicts in Kupwara District, Jammu & Kashmir, India	IND
29	2012_Maheshwari & _LAD_IND	2012	World Wide Fund for Nature – India	Living with Large Carnivores: Mitigating Large Carnivore-Human Conflicts in Kargil, Ladakh, India	IND
30	2013_Sathyakumar & _DNP_J & K	2013	Wildlife Institute of India	Ecology of Asiatic black bear (Ursus thibetanus) in Dachigam National Park, Kashmir, India	IND
31	2013_Maheswari & _UK & HP_IND	2013	Journal of Ecology and the Natural Environment	Snow leopard (Panthera uncia) Surveys in the Western Himalayas, India	IND
32	2013_SuryawanshiK_PhD_Thesis	2013	Manipal University	Human-Carnivore Conflicts: Understanding Predation Ecology and Livestock Damage by Snow Leopards	IND
33	2013_Suryawanshi & _HP_IND	2013	Journal of Applied Ecology	People, Predators and Perceptions: Patterns of Livestock Depredation by Snow Leopards and Wolves leopards and wolves	IND

(continued)

Sr. no.	Year published and author(s) names used as file names	Year published	Journal name/ organisation or institution name	Title of the studies/ publications/ reports/theses/ short papers	Country
34	2013_Steinberg_HP_IND	2013	Albert-Ludwigs-University	Spatial-temporal Dimensions of Human-Wildlife Conflicts: Case Study in the Trans-Himalaya	IND
35	2013_Huyett_LAD_IND	2013	Duke University	Motivations for Conservation: Participation in Community-Based Snow Leopard (Uncia uncia) Conservation in Ladakh, India	IND
36	2013_Shrotriya & _IND	2013	Wildlife Institute of India	Ecology and Conservation of Himalayan Wolf	IND
37	2014_Suryawanshi & _HP_IND	2014	Conservation Biology	Multiscale Factors Affecting Human Attitudes Toward Snow Leopards and Wolves	IND
38	2014_Mohanta & _HP_IND	2014	Asian Journal of Conservation Biology	Anthropogenic Threats and its Conservation on Himalayan Brown Bear (Ursus arctos) habitat in Kugti Wildlife Sanctuary, Himachal Pradesh, India	IND
39	2015_Habib & _J & K	2015	The Journal of Zoology Studies	Human-Wildlife Conflict-Causes, Consequences and Mitigation Measures with Special Reference to Kashmir	IND
40	2015_Mir & _Kashmir_IND	2015	Mountain Research and Development	Attitudes of Local People Toward Wildlife Conservation: A Case Study From the Kashmir Valley	IND
41	2016_Mishra & _HP_IND	2016	Snow Leopard Conservancy	The Role of Village Reserves in Revitalizing the Natural Prey Base of the Snow Leopard	IND

(continued)

Sr. no.	Year published and author(s) names used as file names	Year published	Journal name/ organisation or institution name	Title of the studies/ publications/ reports/theses/ short papers	Country
42	2017_Home & _HP_IND	2017	Ambio	Commensal in Conflict: Livestock Depredation Patterns by Free Ranging Domestic Dogs in the Upper Spiti Landscape, Himachal Pradesh, India	IND
43	2017_Moten & _J & K_IND	2017	International Journal of Research in Medical Sciences	Causalities of human wildlife conflict in Kashmir valley, India; a neglected form of trauma: our 10 year study	IND
44	2018_Jamwal & _HemisNP_LAD_IND	2018	Oryx	Factors contributing to a striking shift in human–wildlife dynamics in Hemis National Park, India: 22 years of reported snow leopard depredation	IND
45	2019_Bagchi & _HP_IND	2019	Wildlife Biology	Change in snow leopard predation on livestock after revival of wild prey in the Trans-Himalaya	IND
46	2019_Hassan & _Kulgam_J & K	2019	Recent Innovations in Biosustainability and Environmental Research-Vol.1	Human-Wildlife Conflict in District Kulgam of Jammu & Kashmir State: Causes, Consequences and Mitigation Measures	IND
47	2019_Lyngdoh & Habib_Spiti_IND	2019	BioRxiv	Predation by Himalayan Wolves: Understanding conflict, culture and co-existence amongst 2 Indo-Tibetan community and large carnivores in High Himalaya	IND
48	2019_Maheswari & _HNP_LAD	2019	Human Dimensions of Wildlife	Snow leopard stewardship in mitigating human–wildlife conflict in Hemis National Park, Ladakh, India	IND

(continued)

Sr. no.	Year published and author(s) names used as file names	Year published	Journal name/ organisation or institution name	Title of the studies/ publications/ reports/theses/ short papers	Country
49	2019_Yadav & _NDBR	2019	Journal of Entomology and Zoology Studies	Occurrence and feeding habit of Asiatic black bear (Ursus thibetanus) in Nanda Devi biosphere reserve, Uttarakhand, India	IND
50	2019_Vanelli & _Ladakh	2019	Human Dimensions of Wildlife	Community participation in ecotourism and its effect on local perceptions of snow leopard (Panthera uncia) conservation	IND
51	2019_Sharief & _Lahaul_HP	2019	Global Ecology and Conservation	Identifying Himalayan brown bear (Ursus arctos isabellinus) conservation areas in Lahaul Valley, Himachal Pradesh	IND
52	2019_Dhendhup & _Himalayas	2019	CAT News	Distribution and status of manul in the Himalayas and China	IND
53	2020_Maheshwari & Sathyakumar_LAD_IND	2020	Journal of Arid Environments	Patterns of Livestock Depredation and Large Carnivore Conservation Implications in the Indian Trans-Himalaya	IND
54	2020_Sharma & _HP_IND	2020	World Wide Fund for Nature – India	Human-Wildlife Conflict Mitigation Strategy in Pangi, Lahaul and Kinnaur Landscapes, Himachal Pradesh	IND
55	2020_Sharma & _LAD_IND_Secure	2020	World Wide Fund for Nature – India	Human-Wildlife Conflict Management Strategy in Changthang Landscape, Ladakh	IND
56	2000_Hussain_BAL_PAK	2000	Mountain Research and Development	Protecting the Snow Leopard and Enhancing Farmers' Livelihoods	PAK
57	2001_Shackleton_PAK	2001	International Union for Conservation of Nature	A Review of Community-Based Trophy Hunting Programs in Pakistan	PAK

(continued)

Sr. no.	Year published and author(s) names used as file names	Year published	Journal name/ organisation or institution name	Title of the studies/ publications/ reports/theses/ short papers	Country
58	2003_HussainS_Bal_PAK	2003	Oryx	The status of the snow leopard in Pakistan and its conflict with local farmers	PAK
59	2006_Mir_NWFPNorthernAreas_PAK	2006	International Union for Conservation of Nature	Impact Assessment of Community Based Trophy Hunting in Mountain Areas Conservancy Project areas of North West Frontier Province and Northern Areas	PAK
60	2007_Nawaz_PAK	2007	Ursus	Status of Brown Bear in Pakistan	PAK
61	2008_Nawaz & _DNP_PAK	2008	Biological Conservation	Pragmatic management increases a flagship species, the Himalayan brown bears, in Pakistan's Deosai National Park	PAK
62	2008_Ali_NWFP_PAK	2008	Wildlife Biology	Conservation and Status of Markhor (Capra falconeri) in the Northern Parts of North West Frontier Province, Pakistan	PAK
63	2009_Dar & _MachiaraNP_PAK	2009	Biological Conservation	Predicting the patterns, perceptions and causes of human–carnivore conflict in and around Machiara National Park, Pakistan	PAK
64	2010_Din & Nawaz_Chitral_PAK	2010	Journal of Animal and Plant Sciences	Status of the Himalayan Lynx in District Chitral, North West Frontier Province, Pakistan	PAK
65	2011_Din & Nawaz_Chitral_PAK	2011	Journal of Animal and Plant Sciences	Status of Snow Leopard and Prey Species in Torkhow Valley, District Chitral, Pakistan	PAK
66	2011_Waseem & Ali_Mansehra_PAK	2011	World Wide Fund for Nature – Pakistan	Assessment Report On Asiatic Black Bear (Ursus theibetanus)-Human Conflict in District Mansehra	PAK

(continued)

Sr. no.	Year published and author(s) names used as file names	Year published	Journal name/ organisation or institution name	Title of the studies/ publications/ reports/theses/ short papers	Country
67	2012_Rosen & _BAL_PAK	2012	Mountain Research and Development	Reconciling Sustainable Development of Mountain Communities with Large Carnivore Conservation	PAK
68	2013_Hameed & _DNP_PAK	2013	International Bear News	Himalayan Brown Bear in Deosai National Park, Pakistan: Current Status and Threats	PAK
69	2013_Parveen & Abid_Koh_PAK	2013	International Journal of Farming and Allied Sciences	Asian black bear, Ursus thibetanus: Human-bear conflict in the Palas Valley, Kohistan, Pakistan	PAK
70	2015_Abbas & _PAK	2015	Journal of Bioresource Management	Bears in Pakistan: Distribution, Population Biology and Human Conflicts	PAK
71	2015_Ali & _BAL_PAK	2015	International Journal of Scientific and Research Publications	Status and Threats of Asiatic Black bear in Gais Valley of Diamer District, Gilgit-Baltistan, Pakistan	PAK
72	2016_Awan & _AJK_PAK	2016	International Journal of Conservation Science	Preliminary survey on Asiatic black bear in Kashmir Himalaya, Pakistan: Implications for preservation	PAK
73	2016_Hameed & _INTLbear news_PAK	2016	International Bear News	The Bears in Musk Deer National Park Gurez, Azad Jammu and Kashmir and their conflict with local people	PAK
74	2016_Ahmad & _MDNP_AJK_PAK	2016	European Journal of Wildlife Research	Carnivores' diversity and conflicts with humans in Musk Deer National Park, Azad Jammu and Kashmir, Pakistan	PAK

(continued)

Sr. no.	Year published and author(s) names used as file names	Year published	Journal name/ organisation or institution name	Title of the studies/ publications/ reports/theses/ short papers	Country
75	2016_Ahmad & _Chitral_PAK_RUFFORDReport	2016	The Rufford Foundation	Community Based Human-Carnivores Conflict Mitigation in Gehrate Gol Community Game Reserve Chitral, Pakistan	PAK
76	2016_Ali & _AJK_PAK	2016	Pakistan Journal of Zoology	Human-Grey Wolf (Canis lupus) Conflict in Shounter Valley, District Neelum, Azad Jammu and Kashmir, Pakistan	PAK
77	2017_Bocci & _CKNP_PAK	2017	European Journal of Wildlife Research	Sympatric snow leopards and Tibetan wolves: coexistence of large carnivores with human-driven potential competition	PAK
78	2017_Wahid & _Kumrat_PAK	2017	Bioscience Research	Conservation status of Black Bear in Kumrat Valley, Pakistan	PAK
79	2018_Khan & _HusheyV_PAK	2018	Oryx	Livestock depredation by large predators and its implications for conservation and livelihoods in the Karakoram Mountains of Pakistan	PAK
80	2018_Ali & _Kaghan_PAK	2018	Ethology, Ecology & Evolution	Human–Asiatic black bear (Ursus thibetanus) interactions in the Kaghan Valley, Pakistan	PAK
81	2019_Kazmi & _MNP_PAK	2019	The Journal of Animal & Plant Sciences	Crop Raiding by Himalayan Black Bear: A Major Cause of Human-Bear Conflict in Machiara National Park, Pakistan	PAK

(continued)

Sr. no.	Year published and author(s) names used as file names	Year published	Journal name/ organisation or institution name	Title of the studies/ publications/ reports/theses/ short papers	Country
82	2019_Khan & _PAK	2019	Animals	Status and Magnitude of Grey Wolf Conflict with Pastoral Communities in the Foothills of the Hindu Kush Region of Pakistan	PAK
83	2019_Khan & _LowerDir_PAK	2019	Applied Ecology and Environmental Research	Status and Attitude of Local Communities Towards the Grey Wolf (Canis lupus) in Lower Dir District, Khyber Pakhtunkhwa, Pakistan	PAK
84	2020_Zahoor & _AJK_PAK	2020	Pakistan Journal of Zoology	Damages Done by Black Bear (Ursus thibetanus) in Moji Game Reserve and its Surroundings, Leepa Valley, Azad Jammu and Kashmir (Pakistan)	PAK
85	2020_Waseem & _Manshera_PAK	2020	Pakistan Journal of Zoology	Ecology and Human Conflict of Asiatic Black Bear (Ursus thibetanus laniger) in Mansehra District, Pakistan	PAK
86	2020_Khan & _MahoodandV_PAK	2020	Pakistan Journal of Zoology	Distribution, Diet Menu and Human Conflict of Grey Wolf Canis lupus inMahodand Valley, Swat District, Pakistan	PAK
87	2020_Ullah & _2PAKValleys	2020	Global Ecology and Conservation	Poaching of Asiatic black bear: evidence from Siran and Kaghan valleys, Pakistan	PAK
88	2021_Ali_KalamValley_PAK	2021 (published online in 2020)	Pakistan Journal of Zoology	Conflicts Involving Brown Bear and Other Large Carnivores in the Kalam Valley, Swat, Pakistan	PAK

Appendix 3: Data Extraction Form Utilised in the Current Review Chapter

Categories	Sub-categories
Section 1 – Basic information regarding the studies	Serial number used for sorting and identification
	Year published
	Author(s) names and year published
	Study duration and sampling Period
	Study started in year
	Journal name
	Title of the study/publication
	Country/countries
	State(s)/district(s)
	Species names
	Article location (folder name on local computer)
	Organisation(s)/institution(s) that conducted the studies
	GPS co-ordinates as per area names mentioned in the studies/publications
Section 2 – Purpose identified or defined through objectives of the study	Purpose was to assess extent of conflict
	Purpose was to quantify impact on animals
	Purpose was to quantify impact on humans
	Purpose was to document interventions
	Purpose was to implement interventions
	Purpose was to evaluate interventions
	Any other purpose based on objectives mentioned in the study
Section 3 – Type of data generated by the study, irrespective of the methodology/study instrument utilised	Data type: ecological
	Data type: social
	Data type: hybrid
	Data type: other
Section 4 – Drivers of HWC – identified in the study through the data collated/prior information referenced in the study	Improper waste disposal
	Lack of wild prey
	Expanding areas
	Herding practices
	Increased livestock
	Easy access to crop or livestock
	Forest dependency
	Forest fragmentation and/or habitat loss
	Expanding settlements and/or increased human disturbance

(continued)

Categories	Sub-categories
	Lack of awareness
	Other factors
Section 5 – Impacts of HWC observed in the study – through the data collated	Hunting of wildlife and/or death of wildlife
	Reduction of forest resources
	Human injury or death
	Livestock loss
	Impact to human livelihood or economic security
	Tourism
	Prey base depletion
	Crop depredation
	Impact on emotional well-being
	Attitude towards species
	Property damage
	Others impacts
Section 6 – Interventions suggested/implemented/evaluated-based on the data collated through the study	Land management and zoning, Land use change
	Prey management
	Relocation or reduction of prey species
	Livestock corrals and/or crop fencing
	Alternate livelihood opportunities
	Use of guard dogs, loud noise, increased vigilance, skilled shepherds, chasing
	Trophy hunting of ungulates
	Relocation or reduction of livestock
	Monetary compensation
	Community conservation and ecotourism
	Insurance programs
	Religion beliefs propaganda
	Education programs
	Women empowerment
	Habitat restoration
	Crop rotation, crop change
	Livestock husbandry, veterinary services
	Tourism development
	Legal management
	Rangeland management
	Other sustainable solutions, technology etc.
	Capacity building
	Additional monitoring
	Increase in current intervention levels
Section 7 – Details on the interventions and its evaluations if any mentioned in the study	Multiple interventions suggested
	Interventions implemented
	Intervened by whom?

(continued)

Categories	Sub-categories
	Type of interventions
	Interventions evaluated
	Comments on effectiveness of interventions
	Comments on the drivers
	Indicative trends of HWC
	Additional notes/comments for Authors to consider

References

Ahmad, A. (2016). *Community based human-carnivores conflict mitigation in Gehrate Gol Community Game Reserve Chitral*. Final Report, The Rufford Foundation.

Ahmed, J., Tareen, N., & Khan, P. (2003). Conservation of Sulaiman Markhor and Afghan Urial by local Tribesmen in Torghar, Pakistan, Case Study.

Alexander, J., et al. (2015). Human wildlife conflict involving large carnivores in Qilianshan, China and the minimal paw-print of snow leopards. *Biological Conservation, 187*, 1–9. https://doi.org/10.1016/j.biocon.2015.04.002

Ali, A. (2008). *Conservation and status of Markhor (Capra falconeri) in the northern parts of North West Frontier Province, Pakistan*. University of Montana.

Ali, A., et al. (2018). Human–Asiatic black bear (Ursus thibetanus) interactions in the Kaghan Valley, Pakistan. *Ethology Ecology and Evolution, 30*(5), 399–415. https://doi.org/10.1080/03949370.2017.1423113

Allen, C. R., & Garmestani, A. S. (2015). Adaptive management. In C. R. Allen & A. S. Garmestani (Eds.), *Adaptive management of socio-ecological systems*. Springer. Available at: https://doi.org/10.1007/978-94-017-9682-8

Anand, S., & Radhakrishna, S. (2017). Investigating trends in human-wildlife conflict: Is conflict escalation real or imagined? *Journal of Asia-Pacific Biodiversity, 10*(2), 154–161. https://doi.org/10.1016/j.japb.2017.02.003

Bagchi, S., Sharma, R. K., & Bhatnagar, Y. V. (2019). Change in snow leopard predation on livestock after revival of wild prey in the Trans-Himalaya. *Wildlife Biology, 2020*(1), 2019. https://doi.org/10.2981/wlb.00583

Bhatnagar, Y. V., Mathur, V., & Mccarthy, T. (2001, September). *A regional perspective for snow leopard conservation in the Indian Trans-Himalaya* (pp. 25–47). National Parks.

Bhatnagar, Y. V., Stakrey, R. W., Jackson, R. (1999). A survey of depredation and related wildlife-human conflicts in Hemis National Park, Ladakh, Jammu & Kashmir, India. Unpublished Report. International Snow Leopard Trust. Seattle, WA. 20 pages

Bombieri, G., et al. (2019). Brown bear attacks on humans: A worldwide perspective. *Scientific Reports, 9*(1), 1–10. https://doi.org/10.1038/s41598-019-44341-w

Bull, J. W., et al. (2014). Importance of baseline specification in evaluating conservation interventions and achieving no net loss of biodiversity. *Conservation Biology, 28*(3), 799–809. https://doi.org/10.1111/cobi.12243

Can, Ö. E., et al. (2014). Resolving human-bear conflict: A global survey of countries, experts, and key factors. *Conservation Letters, 7*(6), 501–513. https://doi.org/10.1111/conl.12117

Chandra, K., et al. (2018). Faunal diversity of Indian Himalaya: An overview. In K. Chandra, D. Gupta, K. C. Gopi, B. Tripathy, & V. Kumavr (Eds.), *Faunal diversity of Indian Himalaya*. Zoological Survey of India. Available at: https://nmhs.org.in/pdf/BOOKS/03.Faunal%20Diversity%20of%20Indian%20Himalaya.pdf

Chettri, N., & Sharma, E. (2016). Reconciling the mountain biodiversity conservation and human well-being: Drivers of biodiversity loss and new approaches in the Hindu-Kush Himalayas. *Proceedings of the Indian National Science Academy, 82*(1), 53–73. https://doi.org/10.16943/ptinsa/2016/v81i1/48378

Dar, N. I., et al. (2009). Predicting the patterns, perceptions and causes of human-carnivore conflict in and around Machiara National Park, Pakistan. *Biological Conservation, 142*(10), 2076–2082. https://doi.org/10.1016/j.biocon.2009.04.003

Dhee, et al. (2019). The leopard that learnt from the cat and other narratives of carnivore–human co-existence in northern India. *People and Nature, 1*(3), 376–386. https://doi.org/10.1002/pan3.10039

Dickman, A. J., Macdonald, E. A., & Macdonald, D. W. (2011). A review of financial instruments to pay for predator conservation and encourage human-carnivore co-existence. *Proceedings of the National Academy of Sciences of the United States of America, 108*(34), 13937–13944. https://doi.org/10.1073/pnas.1012972108

Distefano, E. (2005). *Human-wildlife conflict worldwide: Collection of case studies, analysis of management strategies and good practices*. Available at: http://www.fao.org/3/a-au241e.pdf

Eklund, A., et al. (2017). Limited evidence on the effectiveness of interventions to reduce livestock predation by large carnivores. *Scientific Reports, 7*, 2097. https://doi.org/10.1038/s41598-017-02323-w

Ferraro, P. K., & Pattanayak, S. K. (2006). Money for nothing? A call for empirical evaluation of biodiversity conservation investments. *PLoS Biology, 4*(4), 0482–0488.

Gibbs, G. (2012). Thematic coding and categorizing. In *Analyzing qualitative data* (pp. 38–55). SAGE. https://doi.org/10.4135/9781849208574.n4

Gusenbauer, M., & Haddaway, N. R. (2020). Which academic search systems are suitable for systematic reviews or meta-analyses? Evaluating retrieval qualities of Google Scholar, PubMed, and 26 other resources. *Research Synthesis Methods, 11*(2), 181–217. https://doi.org/10.1002/jrsm.1378

Hill, C. M., Webber, A. D., & Priston, N. E. C. (2017). Understanding conflicts about wildlife. In C. M. Hill, A. D. Webber, & N. E. C. Priston (Eds.), *Understanding conflicts about wildlife* (1st ed.). Berghahn Books. https://doi.org/10.2307/j.ctvw04h12

Holland, K. K., Larson, L. R., & Powell, R. B. (2018). Characterising conflict between humans and big cats Panthera spp: A systematic review of research trends and management opportunities. *PLoS One, 13*(9), 1–19. https://doi.org/10.1371/journal.pone.0203877

Hsieh, H. F., & Shannon, S. E. (2005). Three approaches to qualitative content analysis. *Qualitative Health Research, 15*(9), 1277–1288. https://doi.org/10.1177/1049732305276687

Hussain, S. (2000). Protecting the snow leopard and enhancing farmers' livelihoods. *Mountain Research and Development, 20*(3), 226–231. https://doi.org/10.1659/0276-4741(2000)020[0226,ptslae]2.0.co;2

Hussain, S. (2003). The status of the snow leopard in Pakistan and its conflict with local farmers. *Oryx, 37*(1), 26–33. https://doi.org/10.1017/S0030605303000085

Huyett, A. M., (2013). *Motivations for conservation: Participation in community-based snow leopard (Uncia Uncia) conservation in Ladakh, India*. Dissertation, Duke University. Available at: https://dukespace.lib.duke.edu/dspace/bitstream/handle/10161/6818/Huyett_MP_May2013.pdf?sequence=1

Ijaz, S., et al. (2020). Human interaction, conflict, threats and role of mammals-A review. *Journal of Ethnomedicine and Ethnoecology, 1*(1), 1–11.

Inskip, C., & Zimmermann, A. (2009). Human-felid conflict: A review of patterns and priorities worldwide. *Oryx, 43*(1), 18–34. https://doi.org/10.1017/S003060530899030X

IUCN. (2020). *IUCN SSC position statement on the management of human-wildlife conflict*. IUCN Species Survival Commission (SSC) Human-Wildlife Conflict Task Force.

Jackson, R. M. (2015). HWC ten years later: Successes and shortcomings of approaches to global snow leopard conservation. *Human Dimensions of Wildlife, 20*(4), 310–316. https://doi.org/10.1080/10871209.2015.1005856

Jackson, R. M., & Wangchuk, R. (2001). Linking snow leopard conservation and people-wildlife conflict resolution: Grassroots measures to protect the endangered snow leopard from herder retribution. *Endangered Species Update, 18*(4), 138–141. Available at: https://snowleopardconservancy.org/pdf/Snow%20Leopard%20138-141.pdf

Jackson, R. M., & Wangchuk, R. (2004). A community-based approach to mitigating livestock depredation by snow leopards. *Human Dimensions of Wildlife, 9*(4), 1–16. https://doi.org/10.1080/10871200490505756

Jamwal, P. S., Takpa, J., & Parsons, M. H. (2019). Factors contributing to a striking shift in human-wildlife dynamics in Hemis National Park, India: 22 years of reported snow leopard depredation. *Oryx, 53*(1), 58–62. https://doi.org/10.1017/S0030605317001892

Kansky, R., Kidd, M., & Fischer, J. (2021). Does money "buy" tolerance toward damage-causing wildlife? *Conservation Science and Practice, 3*, e262. https://doi.org/10.1111/csp2.262

Karanth, K. K., & Vanamamalai, A. (2020). Wild seve: A novel conservation intervention to monitor and address human-wildlife conflict. *Frontiers in Ecology and Evolution, 8*(July), 1–11. https://doi.org/10.3389/fevo.2020.00198

Kazmi SS-Ul-Hassan, Minhas, R. A., Ahmad, B., Awan, M. S., Abbasi, S., Ali, U., Shakeel, U., & Dar, N. I. (2019). Crop raiding by Himalayan black bear: A major cause of human. *The Journal of Animal & Plant Sciences, 29*(3), 854–863.

Khan, H., & Baig, S. U. (2020). Biodiversity conservation in the Hindu Kush-Karakoram-Himalayan mountain region of northern Pakistan: Overview of big mammal protection. *Journal of Mountain Science, 17*(6), 1360–1373. https://doi.org/10.1007/s11629-018-5113-0

Khan, M. Z., et al. (2014). Abundance, distribution and conservation of key ungulate species in Hindu Kush, Karakoram and Western Himalayan (HKH) mountain ranges of Pakistan. *International Journal of Agriculture and Biology, 16*(6), 1050–1058.

Krafte, K. E. (2019). *Investigating the cultural context for big cat conservation around the world.* Dissertation, Clemson University. Available at: https://tigerprints.clemson.edu/all_dissertations/2348/

Krippendorff, K. (2004). *Content analysis: An introduction to its methodology.* SAGE Publications. https://doi.org/10.1103/PhysRevB.31.3460

Kumar, A., Adhikari, B. S., & Rawat, G. (2015). Ecology and management of grassland habitats in India. In G. S. Rawat & B. S. Adhikari (Eds.), *ENVIS bulletin wildlife and protected areas* (Vol. 17, pp. 28–41). Wildlife Institute of India.

Lozano, J., et al. (2019). Human-carnivore relations: A systematic review. *Biological Conservation, 237*(July), 480–492. https://doi.org/10.1016/j.biocon.2019.07.002

Maheshwari, A., & Sathyakumar, S. (2019). Snow leopard stewardship in mitigating human–wildlife conflict in Hemis National Park, Ladakh, India. *Human Dimensions of Wildlife, 24*(4), 395–399. https://doi.org/10.1080/10871209.2019.1610815

Maheshwari, A., & Sathyakumar, S. (2020). Patterns of livestock depredation and large carnivore conservation implications in the Indian Trans-Himalaya. *Journal of Arid Environments, 182*(May), 104241. https://doi.org/10.1016/j.jaridenv.2020.104241

Maheshwari, A., et al. (2012). *Living with large carnivores: Mitigating large carnivore-human conflicts in Kargil, Ladakh, India.* Final Report. The Rufford Foundation. Available at: https://ruffordorg.s3.amazonaws.com/media/project_reports/8296-2%20Detailed%20Final%20Report.pdf.

Maikhuri, R. K., et al. (2001). Conservation policy-people conflicts: A case study from Nanda Devi Biosphere Reserve (a World Heritage Site), India. *Forest Policy and Economics, 2*(3–4), 355–365. https://doi.org/10.1016/S1389-9341(01)00037-5

Mani, M. S. (1974). Biogeography of the Himalaya. In M. S. Mani (Ed.), *Ecology and biogeography in India* (pp. 664–681). Springer Netherlands. https://doi.org/10.1007/978-94-010-2331-3_21

Manral, U., et al. (2016). Human wildlife conflict in India: A review of economic implication of loss. *Indian Forester, 142*(10), 928–940. Available at: http://www.indianforester.co.in/index.php/indianforester/article/view/93647

Marchini, S., et al. (2019). Planning for co-existence in a complex human-dominated world. In B. Frank, J. A. Glikman, & S. Marchini (Eds.), *Human-wildlife interactions: Turning conflict into co-existence* (pp. 414–438). Cambridge University Press. https://doi.org/10.1017/9781108235730.022

Meiyappan, P., P. S. Roy, A. Soliman, T. Li, P. Mondal, S. Wang, & A. K. Jain. (2018) India Village-Level Geospatial Socio-Economic Data Set: 1991, 2001. Palisades, NY: NASA Socio-economic Data and Applications Center (SEDAC). https://doi.org/10.7927/H4CN71ZJ. Accessed May 2021.

Mir, A. (2006). *Impact assessment of community based trophy hunting in MACP areas of NWFP and Northern Areas, Mountain Areas Conservancy Project (MACP).* IUCN Islamabad Program Office. Available at: https://www.iucn.org/sites/dev/files/import/downloads/pk_macp_iacbth_cs.pdf

Mishra, C. (1997). Livestock depredation by large carnivores in the Indian Trans-Himalaya: Conflict perceptions and conservation prospects. *Environmental Conservation, 24*(4), 338–343. https://doi.org/10.1017/S0376892997000441

Mishra, C., Prins, H. H. T., & van Wieren, S. E. (2001). Overstocking in the Trans-Himalayan rangelands of India. *Environmental Conservation, 28*(3), 279–283. https://doi.org/10.1017/s0376892901000297

Mishra, C., Allen, P., McCarthy, T., Madhusudan, M. D., et al. (2003). The role of incentive programs in conserving the snow leopard. *Conservation Biology, 17*(6), 1512–1520. https://doi.org/10.1111/j.1523-1739.2003.00092.x

Mishra, C., Bagchi, S., Namgail, T., & Bhatnagar, Y. V. (2009). Multiple use of Trans-Himalayan Rangelands: Reconciling Human Livelihoods with Wildlife Conservation. *In Wild Rangelands: Conserving Wildlife While Maintaining Livestock in Semi-Arid Ecosystems* (Issue December, pp. 291–311). https://doi.org/10.1002/9781444317091.ch11

Mishra, C., Bhatnagar, Y. V., Trivedi, P., et al. (2016). The role of village reserves in revitalising the natural prey base of snow leopard. In T. McCarthy & D. Mallon (Eds.), *Snow leopards: Biodiversity of the world: Conservation from genes to landscapes* (pp. 184–186). Academic Press.

Mishra, C., et al. (2017). Building partnerships with communities for biodiversity conservation: Lessons from Asian mountains. *Journal of Applied Ecology, 54*(6), 1583–1591. https://doi.org/10.1111/1365-2664.12918

Miteva, D. A., Pattanayak, S. K., & Ferraro, P. J. (2012). Evaluation of biodiversity policy instruments: What works and what doesn't? *Oxford Review of Economic Policy, 28*(1), 69–92. https://doi.org/10.1093/oxrep/grs009

Namgail, T., Fox, J. L., & Bhatnagar, Y. V. (2007). Carnivore-caused livestock mortality in Trans-Himalaya. *Environmental Management, 39*(4), 490–496. https://doi.org/10.1007/s00267-005-0178-2

Namgail, T., Wieren, S. E., & Prins, H. H. T. (2010). Pashmina production and socio-economic changes in the Indian Changthang: Implications for natural resource management. *Natural Resources Forum, 34*(3), 222–230. https://doi.org/10.1111/j.1477-8947.2010.01303.x

Nawaz, M. A., Swenson, J. E., & Zakaria, V. (2008). Pragmatic management increases a flagship species, the Himalayan brown bears, in Pakistan's Deosai National Park. *Biological Conservation, 141*(9), 2230–2241. https://doi.org/10.1016/j.biocon.2008.06.012

Nyhus, P. J. (2016). Human-wildlife conflict and co-existence. *Annual Review of Environment and Resources, 41*(1), 143–171. https://doi.org/10.1146/annurev-environ-110615-085634

Pahuja, M., & Sharma, R. K. (2021). Wild predators, livestock, and free ranging dogs: Patterns of livestock mortality and attitudes of people toward predators in an urbanising Trans-Himalayan landscape. *Frontiers in Conservation Science, 2*(December), 1–13. https://doi.org/10.3389/fcosc.2021.767650

Peaty, D. (2009). Community-based tourism in the Indian Himalaya: Homestays and lodges. *Journal of Ritsumeikan Social Sciences and Humanities, 2*, 25–44.

Pooley, S., et al. (2017). An interdisciplinary review of current and future approaches to improving human–predator relations. *Conservation Biology, 31*(3), 513–523. https://doi.org/10.1111/cobi.12859

Quinye, Y., & Du, Z. (2004). *Tibetan geography* (Series of basic information of Tibet of China). China Intercontinental Press.

Rashid, W., et al. (2020). Research trends and management options in human-snow leopard conflict. *Biological Conservation, 242*(January), 108413. https://doi.org/10.1016/j.biocon.2020.108413

Redpath, S. M., et al. (2013). Understanding and managing conservation conflicts. *Trends in Ecology & Evolution, 28*(2), 100–109. https://doi.org/10.1016/j.tree.2012.08.021

Redpath, S. M., Bhatia, S., & Young, J. (2015). Tilting at wildlife: Reconsidering human-wildlife conflict. *Oryx, 49*(2), 222–225. https://doi.org/10.1017/S0030605314000799

Reid, R. S., et al. (2016). Evolution of models to support community and policy action with science: Balancing pastoral livelihoods and wildlife conservation in savannas of East Africa. *Proceedings of the National Academy of Sciences of the United States of America, 113*(17), 4579–4584. https://doi.org/10.1073/pnas.0900313106

Riley, S. J., et al. (2002). The essence of wildlife management. *Wildlife Society Bulletin, 30*(2), 585–593.

Rosen, T., et al. (2012). Reconciling sustainable development of mountain communities with large carnivore conservation: Lessons from Pakistan. *Mountain Research and Development, 32*(3), 286–293. https://doi.org/10.1659/MRD-JOURNAL-D-12-00008.1

Sandhu, H., & Sandhu, S. (2015). Poverty, development, and Himalayan ecosystems. *Ambio, 44*(4), 297–307. https://doi.org/10.1007/s13280-014-0569-9

Sanwal, C. S., & Lone, R. A. (2012). An assessment of the Asiatic black bear-human conflicts in Kupwara district, Jammu & Kashmir, India. *The Indian Forester, 138*(10), 1–6.

Satterfield, L. (2009). *Trailing the snow leopard: Sustainable Wildlife Conservation in Ladakh (India)*. Dissertation, Mount Holyoke College. Available at: http://www.jkwildlife.com/wild/pdf/pub/shawl/sustainable_wildlife_conservation_in_ladakh_Satterfield_2009.pdf

Seoraj-Pillai, N., & Pillay, N. (2017). A meta-analysis of human-wildlife conflict: South African and global perspectives. *Sustainability (Switzerland), 9*(1), 1–21. https://doi.org/10.3390/su9010034

Shackleton, D. M. (2001). *A review of community-based trophy hunting programs in Pakistan, Mountain Areas Conservancy Project (MACP)*. IUCN Islamabad Program Office. Available at: https://www.iucn.org/sites/dev/files/import/downloads/pk_macp_trophyhunting_review.pdf

Sharma, R. K., & Singh, R. (2020). *Over 100 years of snow leopard research, a spatially explicit review of the state of knowledge in the snow leopard range*. Report. WWF.

Sharma, P., Chettri, N., & Wangchuk, K. (2021). Human–wildlife conflict in the roof of the world: Understanding multidimensional perspectives through a systematic review. *Ecology and Evolution*, 11, 11569– 11586. https://doi.org/10.1002/ece3.7980

Singh, R., et al. (2020). Traditional ecological knowledge and contemporary changes in the agro-pastoral system of Upper Spiti Landscape, Indian Trans-Himalayas. *Pastoralism, 10*, Article Number: 15. https://doi.org/10.1186/s13570-020-00169-y.

Suryawanshi, K. R., et al. (2014). Multiscale factors affecting human attitudes toward snow leopards and wolves. *Conservation Biology, 28*(6), 1657–1666. https://doi.org/10.1111/cobi.12320

Sutherland, W. J., et al. (2004). The need for evidence-based conservation. *Trends in Ecology & Evolution, 19*(6), 305–308. https://doi.org/10.1016/j.tree.2004.03.018

Svancara, L. K., et al. (2005). Policy-driven versus evidence-based conservation: A review of political targets and biological needs. *BioScience, 55*(11), 989–995. https://doi.org/10.1641/0006-3568(2005)055[0989,PVECAR]2.0.CO;2

Talbert, L., Leslie, S. C., & Black, S. A. (2020). Use of predator controls to address human wildlife conflict. *International Journal of Avian & Wildlife Biology, 5*(1), 9–13. https://doi.org/10.15406/ijawb.2020.05.00167

Tiwari, P., Joshi, A., & Joshi, B. (2018). Urban growth in the Himalayas: Understanding the process and options for sustainable development. *Journal of Urban and Regional Studies on Contemporary India, 4*(2), 15–27.

Torres, D. F., Oliveira, E. S., & Alves, R. R. N. (2018). Conflicts between humans and terrestrial vertebrates: A global review. *Tropical Conservation Science, 11*. https://doi.org/10.1177/1940082918794084

Treves, A., et al. (2006). Co-managing human–wildlife conflicts: A review. *Human Dimensions of Wildlife, 11*(6), 383–396. https://doi.org/10.1080/10871200600984265

Treves, A., Wallace, R. B., & White, S. (2009). Participatory planning of interventions to mitigate human-wildlife conflicts. *Conservation Biology, 23*(6), 1577–1587. https://doi.org/10.1111/j.1523-1739.2009.01242.x

UNEP-WCMC and IUCN (2021), Protected Planet: The World Database on Protected Areas (WDPA) and World Database on Other Effective Area-based Conservation Measures (WD-OECM) [Online], June 2021, Cambridge, UK: UNEP-WCMC and IUCN. Available at: https://www.protectedplanet.net.

Valentová, K. A. (2017). Abundance and threats to the survival of the snow leopard – A review. *European Journal of Environmental Sciences, 7*(1), 73–93. https://doi.org/10.14712/23361964.2017.7

van Eeden, L. M., et al. (2018). Carnivore conservation needs evidence-based livestock protection. *PLoS Biology, 16*(9), 1–8. https://doi.org/10.1371/journal.pbio.2005577

Vannelli, K., et al. (2019). Community participation in ecotourism and its effect on local perceptions of snow leopard (Panthera uncia) conservation. *Human Dimensions of Wildlife, 24*(2), 180–193. https://doi.org/10.1080/10871209.2019.1563929

Wahid, A., Muhammad, J., Ditta, A., Khan, A., Murtaza, A., & Saeed, S. (2017). Conservation status of black bear (Ursus thibetanus) in the Kumrat valley, Pakistan. *Bioscience Research, 14*(October), 1230–1237.

Woodford, M. H., Frisina, M. R., & Awan, G. A. (2004). The Torghar conservation project: Management of the livestock, Suleiman Markhor (Capra Falconeri) and Afghan Urial (Ovis Orientalis) in the Torghar Hills, Pakistan. *Game and Wildlife Science, 21*(213), 177–187.

Woodhouse, E., et al. (2015). Guiding principles for evaluating the impacts of conservation interventions on human well-being. *Philosophical Transactions of the Royal Society B: Biological Sciences, 370*(1681). https://doi.org/10.1098/rstb.2015.0103

Chapter 15
Groundwater Potential Assessment Using an Integrated AHP-Driven Geospatial Techniques in the High-Altitude Springs of Northwestern Himalaya, India

Dharam Chand, Renu Lata, Rajat Dhiman, and Kireet Kumar

Abstract The Himalayas, also known as the "water tower of the world", acts as a source for a numerous rivers and rivulets. In mountainous region, population is mainly dependent on springs. In the present study, detailed investigations of the springs were done and their water quality was analysed, including all the basic parameters, and the prime objective of this study is to identify the groundwater potential zones. The springs in the study areas doesn't contain any contaminants in its origin. All the physiochemical parameters, like pH, TDS, EC, Ca^{2+}, Mg^{+}, Cl^{-}, F^{-}, SO_4^{2-}, NO_3^{-}, Na^{+}, K^{+}, total hardness, and alkalinity, were well within the permissible limit prescribed by BIS 2012 and WHO 2011. The current study shows that Na^{+} is the dominant cation, whereas Cl^{-} are the most dominant anion in springs water. The identification of the recharge zones was done with the help of satellite data using Analytical Hierarchical Process (AHP) by incorporating the seven factors, viz. geomorphology (highly dissected structural hills and valleys), lineament density (very low in study area), lithology, slope gradient (maximum of study area comes within 30–45° slope), soil (dystric cambisols, eutric cambisols, lithosols) change in land use/land cover (LULC), and drainage density (high in study area) of study areas were considered in assessing the groundwater potential recharge zones. Results indicated that the maximum occurrences of the potential recharge zones were observed along the major river and its tributaries in the study area. Study area are categorized into five different groundwater potential recharge zones: very high, high, moderate, low, and very low. About 28%, 33%, 37%, 2%, and 0.4% of the total area 226 km^2 lie in very high, high, moderate, low, and very low recharge zones, respectively, in Barot valley, while in Thunag valley, 8% of the total area 276 km^2 lies in very high potential recharge zone and 41% lies in high, 49% in moderate 1.1%

D. Chand · R. Lata (✉) · R. Dhiman
G. B. Pant National Institute of Himalayan Environment, Himachal Regional Centre, Mohal-Kullu, Himachal Pradesh, India

K. Kumar
G. B. Pant National Institute of Himalayan Environment, Almora, Uttarakhand, India

S. Sharma et al. (eds.), *Climate Change Adaptation, Risk Management and Sustainable Practices in the Himalaya*,
https://doi.org/10.1007/978-3-031-24659-3_15

in low, and 0.7 in very low groundwater potential recharge zone. Hence, the groundwater recharge potential zone map will help to formulate better groundwater recharge planning by suggesting appropriate recharge structures, like check dams, trenches, ditches, and percolation tanks.

Keywords Analytical Hierarchical Process (AHP) · Climate change · Anthropogenic activities

1 Introduction

The Himalaya is known as the water tower of Asia. It is the source of many perennial rivers. Peoples of the mountains area are mostly dependent on the spring water (Tambe et al. 2012). In Indian Himalayan regions (IHR), the primary source of water is the spring which provides water for both irrigation and for drinking purpose. Springs are the key features for the surrounding environment. Various studies have been carried out for focusing on the relationship between spring discharge rate, rainfall patterns, and catchment area degradation and quality of springs. Due to high demand of water in hilly regions for drinking, agricultural practices and construction springs are facing the threat of drying up (Singh and Rawat 1985). The problem of drying up of springs arising out of changing hydrological cycle and increasing and competing demands for water are becoming more acute in terms of availability and quality. Drying up of springs attributed to increasing in water demand, changing land use patterns, and ecological degradation of the area (Report of Working Group I 2018). Further, this problem of drying up of springs or decreasing discharge from springs aggravated mostly due to the lack of structures or management work to conserve the discharge from the spring sources. If these springs are properly developed, they will serve as a permanent source of drinking water to the people who are living in the remote areas. Recent climate change affects the spatial and temporal pattern of the rainfall in the Himalayan region, which also manifested the problem of drying of the springs, and this is increasingly felt across the IHR (Tambe et al. 2012). Therefore, drying of springs and water scarcity issues underscore the need to increase the understanding of spring hydrology, especially in Himalayan region (Chinnasamy and Prathapar 2016). Unlike watershed, which is larger extent of geographical area, springs have smaller extent of geographical area through which it receives water. Therefore, spring resources needs to be studied and managed at springshed level. Springshed management is less difficult than watershed management, because the area of springshed is less than the watershed, and springshed management in comparison to watershed shows its effects earlier and is cost-effective. Springshed management approach for groundwater recharge and revival of springs holds promise for the Himalayan region (Tambe et al. 2012). Efforts to preserve and save springs from drying up and consequently to recharge them are gaining momentum, but still a lot more studies are necessary to carry out detailed inventory of the springs, understanding the hydro-geology of the spring, its recharge and discharge pattern, and recharging strategy in the context of changing climate.

The serious aspects of the water crisis are thought to be the misconceptions about solutions, which are nowadays being proposed (Frederiksen 1996). Therefore, it is very important to understand how springs function in order to address the problem of drying of springs (Sharma et al. 2016). Further, in IHR, scarcity of continuous and reliable data on spring resources puts challenges to water resource managers. Information on seasonal changes in flows and quality of spring water, elevation, geology, and soil and land use in spring catchment provides a major step towards understanding the characteristics of hydrological regime of spring and helps in preparing strategies to sustain spring discharge (Rani et al. 2018). Keeping these issues in mind and knowing the importance of the springs in IHR, we have prepared a detailed inventory of the spring sources and demonstrated the spring revival plan using the science-based spring sanctuary concept (SSC) (Negi et al. 2001). Spring sanctuary technology aims to improve rainwater infiltration and retention in the spring recharge zone through vegetative and engineering measures such as implementing groundwater augmentation structures in the spring catchment area, such as digging 45 cm deep and 60 cm wide trenches along contours, and recharge pits over geologically permeable soils. Groundwater augmentation techniques in the treatment zone may include gully plugging, rising stone, and mud-built water holding structures. According to several researchers, analytical hierarchical process (AHP) shows the good functionality for mapping the groundwater potential zones. In this study baseline data, spring sample, spring discharge, and social survey conducted during field visits are reported. The main objective of the study is to generate the groundwater potential zones by using GIS techniques (AHP). The present study was carried out in Barot valley and Thunag valley of Mandi district, Himachal Pradesh. Field surveys, Laboratory analysis and potential groundwater recharge zone maps will enhance groundwater recharge planning by suggesting appropriate recharge structure like check dams, trenches, ditches and percolation tanks to save this precious natural resource. Since the last decade, Himachal Pradesh has been facing drought-like conditions. Drought intensity has been rising year after year for the past 6 years. During the summer months, Himachal faces a severe drinking water deficit. The impact of the drought on the state's drinking water sources has been assessed. To address the water deficit during a water scarcity, several existing water management practices must be changed (Singh et al. 2010). Excessive abstraction has been recognized as a key hotspot of groundwater depletion in the Northwestern India, with serious consequences for groundwater sustainability (Kumar Joshi et al. 2021). By using high-resolution satellite imaging, locating groundwater potential recharge zone became very convenient and cost-effective than invasive methods. In this study, remote sensing and GIS systems were combined to construct multiple thematic layers, utilizing weighted overlay analysis for the identification of groundwater potential zones in the Jammu Himalayas (Qadir et al. 2019). The Basantar watershed is part of the Jammu plains and hilly stretch of Siwaliks in Jammu & Kashmir; the scarcity of water is quite hilly nature of terrain and sloppy surface mostly rain-fed, and the lack of proper guidance and management has led to the dying of many springs (Taloor et al. 2020). Because of the underlying trend of decreasing stream flow patterns in the western Himalaya, groundwater potential modelling has become more important for a sustainable and accurate assessment of groundwater resources in the region (Rashid et al. 2020)

2 Study Area

2.1 *Barot and Thunag Valley*

The Mandi district is centrally located district of Himachal Pradesh. The district is entirely hilly, except for a few isolated patches of small and fertile valleys. The district, with its headquarter at Mandi town, lies between 31°13′ and 32° 05′ north latitudes and 76°37′ and 77°25′ east longitudes. The district has a total geographical area of 3950 km^2. There are six towns and 3338 villages in the district. There are nine tehsils (Sadar, Thunag, Sundernagar, Sarkaghat, Padhar, Jogindernagar, Lad Bhraol, Karsog, and Chachyot). The present study has been conducted in two valleys – Barot valley and Thunag valley of Padhar and Thunag Tehsil, respectively – of Mandi district. The study area (Barot and Thunag valley) consists of dams and reservoirs, highly dissected structural hills and valley, flood plains, and alluvial plains. Dams are the structures that regulate the flow of a river or stream. The structures that restrict the flow of river or stream is known as reservoir. Alluvial plains are also known as floodplains are the flat area near the river and stream. Highly dissected hills and valleys are formed by erosion and especially tectonic movement (Fig. 15.1). The district has a population of 9, 99,777 persons with a population density of 253 persons km^2 (Census 2011). Generally, in Mandi district, the least amount of rainfall occurs in the month of November (24 mm), and with an average of 515 mm, the most precipitation falls in the month of July. In Mandi district, geomorphologic features like highly dissected structural hills and valley have covered most of the area. The minimum and maximum temperature ranges from −20°C to 40°C. The major crops grown in the study area are corn maize, wheat, rice, and vegetables

3 Material and Method

Spring water samples were collected from 54 sampling sites in Barot valley and Thunag valley of Mandi district. Water samples were collected in polyethylene bottles. Some physical parameters of water include temperature, pH, TDS, and EC. Elevation of spring from mean sea level, discharge rate, and geographic coordinates of springs were taken at the time of sampling. The discharge rate was measured with the help of stopwatch bucket method. After sampling, water samples were analysed in the laboratory for detection of cation (Na^+, K^+, Ca^+, Mg^+) and anions (Cl^-, SO_4^{2-}, NO_3^-, F^-) by following the standard method given by APHA 2017.

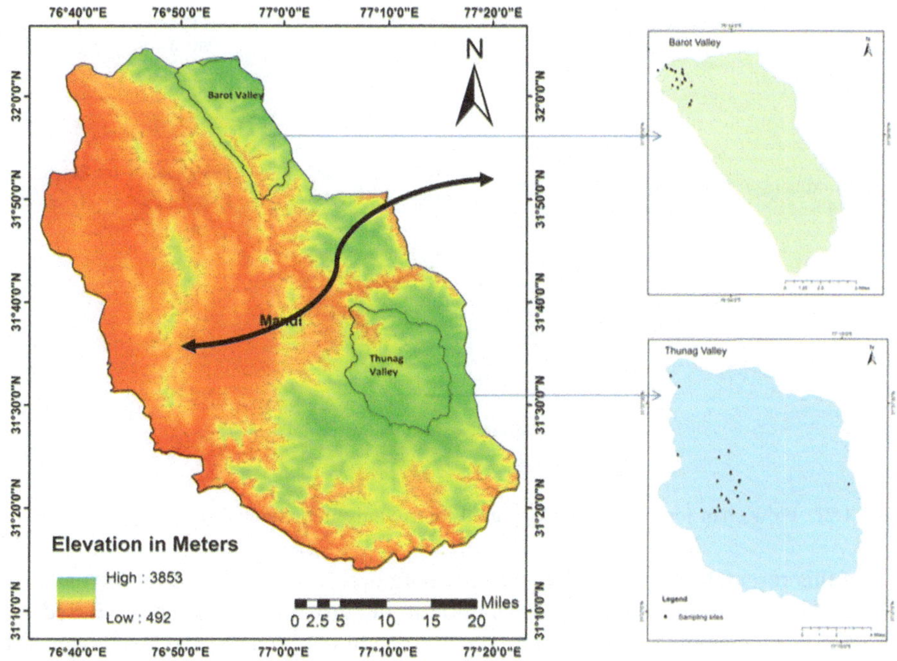

Fig. 15.1 Location of the study area

GIS Techniques Identification of groundwater potential zones was carried out by integrated satellite-derived thematic maps in GIS platform. The GIS analysis was done by using Arc GIS 10.8 of ESRI Inc. Geomorphology, lithology, drainage density, lineament density, slope, soil and land use/land cover maps were rectified and projected in World Geodetic System, UTM zone 1984. All these maps were taken to delineate the groundwater potential recharge zone. By using density analysis tool in software Arc GIS 10.8, lineament density and drainage density maps were prepared. Preprocessing analysis of remote sensing data was carried out using a GIS software to delineate the groundwater potential recharge zone. The flow chart (Fig. 15.2) shows the sequential steps involved in this study.

3.1 Analytical Hierarchy Process

This analytical process model is based on hierarchical process. It is a tool for qualitative and quantitative multi-criteria element used for decision-making. For decision-making AHP having three parts: First, we create the hierarchy of the problem; the second is to assign the value to each level in the hierarchy; the third step is to create the matrix of pair-wise comparison.

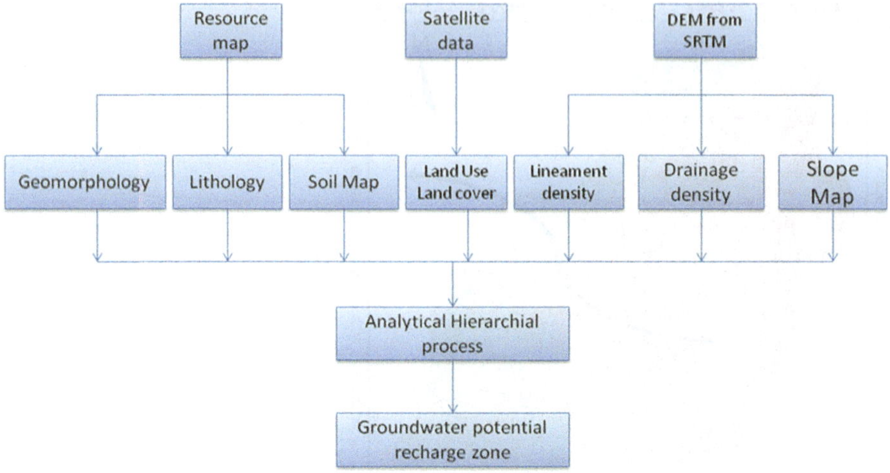

Fig. 15.2 Flow chart for the methodology used for groundwater potential zone mapping

In this research, work weightage of the parameters was assigned on the basis of their relative relevance or their contribution to the groundwater recharge. Based on the expert's viewpoints and literature study, several researchers give weightage and consider more to geomorphology, lineament density, and lithology for groundwater resources (Saaty 1990; Khali et al. 2020; Pradhan 2009) (Tables 15.1 and 15.2).

Ranks and overall weightages of every thematic layer were assigned to it (Tables 15.3 and 15.4).

4 Results and Discussions

Location and their elevation from mean sea level of sampling sites are shown in (Tables 15.5 and 15.6).

4.1 *Physicochemical Parameters*

Temperature value of water samples in Barot valley ranged from 13.6 to 18.4°C and in Thunag valley from 10 to 18.5°C.

In drinking water, pH doesn't have any direct impact, but it's the key water quality parameter. The value of pH in drinking water should be less than 8.5. The range of pH in Barot valley and Thunag valley lies between (5.18 and 6.83) and (6.12 and 6.9), respectively (Table 15.5).

Electrical conductivity in study area ranges between 59.8 and 911 μS/cm in Barot valley and 108 and 675 μS/cm in Thunag valley.

Table 15.1 Pairwise comparison matrix, weights of various thematic classes

Factors	Geomorphology	Lineament density	Lithology	Slope	Soil map	LULC	Drainage density	Weightage
Geomorphology	7	6	5	4	3	2	1	0.38
Lineament density	7/2	6/2	5/2	4/2	3/2	2/2	1/2	0.19
Lithology	7/3	6/3	5/3	4/3	3/3	2/3	1/3	0.12
Slope	7/4	6/4	5/4	4/4	3/4	2/4	1/4	0.10
Soil map	7/5	6/5	5/5	4/5	3/5	2/5	1/5	0.08
Land use/land cover	7/6	6/6	5/6	4/6	3/6	2/6	1/6	0.066
Drainage density	7/7	6/7	5/7	4/7	3/7	2/7	1/7	0.064

Table 15.2 Weights of the thematic maps of the potential groundwater

Parameter	Assigned weight
Geomorphology	32
Lineament density	19
Lithology	12
Slope	10
Soil	8
Land use/land cover	6.6
Drainage density	6.4

Table 15.3 Factors influencing groundwater potential zones (Barot)

Barot valley				
Factors	Parameter	Weight	Rank	Overall
Dam and reservoir	Geomorphology	38	5	190
Highly dissected structural hills and valley			2	76
Older flood plain			3	114
Piedmont alluvial plain			4	152
Younger alluvial plain			4	152
Very high	Lineament density	19	5	95
High			4	76
Moderate			3	57
Low			2	38
very low			1	19
Kulu	Lithology	12	2	24
New alluvium			3	36
Shali			2	24
Sundernagar			4	48
Vaikrita			3	36
0–15	Slope	10	5	50
15–30			4	40
30–45			3	30
45–60			2	20
60–72			1	10
Dystric cambisols	Soil	8	2	16
Eutric cambisols			3	24
Lithosols			1	8
Water bodies	Land use and land cover	6.6	5	33
Forest			4	26.4
Vegetation			4	26.4
Agricultural land			4	26.4
Shrubs			3	19.8
Built area			1	6.6
Barren land			2	13.2

(continued)

Table 15.3 (continued)

Barot valley				
Factors	Parameter	Weight	Rank	Overall
Very high	Drainage density	6.4	1	6.4
High			2	12.8
Moderate			3	19.2
Low			4	25.6
Very low			5	32

Table 15.4 Factors influencing groundwater potential zones (Thunag)

Thunag valley				
Factors	Parameter	Weight	Rank	Overall
Highly dissected structural hills and valleys	Geomorphology	38	2	76
Piedmont alluvial plain			4	152
Younger alluvial plain			4	152
Very high	Lineament density	19	5	95
High			4	76
Moderate			3	57
Low			2	38
Very low			1	19
Kulu	Lithology	12	2	24
Vaikrita			3	36
0–15	Slope	10	5	50
15–30			4	40
30–45			3	30
45–60			2	20
60–72			1	10
Dystric cambisols	Soil	8	2	16
Eutric cambisols			3	24
Water bodies	Land use and land cover	6.6	5	33
Forest			4	26.4
Vegetation			4	26.4
Agricultural land			4	26.4
Shrubs			3	19.8
Built area			1	6.6
Barren land			2	13.2
Snow/ice			3	19.8
Very high	Drainage density	6.4	1	6.4
High			2	12.8
Moderate			3	19.2
Low			4	25.6
Very low			5	32

Table 15.5 Location description of sampling sites in Barot valley

Sample ID	Village	Panchayat	Longitude (E)	Latitude (N)	Elevation (m) amsl	Discharge rate (lpm)
B001	Kahog	Barot	76° 50′ 39.719″	32° 01′ 51.857″	1900.89	2.02
B002	Dug Nallah (Barot)	Barot	76° 50′ 22.989″	32° 02′ 26.179″	1506.26	1.93
B003	Ger Nallah	Barot	76° 50′ 04.975″	32° 02′ 35.296″	1901.97	37.5
B004	Mayot	Barot	76° 47′ 43.406″	32° 03′ 50.828″	2220.2	60
B005A	Bari Jharwaar (IPH Scheme)	Barot	76° 48′ 54.052″	32° 02′ 46.459″	2157.7	300
B005B	Bari Jharwaar	Barot	76° 49′ 02.752″	32° 02′ 50.153″	2133	36.36
B006	Kathyaru	Barot	76° 49′ 30.703″	32° 02′ 26.051″	2102	277
B007	Barot village	Barot	76° 50′ 18.878″	32° 02′ 30.154″	1832.7	3.33
B008	Khadiyaar Nallah spring Thuji	Barot	76° 51′ 1.44943″	32° 2.28.5963″	2100	60
B009	Thuji	Barot	76° 51′ 00.598″	32° 02′ 17.823″	1709.89	3.03
B010	Kharak Nallah	Barot	76° 51′ 03.030″	32° 02′ 12.760″	1935	23.07
B011	Dharangda	Barot	76° 51′ 09.324″	32° 01′ 51.990″	1856	6.52
B012	Dharangda	Barot	76° 51′ 00.910″	32° 01′ 39.295″	1845	1.25
B013	Boching	Barot	76° 51′ 34.193″	32° 00′ 33.767″	1745	1.31
B014	Lapaas	Lapaas	76° 51′ 27.889″	32° 01′ 17.792″	1906	2.23
B015	Lapaas	Lapaas	76° 51′ 30.203″	32° 01′ 20.242″	1930	3.0
B016	Kasambal	Lapaas	76° 51′ 34.975″	32° 01′ 29.445″	1930	2.0
B017	Barot	Barot	76° 50′ 24.89″E	32° 02′ 30.63″N	1822	1.15
B018	Barot	Barot	76° 50′ 23.17″E	32° 02′ 26.44″N	1850	2.01
B019	Turran	Barot	76° 50′ 44.85″E	32° 01′ 21.27″N	1950	4.01
B020	Khlail	Khalail	76° 48′ 03.70″E	32° 03′ 29.06″N	2210	00
B021	Choti Jharwaar	Khalail	76° 48′ 31.14″E	32° 03′ 23.99″N	2124	1.2
B022	Barot	Barot			1869	3.2

(continued)

Table 15.5 (continued)

Sample ID	Village	Panchayat	Longitude (E)	Latitude (N)	Elevation (m) amsl	Discharge rate (lpm)
			76° 49′ 59.06″E	32° 02′ 45.45″N		
B023 (A)	Galu	Lapaas	76° 51′ 12.96″E	32° 01′ 43.5″N	1960	2.33
B023 (B)	Galu	Lapaas	76° 51′ 12.96″E	32° 01′ 43.5″N	1957	3.12
B024	Ruling	Ruling	76° 50′ 35.96″E	32° 02′ 20.05″N	2010	5.45
B025	Dharangan	Barot	76° 51′ 12.05″E	32° 01′ 50.74″N	1894	0.89

Table 15.6 Location description of sampling sites in Thunag valley

Sample ID	Village	Panchayat	Longitude (E)	Latitude (N)	Elevation (m) amsl	Discharge rate (lpm)
S001	Jajar	Thunag	077° 09.370′	31° 32.519′	1975	2.8
S002	Jajar	Thunag	077° 09.269′	31° 32.677′	1988	3.58
S003	Judd	Thunag	077° 09.258′	31° 32.771′	1980	4.2
S004A	Khatiyar Nallah Spring	Thunag	077° 09.023′	31° 32.901′	2149	30
S004B	Khatiyar Nallah Spring	Thunag	077° 09.029′	31° 32.922′	2168	40.5
S005	TankiNala	Thunag	077° 08.883′	31° 32.330′	2233	30
S006	Chhindi Nallah	Thunag	077° 08.714′	31° 32.063′	2275	15
S007	Shagethata	Thunag	077° 08.671′	31° 32.055′	2356	00
S008	Chandi	Thunag	077° 09.007′	31° 32.105′	2221	170
S009	Atinali Baudi	Thunag	077° 08.938′	31° 32.094′	2238	20
S010	Chizedor	Thunag	077° 09.732′	31° 33.231′	1895	12.5
S011	Khotapaani	Thunag	077° 09.463′	31° 34.038′	1855	25.1
S012	Khabadhaar	Thunag	077° 09.372′	31° 35.104′	1922	12.5
S013	Bynallah	Muraag	077° 08.880′	31° 34.801′	1911	3.2
S014	Tutti Nallah spring (Reinn)	Muraag	077° 08.801′	31°.33.586′	2145	15

(continued)

Table 15.6 (continued)

Sample ID	Village	Panchayat	Longitude (E)	Latitude (N)	Elevation (m) amsl	Discharge rate (lpm)
S015	Tandi	Muraag	077° 09.418′	31° 33.982′	2314	25.2
S016	Shaapdi source	Pakhrer	77° 16′ 49.129″	31° 44′ 54.65″	1631	8.9
S017	Bai Nallah spring (Jhakdi)	Pakhrer	077° 10.164′	31° 31.904′	2407	40
S018	Desi (chellamkhad)	Pakhrer	077° 09.624′	31° 32.020′	2115	10.5
S019	Thunaag khad	Thunaag bazaar	077° 10.120′	31° 33.535′	1840	350.6
S020	Halinu Nallah spring	Saroah	077° 06.705′	31° 34.918′	1891	30.6
S021	Khiyogi Nallah spring	Saroah	077° 06.029′	31° 38.219′	1586	25
S022	Naled spring in Kudi Nallah	Saroah	077° 04.958′	31° 38.833′	1252	15.2
S023	Suragi	Thunag	77° 09′ 54.00″E	31° 33′ 37.63′N	1935	5.3
S024	Suragi	Thunag	77° 09′ 54.79″E	31° 33′ 37.63′N	1936	6.4
S025	Thunag	Thunag	77° 09′ 53.66″E	31° 33′ 31.08′N	1883	7.2
S026	Tipra	Thunag	77° 10′ 21.01″E	31° 32′ 42.30′N	1860	5.68

The concentration of total dissolved solid are within the permissible limits of WHO 2011 and BIS 2012. In Barot valley TDS ranges between 40 and 610.4 mg/L and in Thunag valley between 72.4 and 452.3 mg/L (Table 15.5).

4.2 *Major Ion Chemistry*

Cations The cation concentration in springs of Barot valley shows variation as $Na^+ > Ca^{2+} > Mg^{2+} > K^+$. The average concentration of c cations in spring water of Barot valley is as follows: sodium 24.18, potassium 8.54, calcium 12.97, and magnesium 12.9 ppm (Fig. 15.3). In Thunag valley, cations in springs show variation as $Na^+ > K^+ > Ca^{2+} > Mg^{2+}$. On the other hand, the average concentration in Thunag valley is as follows: sodium 31.1, potassium 27.3, calcium 24.2, and magnesium 5.34. Among the cations in Barot valley and Thunag valley, Na^+ is the dominant cation, but in Barot valley, K^+ has the lowest constituent, whereas in Thunag valley, it is Mg^{2+} (Table 15.7).

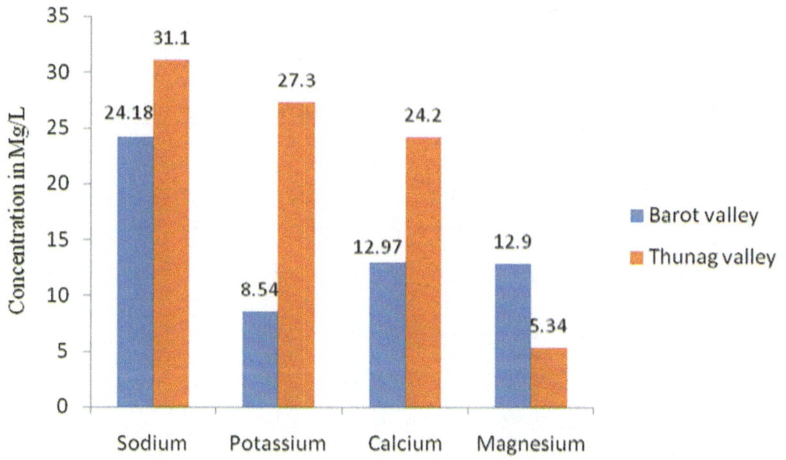

Fig. 15.3 Average concentration of cation in the study area (Barot and Thunag valley)

Anions The anion concentration in springs of Barot valley shows variation as $Cl^- > NO_3^- > SO_4^{2-} > F^-$, with an average concentration of anions in spring water of Barot valley of chloride 47.9, sulphate 1.99, nitrate 8.94, and fluoride 0.03. Anion concentration in springs of Thunag shows variation as $Cl^- > NO_3^- > SO_4^{2-} > F^-$. The average concentration of anions in springs of Thunag is as follows: chloride 90.1, sulphate 2.19, nitrate 9.6, and fluoride 0.03 (Fig. 15.4). Among the anion concentration in Barot valley and Thunag valley, Cl^- is the most abundant and F^- is the minor constituents in anions.

4.3 Factors Influencing Groundwater Potential Zones

4.3.1 Geomorphology

In this study, the maps show five geomorphological features in Barot valley (Fig. 15.5) and three geomorphological features in Thunag valley (Fig. 15.5) from the remote sensing (RS) imagery in order to know about the water resource area. It is one of the main factors used to delineate the groundwater potential zones. Geomorphology includes detailed study of landform their processes and topography of the area. Geomorphic processes occur at a very slow rate, but because of flood and landslide, geomorphology of that area changes rapidly (Rajaveni et al. 2015). More than 95% of study areas is occupied by the highly dissected structural hills and valley in Barot valley, and in Thunag valley, 93% of the area is occupied by highly dissected structural hills and valley. By using the analytical hierarchy process, different weightages are assigned to different classes on the basis of occurrence of groundwater in that geomorphic feature.

Table 15.7 Brief statistics of hydrochemical parameters

S. no.	Parameters	Barot valley				Thunag valley			
		Range		Mean	Std. Dev.	Range		Mean	Std. dev.
		Min	Max			Min	Max		
1.	Temperature (°C)	13.6	18.4	16.7	1.4	10	18.5	15.2	2.07
2.	pH	5.18	6.83	6.41	0.42	6.12	6.9	6.65	0.21
3.	EC (μS/cm)	59.8	911	320	250	108	675	267.3	156
4.	Total dissolved solids (TDS) mg/L	40	610.4	215	168	72.4	452.3	179.08	104.4
5.	Total hardness (as $CaCO_3$) mg/L	4	156	55.5	45.2	2	148	79.5	45.6
6.	Sodium (Na^+) mg/L	0.4	41.4	24.18	9.79	19.4	44.9	31.1	6.81
7.	Potassium (K^+) mg/L	0.1	32.9	8.54	9.59	9.3	47.2	27.3	11.2
8.	Calcium (Ca^{2+}) mg/L	1.68	30.3	12.97	9.12	5.88	83.2	24.2	17.8
9.	Magnesium (Mg^{2+}) mg/L	0.00	30.3	12.9	9.12	0.00	24.9	5.34	10.7
10.	Chloride (Cl^-) mg/L	21.3	71	47.9	15.2	49.7	120.7	90.1	18.3
11.	Sulphate (SO_4^{2-}) mg/L	0.41	12.5	1.99	2.90	0.4	12.5	2.19	3.32
12.	Nitrate (NO_3^-) mg/L	4.49	33.4	8.94	8.36	4.39	40.5	9.6	10.5
13.	Fluoride (F^-) mg/L	0.00	1.2	0.03	0.85	0.00	1	0.02	0.71
	Alkalinity (as $CaCO_3$) mg/L	10	50	25.9	10.8	10	50	24.6	9.7

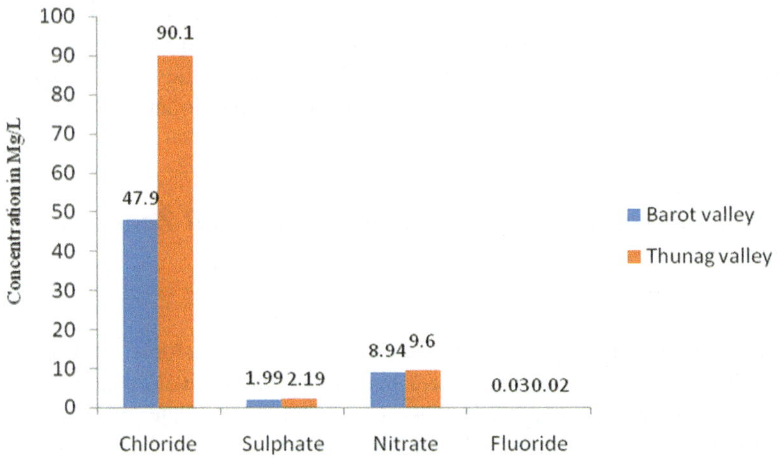

Fig. 15.4 Average concentration of anions in the study area (Barot and Thunag valley)

4.3.2 Lineament Density

Linear features in a landscape are known as lineament. Lineaments are used for geological structures such as fault fracture, or joint. Lineament shows us zones of faults and fractures (Abdullah et al. 2015). Potential for groundwater recharge is

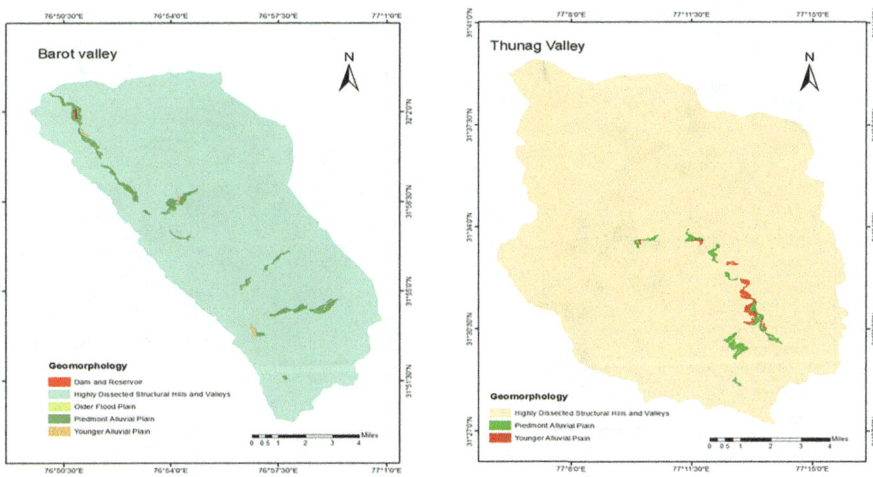

Fig. 15.5 Geomorphology map of the study area

maximum near the lineaments, because it develops secondary porosity, which promotes the groundwater recharge. It can be identified by their linear alignments in satellite imagery. With the help of the Shuttle Radar Topography Mission (SRTM) data and GIS software (Arc GIS 10.8), the lineament density map of study area was prepared. For lineament density map preparation, SRTM data was downloaded from the platform USGS Earth Explorer. With the help of the hillshade tool, the lineament for the study area was extracted. The lineament length varies form few meters to kilometres. Then, the lineament density map were reclassified into five categories – very low, low, moderate, high, and very high. For groundwater potential zone mapping, lineament density is considered as a key parameter (Fig. 15.6). Due to their high potential to recharge groundwater, zones with high lineament density were assigned the highest rank of 5.

4.3.3 Geology/Lithology

Lithology throws light on the general physical characteristics of rocks, composition, and types of rocks. Lithology controls the infiltration and flow process of water. Rock type can influence the groundwater recharge potential (Allafta et al. 2020). In the present study, major five types of lithological characteristics were found to understand the goundwater distribution in the study area of Barot valley and Thunag valley (Fig. 15.7). The formation of the study area was mainly differentiated into five major categories, viz. Kulu, new alluvium, Shali, Sundernagar, and Vaikrita. The study area contains a variety of rocks including streaky and banded gneiss, dolomite, sporadic quartzite, grey sand, silt and clay, shale, slate, a few basic flow, tholeiitic

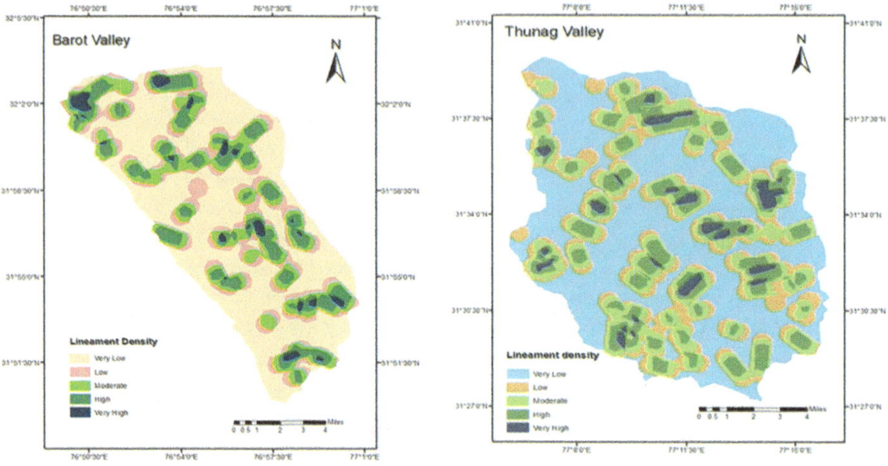

Fig. 15.6 Lineament density map of the study area

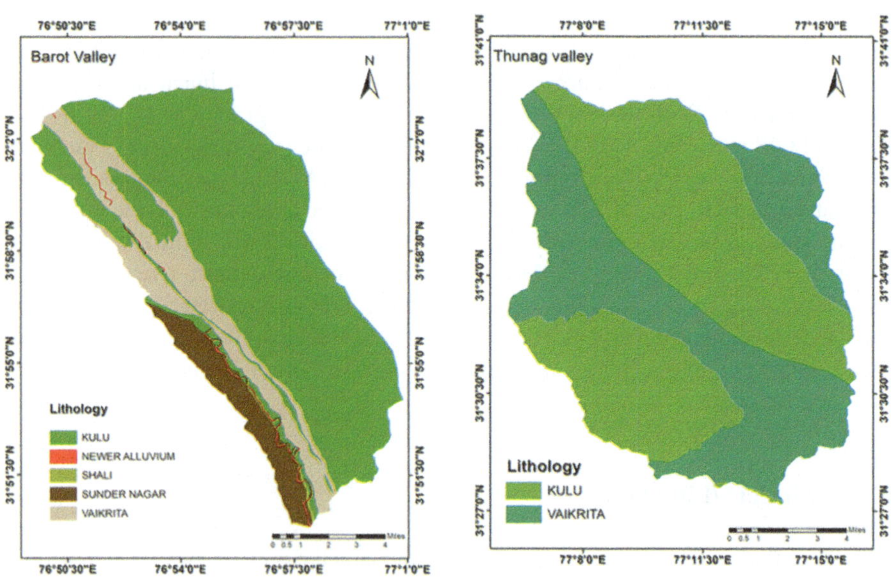

Fig. 15.7 Lithology map of the study area

basalt, minor quartzarenite, sillimanite-kyane-bearing schist, quartzite etc. Kulu-streaky and banded gneiss grey phyllite, schist, and quartzite, as well as Vaikrita-sillimanite-kyanite-bearing schist and quartzite, are the major lithological groups in the study area.

4.3.4 Slope

Slope is the rise or fall of a land surface. Stability of a surface is directly related to slope. In both regions, tropical or subtropical rainfall is the prime source of groundwater recharge. The slope is often ignored in plain surface. The maximum study areas are found between 30° and 45° slope. Slope is one of the major features to delineate the groundwater potential zones. The slope of an area affects the rainwater to infiltrate and overflow, and the slope also determines the surface flow of rainwater. In low-slope areas, the residence time of water is high so that the rate of infiltration of water is high in low-slope areas as compared to high-slope area. The steeper the slope, the greater will be the surface flow. Mostly the study areas are having steep slope due to their geomorphologic feature (highly dissected structural hills and valley; Fig. 15.8). In high-slope gradient area, groundwater potential is low due to high surface run-off. The slope map of the study area is generated by spatial analyst tool in ArcGIS 10.8 (Allafta et al. 2020). The slope map is classified into five categories, viz. very low (0–15°), low (15–30°), moderate (30–45°), high (45–60°), and very high (>60°).

4.3.5 Soil

Soil plays an important role in groundwater recharge. In Food and Agricultural Organization (FAO) classification system, cambisol is one of the 30 soil groups. Cambisols are different from the parental material in their aggregate structure,

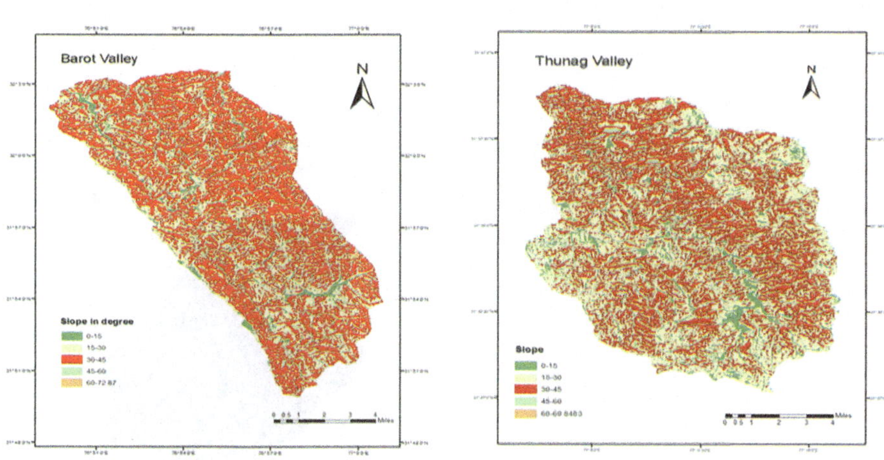

Fig. 15.8 Slope variation map of the study area

colour, clay content, etc that give some evidence of the soil formation. Cambisols lack humus, iron, and aluminium oxide. Lands having dystric cambisols are used for mixed farming because they are less fertile. Eutric cambisols are the productive soil. Lithosol is the group of shallow azonal soils consisting of imperfectly weathered rock fragments (Berhanu and Hatiye 2020). Of the area in Barot valley, dystric cambisols cover almost 85% or eutric cambisols cover 7.7% and lithosols covers almost 7.3% (Fig. 15.9). But in the case of Thunag valley (Fig. 15.9), map shows only two classes of soil, viz. dystric cambisols and eutric cambisols, which cover almost 87% and 13% of the study area, respectively.

4.3.6 Land Use/Land Cover

The area is characterized by water bodies, forest, vegetation, agricultural land, shrub-built area, and barren land (Fig. 15.10). Out of all classes, forest land dominates over the other classes. Occurrence of groundwater is maximum in the forest and lowest in the barren land. Trees facilitate recharge by reducing the surface run-off. For the identification of land use pattern through image interpretation of satellite data, land use/land cover classes were delineated. It includes water bodies, forest, vegetation cover, agricultural lands, built area, and barren land. LUCC maps were prepared by using Sentinel-2, Land use/land cover map produced by ESRI, Microsoft and impact observatory in GIS software. Land use/Land cover is an important feature to know about biogeochemical cycle, biodiversity, and loss of wildlife habitat (Mishra et al. 2020). Land use means how land is used by human (settlements, agricultural land). Most of the study area is covered by the forest and shrubs. Forest cover is the dominant land cover type in the study area.

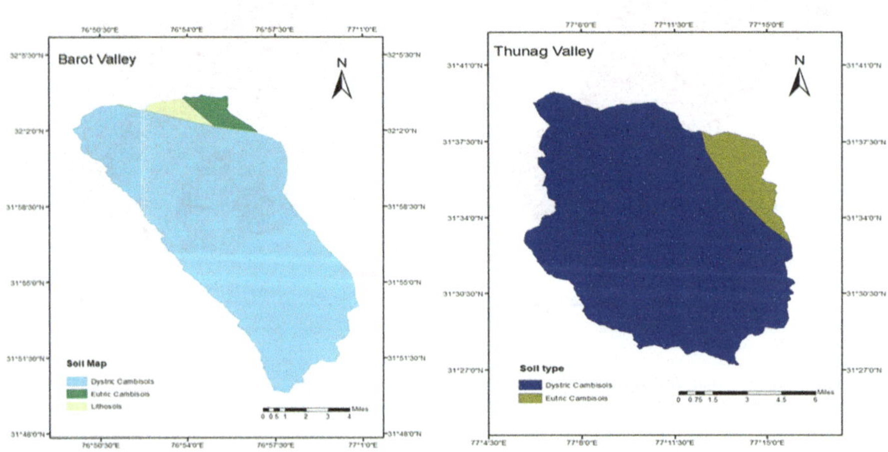

Fig. 15.9 Soil map of the study area

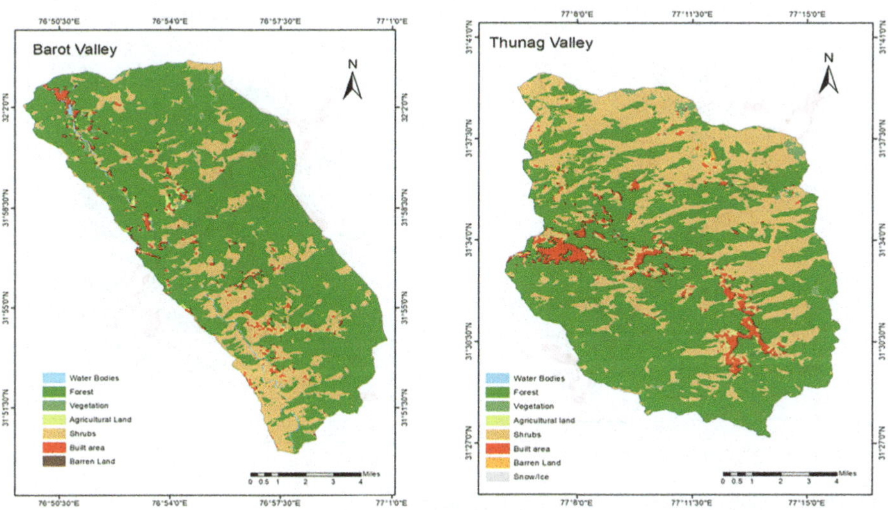

Fig. 15.10 Land use/land cover map of the study area

4.3.7 Drainage Density

Drainage density was introduced by Horton in 1945. Drainage density is the total length of drainage channels per unit area and determines the availability of groundwater and contamination in groundwater. Drainage density is expressed as:

$$D_d = \frac{\sum L}{A_{\text{Basin}}}$$

where $\sum L$ is the total length of drainage and A_{Basin} is the unit of drainage basin. The physical properties of drainage basin are directly related to the drainage density. The higher the closeness in the water channels, the higher will be the drainage density. Drainage density plays a vital role in groundwater recharge. Less water infiltration occurs in high drainage density areas, whereas low density area having high water infiltration to the aquifers. High weightage is assigned to the lower drainage density class than higher drainage density classes. Because low drainage density zones are considered as excellent groundwater potential zones (Ahmed et al. 2021), drainage density map was prepared using ArcGIS10.8. The resultant drainage density map was reclassified and categorized into very low, low, moderate, high, and very high (Fig. 15.11).

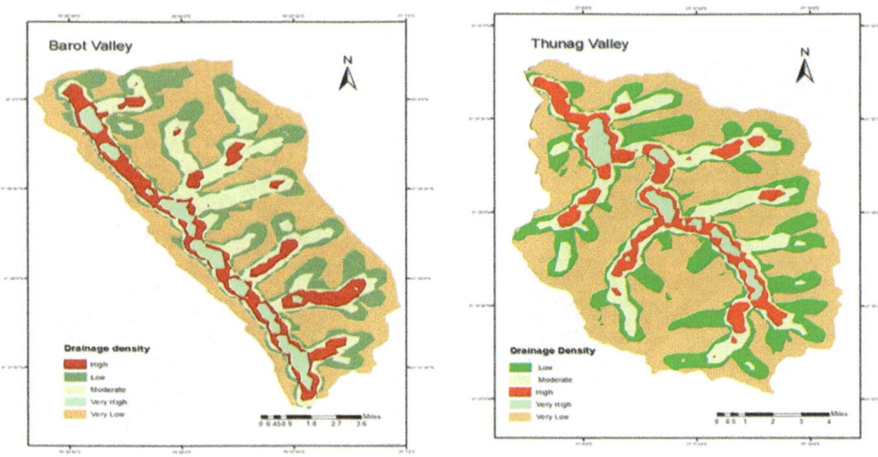

Fig. 15.11 Drainage density map of the study area

4.3.8 Groundwater Potential Zones

Groundwater is an important resource for human life, but due to excessive anthropogenic activities, the level of groundwater decreases day by day. Delineation of groundwater potential recharge zone is done with the help of AHP and GIS. The potential zone map was generated on the basis of various ranks and weightages assigned to the various parameters of the thematic maps. The study area is categorized into five different potential zones – very high, moderate, low, and very low (Fig. 15.12). The parameters considered here are geomorphology, lithology, lineament density, slope gradient, LULC, drainage density, and soil. The analytical hierarchical method has been applied to generate the groundwater potential zones in Barot valley and Thunag valley.

$$\mathrm{GWPZ}=\frac{\mathrm{GM_{Wt}GM_R+LD_{Wt}LD_R+Li_{Wt}Li_R+Sp_{Wt}Sp_R+Soi_{Wt}Soi_R+LULC_{Wt}LULC_R+DD_{Wt}DD_R}}{7}$$

Wt. and R represent the weight of the parameters and rank of the parameters respectively. The results show that the percentage of the regions falls under very low and low potential zones, which are 0.4% and 2% in Barot valley and 0.7% and 1.1% in Thunag valley of Mandi district, respectively (Fig. 15.13). The overall zones with moderate groundwater potential zones are dominant. Groundwater potential recharge zone is important for designing and implementation of groundwater recharge processes.

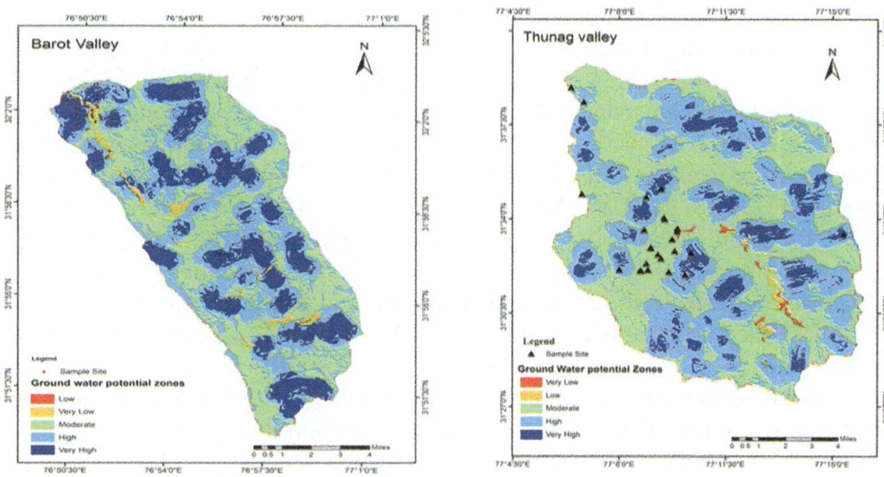

Fig. 15.12 Groundwater potential zone map

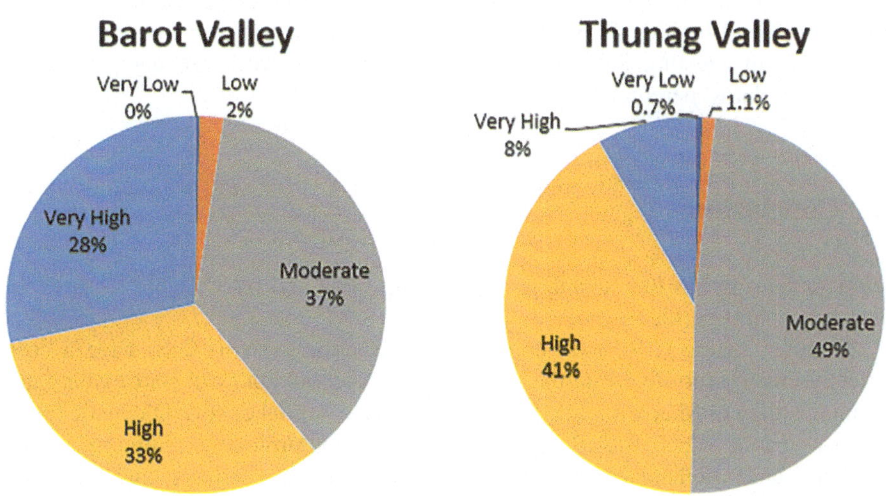

Fig. 15.13 Classifications of groundwater potential zones

5 Discussion

The study established the interrelationship between groundwater recharge potential factors and groundwater recharge potential weights from the general hydrology characteristics of Mandi district. The Spring GPS inventory showing 54 spring locations was prepared from the field survey (27 in Barot valley and 27 in Thunag valley). These groundwater springs mainly originated from the fractured rock aquifers in high-altitude areas of Barot and Thunag valley, Mandi. As per spring water analysis, Na^+ is the dominant cation and Cl^- is the dominant anion in the springs of Barot and Thunag. The average relative abundance of ions was observed as $Cl^- > Na^+ > Ca^{2+} > Mg^{2+} > SO_4^{2-} > K^+ > NO_3^- > F^-$. The springs in the study areas don't contain any contaminants in its origin. All the physiochemical parameters, like pH, TDS, EC, Ca^{2+}, Mg^+, Cl^-, F^-, SO_4^{2-}, NO_3^-. The springs in the study areas doesn't contain any contaminants in its origin. All the physiochemical parameters like pH, TDS, EC, Ca^{2+}, Mg^+, Cl^-, F^-, SO_4^{2-}, NO_3^-, Na^+, and K^+, were well within the permissible limit prescribed by BIS 2012 and WHO 2011. The major geomorphic feature of the study area is highly dissected structural hills and valleys, which act as major run-off zones of that area due to the high degree of slope (30–45°). The use of remote sensing and GIS techniques was done to delineate groundwater potential recharge zones. A total of seven thematic layers were used, and weights were assigned to each layer with the help of the AHP technique. The final map was prepared for each area and classified into five zones: low, very low, moderate, high, and very high. Thus, the construction of artificial recharge structures at identified high potential recharge zones shall help to rejuvenate the aquifer system of the study area. This research would encourage the protection and restoration of spring ecosystem in high-altitude regions of Himachal Pradesh. Based on the final maps obtained from this study, the government and water policymakers can use it as a reference in selecting suitable sited for the groundwater recharge, e.g. percolation pits, check dams, percolation trenches, etc.

6 Conclusion

Remote sensing has the advantage of covering large and inaccessible areas of the Earth in a short period of time.

Remote sensing (RS) and geographic information system (GIS) have a cost-effective method to delineate the groundwater potential zones in Barot valley and Thunag valley of Mandi district, Himachal Pradesh, A total of seven thematic layers were prepared which included the following: geomorphology, lithology, slope gradient, land use/land cover, slope of the area, drainage density, and lineament density were considered due to their association with the groundwater. The weights were assigned to each layer based on its influence on groundwater with the help of AHP matrix. The resultant map was prepared and classified into five categories (very low, low, moderate, high, and very high). About 61% and 49% of the study area in Barot valley and Thunag valley, respectively, have very high and high groundwater recharge potential area.

References

Abdullah, T. O., Ali, S. S., Al-Ansari, N. A., & Knutsson, S. (2015). Groundwater vulnerability mapping using lineament density on standard DRASTIC Model: Case study in Halabja Saidsadiq Basin, Kurdistan Region, Iraq. *Engineering, 07*(10), 644–667. Scientific Research Publishing, Inc.. https://doi.org/10.4236/eng.2015.710057

Ahmed, A., Alrajhi, A., & Alquwaizany, A. S. (2021). Identification of groundwater potential recharge zones in flinders ranges, South Australia using remote sensing, GIS, and MIF techniques. *Water, 13*(18), 2571. MDPI AG. https://doi.org/10.3390/w13182571

Allafta, H., Opp, C., & Patra, S. (2020). Identification of groundwater potential zones using remote sensing and GIS techniques: A case study of the Shatt Al-Arab Basin. *Remote Sensing, 13*(1), 112. MDPI AG. https://doi.org/10.3390/rs13010112

Berhanu, K. G., & Hatiye, S. D. (2020). Identification of groundwater potential zones using proxy data: Case study of Megech Watershed, Ethiopia. *Journal of Hydrology: Regional Studies, 28*, 100676. Elsevier BV. https://doi.org/10.1016/j.ejrh.2020.100676

Chinnasamy, P., & Prathapar, S. A. (2016). *Methods to investigate the hydrology of the Himalayan springs: A review* (IWMI Working Paper 169). International Water Management Institute (IWMI). 28p. https://doi.org/10.5337/2016.205

Frederiksen, H. D. (1996). Water crisis in the developing world: Misconception about solutions. *Journal of Water Resources Planning and Management., 122*(2), 79–87.

Khali, H., Mishra, A., & Singh, R. (2020). Delineation of groundwater potential zone: Remote sensing and GIS approach in Doon Valley, Dehradun, Uttarakhand. *Indian Journal of Ecology, 47*, 910–916.

Kumar Joshi, S., Gupta, S., Sinha, R., Logan Densmore, A., Prakash Rai, S., Shekhar, S., Mason, P. J., & van Dijk, W. M. (2021). Strongly heterogeneous patterns of groundwater depletion in Northwestern India. *Journal of Hydrology, 598*, 126492. Elsevier BV. https://doi.org/10.1016/j.jhydrol.2021.126492

Mishra, P. K., Rai, A., & Rai, S. C. (2020). Land use and land cover change detection using geospatial techniques in the Sikkim Himalaya, India. *The Egyptian Journal of Remote Sensing and Space Science, 23*(2), 133–143. Elsevier BV. https://doi.org/10.1016/j.ejrs.2019.02.001

Negi, G. C. S., Kumar, K., Panda, Y. S., & Satyal, G. S. (2001). Water yield and water quality of some aquifers in the Himalaya. *International Journal of Ecology and Environmental Sciences, 27*, 55–59.

Pradhan, B. (2009). Groundwater potential zonation for basaltic watersheds using satellite remote sensing data and GIS techniques. *Central European Journal of Geosciences, 1*(1), 120–129.

Qadir, J., Bhat, M. S., Alam, A., & Rashid, I. (2019). Mapping groundwater potential zones using remote sensing and GIS approach in Jammu Himalaya, Jammu and Kashmir. GeoJournal. Springer 85, 2, 487–504). doi: https://doi.org/10.1007/s10708-019-09981-5

Rajaveni, S. P., Brindha, K., & Elango, L. (2015). Geological and geomorphological controls on groundwater occurrence in a hard rock region. In Applied water science 7, 3, pp. 1377–1389). Springer https://doi.org/10.1007/s13201-015-0327-6

Rani, M., Joshi, H., Kumar, K., Joshi, R., & Mukherjee, S. (2018). *Inventory of springs of Kosi river basin. Technical report – I.* GBPNIHESD.

Rashid, I., Majeed, U., Aneaus, S., & Pelto, M. (2020). Linking the recent glacier retreat and depleting streamflow patterns with land system changes in Kashmir Himalaya, India. *Water, 12*, 1168. https://doi.org/10.3390/w12041168

Report of Working Group I. (2018). *Inventory and revival of springs in the Himalayas for water security*. NITI Aayog.

Saaty, T. L. (1990). How to make a decision the analytic hierarchy process. *European Journal of Operational Research, 48*, 9–26.

Sharma, B., Nepal, S., Gyawali, D., Pokharel, G. S., Wahid, S. M., Mukherji, A., Acharya, S., & Shrestha, A. B. (2016). *Springs, storage towers, and water conservation in the midhills of Nepal*

(ICIMOD Working Paper 2016/3). Nepal Water Conservation Foundation and International Center for Mountain Development.
Singh, A. K., & Rawat, D. S. (1985). Depletion of oak forests threatening springs: An exploratory study. *The National Geographic Journal of India, 31*(1), 44–48.
Singh, P. H., Sharma, R. M., Hassan, Q., & Ahsan, N. (2010). Impact of drought on drinking water resources of Himachal Pradesh. *International Journal, 2*(1), 73–77.
Taloor, A. K., Pir, R. A., Adimalla, N., Ali, S., Manhas, D. S., Roy, S., & Singh, A. K. (2020). Spring water quality and discharge assessment in the Basantar watershed of Jammu Himalaya using geographic information system (GIS) and water quality Index (WQI). *Groundwater for Sustainable Development, 10*, 100364. Elsevier BV. https://doi.org/10.1016/j.gsd.2020.100364
Tambe, S., Kharel, G., Arrawatia, M. L., Kulkarni, H., Mahamuni, K., & Ganeriwala, A. K. (2012). Reviving dying springs: Climate change adaptation experiments from the Sikkim Himalaya. *Mountain Research and Development, 32*(1), 62–72.

Chapter 16
Climate-Induced and Geophysical Disasters and Risk Reduction Management in Mountains Regions

Ambikapathi Ramya, Ramesh Poornima, Ganesan Karthikeyan, Sengottiyan Priyatharshini, Kalyanasundaram Geetha Thanuja, and Periyasamy Dhevagi

Abstract Mountains, being fragile, act as a vital repository for water and biodiversity. The Himalaya, in specific, the "roof of the world" is endowed with a magnificent and scenic view with temperate green forests, alpine meadows, agricultural fields, gorges, waterfalls, cascades of river valleys, and a human settlement in the unstable slopes or at the perennial streams of major rivers. The vulnerability of mountain ecosystems is being disproportionately influenced by climate change-induced disasters and is poorly understood as well. Cascading effect of temperature change can melt the snow and ice, thereby exhibiting a noticeable impact on the availability of water, biodiversity, boundary shift in ecosystem, agriculture, and on human well-being. Furthermore, several climate-induced disasters, like flash floods, mass movements, debris flows, and landslides, have occurred in the Himalayas. Specifically, this has happened a lot in the recent past, resulting in numerous deaths and property damage. This insecurity is due to the region's unplanned, unscientific and unregulated practices as well as a massive rise in population. This underlines the necessity for a Mountain Specific Risk Management Framework (MSMRMF) and the incorporation of spatial specificities for risk reduction. The three dimensions of vulnerability, namely, adaptive capacity, exposure, and sensitivity, are greatly governed by livelihood strategies, access to water, food, and hygiene. The best available research on disaster risk reduction (DRR) and climate change adaptation must be incorporated in deciding disaster resilience. This chapter sheds light on various climate-induced and geological disasters in mountain regions, their impact,

A. Ramya · R. Poornima · G. Karthikeyan · P. Dhevagi (✉)
Department of Environmental Sciences, Tamil Nadu Agricultural University, Coimbatore, Tamil Nadu, India

S. Priyatharshini
Vanavarayar Institute of Agriculture, Coimbatore, Tamil Nadu, India

K. G. Thanuja
Department of Agricultural Microbiology, Tamil Nadu Agricultural University, Coimbatore, Tamil Nadu, India

S. Sharma et al. (eds.), *Climate Change Adaptation, Risk Management and Sustainable Practices in the Himalaya*,
https://doi.org/10.1007/978-3-031-24659-3_16

and risk management strategies. The significance of regional climate models, development of alternative technologies, people's understanding regarding the social construction of risk, the role of local stakeholders, and enhancing the governance capacity and participation to manage the disaster risk is as well briefly discussed.

Keywords Climate change · Geophysical disaster · Risk reduction · Management · Himalayas

1 Introduction

The Himalayas, the world's tallest, youngest, and most active continental mountain range, have a variety of remarkable geological and tectonic features that began to emerge some 50 million years ago and are still evolving now. The Himalayas are a massive crescent-shaped mountain range that stretches over 2500 km from the Indus Valley's southern end, Nanga Parbat in the west, to Namcha Barwa in the east. The breadth of the range varies from 350 km in the west to 150 km in the east. The beautiful mountain ranges with considerable southerly convexity forms a wall that runs the length of the Indian subcontinent's northern boundary. The Himalayas' rising height, capped with several of the world's prime snow-covered summits, including 10 of the world's 14 above 8000 m peaks, is the Himalayas' most distinctive geomorphological characteristic (Roy & Purohit, 2018). The mountain range is at the crossroads of four different zoogeographic zones (Xu et al., 2009; Kotru et al., 2020). Climate change, along with a diverse range of habitats occupied by animals and plants from many worlds, has resulted in a global biodiversity hotspot (Myers et al., 2000; Wambulwa et al., 2021). The region is prone to natural catastrophes due to strong monsoon rains and occasional earthquakes (Vaidya et al., 2019), which has been worsened by recent human pressures, including deforestation and land-use changes (Sharma et al., 2007; Pandit et al., 2014; Paudel et al., 2020; Dhimal et al., 2021). Grassland makes up around 39% of the HKH, whereas woodland makes up 20%, shrubland makes up 15%, and agricultural land makes up 5%. Other forms of land use, such as barren terrain, rock outcrops, built-up regions, snow cover, and water bodies, account for the remaining 21%. The HKH's elevation zones range from tropical (500 m) to alpine ice-snow (>6000 m), with temperate broadleaf deciduous or mixed forest, temperate coniferous forest, and tropical and subtropical rainforest, as well as high-altitude cold shrub or steppe and cold desert, as the primary vertical vegetation regimes (Pei, 1995; Chettri et al., 2007). The climate ranges from subtropical to very cold. The market-led logic of growth through economic integration states that marginalized communities in India's border zones require infrastructure, particularly electricity grids and transportation networks, to connect them to fast-growing regions. On the other hand, anthropogenic intervention with the Himalayan slopes has been a major cause for disasters. For instance, Uttarakhand is home to various holy places, beautiful landscapes, and a diverse environment, attracting a large number of visitors and pilgrims. The central government has proposed a plan to enhance the Char Dham pilgrimage route, which connects Badrinath, Kedarnath, Yamunotri, and Gangotri. In late 2017, excavation for the project began in small portions along National Highway (NH)-58. The roads in such dangerous terrain are under a lot of strain. Large-scale excavations that alter equilibrium circumstances might result in catastrophic calamities in the near future (Siddique & Pradhan, 2019). Apart from this, climate-related calamities are also widespread in the Himalayan area. It is very vulnerable to climate change and calamity. It is characterized by terrain fragility, instability, tectonic activity, and seismic sensitivity. The severity of the climatic calamity was amplified by monsoon rain. The fast retreat of Himalayan glaciers has ramifications for water-related risks, such as glacier lake outburst floods, and water stress, as freshwater supplies diminish during the lean season (Sati & Litt, 2011; Mir et al., 2021). Rapid environmental and social changes are affecting the HKH region's landscapes and communities at the same time. Identification and understanding of essential biological and socio-economic aspects of mountain ecosystems, especially their sensitivity to and susceptibility to climate change (Yadav et al., 2021), has become critical for environmental management and sustainable development of mountain regions and downstream areas.

2 Climate Change in Mountain Regions

Mountain is a powerful system, which maintains the ecological balance and climate related weather pattern. It accounts for 26.5% of the world's total continental land surface and are highly susceptible to change in climatic conditions (Khan & Qureshi, 2019; Shugar et al., 2021). The major mountain regions like Himalayan region, Andes, and European Alps and the mountainous parts of Africa and Central Asia are prone to various climate-induced and geophysical disasters. It includes earthquakes, landslides, glacial lake, outburst floods, subsidence, snow avalanches, heat extremes, droughts and wildfires (Carrivick & Tweed, 2016). In recent years, climate-induced and geophysical disasters have increased in the Himalayan regions, including northern zone, central zone, and western zone, making it one of the world's most vulnerable and fragile locations. The Himalayas are sensitive to even minor changes in the climate due to geologically young and fragile nature (Maikhuri et al., 2017).

In Himalayan region, increased disaster occurrence was mostly related to climate-induced and geophysical changes (Fig. 16.1). Flood disasters accounted for 52.6% of disasters in the Himalayan region, while mass movements, such as landslides and avalanches, account for 28.5% and storm disasters account for roughly 13%. Moreover, during the monsoon season, heavy rainfall caused landslides, glacial lake outburst floods, and soil erosion throughout the region (Stäubli et al., 2018).

According to the IPCC AR6 report (2021), the following projected changes over Himalayan mountain region were reported

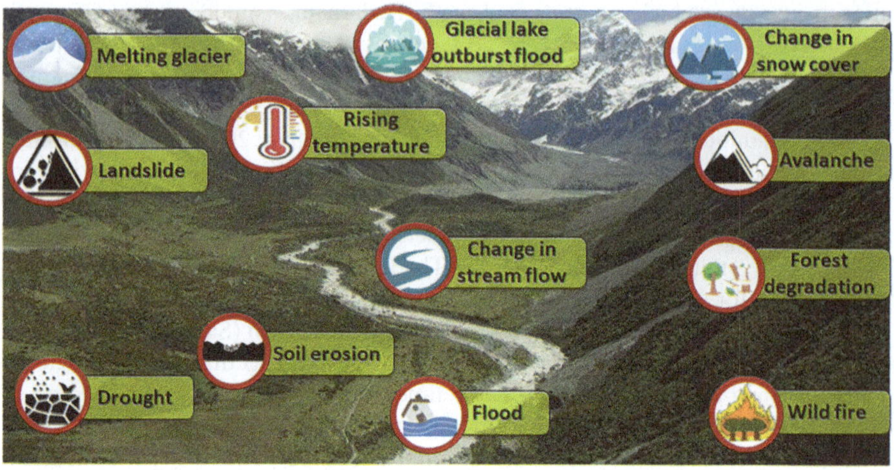

Fig. 16.1 Climate-induced and geophysical disasters in Himalayan mountain region

- Since 1970s, snow cover has decreased and glaciers have thinned and lost mass, though the Karakoram glaciers have marginally gained mass or in a balanced state.
- During the twenty-first century, snow-covered areas and volumes will decrease, snowline heights will rise, and glacier mass will likely drop with larger mass loss due to higher greenhouse gas (GHGs) emission.
- The glacial lake outburst floods and landslides have occurred over moraine-dammed lakes, which may increase with the rise in temperature and precipitation.
- In the twenty-first century, a general wetting of the Himalaya is expected with increase in heavy precipitation.

2.1 Climate-Induced Disasters

The annual mean temperature trends in the Hindu Kush Himalayas showed broad agreement with worldwide land surface temperature trends, with minor differences. Between 1901 and 2020, it was anticipated that the annual mean warming rates would be 0.19 °C per decade; however, it was observed to increase by 0.20 °C per decade during 1951–2014 (Ren et al., 2017; Sun et al., 2017). According to Global Circulation Model (GCM) simulations, a 1.5 °C global temperature increase at the end of twenty-first century would result in a temperature increase of 1.8 ± 0.4 °C averaged over the whole Hindu Kush Himalayan (HKH) areas. At high mountain ranges, it would result in temperature increase of 2.2 ± 0.4, 2.0 ± 0.5, and 2.0 ± 0.5 °C, respectively for the Karakoram, central Himalaya, and south-eastern Himalaya (Krishnan et al., 2019). At higher elevation of mountain regions, the air and surface temperatures have been rising at higher rate (Krishnan et al., 2020).

The climate change-mediated Himalayan glacier shrinkage is increasing throughout the region, ultimately resulting in drastic alteration in the underlying rock's hydrology and thermal regimes (Maurer et al., 2019). High-mountain glaciers with steep slopes and perennially frozen glaciers were extremely vulnerable to climate change (Hock et al., 2019). The regional climate and related cryospheric change interacted with the geology and topography caused huge slope failures. Changes in rock temperature and mechanical properties would have resulted in snow avalanches (Gruber & Haeberli, 2007; Hock et al., 2019). Temperatures in permafrost are rising over the world, especially in cold permafrost, causing long-term thermal anomalies as well as permafrost degradation (Noetzli et al., 2020).

Increased temperatures in the Himalayas will result in higher glacier melting, glacier contraction, and a reduction in summer runoff (Sorg et al., 2014). As a result of continuous glacier retreat in the Himalaya-Karakoram areas, several new lakes are expected to form (Linsbauer et al., 2016). The eroding nature of permafrost with decreasing slope stability, the risk of glacial lake outburst floods (GLOFs), and also landslides into moraine-dammed lakes are increasing. In Hindu Kush Himalayas, there are 204 glacial lakes that could break at any time (Narama et al., 2018; Wang et al., 2020). Glacier runoff in Asia's high mountain regions will increase until 2050, after which runoff would likely decline due to the glacier loss (Huss & Hock, 2018). In the next century, glacier mass and area will continue to decline in North Asia, High Mountains of Asia, Caucuses, and the Middle East. Glacier loss slowed under RCP2.6, but accelerates under RCP8.5, and peak in the mid- to late-twentieth century. According to GlacierMIP (Glacier Model Intercomparison Project) forecasts, glaciers in Asia's High Mountains will lose 42.5%, 56.4%, and 71.2% of their mass (at 2015) by the end of the century for RCP of 4.5, 6.0, and 8.5 scenarios, respectively (Rounce et al., 2020).

The altitudinal shift is projected to be ~20–80 m/decade in the eastern Himalaya based on the projections of temperature increase of 0.01–0.04 °C per year (Hussain et al., 2021; Liu & Rasul, 2007). Under the RCP8.5 scenario, snowline elevations in the Indus basins, Ganges, and Brahmaputra basins are expected to rise between 400 and 900 m, i.e., 4.4–10.0 m $year^{-1}$ by 2100 (Viste & Sorteberg, 2015). There was no significant interannual variation in snow cover in Eurasia between 2000 and 2016 (Sun & Zhou, 2020). Since 1970s, there have been considerable seasonal changes of Eurasian snow cover extent (particularly for faster spring snowmelt), and these changes are projected to continue in the future (Zhong et al., 2021).

The amount of precipitation did not vary much in the eastern Himalaya, but the number of rainy days reduced, resulting in more rainfall in a shorter period of time. In the eastern Himalaya and mountainous places, this intense rain may trigger flash floods and landslides (Syed & Al Amin, 2016). High-resolution climate model projections showed that the variations of the western disturbances might increase the winter rain over the Karakoram and Western Himalayas (Forsythe et al., 2017). Flood occurring more frequently at higher altitudes might hasten glacier melting and flood events, posing considerable disaster risks in the region (Dalezios et al., 2017). During the last several decades, the Himalayan regions have seen a dramatic increase in the extreme rain events (Krishnan et al., 2019). A western disturbance is denoted as extratropical storm that originates in the Mediterranean region that would cause heavy wintertime precipitation to the Indian subcontinent especially around north-western regions (Krishnan et al., 2019).

Droughts are caused by an imbalance in precipitation pattern, duration, and frequency, thereby creating an irreversible damage to society and are directly related to climate change (Wilhite et al., 2014). Precipitation variability and drought monitoring have been observed over many regions; however, the interrelationship between the climate change and droughts has received little attention due to its slow onset (Ahmed et al., 2019; Keshavarz et al., 2013; Nikravesh et al., 2020). In Himalayan region, low precipitation was noticed in Karakoram compared to the Hindukush region (Khan et al., 2020).

Forests are more prone to fires as a result of global warming caused by increased climate variability. Remote sensing studies also showed the occurrence of forest fire due to extended dry winter period and rising mean temperature (Sannigrahi et al., 2020). In the Himalayas, wildfire season varied from November to May, with a peak in March to April. Approximately, 69% of the forest area was susceptible to forest fire in Uttarakhand, followed by 50% in Himachal Pradesh and 40% in Jammu and Kashmir and Sikkim (Sharma et al., 2014). Forest fires in the Himalayan regions are becoming more common due to rising temperatures, decreasing precipitation, and drought (Abrha & Adhana, 2019). The majority of forest fires in Himalayan regions are linked to meteorological parameters, such as temperature, dryness, rainfall, and humidity (Wang et al., 2021).

2.2 *Geophysical Disasters*

Geologically, Himalayan mountain ranges are broadly classified into four regions across its length: (1) Foothill/Outer Himalaya, (2) Lesser Himalaya, (3) Higher Himalaya, and (4) Trans-Himalaya (Chakrabarti, 2016). Climate change is attributed to long-term patterns of increasing slope collapse occurrence in mountain areas. Mountain locations are prone to mass movements due to steep slopes, considerable topographic relief, and seismic activity, which are intensified by the sensitivity of glaciers and permafrost (Gruber et al., 2017).

The Himalayan mountain range is tectonically active and prone to earthquakes and landslides (Sastry & Mondal, 2013; Velayudham et al., 2021). Landslides can occur both in Lesser and Higher Himalayas as well as in cold desert regions (Sastry & Mondal, 2013). Intense landslide activity is influenced by undeveloped geology of the Himalayan region, fragile geological structures and uneven topography, as well as the pressures generated by an earthquake. Some of the key attributes in the geology of the failed rocks are the following: (a) the failed mass detached and cut by several directions of planar weakness; (b) the rock mass located near a major thrust fault contains several local shear fractures, which would have aided in chemical weathering; and (c) the unweathered rock types, i.e., schist and gneiss contains soft, platy, and minerals like phyllosilicates and kyanite; further, weathering of these rocks will be more weakened, and making them to disintegrate into fine material would enhance the mass flow mobility (Shugar et al., 2021).

The major rivers, Indus, Ganges, and their respective tributaries are originated from the Himalayan regions. Hazard cascades, in which an initial incident triggered a downstream chain reaction, can have far-reaching consequences, especially when a huge quantity of water is involved (Kirschbaum et al., 2019). Moreover, rainfall-induced landslides are high risk in the Western Himalayan region due to excessive rainfall during the monsoon season (June to September) that would moisten the terrain by 77.7% (Dikshit et al., 2020).

The majority of the landslides seen in the field around 65% fell into the debris slide category. This clearly demonstrates that the overburden or debris material was saturated by extended heavy rains causing catastropic landslides. A steep, glacierized north-facing slope triggered the rock-ice avalanche in Chamoli region, Uttarakhand, during February, 2021, causing debris flow (Mergili et al., 2021). Moreover, Chamoli event might be viewed in the circumstance of a shift in geomorphological sensitivity due to continuous global warming events in recent days (Shugar et al., 2021).

2.3 *Implications on Environment*

Disasters in the mountain areas affect the wildlife and human society by declining water resources, wild food resources, fuel wood, and biodiversity. Furthermore, changes in land use patterns caused by climate change have an equal impact on animal and human environment (Khanduri et al., 2018). Climate change impact with signs of phenological changes and reduced agricultural productivity in the Himalayan regions progressed during recent decades (Hart et al., 2014; Krishnan et al., 2020).

Changes in temperature and precipitation over the mountain regions could have significant implications on biodiversity, ecosystem goods, and services (Chettri & Sharma, 2016). High evaporation will cause wetlands to diminish, which will be aggravated by the spread of towns and human activities. Water bodies are becoming more suited as habitats for invading species, which competes with native species and endangers native life. As the temperature rises, more evaporation occurs, resulting in an increase in atmospheric moisture content, and, as a result, future geographical and temporal precipitation patterns will shift. Due to the region's diverse topography, localized weather phenomena, such as cloudbursts, flash floods, and landslides, may increase, eventually leading to greater risk (Hussain et al., 2021).

The MODIS satellite results showed that the significant decline in wintertime snow across the HKH region in present decade affects the river flow regimes and water resource availability (Maskey et al., 2011). Heavy rains and rapid ice melting in high-mountain glaciated areas have increased the risk of dammed lake outburst floods, which have harmed the human population and infrastructure (Khanduri et al., 2018; Mergili et al., 2021). The occurrence of disasters, like floods and landslides on a regular basis, changes the distribution of groundwater recharge and runoff, which controls the soil moisture indirectly (Pangali Sharma et al., 2020).

Rapid loss of forest to fires can aggravate the extent and magnitude of socio-ecological consequences, including degradation of ecosystem services, shortage of fuel wood, and displacement of people (Han & Han, 2020). Forest fires destroyed the endemic forest plants and species and devastated the dead vegetation and detritus, thereby exposing the top soil to erosion (Wang et al., 2021). This results in massive amounts of carbon dioxide released into the atmosphere and enhancing the effect of global warming (Hanson & Ranganathan, 2017). Due to biomass burning and forest fires, almost 100 million tons of smoke aerosols and black carbon are released into the atmosphere, and an increase in surface albedo and water runoff were observed (Fearnside, 2019). On the other hand, absorbing aerosols at high elevations enhanced the warming rate on snow, indirectly amplifying the melting of glaciers.

3 Occurrence of Climate-Induced and Geophysical Disasters in Himalayan Regions

3.1 Earthquakes

The Himalayan region is known for the maximum seismic activity in southern Asia, where multiple massive earthquakes have caused thousands of deaths in the past (Table 16.1). The geodetic strain build-up rate and the stress release rate through earthquakes are two important factors in determining a seismogenic area's energy budget. As a result, the comparison of geodetic rate and seismic moments along the Himalaya can be viewed as a main predictor of seismic hazard. It was projected that the rate of geodetic moment varied from 1.7×10^{18} to 10.2×10^{18} Nm per year, while the rate of seismic moment ranged from 3.7×10^{16} to 5.1×10^{19} Nm per year (Sharma et al., 2020). This difference in geodetic and seismic moment rates was equal to a moment deficit rate ranging from $\sim 1.15 \times 10^{17}$ to 7.97×10^{18} throughout the northwest to northeast Himalaya, which corresponds to a magnitude of 5.7–8.2 earthquake potential. Even though the regions of Kashmir and Kathmandu are thought to have low seismic potential, the northeast half of Himalaya have a larger earthquake potential ($Mw \geq 8.0$) (Bilham, 2019).

3.2 Landslides

The landslide, being one of the primary disasters in the steep and hilly region, is a downward movement of dislodged rocks, debris, and earth materials caused by gravity and one of the most destructive geo-hazards (Mandal & Maiti, 2015). In Asia, over 66 million people live in landslide-prone zones (Tien Bui et al., 2019).

Table 16.1 Occurrence of earthquake in Himalayan mountain regions

Region	Year of occurrence	Place	Casualties	Richter magnitude	Impact	References
Northwest Himalayan region	November, 1997	Bangladesh	23	5.8	Infrastructure damage and property loss	Islam et al. (2011)
	October, 2005	Muzaffarabad, Kashmir	>80,000	7.7	Life loss and infrastructure damage	Kamp et al. (2008)
	January, 2016	Manipur	–	6.7	Infrastructure damage and transport system collapsed	Gahalaut and Kundu (2016)
	January, 2017	Ambasa earthquake, Tripura	10	5.7	Infrastructure damage	Halder et al. (2021)
	June, 2020	Mizoram earthquake	–	5.5	Transport system collapsed	
Western Himalayan region	October, 1991	Uttarakhand	2000	7.0	Infrastructure damaged and human life loss	Chopra et al. (2012)
	March, 1999	Uttarakhand (Chamoli)	110	6.8	Infrastructure damage, property loss, and transport system collapsed	
	December, 2005	Uttarakhand	–	5.4	Transport system collapsed	Mittal et al. (2019)
Eastern Himalayan region	September, 2011	Sikkim	97	6.8	Transport system collapsed	Halder et al. (2021)
	April, 2015	Gorkha, Nepal	8964	7.8	Life loss, infrastructure damage, and property loss	
	May, 2015	Nepal	218	7.3	Property loss, infrastructure damage, and transport system collapsed	
	April, 2021	Assam	2	6.0	Property loss and infrastructure damage	
Southeast Himalayan region	August, 2016	India and Myanmar region	11	6.7	Infrastructure damage	Aung et al. (2019)

According to the Geological Survey of India (GSI), landslides threaten 4,20,000 km^2 or 12.6% of India's entire geographical land area. Rainfall and earthquakes are the two most common causes of landslides in Himalayan terrain (Table 16.2). India, as a monsoon-affected country, experiences the majority of landslides during the monsoon season, notably from June to August (Mondal & Mandal, 2020). Rain-induced landslides, contributing for nearly 76% of landslide events, occur during or after the monsoon season (Ghosh et al., 2020), whereas earthquake-induced landslides happen right after the earthquake (Dikshit et al., 2020; Kumar et al., 2021). The Himalayan region accounts for roughly 30% of all landslide occurrences worldwide, while 42% of India's landslide occurred around north-west Himalayas, including the Darjeeling-Sikkim Himalayas (Dikshit et al., 2018).

Across Goriganga valley, Kumaun Himalaya region, ~20–25% of the area along the Goriganga river and the Lesser Himalaya lies in the high and very high landslide susceptible zones, whereas ~50–63% of the areas located in the Higher Himalaya and Tethys Himalaya lies in the low and very low susceptible zone (Kumar & Gupta, 2021). Across Mussoorie Township and around Lesser Himalaya, ~44% of the area are prone to very high landslide susceptible zones while ~56% of the area comes under the low susceptible zones (Ram et al., 2020).

3.3 Flood

Intense precipitation events enhance the risk of disasters that can cause unexpected events like flooding and flood-related disasters (Aryal et al., 2020; Zhao et al., 2019). Continuous and varied precipitation eventually leads to an intense river flow to exceed a threshold, causing flood by breaching the river bank or previously constructed flood restoration regions (Aryal et al., 2020; Dulal et al., 2007).

With warming rate higher in Himalayan region than in other parts of the world, the availability of water vapor increases, eventually leading to intense precipitation resulting in increased risk of flooding (IPCC, 2021; Simmons et al., 2010). Nearly 63% of basins under the zone of Sikkim–Darjeeling and Bhutanese Himalayas falls under high to very high flash flood risk, whereas 28% comes under medium and only 9% in the low categories of flash flood risk (Karmokar & De, 2020). Furthermore, nearly 29% area of Himalayan foothills of Jalpaiguri district lies under a high threat level (Roy et al., 2021). Recent flood events over Himalayan mountain regions are detailed in Table 16.3.

3.4 Glacial Lake Outburst Flood (GLOF)

Due to climate change and the acceleration of glacier melt in recent decades, glacial lakes in the Himalayas are rapidly increasing (Khadka et al., 2018). Glacial lakes are considered as an important source of water reservoirs that would highly vary in

Table 16.2 Occurrence of landslides in Himalayan mountain regions

Region	Factors	Year of occurrence	Place	Casualties	Economic loss	References
Northeastern Himalayan region	Cloudburst-landslide	August, 1993	Nagaland (Kohima)	500	Infrastructure damage and transport system collapsed	Nokendangba Chang et al. (2021)
	Earthquake-induced landslide	May, 1995	Mizoram (Aizawl)	25	Transport system collapsed	Vinoth et al. (2022)
	Flood-induced landslide	June, 2017	Bangladesh	152	Life loss	Uddin et al. (2021)
	Rainfall-induced landslide	July, 2020	Manipur (Noney, Mao Town, and Senapati District)	–	Infrastructure damage	Geological survey of India (2020)
Western Himalayan region	Rainfall-induced landslide	July, 1970	Belakuchi, Uttarakhand	–	Infrastructure damage	Parkash (2015)
		July, 1990	Uttarakhand	25	Property loss	Dikshit et al. (2019)
	Landslide-flash flood	July, 1991	Himachal Pradesh	300	Human life losses and transport system collapsed	Meena and Mishra (2018)
	Cloudburst-landslide	June, 1993	Himachal Pradesh	–	Transport system collapsed	Haritashya et al. (2006)
	Rainfall-induced landslide	September, 1995	Himachal Pradesh (Kullu)	65	Property loss and transport system collapsed	Meena and Mishra (2018)
		August, 1998	Uttarakhand (Malpa)	380	Human life loss and transport system collapsed	Onagh et al. (2012)
		March, 1999	Uttarakhand (Chamoli)	–	Infrastructure damage	Gupta et al. (2021)
		August, 2002	Tehri, Uttarakhand	29	Transport system collapsed	Parkash et al. (2015)

(continued)

Table 16.2 (continued)

Region	Factors	Year of occurrence	Place	Casualties	Economic loss	References
		July, 2003	West Bengal	11	Human life loss, infrastructure damage	
			Kullu, Himachal Pradesh	–	Transport system collapsed	Dubey et al. (2013)
		September, 2003	Uttarkashi, Uttarakhand	–	Infrastructure damage	Parkash et al. (2015)
	Earthquake-induced landslide	October, 2005	Kashmir (Muzaffarabad)	26,500	Human life loss, transport system collapsed, and property loss	Mahmood et al. (2015)
	Cloudburst-landslide	July, 2007	Uttarakhand	28	Infrastructure damage and transport system collapsed	Kumar et al. (2017)
	Avalanche-induced landslide	February, 2008	Jammu and Kashmir	25	Infrastructure damage and transport system collapsed	Parkash et al. (2015)
	Rainfall-induced landslide	September, 2008	Shimla, Himachal Pradesh	42	Infrastructure damage and transport system collapsed	Parkash et al. (2015)
		June, 2009	Munsiyari, Uttarakhand	–	Transport system collapsed	Sati (2020)
	Avalanche-induced landslide	February, 2010	Jammu and Kashmir	17	Transport system collapsed	Banerjee and Dimri (2019)
	Cloudburst-landslide	July, 2011	Jammu and Kashmir	3	Transport system collapsed	Parkash et al. (2015)
	Flood-induced landslide	June, 2013	Uttarakhand and Himachal Pradesh	4094	Transport system collapsed and property loss	Shekhar et al. (2015)
	Rainfall-induced landslide	August, 2021	Uttarakhand (Pithoragarh)	2	Transport system collapsed	ECHO (2021)
		October, 2021	Uttarakhand	52	Life loss and transport system collapsed	

Eastern Himalayan region	Rainfall-induced landslide	August, 1948	Assam	500	Infrastructure damage	Parkash (2015)
	Landslide and subsidence zone	August, 1984	Gangtok, Sikkim	9	Transport system collapsed	Murgese et al. (2015)
	Rainfall-induced landslide	August, 1993	West Bengal (Kalimpong)	40	Infrastructure damage and life loss	Roy et al. (2019a, b)
		July, 1993	Arunachal Pradesh (Itanagar)	25	Property loss	Rupa (2015)
		June, 1997	Gangtok City, Sikkim	–	Transport system collapsed	Ramanathan et al. (2020)
		September, 2003	Sikkim	4	Human life loss and transport system collapsed	
		August, 2005	Assam	12	Human life loss, property loss, and infrastructure damage	Parkash et al. (2015)
		September, 2005	Sikkim	7	Human life loss and property loss	
		July–September, 2007	Sikkim, Darjeeling	–	Transport system collapsed	Mondal and Mandal (2018)
		May, 2009	Kurseong, Darjeeling	27	Property loss	Parkash et al. (2015)
		September, 2010	Dhemaji, Golaghat, and Bongaigaon, Assam	2	Infrastructure damage and property loss	
		June, 2011	Sikkim	16	Transport system collapsed and infrastructure damage	Sridharan and Gopalan (2019)
		May, 2020	Sikkim	–	Transport system collapsed	Geological survey of India (2020)
			Meghalaya	–	Transport system collapsed	

(continued)

Table 16.2 (continued)

Region	Factors	Year of occurrence	Place	Casualties	Economic loss	References
		June, 2020	Assam	29	Transport system collapsed, life loss, and property loss	
		July, 2020	Arunachal Pradesh (Papum Pare)	–	Transport system collapsed	
		October, 2021	Nepal	88	Life loss and transport system collapsed	ECHO (2021)
Southeast Himalayan region	Rainfall-induced landslide	July, 2019	Myanmar	75	Infrastructure damage	Panday and Dong (2021)
		August, 2019	Myanmar (Kpakant)	115	Infrastructure damage and property loss	

Table 16.3 Occurrence of flood events in Himalayan mountain regions

Region	Factors	Year of occurrence	Place	Casualties	Impact	References
Western Himalayan region	Cloudburst-flash flood	September, 1988	Himachal Pradesh	65	Property loss and infrastructure damage	Prasad et al. (2016)
		August, 1997	Himachal Pradesh (Chirgaon)	–	Property loss	
		June, 2000	Uttarakhand (Gangotri Glacier)	–	Property loss	Kumar et al. (2018)
		July, 2003	Kullu, Himachal Pradesh	40	Infrastructure damage and transport system collapsed	Prasad et al. (2016)
		June, 2005	Himachal Pradesh (Sutlej Valley)	19	Property loss and transport system collapsed	Pandey et al. (2015)
		June, 2005–August, 2009	Jammu and Kashmir	–	Infrastructure damage and transport system collapsed	Kumar et al. (2018)
		August, 2010	Leh, Jammu and Kashmir	25	Transport system collapsed, life loss, and property loss	
	Rainfall-induced flash flood	August, 2010	Uttarakhand	65	Property loss	
		August, 2011	Uttarakhand (Dokriani Glacier)	–	Property loss	Khanduri (2020)
	Cloudburst-flash flood	September, 2012	Uttarakhand	–	Property loss and infrastructure damage	
	Uttarakhand floods	August–September, 2012	Uttarkashi, Rudraprayag, and Bageshwar	52	Property loss and infrastructure damage	Parkash et al. (2015)
	Cloudburst-flash flood	September, 2012	Uttarakhand (Asi ganga, Uttarkashi)	–	Transport system collapsed	

(continued)

Table 16.3 (continued)

Region	Factors	Year of occurrence	Place	Casualties	Impact	References
		June, 2013	Uttarakhand (Kedarnath)	5000	Transport system collapsed and property loss	Singh et al. (2016)
			Uttarakhand (Gangotri Glacier)	–	Transport system collapsed and property loss	Kumar et al. (2018)
			Himachal Pradesh (Kinnaur)	23	Property loss, life loss, and infrastructure damage	Pandey et al. (2015)
	Floods	October, 2014	Jammu and Kashmir	8	Infrastructure damage and transport system collapsed	Bhatt et al. (2020)
	Cloudburst-flash flood	July, 2016	Uttarakhand (Bastadi Narula)	–	Transport system collapsed	Kumar et al. (2018)
	Avalanche-flash flood	February, 2021	Uttarakhand	200	Property loss and infrastructure damage	Pandey et al. (2021)
Eastern Himalayan region	Cloudburst-flash flood	July, 2016	Assam	28	Life loss, property loss, infrastructure damage, and transport system collapsed	Paik (2018)
		May, 2020	Assam	149	Property loss, infrastructure damage, and transport system collapsed	Sachdev and Kumar (2021)

number, size, and volume with respect to climate change. Glacial lake outburst floods (GLOF) occur when significant quantities of water from glacial lakes are released suddenly, either by dam failure, self-destruction, ice/rock avalanche, or excessive precipitation into the lake (Harrison et al., 2018).

Dam self-destruction was responsible for more than two-fifth of GLOF incidents in the Himalayan region (Emmer & Cochachin, 2013). Across 38 cases of GLOF events assessed over Himalayas, 34% was due to ice avalanche falling into the lake, 20% by hydrostatic pressure mediated water level rise, and 14% by melting of dead ice (Falátková, 2016). According to Mal et al. (2021), Jammu and Kashmir regions have higher number of lakes at high risk ($n = 556$), followed by Arunachal Pradesh (388) and Sikkim (219). The top three triggering mechanisms of 21 historical GLOF episodes in Bhutan were glacier ice falling into the lake (44%), glacier calving (33%), and ice deformation (10%) (Komori et al., 2012). Furthermore, 36 out of 329 glacial lakes analyzed were vulnerable to snow avalanche, and 23 glacial lakes were found to be very high-risk lakes and 50 as high-risk lakes in Indian Himalayas (Dubey & Goyal, 2020). The historical glacial lake outburst flood events over Himalayan mountain regions are given in Table 16.4.

3.5 Snow Avalanche

Avalanches are a common hazard in snow-covered mountainous area all over the world. Two union territories and three states in the Indian Himalaya are more prone to avalanches: Jammu and Kashmir, Ladakh, Himachal Pradesh, Uttarakhand, and Sikkim (Singh et al., 2020a, b). Greater number of snow avalanches occurred in Asia's high mountains, such as the Himalaya, Karakoram, Pamir, Hindu Kush, Tien Shan, and Dazu Shan of India, China, Pakistan, and Nepal (McClung, 2016). Due to its unique terrain, snow characteristics, and seismic activity, the Karakoram Mountains in the north-western Himalaya experiences earthquake-induced and delayed-action avalanches throughout the year (Singh & Ganju, 2002; Kumar et al., 2019; Singh et al., 2021a, b, c). In general, the zone heights of hazardous avalanche slopes in the Upper Himalayan zone are around 5000 m or greater. About 33% of snow avalanche paths have slope angles of 32°–40°, and majority of its track faces north-west (Kanna et al., 2018). The details of snow avalanche occurred over Himalayan mountain regions are described in Table 16.5.

3.6 Subsidence

Himalayan terrain is more prone to ground subsidence due to its seismic activity and soil erosion. The river Kali constantly accumulated sediments by depositing mass from the north to down regions of Garbyang basin. Moreover, the stream passed through stabilized landslide mass, and water percolated through the destabilized

Table 16.4 Occurrence of glacial lake outburst flood events in Himalayan mountain regions

Region	Glacier	Year	Location	Casualties	Impact	References
Northern Himalayan region	Ghulkin Glacial lake	2008	Pakistan	–	Property loss and transport system collapsed	Nie et al. (2018)
Central Himalayan region	Cirenmaco	1964	Central Himalaya (Zhangzangbo valley)	–	Property loss and transport system collapsed	Gurang et al. (2017a, b)
	Lahaul Valley	1979	India	200	Life loss and infrastructure damage	Gurang et al. (2017a, b)
	Dig Tsho	1985	Nepal	–	Life loss and infrastructure damage	Gurang et al. (2017a, b)
	Chubung lake	1991	Nepal	–	Property loss and transport system collapsed	Westoby et al. (2014)
	Khumbu Himal	1998	Sabai Tsho, Nepal	–	Infrastructure damage	Nie et al. (2018)
	Kara, Hussain, Yigeong	2000	Glacial lake, China and India	–	Property loss	
	Parechu Zhu	2005	China and India	–	Infrastructure damage, property loss, and transport system collapsed	
	Chorabari lake	2013	India	1600	Infrastructure damage and transport system collapsed	Allen et al. (2016)
	Lhotse glacier	2015	Supraglacial lake, Nepal	–	Infrastructure damage and transport system collapsed	Rounce et al. (2017)
	Lhotse glacier	2016	Supraglacial lake	10	Infrastructure damage	
	Gongbatongsha Tsho	2016	Tibet	5	Infrastructure damage and property loss	

Western Himalayan region		2021	Chamoli, Uttarakhand	61	Infrastructure damage and property loss	Vishwanath and Sharma (2021)
Eastern Himalayan region	Lunana glacier	1960	Northern Bhutan	–	Property loss and transport system collapsed	Nie et al. (2018)
	Zharico lake	1981	Tibet	–	Life loss and infrastructure damage	Yao et al. (2018)
	Luggye Tsho lake	1994	Tibet	20	Infrastructure damage	Gurang et al. (2017a, b)
	Gangri Tsho	1998	Tibet	–	Property loss	Nie et al. (2018)
	Lemthang Tsho	2015	Tibet	–	Infrastructure damage	Gurung et al. (2017a, b)

Table 16.5 Occurrence of snow avalanche in Himalayan mountain regions

Region	Factors	Year	Location	Casualties	Impact	References
East central Himalayan region	Manaslu Nepal avalanche	April, 1972	Nepal	15	Transport system collapsed and property loss	Thakuri et al. (2020)
	Gokyo avalanche	November, 1995	Nepal	42 people died	Transport system collapsed and property loss	
	Mount Everest	May, 1996	Nepal	3 people died	Transport system collapsed	
	Manaslu	September, 2012	Nepal	11 people died	Life loss and transport system collapsed	
	Mount Everest avalanche	April, 2014	Nepal	16 people died	Life loss and transport system collapsed	
	Mount Everest avalanches	April, 2015	Nepal	22 people died	Life loss, property loss, and transport system collapsed	
Northeastern Himalayan region	Mokokchung debris avalanche	May, 2005	Nagaland	14 people died	Life loss and transport system collapsed	Singh (2018)
	Kohistan avalanche	February, 2010	Pakistan	102 people died	Life loss and transport system collapsed	Khan and Qureshi (2019)
	Gayari sector avalanche	April, 2012	Pakistan	138 people died	Life loss and transport system collapsed	
Western Himalayan region	Gurez sector avalanche	January, 2017	Jammu and Kashmir	28 people died	Life loss and transport system collapsed	Khan and Qureshi (2019)
	Massive rock and ice avalanche	February, 2021	Uttarakhand	200 people died	Life loss, property loss, and transport system collapsed	Shugar et al. (2021)

mass would reflect in ground subsidence over the regions of Umli-Bhandarigaon, Talla Dhumar, and Bagi (Rautela, 2005). The Indian continent's northward movement have been seen at the surface as a southward migrating wave of subsidence and uplift, which has been observed from the Himalaya to the Ganges basin (Webb et al., 2017). The Kailas basin is a one kind of perched basin that was created by dynamic deflection inside a mountain belt (Shen et al., 2020).

3.7 Climate Extreme Events: Temperature, Droughts, and Wildfires

Extreme temperature events were more prevalent in Himalayan mountain compared to the other regions of the world (Singh et al., 2011). Warm spells are becoming more common in the Himalayan highlands. The cold spell is decreasing primarily in valley areas, such as Kathmandu and Karnali, as well as the Himalayan Chisapani region. On an annual basis, the maximum temperature is increasing (0.05 °C/year) than the minimum temperature (0.003 °C/year) (Awasthi & Owen, 2020).

Extreme drought occurrences in the Himalayas have become common and intense over the last three to four decades, during both winter and summer seasons (Dahal et al., 2016). Trans-Himalaya exhibited the highest temperature variation during summer (7 °C) and a low during winter (−13 °C). The mountain plains (the Terai in Nepal and the Indo-Gangetic Plain in India) had the highest temperature varying from 16 to 30 °C, whereas the middle mountains (north of the Terai up to the Himalayan peaks in Nepal) had a range of 7–20 °C. More than 75% of the area in the mountains had a soil moisture deficit index (SMDI) below −1, while drought (SMDI < -3) prevailed over the Koshi River basin area for >40% during 2005–2006 (Nepal et al., 2021). The extreme heat and wildfire events that occurred over Himalayan mountain regions are detailed in Table 16.6.

4 Observation and Monitoring Systems

The cascading disasters occurring in Himalayas increase every day and trigger one another, creating a cumulative hazard. The Himalayan mountains are geologically fragile, vulnerable to various climate change impacts, including erosion, landslides, glacial floods, avalanches, etc. There is a pressing requisite for dramatic step in the monitoring processes for growing environmental concerns. Such highly bio-diverse and multi-hazard zones demand focused risk reduction strategies, considering bio-physical and sociocultural certainties of the mountains.

For the wide range analysis of earth surface processes, Himalayan topography assessment is a prime requisite to assess where disaster events cause threat to population and infrastructure. Various techniques are available to observe and monitor the pre- and post-disaster, which are detailed under each climate risks and

Table 16.6 Recent occurrence of heat extremes and wildfire events in Himalayan mountain regions

Types of disaster	Factors	Year	Location	Casualties	Environmental loss	References
Asian heatwave	Heatwave	2007	India, Pakistan, Bangladesh, and Nepal	More than 200 people died	Life loss, property loss, and transport system collapsed	Naveena et al. (2021)
Uttarakhand wildfire	Heatwave	2016	Uttarakhand	7 people died	Infrastructure damage	Biswas et al. (2018)
Northern Indian cold wave	Cold wave	2017	Jammu and Kashmir, Himachal Pradesh	7 people died	Infrastructure damage and transport system collapsed	Das et al. (2020)
Shimla forest fire	Crown fire	April,1999	Himachal Pradesh	–	Biodiversity loss and reduced forest area	Kanga et al. (2017)
Uttarakhand forest fire	Crown fire	April, 2020	Uttarakhand	2 people died	Biodiversity loss and infrastructure damage	Latestly (2021)
Uttarakhand forest fire	Crown fire	April, 2021	Uttarakhand	–	Biodiversity loss and infrastructure damage	Latestly (2021)

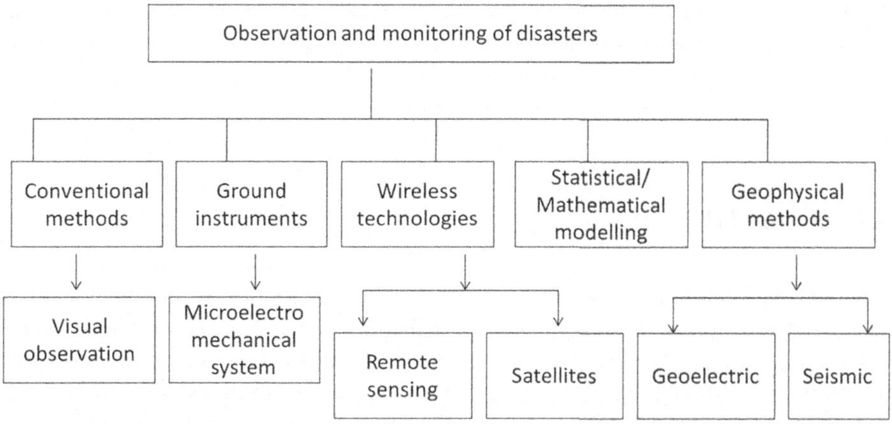

Fig. 16.2 Disaster observation and monitoring techniques over Himalayan mountain regions

briefed in Fig. 16.2. The processes find ease after the advent of remote sensing technologies.

4.1 *Earthquakes*

Understanding the earthquake mechanism, structural complexities of fault zones, and its recurrence interval has greatly enhanced the need and interest on earthquake prediction techniques. Sudden disturbance on the earth manifests the surface causing interruption on seismic waves and ground shaking. Prediction of expected magnitude, geographical location, and time of occurrence with accurate precision would be indispensable. Macro-level zonation mapping based on peak ground acceleration (PGA) was implied to be imprecise compared with micro-level zonation. The major drawback associated with such map is that it was designed depending only on past events. Hence, a scientific seismic hazard analysis and reliable zoning are essential for effective monitoring. Various models have been devised to forecast the seismic activity rates. A well-acknowledged "regional time and magnitude predictable model" was proposed by Papazachos (1989) to assess the probability of occurrence of forth-coming main shock at specified time and magnitude in seismogenic source zone. Radon, being sensitive to tectonic disturbances, is used as seismotectonic indicators (Ramola et al., 2008; Barkat et al., 2017). Ramkrishnan et al. (2021) have developed ground motion prediction equation based on moment magnitude and hypocentral distances, which could accurately predict PGA than previous reports.

Freeman (1975) has evaluated the capacity spectrum technology for damage and loss assessment of earthquake and attained transition from employing intensity parameters to physical parameters like spectral accelerations and spectral displacements. Bradley et al. (2015) have attempted the comparisons of ground motion equation and simulation-based technique. The result of the study recommends that rupture kinematics allows better-constrained damage and loss estimates. However, the difficulty and uncertainties associated with it reveal stochastic methods as better compromise in observing variability of interest.

4.2 Landslides

Earthquakes or rainfall-triggered landslides cause devastation in highlands, either causing severe threat to human lives or damaging roads, forests, and agricultural land. The prediction of landslides is unattainable as it involves continuous monitoring of minor likely regions within large area and relies on complex parameters, such as topography, geology, environment, and anthropogenic activities. Various techniques have been adapted for landslide susceptibility assessments, including expert knowledge and statistical methods (linear, logistic, multiple and geographically weighted regression, frequency ratio, and function of past evidences). In general, landslide hazards are initially assessed by landslide susceptibility mapping (hazard zonation mapping). Such maps and photo interpretation provide widespread knowledge on location and type of past landslide occurrences, frequency of occurrence, casual factors, volume, and damages that have been associated with previous reports (Raetzo & Loup, 2016; Ambrosi & Scapozza, 2015). It is then performed based on the state of activity of mapped phenomena acquired from synthetic aperture radar (SAR) (Calvello et al., 2017). The severity is determined by measuring the volume and velocity and debris thickness from field survey. Several GIS-based soft-computing models being used for envisaging landslide susceptibility include super vector machines (Nhu et al., 2020), artificial neural networks (Tien Bui et al., 2019), adaptive network-based fuzzy interference system (Polykretis et al., 2019), and decision tress (Hong et al., 2015). Digital elevation model (DEM) generated using GIS delivers topographic attributes for hazard assessment (Sarkar et al., 2015).

4.3 Floods

The extreme precipitation events have resulted in flash floods. The acceleration of climate change in Himalayas has resulted in glacial lakes, and recent decades have witnessed glacier melt (Maurer et al., 2019). Management of Himalayan flood is critical as it results from the combination of land and atmospheric processes. The prime step involved is hazard mapping, risk zonation, and enhancing early warning systems and forecasting. Flood forecasting serves as a more efficient and cost-effective technique than flood control measures. The ground details of watershed morphometrics conjoined with land and slope aid in assessing the hazard vulnerability. Various stochastic methods have been used widely for simulating and forecasting rainfall, temperature, water level, and hydrological time series, such as

Box-Jenkins or ARIMA (autoregressive integrated moving average) (Parvaze et al., 2021), ARMA (autoregressive moving average), and SARIMA (seasonal autoregressive integrated moving average) (Valipour and Eslamian, 2014). These models require key features, such as mechanism of precipitation formation, snow-melt modelling, catchment runoff, and flood routing. The occurrence of glacial lake outburst floods has been recognized by satellite imagery and first-hand observations. The frontal topographical analysis could be the first-order approximation of future disasters. Devrani et al. (2015) have determined spatial pattern of Mandakini river basin with the topographical analysis as first-order prediction measuring normalized channel steepness with chi analysis. The integrated use of linear imaging self-scanner satellite data and advanced space-borne thermal emission and reflection radiometer digital elevation model (DEM) by Meeraj et al. (2015) have assessed the influence of morphometry, landcover, and slope on flood vulnerability, which strongly governs hydrological functionality and response. In the absence of finer-resolution data, global DEM are often used (Watson et al., 2015).

4.4 Avalanches

The observations on activity of avalanche are significant, since they are strong and trustworthy indicator of snow instability and its detection is presently carried out by aerial reconnaissance by experts. Traditional field-based method to monitor the snow cover holds hurdles as it is time-consuming and labour-intensive and becomes impossible when harsh climate exists. Several physical and statistical models are used in avalanche forecasting, which encompasses its own limitations with the mathematical formulations. Snowmelt runoff model developed by Martinec (1975) has been used in over 100 basins, which simulate daily runoff ensuing from snowmelt and rainfall using daily temperature, precipitation, and daily snow area as input parameters. An attempt by Larsen et al. (2013) have evaluated the deposits of avalanche using object-oriented approach and validated with manually generated debris maps. The fresh avalanches in the image could be detected by conducting texture classification of image and removing the false detection. Besides, manual monitoring ground-based radar or laser scanning is also being used in monitoring the avalanche occurrence. The high spatial variability endowed causes difficulty in representing the actual snow cover (Rathore et al., 2018).

The automatic detection and mapping of snow avalanche debris in regions of Western Himalayas was conducted by Singh et al. (2020a, b) using satellite data and to formulate an index. The spectral signature of avalanche and non-avalanche snow detected using near-infrared wavelength and object-based image analysis was suggested to be accurate than pixel-based methods. Kaur et al. (2021) have used a multi-model decision support system for predicting avalanche in north-west Himalaya, which integrates four avalanche forecasting models. Singh et al. (2021a, b, c) have assessed the snow cover change using MODIS in basins of

Western Himalayas, whose results revealed ~95% mean overall accuracy of cloud removal.

4.5 Subsidence

The monitoring techniques of subsidence are point-based including leveling, GPS, and ground-based observation field data, which are now substituted by space-based observations. Interferometric synthetic aperture radar (InSAR) was the first method to monitor ground subsidence, which covers hundreds to thousand square kilometers per interferogram (Prati et al., 2010). Generation of high spatial resolution deformation maps covering centimeters to millimeters (Xu et al., 2016) was monitored using differential interferometric synthetic aperture radar (DInSAR) and greatly used to assess the temporal evolution of land subsidence and compaction of subsurface layers. Bali et al. (2021) have studied mining-induced subsidence assisted by ground-penetrating radar in structural and engineering aspects of sinkhole kinematics of subsidence phenomena, geometrical pattern of deformation structures, sags, paleo collapses, groundwater table, etc., Such prospects using high-resolution antenna configuration would successfully map a network of structures responsible for creep process.

4.6 Drought and Wildfires

Several indices are available for assessing, monitoring, and quantifying drought that evaluates its intensity and duration like modified Palmer drought index, surface water supply index, soil moisture anomaly index, drought severity index, etc. Khatiwada and Pandey (2019) have studied the drought dynamics evaluating various indices for characterizing hydrometeorological drought. Standardized precipitation index serves as a powerful tool, which requires only data on precipitation for evaluating the intensity, duration, magnitude, severity, and frequency of drought predicting the drought risk (Khan et al., 2020).

Understanding of anthropogenic, ecological, and environmental drivers of wildfires is the chief component in assessing the occurrence of wildfires. The active fire detection and monitoring capabilities were assessed using moderate resolution imaging spectroradiometer (MODIS) satellite sensors (Cochrane, 2003). A specific biomarker for fire emission is levoglucosan (You & Xu, 2018), and its detection has been reported in regions surrounding Himalayas (You et al., 2016). The fine-scale analysis of fire incidences at Himalayan foothills was demonstrated by Murthy et al. (2019).

5 Risk Reduction and Management in Himalayan Region

People in the region have evolved to cope with the unpredictability of climate change, variability, and extremes by modifying their livelihoods, agricultural techniques, and cultural traditions; they have even endured great loss. Despite the fact that cultural norms influence people's adaptation options, new vocations and livelihood practices that are socially and culturally undesirable have emerged in response to climate-related changes and stress in some circumstances. Enhancing adaptive capacities for collective water management, equitable distribution of irrigation facilities, livestock management, communal grazing, and development of infrastructure for flood management and water security requires social networks and local institutional support with good governance and planning (Hua, 2009). The following are a few steps that can be initiated to reduce the risk and damage caused due to climate-induced and geological disasters.

5.1 Understanding Risk

The impact of climate change induced and geological disaster at the Himalaya region is not well understood to predict the full-scale downstream impact. Hence, in-depth studies of snow, glaciers, landslides, earthquakes, floods, snow avalanches, and drought need to be carried out in the vulnerable region. There lacks a baseline study for most of the areas especially in regions greater than 4000 meters above mean sea level (masl), and very few long-term monitoring studies of climate variables have been conducted (Liu & Chen, 2000; Rees & Collins, 2006). Furthermore, the complexity due to its topography is one common feature for all mountain areas, and hence, the precipitation and temperature vary for every short distance eventually making the projection studies difficult. Hence, understanding the risk through forecasting and modeling studies is vital to reduce the risk and managing them in Himalayan region. In addition, the link between climate change, geological disasters, and built environment is multifaceted and hence has to be understood through an interdisciplinary perspective (Nibanupudi & Shaw, 2015).

In order to build adaptation and mitigation strategies, up to date scientific data on climatic parameters and their implications should be collected in a systematic way in partnership with government agencies, academia, and local-level partners. By exchanging data across partner nations in the area, data banks may be maintained up to date. In this method, regional climatic models may be built to better understand the climate process. The initial step in developing successful adaptation strategies should be to create a list of acceptable factors and analyze them. For fast risk evaluations of development projects, a mechanism for screening climate threats should be developed. The program should be able to determine if expected climatic changes will increase or decrease hazards, allowing planners to make informed decisions. This is especially true in case of hydropower projects, which have a lot of potential in mountainous nations and may help with national development. Choosing the right type of hydropower project (run-of-river or storage) is an important step toward ensuring the long-term viability of hydropower projects. Other water storage solutions (ponds, lakes, and aquifer recharges) should be selected based on geographical regions and implemented based on hydrological, geological, and climatic criteria (Tse-ring et al., 2010).

Furthermore, knowledge regarding the risk's potential and its communication, which includes a solution and direction to risk discourse, may help to successfully mitigate catastrophic adversity. Mountain is riskier due to its unique characteristics and proclivity to several risks. As a result, landscape-level complexity evaluation is valuable for a range of practical applications and policy planners (Papadimitriou, 2012). The risk assessment would educate on the potentiality of catastrophe risk based on the system under consideration (Das et al., 2018) as well as develop future dialogue to mitigate the potential danger. The risk assessment takes into account the system's adaptive capacity and sensitivity and may thus be used to improve the inherent and external features that are causing the system to deteriorate (Sekhri et al., 2020; IPCC, 2021).

Due to their nature and methodology, risk assessment and discourse frameworks for multi-hazards are unique (Gill & Malamud, 2016). In the mountains, particularly in the Himalaya, a lack of a consistent strategy to risk management limits the integration of risk assessments (Wester et al., 2019). The existing gap was widened by the lack of uniform and varied risk assessments, which hampered the vulnerable region's comprehensive mountain-specific development. Danger discourse would be guided by two simultaneous techniques based on actual evidence of possible risk analyzed from risk information and assessment. These two action-based solutions would make it easier to cope with risk and give options for mitigation if the risk is greater than the coping capacity. The study includes a coping capacity component in the risk assessment formulation, making it possible to give metrics to comprehend various aspects of risk. As a result, the coping capacity may be changed to offset the weaknesses of the system's exposed units based on the probable risk during a crisis. Furthermore, the mitigation plan must be integrated with the coping strategy, which includes preventative and inbuilt remedial action based on local resources and tactics, as well as lessons learnt from previous catastrophes.

5.2 Mountain Specific Risk Management Framework (MSMRMF)

The Mountain Specific Risk Management Framework (MSMRMF) is a tool for analyzing risk reduction methods by altering the system's internal strength and risk mitigation capability, as well as recommending feasible adaptation techniques based on multi-hazard risk. The MSMRMF is divided into two parts and is based on a multidisciplinary approach. The first component, termed as "risk information," is to qualify and create information on the system's risk. The second component of risk is known as "risk discourse," and it deals with addressing and mitigating the risk. Risk information includes components of hazard, climate exposure, and susceptibility, and it aids risk assessment of the mountain region's multi-hazards. The methods section has further information on each component. Risk information is used to manage and mitigate system risk, as well as to steer risk discourse, which leads to an explanation of how to address the risk (Sekhri et al., 2020). Hence it is imperative to identify risk drivers (Cardona & Dario, 2009) in the context of spatially explicit hazards, developing proactive measures (UNISDR, 2017), managing risk measures (Azadi et al., 2020) through various individual, community, and institutional supports (UNISDR, 2017), and finally confronting multi-hazard risks with all physiological, psychological, natural, and social hazards (Twigg, 2004; UNISDR, 2017).

5.3 *Reducing Risk and Increasing Resilience to Disasters*

Resilience demonstrates the ability to adapt, alter, and recognize the existence of change (Folke, 2006). Furthermore, the capacity of a system to resist change, preserve structure, demonstrate consistency, and facilitate strength to persevere through adversity without affecting the system's overall welfare is referred to as resilience (Pachauri et al., 2014). Due to the availability of space and expansion potential, mountains containing smaller towns have a higher capability to adapt and enhance resilience than many major metropolitan centers (Sakai et al., 2019). Communities with well-planned infrastructure, well-formulated emergency responses, a healthy population, hereditary and taught skill sets, and greater literacy rates are more resilient (Hatai & Sen, 2008). Moreover, communities from the Himalayan region's traditional wisdom aid in improving community resilience (Wester et al., 2019). Local community members' participation in developmental initiatives may put forth concepts and ideas based on a community's livelihood, observations, viewpoints, and wants (Ahmed et al., 2019), as well as provide locals more authority to interact with development. Additionally, the method would help to improve resilience and would be in line with disaster risk reduction (DRR) (Ahmed et al., 2019) and community-driven development (Ahmed et al., 2019). (CDD).

The basic goal of adaptation techniques for wildlife and habitat protection is to maintain ecosystem resilience (IPCC, 2021). Ecosystems will be better able to withstand anthropogenic environmental pressures, such as climate change, if they retain biological variety, decrease habitat fragmentation and degradation, and promote functional connectedness across habitat fragments. Reducing or eliminating current pressures is an excellent technique for doing this. Ecosystem adaptation alternatives, on the other hand, are limited, and their efficacy is unknown. Climate change-specific measures are unlikely to be extensively implemented sufficiently to safeguard the variety of ecological services on which civilizations rely. Fortunately, minimizing the negative effects of non-climate pressures on ecosystem would also protect ecosystem against climate change's negative consequences.

5.4 Adaptation Strategies

Adaptation to climate-induced and geological disasters is related to both vulnerability and potential impacts in the near future. Effective adaptation includes the formation of adaptive capacity (knowledge, awareness, and governance) and adaptation (changing the lifestyle, practices, and behavior according to the new conditions) (Mirza, 2007). Various factors like scale (local, regional, or national), context (management of water demand and new farming practices), and approach (poverty alleviation, better transparency in decision-making, women empowerment) strongly influence the adaptation strategies. One of the prime factors that makes adaptation strategy difficult for poor people is the structural inequality. Hence, measures have to be taken in moving all the people out of poverty and must enhance the capacity to adopt. Other adaptation measures include good governance to transform the climate change into institutional, general political, and developmental reforms, health education programs, and advancement in early warning systems.

5.5 Decision-Making

The vision of national and global institution guides the bulk of the catastrophe response strategy. The Sendai Disaster Framework 2015–2030 of the UNDP, the Resilient Development Framework of USAID, and Nepal's NAPA, for example, continue to be an important driver of local activities. While each framework emphasizes stakeholder interaction, the process begins with the frameworks' pre-defined corporate objectives. As a result, external institutions remain in charge of the planning process that forms the entire risk reduction arena. Hence, although local institutions frequently determine priority, external institutional rules determine the knowledge acquired and the methods for execution. Furthermore, the NGOs may cause problems when it comes to developing long-term, sustainable decision-making procedures for disaster risk reduction. Rounce et al. (2017) observed multiple situations in which government engagement in program management was reduced as a result of the existence of NGO programs (Thompson et al., 2020).

5.6 Sustainable Development Goal

The increased intensity and frequency of disasters in Himalayan regions is in one way due to high population density and increasing developmental activities. Several unsustainable constructions due to increased tourism have replaced the farmlands and forests, thereby violating the laws. Though tourism is one of the prime revenues in these regions, the importance of agriculture and agro-based industries should not be overlooked. In this regard, there is an urge in integrating the DRR with sustainable development goals. Models, land zonation maps demarcating the regions prone to geological disasters, and hazard scenarios have to be developed. Human habitation must be confined to safe zones; the mitigation plan should have a balance between the acceptable and development level of risk (Singh et al., 2021a, b, c). The key principles in the Himalayan regions for sustainable development are as follows:

- Building of a sustainable and viable forest-based economy.
- The approaches for water development must balance the threat to living being and the choice for energy.
- Enhanced early warning system and immediate response of local communities to disasters.
- Enhanced access to emergency evacuation, shelter, and protection against disasters.
- Strengthening the disaster risk management (DRM).
- The energy requirement for the remote villages must be secured.

5.7 *Risk Reduction by Technological Advancement*

Deeper human causes and impacts of catastrophes, such as inequality and poverty, which define the exposed and impacted entities owing to risks, are not usually addressed by technological answers. This implies the need to enhance effective links between research, technology, and decision-makers in order to develop long-term technological catastrophe risk reduction solutions (Busayo et al., 2020). People-centered technology may aid in effectively combating several risks in the society (Huang et al., 2017). The use of technology to collect data directly from the impacted population gives clear proof for damage estimates and makes catastrophe risk mitigation easier (Sakurai & Murayama, 2019). To reduce the effect of dangers, pre-disaster monitoring systems, such as early warning systems, can be built (Gupta, 2018). For example, the application of sensors and artificial intelligence for the detection of water levels in a river to assess rising water levels demarcates the pre-disaster technological advancements and can be used to generate timely warnings (Gupta, 2018). In a multi-hazard environment like the Himalayan region, effective technology for early warning systems for multi-hazard hazards, like cloud burst, strong rains, and landslides, might be created and used for disaster response planning (Wester et al., 2019). Overall, need-based disaster technology, including indigenous and traditional techniques in terms of early warnings, catastrophe facilitation through communication, and other necessary resources, would be highly effective in combating the damage caused due to these climate-induced and geological disasters. The various risk reduction techniques that can be adopted for Himalayan region are given in Fig. 16.3.

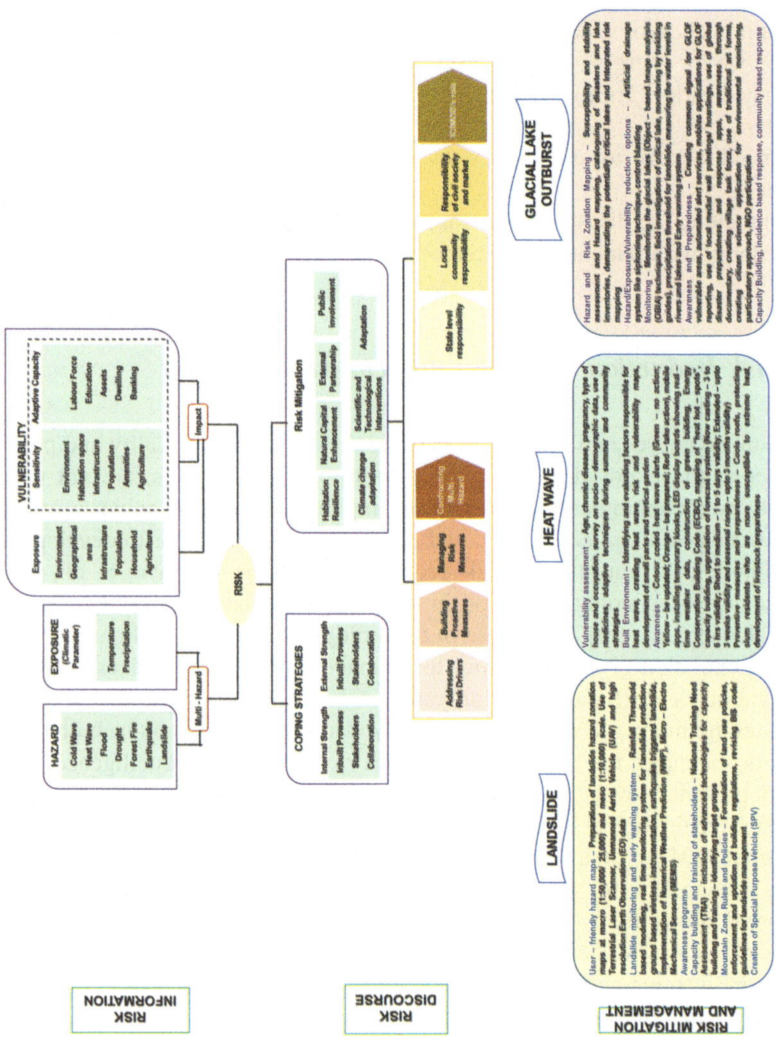

Fig. 16.3 Risk reduction strategies for Himalayan regions

5.8 Policies to Reduce Risk

Since certain disasters have worldwide implications, mitigating the risks associated with such extreme hazards necessitates intergovernmental collaboration. Human health, water supply and sanitation, energy, transportation, industry, mining, construction, commerce and tourism, agriculture, forestry, fisheries, environmental protection, and disaster management are the primary development sectors that are directly impacted by climate change (World Bank, 2006). Government and non-government organizations should work together to safeguard the ecosystem in the mountains and uplands as well as assist local inhabitants in adapting to changing ecological, economic, and health-related consequences. Policies should also seek to persuade important global players, such as the World Trade Organization (WTO), to include mountain concerns in future trade agreements and commercial practices. Governments in the nations have chosen the approach of recognizing the negative effects of climate change and severe events on future development and incorporating this understanding into policy and strategy papers. Adaptation actions should be woven into current development plans (Alam & Regmi, 2004). Different stakeholders, such as government policymakers, implementing agencies, development partners, the commercial sector, and communities, must participate and cooperate. Policies should incorporate adaptation actions into sectoral activities: policies, initiatives, and strategies used to deal with the effects of hazards should be changed to reflect the combined effects of climate change.

6 Conclusion

The Intergovernmental Panel on Climate Change (IPCC) referred to the Himalayan area as a “white spot” in the context of climate change, stressing the paucity of data. However, in the last 5 years or so, data gaps have begun to close. According to recent statistics, the Himalayas are expected to warm faster than the world average, with extreme temperature rise forecast in some areas. Glaciers are expected to continue to lose mass, and exceptional events are anticipated to grow more violent, if not more often. Drought during the premonsoon season may worsen, affecting key processes, such as spring flow, seed germination of important tree species, forest fires, and tourism. Some initiatives that directly benefit local people may result in favorable changes in forest carbon sequestration and biodiversity. Unless they occur in a densely populated region and result in the loss of life and property, all dangers are not catastrophes. Hazards, on the other hand, are only declared catastrophes when they devastate the landscape, kill people, and cause damage to residential areas, industry, and agricultural land. The Uttarakhand catastrophe was exacerbated by the building of villages along streams and large rivers, which were discovered to be the sites of the majority of victims. Many towns and villages are built on debris deposition of floodplains. The pilgrimage to Kedarnath and Srinagar town are two classic examples. The most susceptible regions are the growing number of hotels, guest houses, and towns along the routes leading to the pilgrimage centers. The quality of the landscape and the construction of dwellings may have a significant impact on settlement safety. The building of communities near the riverbank should be discouraged. The necessity of the hour is for a holistic strategy to reduce the number of people killed or injured as a result of natural disasters. People in the community and the government should come out and collaborate. In the context of climate change and mountain specificities, solid science based on credible, salient, and authentic knowledge may frequently lead to appropriate policy, and vice versa. For poverty reduction and long-term development, reducing the hazards of catastrophic calamities is critical. Hydrometeorological data must be shared in a regional trans-boundary, upstream-downstream context in order to build effective early warning systems and catastrophe preparedness.

References

Abrha, H., & Adhana, K. (2019). Desa'a national forest reserve susceptibility to fire under climate change. *Forest Science and Technology, 15*(3), 140–146.

Ahmed, K., Shahid, S., Chung, E. S., Wang, X. J., & Harun, S. B. (2019). Climate change uncertainties in seasonal drought severity-area-frequency curves: Case of arid region of Pakistan. *Journal of Hydrology, 570*, 473–485.

Alam, M., & Regmi, B. R. (2004). *Adverse impacts of climate change on development of Nepal: Integrating adaptation into policies and activities*. Bangladesh Center for Advanced Studies.

Allen, S. K., Linsbauer, A., Randhawa, S. S., Huggel, C., Rana, P., & Kumari, A. (2016). Glacial lake outburst flood risk in Himachal Pradesh, India: An integrative and anticipatory approach considering current and future threats. *Natural Hazards, 84*(3), 1741–1763.

Ambrosi, C., & Scapozza, C. (2015). Improvements in 3-D digital mapping for geomorphological and quaternary geological cartography. *Geographica Helvetica, 70*(2), 121–133.

Aryal, D., Wang, L., Adhikari, T. R., Zhou, J., Li, X., Shrestha, M., Wang, Y., & Chen, D. (2020). A model-based flood hazard mapping on the southern slope of Himalaya. *Water, 12*(2), 540.

Aung, L. T., Martin, S. S., Wang, Y., Wei, S., Thant, M., Htay, K. N., Aung, H. M., Kyaw, T. Z., Min, S., Sithu, K., & Naing, T. (2019). A comprehensive assessment of ground motions from two 2016 intra-slab earthquakes in Myanmar. *Tectonophysics, 765*, 146–160.

Awasthi, R. P., & Owen, J. S. (2020). Observed climate extreme in Nepal. *The Geographic Base, 7*, 1–14.

Azadi, M., Northey, S. A., Ali, S. H., & Edraki, M. (2020). Transparency on greenhouse gas emissions from mining to enable climate change mitigation. *Nature Geoscience, 13*(2), 100–104.

Bali, B. S., Wani, A. A., Bhat, G. R., & Mir, S. A. (2021). GPR investigation of mining induced subsidence and its effects on surface structures: A case study of Srinagar City, J & K, India, NW Himalayas. *Journal of the Geological Society of India, 97*(7), 751–759.

Banerjee, A., & Dimri, A. P. (2019). Comparative analysis of two rainfall retrieval algorithms during extreme rainfall event: A case study on cloudburst, 2010 over Ladakh (Leh), Jammu and Kashmir. *Natural Hazards, 97*(3), 1357–1374.

Barkat, A., Ali, A., Siddique, N., Alam, A., Wasim, M., & Iqbal, T. (2017). Radon as an earthquake precursor in and around northern Pakistan: A case study. *Geochemical Journal, 51*(4), 337–346.

Bhatt, C. M., Rao, G. S., & Jangam, S. (2020). Detection of urban flood inundation using RISAT-1 SAR images: A case study of Srinagar, Jammu and Kashmir (North India) floods of September 2014. *Modeling Earth Systems and Environment, 6*(1), 429–438.

Bilham, R. (2019). Himalayan earthquakes: A review of historical seismicity and early 21st century slip potential. *Geological Society, London, Special Publications, 483*(1), 423–482.

Biswas, S., Mehta, M., & Ghosh, A. (2018, May). Study of aerosol distribution in the vertical air column of atmosphere during Uttarakhand forest fires 2016. In *2018 3rd IEEE International Conference on Recent Trends in Electronics, Information & Communication Technology (RTEICT)* (pp. 98–102). IEEE.

Bradley, B. A., Burks, L. S., & Baker, J. W. (2015). Ground motion selection for simulation-based seismic hazard and structural reliability assessment. *Earthquake Engineering and Structural Dynamics, 44*(13), 2321–2340.

Busayo, E. T., Kalumba, A. M., Afuye, G. A., Ekundayo, O. Y., & Orimoloye, I. R. (2020). Assessment of the Sendai framework for disaster risk reduction studies since 2015. *International Journal of Disaster Risk Reduction, 50*, 101906.

Calvello, M., Peduto, D., & Arena, L. (2017). Combined use of statistical and DInSAR data analyses to define the state of activity of slow-moving landslides. *Landslides, 14*(2), 473–489.

Cardona, A., & Darío, O. (2009). La gestión financiera del riesgo de desastres: Instrumentos financieros de retención y transferencia para la Comunidad Andina. In *La gestión financiera del riesgo de desastres: Instrumentos financieros de retención y transferencia para la Comunidad Andina* (p. 288).

Carrivick, J. L., & Tweed, F. S. (2016). A global assessment of the societal impacts of glacier outburst floods. *Global and Planetary Change, 144*, 1–16.

Chakrabarti, B. K. (2016). Lithotectonic Subdivisions of the Himalaya. In Geology of the Himalayan Belt; Chakrabarti, B.K., Ed.; Elsevier: Amsterdam, The Netherlands, 2016; pp. 1–9.

Chettri, N. A. K. U. L., & Sharma, E. K. L. A. B. Y. A. (2016). Reconciling the mountain biodiversity conservation and human wellbeing: Drivers of biodiversity loss and new approaches in the Hindu-Kush Himalayas. *Proceedings of the Indian National Science Academy, 82*, 53–73.

Chettri, N., Sharma, E., Shakya, B., & Bajracharya, B. (2007). Developing forested conservation corridors in the Kangchenjunga landscape, Eastern Himalaya. *Mountain Research and Development, 27*(3), 211–214.

Chopra, S., Kumar, V., Suthar, A., & Kumar, P. (2012). Modeling of strong ground motions for 1991 Uttarkashi, 1999 Chamoli earthquakes, and a hypothetical great earthquake in Garhwal–Kumaun Himalaya. *Natural Hazards, 64*(2), 1141–1159.

Cochrane, M. A. (2003). Fire science for rainforests. *Nature, 421*(6926), 913–919.

Dahal, P., Shrestha, N. S., Shrestha, M. L., Krakauer, N. Y., Panthi, J., Pradhanang, S. M., Jha, A., & Lakhankar, T. (2016). Drought risk assessment in Central Nepal: Temporal and spatial analysis. *Natural Hazards, 80*(3), 1913–1932.

Dalezios, N. R., Dunkel, Z., & Eslamian, S. (2017). Meteorological drought indices: Definitions. In *Handbook of drought and water scarcity* (pp. 27–44). CRC Press.

Das, B., Paul, A., Bordoloi, R., Tripathi, O. P., & Pandey, P. K. (2018). Soil erosion risk assessment of hilly terrain through integrated approach of RUSLE and geospatial technology: A case study of Tirap District, Arunachal Pradesh. *Modeling Earth Systems and Environment, 4*(1), 373–381.

Das, P. K., Podder, U., Das, R., Kamalakannan, C., Rao, G. S., Bandyopadhyay, S., & Raj, U. (2020). Quantification of heat wave occurrences over the Indian region using long-term (1979–2017) daily gridded (0.5°× 0.5°) temperature data – A combined heat wave index approach. *Theoretical and Applied Climatology, 142*(1), 497–511.

Devrani, R., Singh, V., Mudd, S. M., & Sinclair, H. D. (2015). Prediction of flash flood hazard impact from Himalayan river profiles. *Geophysical Research Letters, 42*(14), 5888–5894.

Dhimal, M., Bhandari, D., Dhimal, M. L., Kafle, N., Pyakurel, P., Mahotra, N., et al. (2021). Impact of climate change on health and well-being of people in Hindu-Kush-Himalayan (HKH) region – A review. *Frontiers in Physiology, 12*, 1139.

Dikshit, A., Satyam, D. N., & Towhata, I. (2018). Early warning system using tilt sensors in Chibo, Kalimpong, Darjeeling Himalayas, India. *Natural Hazards, 94*(2), 727–741.

Dikshit, A., Satyam, N., & Pradhan, B. (2019). Estimation of rainfall-induced landslides using the TRIGRS model. *Earth Systems and Environment, 3*(3), 575–584.

Dikshit, A., Sarkar, R., Pradhan, B., Segoni, S., & Alamri, A. M. (2020). Rainfall induced landslide studies in Indian Himalayan region: A critical review. *Applied Sciences, 10*(7), 2466.

Dubey, S., & Goyal, M. K. (2020). Glacial lake outburst flood hazard, downstream impact, and risk over the Indian Himalayas. *Water Resources Research, 56*(4), e2019WR026533.

Dubey, C. S., Shukla, D. P., Ningreichon, A. S., & Usham, A. L. (2013). Orographic control of the Kedarnath disaster. *Current Science, 105*(11), 1474–1476.

Dulal, K. N., Takeuchi, K., & Ishidaira, H. (2007). Evaluation of the influence of uncertainty in rainfall and discharge data on hydrological modeling. *Proceedings of Hydraulic Engineering, 51*, 31–36.

ECHO 2021. European Commission Mobilises the Civil Protection Mechanism for Victims of the Earthquake and Tsunami in South Asia, Press, Brussels, OCTOBER 2021. ECHO, Oct (2021)- https://reliefweb.int/node/3784525

Emmer, A., & Cochachin, A. (2013). The causes and mechanisms of moraine-dammed lake failures in the Cordillera Blanca, North American Cordillera, and Himalayas. *AUC Geographica, 48*(2), 5–15.

Falátková, K. (2016). Temporal analysis of GLOFs in high-mountain regions of Asia and assessment of their causes. *AUC Geographica, 51*(2), 145–154.

Fearnside, P. M. (2019). Greenhouse gas emissions from land-use change in Brazil's Amazon region. In *Global climate change and tropical ecosystems* (pp. 231–249, 438). CRC Press.

Folke, C. (2006). Resilience: The emergence of a perspective for social–ecological systems analyses. *Global Environmental Change, 16*(3), 253–267.

Forsythe, N., Fowler, H. J., Li, X. F., Blenkinsop, S., & Pritchard, D. (2017). Karakoram temperature and glacial melt driven by regional atmospheric circulation variability. *Nature Climate Change, 7*(9), 664–670.

Freeman, S. A. (1975). Evaluations of existing buildings for seismic risk – A case study of Puget Sound Naval Shipyard. In *Proceedings of 1st U.S. National Conference on Earthquake Engineering, Bremerton, Washington, 1975* (pp. 113–122).

Gahalaut, V. K., & Kundu, B. (2016). The 4 January 2016 Manipur earthquake in the Indo-Burmese wedge, an intra-slab event. *Geomatics, Natural Hazards and Risk, 7*(5), 1506–1512.

Geographical Survey of India (GSI). (2020). *State resource map series: Landslides 2020, India* (p. 1). Published by Geological Survey of India-Explanatory Note.

Ghosh, T., Bhowmik, S., Jaiswal, P., Ghosh, S., & Kumar, D. (2020). Generating substantially complete landslide inventory using multiple data sources: A case study in Northwest Himalayas, India. *Journal of the Geological Society of India, 95*(1), 45–58.

Gill, J. C., & Malamud, B. D. (2016). Hazard interactions and interaction networks (cascades) within multi-hazard methodologies. *Earth System Dynamics, 7*(3), 659–679.

Gruber, S., & Haeberli, W. (2007). Permafrost in steep bedrock slopes and its temperature-related destabilization following climate change. *Journal of Geophysical Research - Earth Surface, 112*(F2). https://doi.org/10.1029/2006JF000547

Gruber, S., Fleiner, R., Guegan, E., Panday, P., Schmid, M. O., Stumm, D., Wester, P., Zhang, Y., & Zhao, L. (2017). Inferring permafrost and permafrost thaw in the mountains of the Hindu Kush Himalaya region. *The Cryosphere, 11*(1), 81–99.

Gupta, A. T. (2018). A software system proposing the processing of crowdsourced data to monitor a flood event: An AI approach. *Open Water Journal, 5*(2), 2.

Gupta, V., Paul, A., Kumar, S., & Dash, B. (2021). Spatial distribution of landslides vis-à-vis epicentral distribution of earthquakes in the vicinity of the Main Central Thrust zone, Uttarakhand Himalaya, India. *Current Science, 120*(12), 1927–1932.

Gurung, D. R., Maharjan, S. B., Shrestha, A. B., Shrestha, M. S., Bajracharya, S. R., & Murthy, M. S. R. (2017a). Climate and topographic controls on snow cover dynamics in the Hindu Kush Himalaya. *International Journal of Climatology, 37*(10), 3873–3882.

Gurung, D. R., Khanal, N. R., Bajracharya, S. R., Tsering, K., Joshi, S., Tshering, P., Chhetri, L. K., Lotay, Y., & Penjor, T. (2017b). Lemthang Tsho glacial Lake outburst flood (GLOF) in Bhutan: Cause and impact. *Geoenvironmental Disasters, 4*(1), 1–13.

Halder, L., Dutta, S. C., Sharma, R. P., & Bhattacharya, S. (2021). Lessons learnt from post-earthquake damage study of Northeast India and Nepal during last ten years: 2021 Assam earthquake, 2020 Mizoram earthquake, 2017 Ambasa earthquake, 2016 Manipur earthquake, 2015 Nepal earthquake, and 2011 Sikkim earthquake. *Soil Dynamics and Earthquake Engineering, 151*, 106990.

Han, S. K., & Han, H. S. (2020). Productivity and cost of whole-tree and tree-length harvesting in fuel reduction thinning treatments using cable yarding systems. *Forest Science and Technology, 16*(1), 41–48.

Hanson, C., & Ranganathan, J. (2017). *Insider: Why burning trees harms the environment* (p. 6). World Resources Institute.

Haritashya, U. K., Singh, P., Kumar, N., & Singh, Y. (2006). Hydrological importance of an unusual hazard in a mountainous basin: Flood and landslide. *Hydrological Processes: An International Journal, 20*(14), 3147–3154.

Harrison, S., Kargel, J. S., Huggel, C., Reynolds, J., Shugar, D. H., Betts, R. A., et al. (2018). Climate change and the global pattern of moraine-dammed glacial lake outburst floods. *The Cryosphere, 12*, 1195. https://doi.org/10.5194/tc-12-1195-2018

Hart, R., Salick, J., Ranjitkar, S., & Xu, J. (2014). Herbarium specimens show contrasting phenological responses to Himalayan climate. *Proceedings of the National Academy of Sciences, 111*(29), 10615–10619.

Hatai, L. D., & Sen, C. (2008). An economic analysis of agricultural sustainability in Orissa. *Agricultural Economics Research Review, 21*(347–2016-16706), 273–282.

Hock, R., Rasul, G., Adler, C., Caceres, S., Gruber, S., Hirabayashi, Y., Jackson, M., Kääb, A., Kang, S., Kutuzov, S., & Milner, A. (2019). Chapter 2: High mountain areas – Special report on the ocean and cryosphere in a changing climate. In *IPCC special report on the ocean and cryosphere in a changing climate*.

Hong, H., Pradhan, B., Xu, C., & Bui, D. T. (2015). Spatial prediction of landslide hazard at the Yihuang area (China) using two-class kernel logistic regression, alternating decision tree and support vector machines. *Catena, 133*, 266–281.

Hua, O. (2009). The Himalayas-water storage under threat. *ICIMOD, Sustainable Mountain Development, 56*, 3–5.

Huang, G., Wang, J., Chen, C. H. E. N., Guo, C., & Zhu, B. (2017). System resilience enhancement: Smart grid and beyond. *Frontiers of Engineering Management, 4*(3), 271–282.

Huss, M., & Hock, R. (2018). Global-scale hydrological response to future glacier mass loss. *Nature Climate Change, 8*(2), 135–140.

Hussain, A., Cao, J., Hussain, I., Begum, S., Akhtar, M., Wu, X., Guan, Y., & Zhou, J. (2021). Observed trends and variability of temperature and precipitation and their global teleconnections in the Upper Indus Basin, Hindukush-Karakoram-Himalaya. *Atmosphere, 12*(8), 973.

IPCC. (2021). Summary for policymakers. In *Climate change 2021: The physical science basis. Impacts, adaptation, and vulnerability. Part A: Global and sectoral aspects. Contribution of Working Group II to the Fifth Assessment Report of the Intergovernmental Panel on Climate Change* (pp. 485–533). Cambridge University Press.

Islam, A. S., Jameel, M., Rahman, M. A., & Jumaat, M. Z. (2011). Earthquake time history for Dhaka, Bangladesh as competent seismic record. *International Journal of Physical Sciences, 6*(16), 3923–3928.

Kamp, U., Growley, B. J., Khattak, G. A., & Owen, L. A. (2008). GIS-based landslide susceptibility mapping for the 2005 Kashmir earthquake region. *Geomorphology, 101*(4), 631–642.

Kanga, S., Kumar, S., & Singh, S. K. (2017). Climate induced variation in forest fire using Remote Sensing and GIS in Bilaspur District of Himachal Pradesh. *International Journal of Engineering and Computer Science, 6*(6), 21695–21702.

Kanna, N., Gupta, S., & Prakasam, K. S. (2018). Micro-seismicity and seismotectonic study in Western Himalaya–Ladakh-Karakoram using local broadband seismic data. *Tectonophysics, 726*, 100–109.

Karmokar, S., & De, M. (2020). Flash flood risk assessment for drainage basins in the Himalayan foreland of Jalpaiguri and Darjeeling Districts, West Bengal. *Modeling Earth Systems and Environment, 6*(4), 2263–2289.

Kaur, P., Joshi, J. C., & Aggarwal, P. (2021). A multi-model decision support system (MM-DSS) for avalanche hazard prediction over North-West Himalaya. *Natural Hazards*, 1–23. https://doi.org/10.1007/s11069-021-04958-5

Keshavarz, M., Karami, E., & Vanclay, F. (2013). The social experience of drought in rural Iran. *Land Use Policy, 30*(1), 120–129.

Khadka, N., Zhang, G., & Thakuri, S. (2018). Glacial lakes in the Nepal Himalaya: Inventory and decadal dynamics (1977–2017). *Remote Sensing, 10*(12), 1913.

Khan, S., & Qureshi, S. F. (2019). Climate change and hazards risk management, community capability, resilience and vulnerability in Swat, Shangla, and Kohistan District, Northwest Pakistan. *International Journal of Economic and Environmental Geology, 10*(4), 69–77.

Khan, I., Waqas, T., & Ullah, S. (2020). Precipitation variability and its trend detection for monitoring of drought hazard in northern mountainous region of Pakistan. *Arabian Journal of Geosciences, 13*(15), 1–18.

Khanduri, S. (2020). Cloudbursts over Indian sub-continent of Uttarakhand Himalaya: A traditional habitation input from Bansoli, District-Chamoli, India. *International Journal of Earth Sciences Knowledge and Applications, 2*(2), 48–63.

Khanduri, S., Sajwan, K. S., Rawat, A., Dhyani, C., & Kapoor, S. (2018). Disaster in Rudraprayag District of Uttarakhand Himalaya: A special emphasis on geomorphic changes and slope instability. *Journal of Geography and Natural Disasters, 8*(218), 2167–0587.

Khatiwada, K. R., & Pandey, V. P. (2019). Characterization of hydro-meteorological drought in Nepal Himalaya: A case of Karnali River Basin. *Weather and Climate Extremes, 26*, 100239.

Kirschbaum, D., Watson, C. S., Rounce, D. R., Shugar, D. H., Kargel, J. S., Haritashya, U. K., Amatya, P., Shean, D., Anderson, E. R., & Jo, M. (2019). The state of remote sensing capabilities of cascading hazards over High Mountain Asia. *Frontiers in Earth Science, 7*, 197.

Komori, D., Nakamura, S., Kiguchi, M., Nishijima, A., Yamazaki, D., Suzuki, S., et al. (2012). Characteristics of the 2011 Chao Phraya River flood in Central Thailand. *Hydrological Research Letters, 6*, 41–46.

Kotru, R. K., Shakya, B., Joshi, S., Gurung, J., Ali, G., Amatya, S., & Pant, B. (2020). Biodiversity conservation and management in the Hindu Kush Himalayan Region: Are transboundary landscapes a promising solution? *Mountain Research and Development, 40*(2), A15.

Krishnan, R., Sabin, T. P., Madhura, R. K., Vellore, R. K., Mujumdar, M., Sanjay, J., Nayak, S., & Rajeevan, M. (2019). Non-monsoonal precipitation response over the Western Himalayas to climate change. *Climate Dynamics, 52*(7), 4091–4109.

Krishnan, R., Sanjay, J., Gnanaseelan, C., Mujumdar, M., Kulkarni, A., & Chakraborty, S. (2020). *Assessment of climate change over the Indian region: A report of the Ministry of Earth Sciences (MOES), Government of India* (p. 226). Springer.

Kumar, S., & Gupta, V. (2021). Evaluation of spatial probability of landslides using bivariate and multivariate approaches in the Goriganga valley, Kumaun Himalaya, India. *Natural Hazards*, 1–28.

Kumar, M., Rana, S., Pant, P. D., & Patel, R. C. (2017). Slope stability analysis of Balia Nala landslide, Kumaun Lesser Himalaya, Nainital, Uttarakhand, India. *Journal of Rock Mechanics and Geotechnical Engineering, 9*(1), 150–158.

Kumar, A., Gupta, A. K., Bhambri, R., Verma, A., Tiwari, S. K., & Asthana, A. K. L. (2018). Assessment and review of hydrometeorological aspects for cloudburst and flash flood events in the third pole region (Indian Himalaya). *Polar Science, 18*, 5–20.

Kumar, A., Negi, H. S., Kumar, K., Shekhar, C., & Kanda, N. (2019). Quantifying mass balance of East-Karakoram glaciers using geodetic technique. *Polar Science, 19*, 24–39.

Kumar, S., Gupta, V., Kumar, P., & Sundriyal, Y. P. (2021). Coseismic landslide hazard assessment for the future scenario earthquakes in the Kumaun Himalaya, India. *Bulletin of Engineering Geology and the Environment*, 1–17.

Larsen, S. Ø., Salberg, A. B., & Solberg, R. (2013). Automatic avalanche mapping using texture classification of Abrha, H. and Adhana, K., 2019. Desa'a national forest reserve susceptibility to fire under climate change. *Forest Science and Technology, 15*(3), 140–146.

Latestly. (2021). *Forest fire in Uttarakhand leaves 71 hectares of land destroyed; PIB says wildfire incidents dropped this year, shares graph*. https://www.latestly.com/india/news/forest-fire-in-uttarakhand-leaves-71-hectares-of-land-destroyed-pib-says-wildfire-incidents-dropped-this-year-shares-graph-1780127.html

Linsbauer, A., Frey, H., Haeberli, W., Machguth, H., Azam, M. F., & Allen, S. (2016). Modelling glacier-bed overdeepenings and possible future lakes for the glaciers in the Himalaya – Karakoram region. *Annals of Glaciology, 57*(71), 119–130.

Liu, X., & Chen, B. (2000). Climatic warming in the Tibetan Plateau during recent decades. *International Journal of Climatology: A Journal of the Royal Meteorological Society, 20*(14), 1729–1742.

Liu, J., & Rasul, G. (2007). Climate change, the Himalayan mountains, and ICIMOD. *Sustainable Mountain Development, 53*, 11–14.

Mahmood, I., Qureshi, S. N., Tariq, S., Atique, L., & Iqbal, M. F. (2015). Analysis of landslides triggered by October 2005, Kashmir Earthquake. *PLoS Currents, 7*. https://doi.org/10.1371/currents.dis.0bc3ebc5b8adf5c7fe9fd3d702d44a99

Maikhuri, R. K., Nautiyal, A., Jha, N. K., Rawat, L. S., Maletha, A., Phondani, P. C., Bahuguna, Y. M., & Bhatt, G. C. (2017). Socio-ecological vulnerability: Assessment and coping strategy to environmental disaster in Kedarnath valley, Uttarakhand, Indian Himalayan Region. *International Journal of Disaster Risk Reduction, 25*, 111–124.

Mal, S., Allen, S. K., Frey, H., Huggel, C., & Dimri, A. P. (2021). Sectorwise assessment of glacial lake outburst flood danger in the Indian Himalayan Region. *Mountain Research and Development, 41*(1), R1.

Mandal, S., & Maiti, R. (2015). *Semi-quantitative approaches for landslide assessment and prediction* (pp. 57–93). Springer.

Martinec, J. (1975). Snowmelt-runoff model for stream flow forecasts. *Hydrology Research, 6*(3), 145–154.

Maskey, S., Uhlenbrook, S., & Ojha, S. (2011). An analysis of snow cover changes in the Himalayan region using MODIS snow products and in-situ temperature data. *Climatic Change, 108*(1), 391–400.

Maurer, J. M., Schaefer, J. M., Rupper, S., & Corley, A. (2019). Acceleration of ice loss across the Himalayas over the past 40 years. *Science Advances, 5*(6), eaav7266.

McClung, D. M. (2016). Avalanche character and fatalities in the high mountains of Asia. *Annals of Glaciology, 57*(71), 114–118.

Meena, S. R., & Mishra, B. K. (2018, December). Landslide risk assessment of Kullu valley using frequency ratio methods and its controlling mechanism, Himachal Himalayas, India. In *Proceedings of the INQUIMUS 2018 workshop: Methods and tools to assess multi-hazard risk, vulnerability and resilience, Venice, Italy* (pp. 3–5).

Meraj, G., Romshoo, S. A., Yousuf, A. R., Altaf, S., & Altaf, F. (2015). Assessing the influence of watershed characteristics on the flood vulnerability of Jhelum basin in Kashmir Himalaya. *Natural Hazards, 77*(1), 153–175.

Mergili, M., Sattar, A., Carrivick, J., Emmer, A., Mandli, K. T., Jacquemart, M., Watson, S., Westoby, M., Clague, J. J., Haritashya, U. K., & Shugar, D. H. (2021, April). Exploring the flow evolution of the Chamoli event (Uttarakhand, India) of 7 February 2021: From geomorphological mapping to multi-model computer simulations. In *EGU General Assembly Conference Abstracts* (pp. EGU21–16591).

Mir, A. Y., Yaqoob, U., Hassan, M., Bashir, F., Zanit, S. B., Haq, S. M., & Bussmann, R. W. (2021). Ethnopharmacology and phenology of high-altitude medicinal plants in Kashmir, Northern Himalaya. *Ethnobotany Research and Applications, 22*, 1–15.

Mirza, M. (2007). *Climate change, adaptation and adaptative governance in the water sector in South Asia*. Adaptation and Impacts Research Division (AIRD), Department of Physical and Environmental Sciences, University of Toronto.

Mittal, H., Wu, Y. M., Lin, T. L., Legendre, C. P., Gupta, S., & Yang, B. M. (2019). Time-dependent shake map for Uttarakhand Himalayas, India, using recorded earthquakes. *Acta Geophysica, 67*(3), 753–763.

Mondal, S., & Mandal, S. (2018). RS & GIS-based landslide susceptibility mapping of the Balason River basin, Darjeeling Himalaya, using logistic regression (LR) model. *Georisk: Assessment and Management of Risk for Engineered Systems and Geohazards, 12*(1), 29–44.

Mondal, S., & Mandal, S. (2020). Data-driven evidential belief function (EBF) model in exploring landslide susceptibility zones for the Darjeeling Himalaya, India. *Geocarto International, 35*(8), 818–856.

Murgese, D., Marchisio, D., Notaro, P., & Vigna, R. (2015). Broad gauge rail link between Sivok in West Bengal and Rangpo in the state of Sikkim with future connectivity to Gangtok (Sikkim) – Shallow-landslide susceptibility map. In *Engineering geology for society and territory – Volume 2* (pp. 979–983). Springer.

Murthy, K. K., Sinha, S. K., Kaul, R., & Vaidyanathan, S. (2019). A fine-scale state-space model to understand drivers of forest fires in the Himalayan foothills. *Forest Ecology and Management, 432*, 902–911.

Myers, N., Mittermeier, R. A., Mittermeier, C. G., da Fonseca, G. A. B., & Kent, J. (2000). Biodiversity hotspots for conservation priorities. *Nature, 403*, 853–858.

Narama, C., Daiyrov, M., Duishonakunov, M., Tadono, T., Sato, H., Kääb, A., Ukita, J., & Abdrakhmatov, K. (2018). Large drainages from short-lived glacial lakes in the Teskey Range, Tien Shan Mountains, Central Asia. *Natural Hazards and Earth System Sciences, 18*(4), 983–995.

Naveena, N., Satyanarayana, G. C., Rao, K. K., Umakanth, N., & Srinivas, D. (2021). Heat wave characteristics over India during ENSO events. *Journal of Earth System Science, 130*(3), 1–16.

Nepal, S., Pradhananga, S., Shrestha, N. K., Kralisch, S., Shrestha, J. P., & Fink, M. (2021). Space–time variability in soil moisture droughts in the Himalayan region. *Hydrology and Earth System Sciences, 25*(4), 1761–1783.

Nhu, V. H., Shirzadi, A., Shahabi, H., Singh, S. K., Al-Ansari, N., Clague, J. J., et al. (2020). Shallow landslide susceptibility mapping: A comparison between logistic model tree, logistic regression, naïve bayes tree, artificial neural network, and support vector machine algorithms. *International Journal of Environmental Research and Public Health, 17*(8), 2749.

Nibanupudi, H. K., & Shaw, R. (Eds.). (2015). *Mountain hazards and disaster risk reduction*. Springer Japan.

Nie, Y., Liu, Q., Wang, J., Zhang, Y., Sheng, Y., & Liu, S. (2018). An inventory of historical glacial lake outburst floods in the Himalayas based on remote sensing observations and geomorphological analysis. *Geomorphology, 308*, 91–106.

Nikravesh, G., Aghababaei, M., Nazari-Sharabian, M., & Karakouzian, M. (2020). Drought frequency analysis based on the development of a two-variate standardized index (rainfall-runoff). *Water, 12*(9), 2599.

Noetzli, J., Christiansen, H. H., Isaksen, K., Smith, S., Zhao, L., & Streletskiy, D. A. (2020). Permafrost thermal state. *Bulletin of the American Meteorological Society, 101*, 34–36.

Nokendangba Chang, C., Ezung, M., Apon, M., Walling, T., & Thong, G. T. (2021). Assessment of landslides along NH 29 in the Kevüza Area, Kohima, Nagaland. *Indian Geotechnical Journal, 51*(4), 841–860.

Onagh, M., Kumra, V. K., & Rai, P. K. (2012). Landslide susceptibility mapping in a part of Uttarkashi district (India) by multiple linear regression method. *International Journal of Geology, Earth and Environmental Sciences, 2*(2), 102–120.

Pachauri, R. K., Allen, M. R., Barros, V. R., Broome, J., Cramer, W., Christ, R., et al. (2014). *Climate change 2014: Synthesis report. Contribution of Working Groups I, II and III to the fifth assessment report of the Intergovernmental Panel on Climate Change* (p. 151). IPCC.

Paik, S. (2018). Impact of natural disasters on schools in India: Current scenario in terms of policies and practices. In *Disaster risk governance in India and cross cutting issues* (pp. 295–312). Springer.

Panday, S., & Dong, J. J. (2021). Topographical features of rainfall-triggered landslides in Mon State, Myanmar, August 2019: Spatial distribution heterogeneity and uncommon large relative heights. *Landslides, 18*(12), 1–15.

Pandey, V. K., Mishra, A., & Mishra, S. S. (2015). Climate change and mitigation measures for the hydrometerological disaster in Himachal Pradesh, India-in light of dams. *International Journal of Scientific & Technology Research, 4*(01), 267.

Pandey, P., Chauhan, P., Bhatt, C. M., Thakur, P. K., Kannaujia, S., Dhote, P. R., Roy, A., Kumar, S., Chopra, S., Bhardwaj, A., & Aggrawal, S. P. (2021). Cause and process mechanism of rockslide triggered flood event in Rishiganga and Dhauliganga River Valleys, Chamoli, Uttarakhand, India using satellite remote sensing and in situ observations. *Journal of the Indian Society of Remote Sensing, 49*(5), 1011–1024.

Pandit, M. K., Manish, K., & Koh, L. P. (2014). Dancing on the roof of the world: Ecological transformation of the Himalayan landscape. *Bioscience, 64*(11), 980–992.

Pangali Sharma, T. P., Zhang, J., Khanal, N. R., Prodhan, F. A., Paudel, B., Shi, L., & Nepal, N. (2020). Assimilation of snowmelt runoff model (SRM) using satellite remote sensing data in Budhi Gandaki River Basin, Nepal. *Remote Sensing, 12*(12), 1951.

Papadimitriou, F. (2012). Modelling landscape complexity for land use management in Rio de Janeiro, Brazil. *Land Use Policy, 29*(4), 855–861.

Papazachos, B. C. (1989). A time-predictable model for earthquake generation in Greece. *Bulletin of the Seismological Society of America, 79*(1), 77–84.

Parkash, S. (2015). Some socio-economically significant landslides in Uttarakhand Himalaya: events, consequences and lessons learnt. *Mountain hazards and disaster risk reduction*, 211–232.

Parvaze, S., Khan, J. N., Kumar, R., & Allaie, S. P. (2021). Temporal flood forecasting for trans-boundary Jhelum River of Greater Himalayas. *Theoretical and Applied Climatology, 144*(1), 493–506.

Paudel, B., Zhang, Y., Yan, J., Rai, R., Li, L., Wu, X., et al. (2020). Farmers' understanding of climate change in Nepal Himalayas: Important determinants and implications for developing adaptation strategies. *Climatic Change, 158*(3), 485–502.

Pei, S. (1995). *Banking on biodiversity: Report on the regional consultations on biodiversity assessment in the Hindu Kush Himalaya*. ICIMOD.

Polykretis, C., Chalkias, C., & Ferentinou, M. (2019). Adaptive neuro-fuzzy inference system (ANFIS) modeling for landslide susceptibility assessment in a Mediterranean hilly area. *Bulletin of Engineering Geology and the Environment, 78*(2), 1173–1187.

Prasad, A. S., Pandey, B. W., Leimgruber, W., & Kunwar, R. M. (2016). Mountain hazard susceptibility and livelihood security in the upper catchment area of the river Beas, Kullu Valley, Himachal Pradesh, India. *Geoenvironmental Disasters, 3*(1), 1–17.

Prati, C., Ferretti, A., & Perissin, D. (2010). Recent advances on surface ground deformation measurement by means of repeated space-borne SAR observations. *Journal of Geodynamics, 49*(3–4), 161–170.

Raetzo, H., & Loup, B. (2016). *Schutz vor Massenbewegungsgefahren: Vollzugshilfe für das Gefahrenmanagement von Rutschungen, Steinschlag und Hangmuren*. Schweizerische Eidgenossenschaft, Bundesamt für Umwelt BAFU.

Ram, P., Gupta, V., Devi, M., & Vishwakarma, N. (2020). Landslide susceptibility mapping using bivariate statistical method for the hilly township of Mussoorie and its surrounding areas, Uttarakhand Himalaya. *Journal of Earth System Science, 129*(1), 1–18.

Ramanathan, K., Vasudevan, N., & Drisya, R. (2020). Analysis of the 1997 Chandmari Landslide, India, and technology-enabled risk reduction. In *2020 8th International Conference on Reliability, Infocom Technologies and Optimization (Trends and Future Directions) (ICRITO)* (pp. 1015–1019). IEEE.

Ramkrishnan, R., Sreevalsa, K., & Sitharam, T. G. (2021). Development of new ground motion prediction equation for the North and Central Himalayas using recorded strong motion data. *Journal of Earthquake Engineering, 25*(10), 1903–1926.

Ramola, R. C., Prasad, Y., Prasad, G., Kumar, S., & Choubey, V. M. (2008). Soil-gas radon as seismotectonic indicator in Garhwal Himalaya. *Applied Radiation and Isotopes, 66*(10), 1523–1530.

Rautela, P. (2005). Indigenous technical knowledge inputs for effective disaster management in the fragile Himalayan ecosystem. Disaster Prevention and Management: An International Journal, 14(2), 233–241.

Rathore, B. P., Singh, S. K., Jani, P., Bahuguna, I. M., Brahmbhatt, R., Rajawat, A. S., et al. (2018). Monitoring of snow cover variability in Chenab Basin using IRS AWiFS sensor. *Journal of the Indian Society of Remote Sensing, 46*(9), 1497–1506.

Rees, H. G., & Collins, D. N. (2006). Regional differences in response of flow in glacier-fed Himalayan rivers to climatic warming. *Hydrological Processes: An International Journal, 20*(10), 2157–2169.

Ren, Y. Y., Ren, G. Y., Sun, X. B., Shrestha, A. B., You, Q. L., Zhan, Y. J., et al. (2017). Observed changes in surface air temperature and precipitation in the Hindu Kush Himalayan region over the last 100-plus years. *Advances in Climate Change Research, 8*(3), 148–156.

Rounce, D. R., Watson, C. S., & McKinney, D. C. (2017). Identification of hazard and risk for glacial lakes in the Nepal Himalaya using satellite imagery from 2000–2015. *Remote Sensing, 9*(7), 654.

Rounce, D. R., Hock, R., & Shean, D. E. (2020). Glacier mass change in High Mountain Asia through 2100 using the open-source python glacier evolution model (PyGEM). *Frontiers in Earth Science, 7*, 331.

Roy, A. B., & Purohit, R. (2018). The Himalayas: Evolution through collision. In *Indian shield; Precambrian evolution and phanerozoic reconstitution* (pp. 311–327). Elsevier.

Roy, A., Chatterjee, A., Ghosh, A., Das, S. K., Ghosh, S. K., & Raha, S. (2019a). Below-cloud scavenging of size-segregated aerosols and its effect on rainwater acidity and nutrient deposition: A long-term (2009–2018) and real-time observation over eastern Himalaya. *Science of the Total Environment, 674*, 223–233.

Roy, J., Saha, S., Arabameri, A., Blaschke, T., & Bui, D. T. (2019b). A novel ensemble approach for landslide susceptibility mapping (LSM) in Darjeeling and Kalimpong districts, West Bengal, India. *Remote Sensing, 11*(23), 2866.

Roy, S., Bose, A., & Chowdhury, I. R. (2021). Flood risk assessment using geospatial data and multi-criteria decision approach: A study from historically active flood-prone region of Himalayan foothill, India. *Arabian Journal of Geosciences, 14*(11), 1–25.

Rupa, T. (2015). Geomorphological study with special reference to landslide in the adjoining areas of National Highway No. 52-A (Banderdewa to Itanagar), Arunachal Pradesh. *IOSR Journal of Environmental Science, Toxicology and Food Technology (IOSR-JESTFT), 9*(4), 20–26.

Sachdeva, S., & Kumar, B. (2021). Flood susceptibility mapping using extremely randomized trees for Assam 2020 floods. *Ecological Informatics*, 101498. https://doi.org/10.46717/igj.55.2C.10ms-2022-08-23

Sakai, S., Poudel, R., Asari, M., & Kirikawa, T. (2019). Disaster waste management after the 2016 Kumamoto Earthquake: A mini-review of earthquake waste management and the Kumamoto experience. *Waste Management & Research, 37*(3), 247–260.

Sakurai, M., & Murayama, Y. (2019). Information technologies and disaster management – Benefits and issues. *Progress in Disaster Science, 2*, 100012.

Sannigrahi, S., Pilla, F., Basu, B., Basu, A. S., Sarkar, K., Chakraborti, S., Joshi, P. K., Zhang, Q., Wang, Y., Bhatt, S., & Bhatt, A. (2020). Examining the effects of forest fire on terrestrial carbon emission and ecosystem production in India using remote sensing approaches. *Science of the Total Environment, 725*, 138331.

Sarkar, S., Kanungo, D. P., & Sharma, S. (2015). Landslide hazard assessment in the upper Alaknanda valley of Indian Himalayas. *Geomatics, Natural Hazards and Risk, 6*(4), 308–325.

Sastry, R. G., & Mondal, S. K. (2013). Geophysical characterization of the Salna sinking zone, Garhwal Himalaya, India. *Surveys in Geophysics, 34*(1), 89–119.

Sati, V. P. (2020). Increasing events of disasters. In *Himalaya on the threshold of change* (pp. 79–99). Springer.

Sati, V. P., & Litt, D. (2011, September). Climate disasters in the Himalaya: Risk and vulnerability. In *International conference on climate change and natural hazards in mountain areas, Dushanbe Sept* (pp. 19–21).

Sekhri, S., Kumar, P., Fürst, C., & Pandey, R. (2020). Mountain specific multi-hazard risk management framework (MSMRMF): Assessment and mitigation of multi-hazard and climate change risk in the Indian Himalayan Region. *Ecological Indicators, 118*, 106700.

Sharma, S., Kuniyal, J. C., & Sharma, J. C. (2007). Assessment of man-made and natural hazards in the surroundings of hydropower projects under construction in the beas valley of northwestern Himalaya. *Journal of Mountain Science, 4*, 221–236.

Sharma, S., Joshi, V., & Chhetri, R. K. (2014). Forest fire as a potential environmental threat in recent years in Sikkim, Eastern Himalayas, India. *Climate Change and Environmental Sustainability, 2*(1), 55–61.

Sharma, Y., Pasari, S., Ching, K. E., Dikshit, O., Kato, T., Malik, J. N., Chang, C. P., & Yen, J. Y. (2020). Spatial distribution of earthquake potential along the Himalayan arc. *Tectonophysics, 791*, 228556.

Shekhar, M. S., Pattanayak, S., Mohanty, U. C., Paul, S., & Kumar, M. S. (2015). A study on the heavy rainfall event around Kedarnath area (Uttarakhand) on 16 June 2013. *Journal of Earth System Science, 124*(7), 1531–1544.

Shen, T., Wang, G., Replumaz, A., Husson, L., Webb, A. A. G., Bernet, M., Leloup, P. H., Zhang, P., Mahéo, G., & Zhang, K. (2020). Miocene subsidence and surface uplift of southernmost Tibet induced by Indian subduction dynamics. *Geochemistry, Geophysics, Geosystems, 21*(10), e2020GC009078.

Shugar, D. H., Jacquemart, M., Shean, D., Bhushan, S., Upadhyay, K., Sattar, A., Schwanghart, W., McBride, S., de Vries, M. V. W., Mergili, M., & Emmer, A. (2021). A massive rock and ice avalanche caused the 2021 disaster at Chamoli, Indian Himalaya. *Science, 373*(6552), 300–306.

Siddique, T., & Pradhan, S. P. (2019). Road widening along National Highway-58, Uttarakhand, India. *Current Science, 117*(8), 1267–1269.

Simmons, A. J., Willett, K. M., Jones, P. D., Thorne, P. W., & Dee, D. P. (2010). Low-frequency variations in surface atmospheric humidity, temperature, and precipitation: Inferences from reanalyses and monthly gridded observational data sets. *Journal of Geophysical Research-Atmospheres, 115*(D1). https://doi.org/10.1029/2009jd012442

Singh, A. (2018). Communities and disasters in Nagaland: Landslides and the cost of development. In A. Singh, M. Punia, N. Haran, & T. Singh (Eds.), *Development and disaster management* (p. 409). Palgrave Macmillan. https://doi.org/10.1007/978-981-10-8485-0_28

Singh, A., & Ganju, A. (2002, December). Earthquakes and avalanches in western Himalaya. In *Proceedings of the 12th symposium on earthquake engineering* (Vol. 16, p. 18).

Singh, S. P., Bassignana-Khadka, I., Karky, B. S., & Sharma, E. (2011). *Climate change in the Hindu Kush-Himalayas: The state of current knowledge*. ICIMOD.

Singh, N., Garbyal, Y., & Kumar, K. (2016). Orographic and tectonic control on extreme events with special reference to Uttarakhand disaster of June 2013 in the Mandakini Valley, India. *Journal of Earth Science and Climatic Change, 7*(4), 1000346.

Singh, D. K., Mishra, V. D., & Gusain, H. S. (2020a). Simulation and analysis of a snow avalanche accident in Lower Western Himalaya, India. *Journal of the Indian Society of Remote Sensing, 48*(11), 1555–1565.

Singh, K. K., Singh, D. K., Thakur, N. K., Dewali, S. K., Negi, H. S., Snehmani, & Mishra, V. D. (2020b). Detection and mapping of snow avalanche debris from Western Himalaya, India using remote sensing satellite images. *Geocarto International*, 1–19.

Singh, A., Juyal, V., Kumar, B., Gusain, H. S., Shekhar, M. S., Singh, P., Kumar, S., & Negi, H. S. (2021a). Avalanche hazard mitigation in East Karakoram mountains. *Natural Hazards, 105*(1), 643–665.

Singh, C., Chauhan, N., Upadhyay, S. K., Singh, R., & Rani, A. (2021b). The Himalayan natural resources: Challenges and conservation for sustainable development. *Journal of Pharmacognosy and Phytochemistry, 10*(1), 1643–1648.

Singh, M. K., Thayyen, R. J., & Jain, S. K. (2021c). Snow cover change assessment in the upper Bhagirathi basin using an enhanced cloud removal algorithm. *Geocarto International, 36*(20), 2279–2302.

Sorg, A., Huss, M., Rohrer, M., & Stoffel, M. (2014). The days of plenty might soon be over in glacierized Central Asian catchments. *Environmental Research Letters, 9*(10), 104018.

Sridharan, A., & Gopalan, S. (2019). Prediction studies of landslides in the Mangan and Singtam areas triggered by 2011 Sikkim Earthquake. In *International conference on advances in computing and data sciences* (pp. 609–617). Springer.

Stäubli, A., Nussbaumer, S. U., Allen, S. K., Huggel, C., Arguello, M., Costa, F., . . ., Zimmermann, M. (2018). Analysis of weather-and climate-related disasters in mountain regions using different disaster databases. In *Climate change, extreme events and disaster risk reduction* (pp. 17–41). Springer.

Sun, C., & Zhou, X. (2020). Characterizing hydrological drought and water scarcity changes in the future: A case study in the Jinghe River Basin of China. *Water, 12*(6), 1605.

Sun, X. B., Ren, G. Y., Shrestha, A. B., Ren, Y. Y., You, Q. L., Zhan, Y. J., et al. (2017). Changes in extreme temperature events over the Hindu Kush Himalaya during 1961–2015. *Advances in Climate Change Research, 8*(3), 157–165.

Syed, M., & Al Amin, M. (2016). Geospatial modeling for investigating spatial pattern and change trend of temperature and rainfall. *Climate, 4*(2), 21.

Thakuri, S., Chauhan, R., & Baskota, P. (2020). *Glacial hazards and avalanches in high mountains of Nepal Himalaya* (p. 87). Nepal Mountain Academy.

Thompson, I., Shrestha, M., Chhetri, N., & Agusdinata, D. B. (2020). An institutional analysis of glacial floods and disaster risk management in the Nepal Himalaya. *International Journal of Disaster Risk Reduction, 47*, 101567.

Tien Bui, D., Shahabi, H., Omidvar, E., Shirzadi, A., Geertsema, M., Clague, J. J., Khosravi, K., Pradhan, B., Pham, B. T., Chapi, K., & Barati, Z. (2019). Shallow landslide prediction using a novel hybrid functional machine learning algorithm. *Remote Sensing, 11*(8), 931.

Tse-ring, K., Sharma, E., Chettri, N., & Shrestha, A. B. (2010). *Climate change vulnerability of mountain ecosystems in the Eastern Himalayas*. International Centre for Integrated Mountain Development (ICIMOD).

Twigg, J. (2004). *Disaster risk reduction: Mitigation and preparedness in development and emergency programming*. Overseas Development Institute (ODI).

Uddin, T., Islam, M. T., & Gosney, J. E. (2021). 2017 Bangladesh landslides: Physical rehabilitation perspective. *Disability and Rehabilitation, 43*(5), 718–725.

UNISDR (United Nations International Strategy for Disaster Reduction). (2017). *UNISDR annual report 2017*. UNISDR.

Valipour, M., & Eslamian, S. (2014). Analysis of potential evapotranspiration using 11 modified temperature-based models. *International Journal of Hydrology Science and Technology, 4*(3), 192–207.

Vaidya, R. A., Shrestha, M. S., Nasab, N., Gurung, D. R., Kozo, N., Pradhan, N. S., & Wasson, R. J. (2019). Disaster risk reduction and building resilience in the Hindu Kush Himalaya. In *The Hindu Kush Himalaya assessment* (pp. 389–419). Springer.

Velayudham, J., Kannaujiya, S., Sarkar, T., Taloor, A. K., Bisht, M. P. S., Chawla, S., & Pal, S. K. (2021). Comprehensive study on evaluation of Kaliasaur Landslide attributes in Garhwal

Himalaya by the execution of geospatial, geotechnical and geophysical methods. *Quaternary Science Advances, 3*, 100025.

Vinoth, M., Prasad, P. S., Mathur, S., & Kumar, K. (2022). Investigation and design of remedial measures for landslide in Hunthar Veng, Mizoram – A case study. In *Stability of slopes and underground excavations* (pp. 79–90). Springer.

Vishwanath, A., & Sharma, R. (2021). *As Himalayan nations deal with glacial floods, cooperation is the key to mitigate disasters*. Available at https://scroll.in/article/1004246/as-himalayan-nations-deal-with-glacial-floods-cooperation-is-the-key-to-mitigate-disasters. Accessed 26 Dec 2021.

Viste, E., & Sorteberg, A. (2015). Snowfall in the Himalayas: An uncertain future from a little-known past. *The Cryosphere, 9*(3), 1147–1167.

Wambulwa, M. C., Milne, R., Wu, Z. Y., Spicer, R. A., Provan, J., Luo, Y. H., et al. (2021). Spatiotemporal maintenance of flora in the Himalaya biodiversity hotspot: Current knowledge and future perspectives. *Ecology and Evolution, 11*(16), 10794–10812.

Wang, S., Che, Y., & Xinggang, M. (2020). Integrated risk assessment of glacier lake outburst flood (GLOF) disaster over the Qinghai–Tibetan Plateau (QTP). *Landslides, 17*(12), 2849–2863.

Wang, S. W., Lim, C. H., & Lee, W. K. (2021). A review of forest fire and policy response for resilient adaptation under changing climate in the Eastern Himalayan region. *Forest Science and Technology*, 1–9.

Watson, C. S., Carrivick, J., & Quincey, D. (2015). An improved method to represent DEM uncertainty in glacial lake outburst flood propagation using stochastic simulations. *Journal of Hydrology, 529*, 1373–1389.

Webb, A. A. G., Guo, H., Clift, P. D., Husson, L., Müller, T., Costantino, D., Yin, A., Xu, Z., Cao, H., & Wang, Q. (2017). The Himalaya in 3D: Slab dynamics controlled mountain building and monsoon intensification. *Lithosphere, 9*(4), 637–651.

Wester, P., Mishra, A., Mukherji, A., & Shrestha, A. B. (2019). *The Hindu Kush Himalaya assessment: Mountains, climate change, sustainability and people* (p. 627). Springer.

Westoby, M. J., Glasser, N. F., Hambrey, M. J., Brasington, J., Reynolds, J. M., & Hassan, M. A. (2014). Reconstructing historic Glacial Lake Outburst Floods through numerical modelling and geomorphological assessment: Extreme events in the Himalaya. *Earth Surface Processes and Landforms, 39*(12), 1675–1692.

Wilhite, D. A., Sivakumar, M. V., & Pulwarty, R. (2014). Managing drought risk in a changing climate: The role of national drought policy. *Weather and Climate Extremes, 3*, 4–13.

World Bank. (2006). *The World Bank annual report 2006*. The World Bank.

Xu, J., Grumbine, R. E., Shrestha, A., Eriksson, M., Yang, X., Wang, Y. U. N., & Wilkes, A. (2009). The melting Himalayas: Cascading effects of climate change on water, biodiversity, and livelihoods. *Conservation Biology, 23*(3), 520–530.

Xu, Y. S., Shen, S. L., Ren, D. J., & Wu, H. N. (2016). Analysis of factors in land subsidence in Shanghai: A view based on a strategic environmental assessment. *Sustainability, 8*(6), 573.

Yadav, R. K., Gahalaut, V. K., & Bansal, A. K. (2021). Tectonic and non-tectonic crustal deformation in Kumaun Garhwal Himalaya. *Quaternary International, 585*, 171–182.

Yao, X., Liu, S., Han, L., Sun, M., & Zhao, L. (2018). Definition and classification system of glacial lake for inventory and hazards study. *Journal of Geographical Sciences, 28*(2), 193–205.

You, C., & Xu, C. (2018). Review of levoglucosan in glacier snow and ice studies: Recent progress and future perspectives. *Science of the Total Environment, 616*, 1533–1539.

You, C., Yao, T., Xu, B., Xu, C., Zhao, H., & Song, L. (2016). Effects of sources, transport, and postdepositional processes on levoglucosan records in southeastern Tibetan glaciers. *Journal of Geophysical Research-Atmospheres, 121*(14), 8701–8711.

Zhao, G., Xu, Z., Pang, B., Tu, T., Xu, L., & Du, L. (2019). An enhanced inundation method for urban flood hazard mapping at the large catchment scale. *Journal of Hydrology, 571*, 873–882.

Zhong, X., Zhang, T., Kang, S., & Wang, J. (2021). Spatiotemporal variability of snow cover timing and duration over the Eurasian continent during 1966–2012. *Science of the Total Environment, 750*, 141670.

Chapter 17
Modelling of Reference Crop Evapotranspiration in Humid-Wet Tropical Region of India

Ujjal Deka Baruah, Anup Saikia, and Nitashree Mili

Abstract Reference crop evapotranspiration is an essential parameter for crop water management and the hydrological cycle. Therefore, selecting methods to predict the reference crop evapotranspiration plays an essential role in agriculture water management and the hydrological cycle. This study attempts to compare different methods for the estimation of reference crop evapotranspiration in the subtropical region of Assam. The Indian Meteorological Department (IMD) gridded temperature, and rainfall data at 1° × 1° spatial resolution was used for the period 1971–2011. Three methods of estimation and comparison of reference crop evapotranspiration were used, which include Thornthwaite's plan (1948), Hargreaves and Samani's method (1985), and Turc's method (1961). Thornthwaite's method was likely to give better results for humid regions for the estimation of reference crop evapotranspiration. The reference crop evapotranspiration, as well as rainfall in eastern Assam, was found to be high in comparison to other districts of Assam.

Keywords Reference crop evapotranspiration · Thornthwaite's method · Hargreaves and Samani's method · Turc's method · Brahmaputra valley · Assam

1 Introduction

Crops exhibit distinctly different responses to rain with high humidity than irrigation with low humidity (Breazeale et al., 1950). Winter crops grown under favourable atmospheric conditions with humidity and temperature show less stress than summer-grown crops. Some crops grow where fog exists, even without irrigation and with less rainfall (Breazeale et al., 1950). Food grains in Assam, India,

U. D. Baruah (✉) · N. Mili
Department of Geography, Cotton University, Guwahati, Assam, India
e-mail: udbmail@cottonuniversity.ac.in

A. Saikia
Department of Geography, Gauhati University, Guwahati, Assam, India

S. Sharma et al. (eds.), *Climate Change Adaptation, Risk Management and Sustainable Practices in the Himalaya*,
https://doi.org/10.1007/978-3-031-24659-3_17

accounted for approximately 66% of the total cultivated area during 2006–2008 and for 2.8–5.1% of the total irrigated area. Any delay in the monsoon or shortfall experienced in the amount of precipitation received aggravates the risk of crop failure or steep reductions in crop yield (Isik & Devadoss, 2006; Lamaoui et al., 2018). The paucity of irrigation adversely affects farmers as well as the contribution to the gross domestic product of the state (Touch et al., 2016). The region is experiencing an increase in temperature at a rate of 0.15 °C per decade and erratic patterns of rainfall (Baruah, 2018; Kothawale et al., 2005).

For developing irrigation systems and estimating crop moisture availability, evapotranspiration is crucial for a sustainable agriculture (Djaman et al., 2015). Evapotranspiration (ET) is an important component in water-balance models and irrigation scheduling and was often estimated in a two-step process (Fisher & Pringle, 2013).

Reference crop evapotranspiration (ET_o) has been defined as the rate of evapotranspiration from a hypothetical reference crop with an assumed crop height of 0.12 m, a fixed surface resistance of 70 s m^{-1}, and an albedo of 0.23, closely resembling the evapotranspiration from an extensive surface of green grass of uniform height, actively growing, completely shading the ground, and not short of water (Allen et al., 1998a). Evapotranspiration rates vary between plant species over different times of the year and during different stages of plant growth. The Food and Agricultural Organisation (FAO) recommended the Penn-Monteith method for computation of reference crop evapotranspiration which requires many variables (Allen et al., 1998b). On the other hand, Thornthwaite and Hargreaves-Samani's method requires only air temperature (Sepaskhah & Razzaghi, 2009). The paucity of weather stations across North-East India (Saikia, 2009) necessitates the use of equations with a minimal number of climate variables. Reference crop evapotranspiration has been calculated by three methods to check the variation in their estimation using Thornthwaite (1948), Turc (1961), and Hargreaves et al. (1985a) methods (Table 17.1).

Although the potential evapotranspiration and reference crop evapotranspiration are two different concepts, the potential evapotranspiration (PET) values approximates the values derived from reference crop evapotranspiration (ET_o), provided the data used in the equations are from same station (Michalopoulou & Papaioannou, 1991). Thus, in the present study, we used Thornthwaite's (1948) potential evapotranspiration equation along with the other two models to estimate reference crop evapotranspiration.

2 Database and Methodology

2.1 *Study Area*

Assam is situated in the north-eastern part of India. According to Human Development Report, India 2011, the Human Development Index (HDI) score for Assam during 2007–2008 was 0.444 (India 0.467), and it ranked 16th out of 23 states in

India. It is categorized as a humid and wet region under the broad classification of humidity province after Thornthwaite (1931) (Baruah, 2018). However, area wise there may be local variation. In the higher Barails due to altitude, the climate tends to be temperate. The climate of Assam can be summed up as 'tropical monsoon type'. Generally, the state enjoys a hot and humid climate during summer, while the winter months are cool and dry. The climate of Assam is slightly different from Cwg (humid mesothermal gangetic type) of Koppen climate classification, because of the development of an orographic low, a spectacular, but complex thermodynamic phenomenon. Perhaps it is more appropriate to designate the climate of Assam as Cwa (monsoon-influenced humid subtropical climate) instead of Cwg.

The layout of the hills and mountains and their height influence the climate of the state. The Himalayas in the north, the Patkai and other Hills in the east, and the Meghalaya Plateau-Barail Range in the central part play a major role in shaping the climate of Assam with its distinctive local characters. These hills and plateaus exert both mechanical and thermodynamic influences on the general and seasonal circulation of winds, distribution of pressure, temperature, and precipitation. The mountain wall of Bhutan and Arunachal Himalayas prevent the invasion of severe cold air masses of Central Asia in winter. On the other hand, it acts as a barrier to the moisture-bearing south-west monsoon winds with the result that the whole area receives a very heavy amount of rainfall in summer.

Of the total geographical region of Assam, only 2.38% (186,806 ha) was irrigated during 2013–2014. Agriculture remains the mainstay of Assam's economy. Of the total work force, 33–39% were farmers, while 13–15% were agricultural labourers during 2001–2011. More than 80% of the rainfall occurs monsoon season (June–September). However, in recent years, even during monsoons and winter season, the state reported drought-like situation as well as floods (Bhattacharjya et al., 2021; Mudi & Das, 2022). The unpredictability of prevailing vivid climatic conditions creates an uncertainity in the mindsets of farmers. However, with proper artificial irrigation system, farmers would be encouraged to perform agricultural activities during adverse climatic condition. For proper management and monitoring of irrigation system, the assessment of reference crop evapotranspiration becomes a prerequisite condition (Fig. 17.1).

2.2 Data Sources

We used daily gridded rainfall and temperature dataset available at $1° \times 1°$ latitude-longitude resolution for the period 1971–2011 developed by the India Meteorological Department (IMD). These datasets were developed using modified Shepard's angular distance weighting algorithm to interpolate the station data into $1° \times 1°$ grids (Rajeevan et al., 2005, 2006; Revadekar et al., 2009; Srivastava et al., 2009). The daily rainfall data were averaged over months and years for each district of Assam.

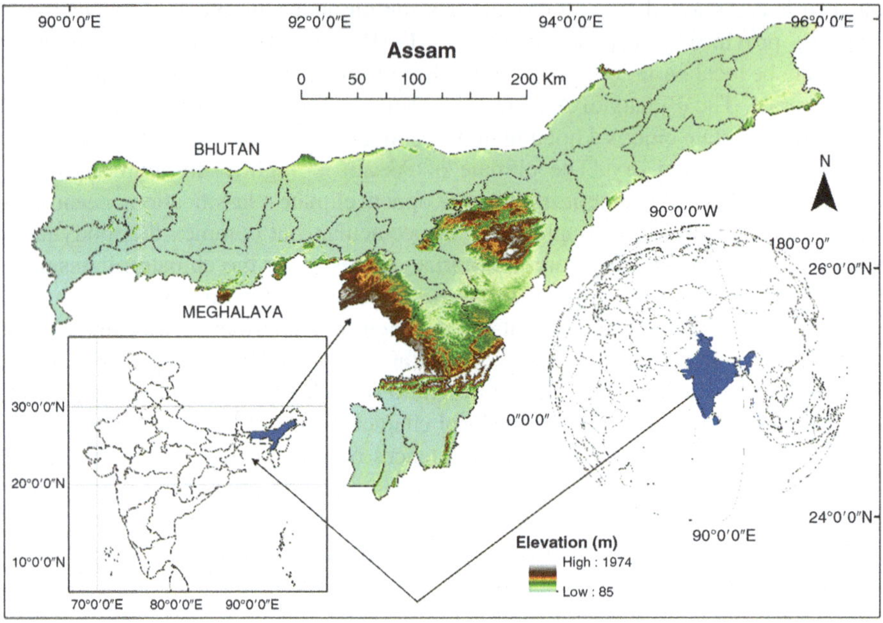

Fig. 17.1 Location of the study area

2.3 *Methodology*

2.3.1 Thornthwaite's Method (1948)

The potential evapotranspiration (Thornthwaite, 1948) in each area and the relationship with the range of temperature involved were expressed by an equation of the form

$$e = ct^a$$

in which '*e*' was monthly evapotranspiration (cm) and '*t*' was mean monthly temperature in °C. The coefficients '*c*' and '*a*' vary from one place to another. The exponent *a* in the above equation varies from 0 to 4.25. Thus, an equation having coefficients derived from observations made in a warm climate does not yield correct values of potential evapotranspiration for an area having a cold climate, and vice versa (Thornthwaite, 1948). Reference Evapotranspiration Was Calculated by the Following Steps

1. The Thornthwaite's heat index was calculated by $i = \left(\frac{t}{5}\right)^{1.514}$, where '*t*' was mean monthly average temperature (°C). The coefficient '*c*' in the equation above varies inversely with '*I*'.

2. Summation of the Thornthwaite's heat index gives the annual heat index (I). This heat index varies from 0 to 160 and is given by

$$I = \sum_{i=1}^{12} i$$

3. The unadjusted PET (UPET) was calculated as

$$\text{UPET} = 1.6 \times \left(\frac{10t}{I}\right)^a$$

4. Where a was calculated by the equation

$$a = 0.000000675 \times I^3 - 0.0000771 \times I^2 + 0.01792 \times I + 0.49239$$

Adjusted PET was given by – $\text{PET} = \text{UPET} \times \frac{N}{12} \times \frac{d}{30}$.

2.3.2 Hargreaves and Samani's Method (1985)

This method is radiation based and uses maximum and minimum temperature for estimating reference crop evapotranspiration (ET_o) (Droogers & Allen, 2002; Hargreaves et al., 1985a, b; Sepaskhah & Razzaghi, 2009).

$$ET_o = 0.0023(0.408)(T_{\text{mean}} + 17.8)(T_{\text{max}} - T_{\text{min}})^{0.5} R_a$$

where T_{mean}, T_{max}, and T_{min} are mean, maximum, and minimum temperature (°C) respectively. R_a was extraterrestrial radiation (M.Jm^{-2}), 0.408 was a factor to convert M.Jm^{-2} d^{-1} to mmd^{-1}, and 0.0023 was an empirical coefficient proposed by Hargreaves and Samani (1985). R_a depends on the Julian day number and latitude and can be computed as described by Allen et al. (1998a) and was calculated by

$$R_a = \frac{24(60)}{\pi} G_{sc} d_r [\omega \ \sin(\varphi) \sin(\delta) + \ \cos(\varphi) \cos(\delta) \sin(\omega_s)]$$

where G_{sc} was solar constant (0.0820 MJ·m^{-2}·min^{-1}); φ was latitude in radians; the term 24(60) was a factor to convert min to day; and [ω sin(φ)sin(δ) + cos(φ)cos(δ)sin (ω_s)].

Based on the calendar day of the year,

$$d_r = 1 + 0.33 \cos\left(\frac{2\pi}{365} J\right)$$

where d_r is the inverse relative distance from earth to sun and J is the calendar day of the year

$$\delta = 0.409 \sin\left(\frac{2\pi}{365} J - 1.39\right)$$

where δ = solar declination (radians), and

$$\omega_s = \arccos(-\tan(\varphi)\tan(\delta))$$

where ω_s = sunset hour angle (radians).

or R_a in daily joules per square meter per day

$$R_a = \frac{24 \times 3600 G_{sc}}{\pi}\left(1 + 0.033 \cos\frac{360n}{365}\right)$$
$$\times \left(\cos\varphi \cos\delta \sin\omega_s + \frac{\pi\omega_s}{180} \sin\varphi \sin\delta\right)$$

where G_{sc} is in (wattm^{-2}).

2.3.3 Turc Method (1961)

Reference crop evapotranspiration (mm/month) was given by $ET_o = 0.40\left(\frac{T_{mean}}{T_{mean}+15}\right)(R_s + 50)$, where R_s = solar radiation (MJ·m^{-2}) and T_{mean} = average air temperature (°C) calculated as $(T_{max} + T_{min})/2$. To estimate ET_o on a daily basis, the factor 0.40 was divided by 30 (average days per month), and the final equation for daily ET_o (mm/day) was given by

$$ET_o = 0.0133\left(\frac{T_{mean}}{T_{mean} + 15}\right)(R_s + 50)$$

where R_s can be estimated by, $R_s = 0.16\,(T_{max} - T_{min})^{0.5} R_a$.

Thornthwaite's reference crop evapotranspiration, rainfall, and water balance for each year in 23 districts (2001 census) of Assam was calculated and averaged during the period 1971–2011.

3 Results and Discussion

Thornthwaite's method yielded the highest reference crop evapotranspiration (ET_o) in comparison to Hargreaves-Samani's method contrary to the previous findings of Lu et al. (2005) and Djaman et al. (2015). Turc's (1961) method underestimated the

Table 17.1 Reference crop evapotranspiration (mmd^{-1}) and rainfall estimated using different methods

	Reference crop evapotranspiration (mmd^{-1})			
	Hargreaves's method	Turc's method	Thornthwaite's method	Rainfall (mmd^{-1})
Jan	1.21	0.38	0.99	0.55
Feb	1.45	0.41	1.36	1.17
Mar	1.81	0.45	2.63	2.27
Apr	1.92	0.46	3.59	6.08
May	2	0.48	4.7	9.65
Jun	1.99	0.49	5.56	14.34
Jul	1.9	0.48	5.75	16.77
Aug	1.93	0.49	5.69	12.76
Sep	1.72	0.48	4.76	10.28
Oct	1.58	0.46	3.78	4.88
Nov	1.4	0.43	2.19	1.25
Dec	1.21	0.39	1.24	0.8

ET_o in case of Assam, similar to the findings of Djaman et al. (2015). Suitability of these methods differs, depending on the situation and location of a region.

The areas which were studied by Djaman et al. (2015) and Lu et al. (2005) are climatologically not similar to Assam; the former are closer to oceans (Southeastern US and West Africa, respectively) and at higher latitudes.

Assam is a monsoon-dominated region; it was inevitable to show large variations in ET_o. Hence, the Thornthwaite's method was likely to give better results, since variation in the ET_o in Hargreaves-Samani method was very less (Table 17.1). Hargreaves-Samani method was best suited for arid and semiarid regions (Tabari, 2010; Tabari & Aghajanloo, 2013) but overestimates in humid areas (Rojas & Sheffield, 2013). Turc's method is the best-suited model in cold humid and arid climates (Tabari, 2010). Turc's method underestimated the reference evapotranspiration in this hot-humid tropical part of the world. Therefore, for the purpose of the actual evapotranspiration estimation and irrigation water need, Thornthwaite's ET_o estimates may be used.

3.1 Estimation of $\mathrm{ET_0}$ *Based on Thornthwaite's Method*

Based on the Thornthwaite's method, potential evapotranspiration was estimated during 1970–2011. It was observed that the highest ET_o occurred during the monsoon season and lowest during the winter season (Fig. 17.2). The effect was plausibly due to heat during the summer/monsoon season, when the temperature levels was generally at a peak. The ET_o during July was 170 mm, while the lowest occurred in January (30 mm), during 1970–2007 in Assam (Fig. 17.2).

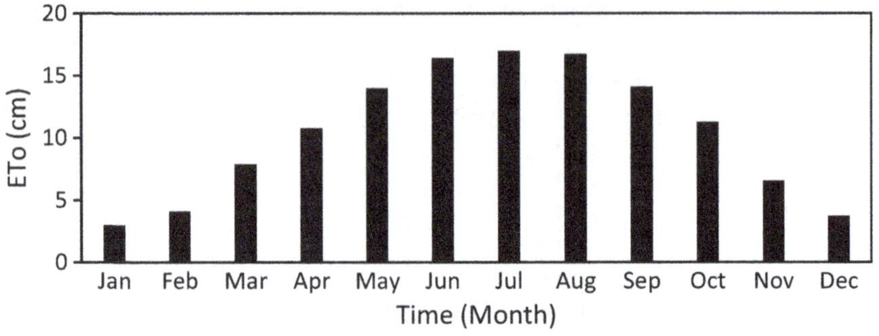

Fig. 17.2 Potential evapotranspiration based on Thornthwaite's method, Assam, 1971–2011

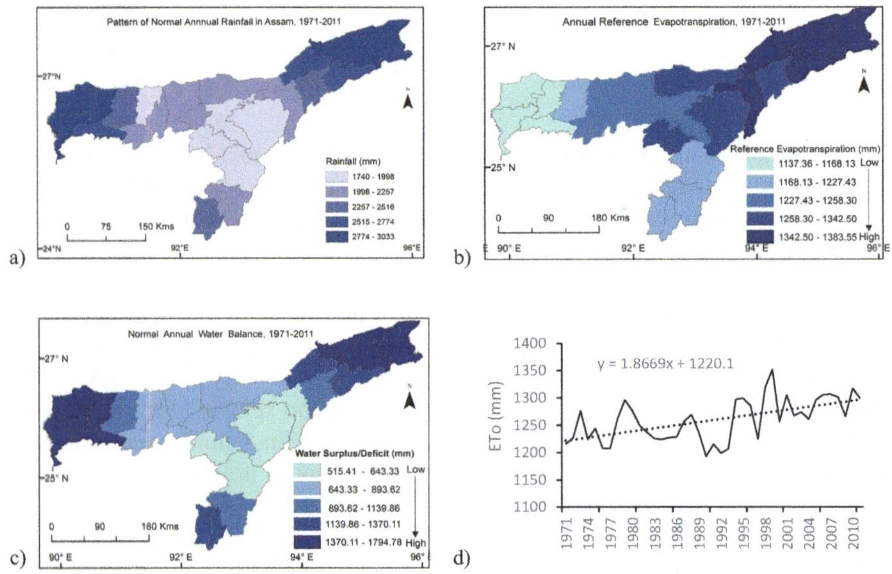

Fig. 17.3 (**a**) Normal annual rainfall, (**b**) annual reference crop evapotranspiration, (**c**) normal annual water balance, and (**d**) trend of reference evapotranspiration in Assam using Thornthwaite's method (1948) during 1971–2011

There was at least 200 mm difference in the annual normal ET_o between districts of Assam during 1971–2011 (Fig. 17.3b). High ET_o concentrations were observed in the eastern parts of Assam and the least concentrations accrued in the western parts of Assam. Although high rainfall in the eastern Assam compensates the loss of water through evapotranspiration in eastern Assam, the amount of water availability in these regions varied according to the ET_o. Assam's crop water availability was found to be naturally compensated by rainfall. Although the water balance amount is not uniform throughout the region, the crop water availability in upper Brahmaputra region (eastern Assam), western (lower Brahmaputra valley (western Assam), and southern (Barak valley) Assam was high in comparison to central Brahmaputra valley (central Assam) and Hill zones (Karbi Anglong and Dima Hasao) (Fig. 17.3c). The ET_o increased at the rate of 1.82 mm/year during 1971–2011, statistically significant at 0.01 level of significance (p value 0.0013) (Fig. 17.3d). The rise in ET_o indicates that sooner or later the moisture requirements for crops would be inadequate for sustainable agriculture primarily in central Assam. The irrigation system needs to be developed in areas of low crop water availability.

4 Conclusion

Three methods were compared for estimating the reference crop evapotranspiration (ET_o). Turc and Hargreaves-Samani methods were found to be less suitable for tropical regions with wet-humid type of climate in contrast to previous study (Tabari, 2010). The estimated values from Thornthwaite's method were best fitted for the study area of Assam, India. However, the accuracy of Thornthwaite's method depends on the type of data used for estimating reference crop evapotranspiration. The data from high density of weather stations would give better results rather than gridded datasets. Shortfall in real-time data availability and reliability along with possible technical errors due to sophisticated techniques may lead to several problems in estimating evapotranspiration (Hargreaves et al., 1985a). Thornthwaite's simple method was tested and found to be suitable for tropical areas, where data is an acute problem.

References

Allen, R. G., Pereira, L. S., Raes, D., & Smith, M. (1998a). *Crop evapotranspiration-guidelines for computing crop water requirements* (FAO Irrigation and Drainage Paper 56) (Vol. 300(9)). FAO.

Allen, R. G., Pereira, L. S., Raes, D., & Smith, M. (1998b). *Crop evapotranspiration: Guidelines for computing crop water requirements* (Irrigation and Drainage Paper) (Vol. 56, p. 300). Food and Agriculture Organization of the United Nations.

Baruah, U. D. (2018). *Crops and farmers' responses to climate change: A case study of Assam.* (PhD), Gauhati University.

Bhattacharjya, B. K., Yadav, A. K., et al. (2021). Effect of extreme climatic events on fish seed production in lower Brahmaputra Valley, Assam, India: Constraint analysis and adaptive strategies. *Aquatic Ecosystem Health & Management, 24*(3), 39–46.

Breazeale, E. L., McGeorge, W. T., & Breazeale, J. F. (1950). Moisture absorption by plants from an atmosphere of high humidity. *Plant Physiology, 25*(3), 413.

Djaman, K., Balde, A. B., et al. (2015). Evaluation of sixteen reference evapotranspiration methods under sahelian conditions in the Senegal River valley. *Journal of Hydrology: Regional Studies, 3*, 139–159. https://doi.org/10.1016/j.ejrh.2015.02.002

Droogers, P., & Allen, R. G. (2002). Estimating reference evapotranspiration under inaccurate data conditions. *Irrigation and Drainage Systems, 16*(1), 33–45.

Fisher, D. K., & Pringle, H. C., III. (2013). Evaluation of alternative methods for estimating reference evapotranspiration. *Agricultural Sciences, 4*(8A), 51–60. https://doi.org/10.4236/as.2013.48A008

Hargreaves, G. H., & Samani, Z. A. (1985a). Reference crop evapotranspiration from temperature. *Applied Engineering Agriculture, 1*(2), 96–99.

Hargreaves, G. L., Hargreaves, G. H., & Riley, J. P. (1985b). Irrigation water requirements for Senegal River basin. *Journal of Irrigation and Drainage Engineering, 111*(3), 265–275.

Isik, M., & Devadoss, S. (2006). An analysis of the impact of climate change on crop yields and yield variability. *Applied Economics, 38*(7), 835–844. https://doi.org/10.1080/00036840500193682

Kothawale, D. R., & Rupa Kumar, K. (2005). On the recent changes in surface temperature trends over India. *Geophys Res Lett, 32*(18), n/a-n/a. https://doi.org/10.1029/2005gl023528

Lamaoui, M., Jemo, M., Datla, R., & Bekkaoui, F. (2018). Heat and drought stresses in crops and approaches for their mitigation. *Frontiers in Chemistry, 6*, 26. https://doi.org/10.3389/fchem.2018.00026

Lu, J., Sun, G., McNulty, S. G., & Amatya, D. M. (2005). A comparison of six potential evapotranspiration methods for regional use in the southeastern United States. *Journal of the American Water Resources Association, 41*(3), 621–633.

Michalopoulou, H., & Papaioannou, G. (1991). Reference crop evapotranspiration over Greece. *Agricultural Water Management, 20*, 209–221.

Mudi, S., & Das, J. P. (2022). Flood hazard mapping in Assam using Sentinel-1 SAR data. In P. K. Shit, H. R. Pourghasemi, G. S. Bhunia, P. Das, & A. Narsimha (Eds.), *Geospatial technology for environmental hazards* (Advances in geographic information science). Springer.

Rajeevan, M., Bhate, J., Kale, J. D., & Lal, B. (2005). *Development of a high resolution daily gridded rainfall data for the Indian region, India* (Meteorological monograph climatology, 22/2005) (p. 26). Meteorological Department.

Rajeevan, M., Bhate, J., Kale, K. D., & Lal, B. (2006). High resolution daily gridded rainfall data for the Indian region: Analysis of break and active monsoon spells. *Current Science (Bangalore), 91*, 296–306.

Revadekar, J. V., Kothawale, D. R., & Rupa Kumar, K. (2009). Role of El Niño/La Niña in temperature extremes over India. *International Journal of Climatology, 29*(14), 2121–2129. https://doi.org/10.1002/joc.1851

Rojas, J. P., & Sheffield, R. E. (2013). Evaluation of daily reference evapotranspiration methods as compared with the ASCE-EWRI Penman-Monteith equation using limited weather data in Northeast Louisiana. *Journal of Irrigation and Drainage Engineering, 139*(4), 285–292. https://doi.org/10.1061/(asce)ir.1943-4774.0000523

Saikia, A. (2009). NDVI variability in North East India. *Scottish Geographical Journal, 125*(2), 195–213. https://doi.org/10.1080/14702540903071113

Sepaskhah, A. R., & Razzaghi, F. (2009). Evaluation of the adjusted Thornthwaite and Hargreaves-Samani methods for estimation of daily evapotranspiration in a semi-arid region of Iran. *Archives of Agronomy and Soil Science, 55*(1), 51–66. https://doi.org/10.1080/03650340802383148

Srivastava, A. K., Rajeevan, M., & Kshirsagar, S. R. (2009). Development of a high resolution daily gridded temperature data set (1969–2005) for the Indian region. *Atmospheric Science Letters*. https://doi.org/10.1002/asl.232

Tabari, H. (2010). Evaluation of reference crop evapotranspiration equations in various climates. *Water Resources Management, 24*(10), 2311–2337. https://doi.org/10.1007/s11269-009-9553-8

Tabari, H., & Aghajanloo, M.-B. (2013). Temporal pattern of aridity index in Iran with considering precipitation and evapotranspiration trends. *International Journal of Climatology, 33*(2), 396–409. https://doi.org/10.1002/joc.3432

Thornthwaite, C. W. (1931). The climates of North America: according to a new classification. *Geographical Review, 21*(4), 633–655.
Thornthwaite, C. W. (1948). An approach toward a rational classification of climate. *Geographical Review, 38*(1), 55–94.
Touch, V., Martin, R. J., Scott, J. F., Cowie, A., & Liu de, L. (2016). Climate change adaptation options in rainfed upland cropping systems in the wet tropics: A case study of smallholder farms in North-West Cambodia. *Journal of Environmental Management, 182*, 238–246. https://doi.org/10.1016/j.jenvman.2016.07.039
Turc, L. (1961). Evaluation de besoins en eau d'irrigation, ET potentielle. *Annals of Agronomy, 12*, 13–49.

Index

A

Adaptation, 52, 55–60, 66–85, 91, 99–101, 104, 165, 188, 208, 246, 387–390, 393

Analytical hierarchical process (AHP), 339, 341, 356, 358

Anthropogenic activities, 55, 85, 209, 216, 240, 356, 384

Assam, 24, 103, 158, 369, 373, 374, 376, 407–409, 412–415

B

Biodiversity, 2, 46, 56, 58, 66, 77, 80, 82, 85, 91, 92, 159, 161–162, 166, 183, 184, 188, 200, 221, 231, 238–240, 243, 244, 246, 247, 255, 259, 268, 354, 362, 367, 382, 393

Brahmaputra valley, 414, 415

C

Climate change, 2–16, 22, 23, 25, 27, 29, 30, 36, 46, 47, 50–60, 66, 67, 69–78, 80–85, 90, 92, 94–101, 103, 131, 158, 164, 165, 184, 188, 203, 238–247, 293, 338, 362, 363, 365–367, 370, 377, 381, 384, 387, 389, 390, 393, 394

Climate finance, 99, 100

Communities, 57, 66, 67, 70, 72, 73, 75–85, 91, 93, 98, 99, 101–103, 110, 159, 161–170, 174, 176, 178, 179, 181–188, 200, 201, 203, 205–209, 214–216, 220–222, 225–229, 231, 233, 239, 240, 254–258, 260, 264, 265, 283, 285, 298, 301–303, 305–308, 319, 322–327, 329, 362, 363, 389, 391, 393, 394

Conservation interventions, 282–284, 288, 291–294, 296, 297, 304, 307, 308, 316

Cultivation, 74, 78, 144, 174–176, 184, 186, 227, 228, 254, 255, 258, 259, 261, 263–265

Cultural diversity, 258

D

Debris covered, 111, 116, 117, 119, 120, 122, 123, 127, 128, 131

Digital elevation model (DEM), 5, 6, 114, 116, 118, 119, 131, 384, 385

Disaster management, 25, 26, 59, 83, 91, 98–99, 155, 201, 207, 209, 210, 393

Drivers of human-wildlife conflicts, 282–285, 287, 291–294, 304–308, 328

E

Ecosystem services, 46, 56, 72, 82, 85, 166, 228, 231, 240, 243, 246, 268, 284, 368

Eddy covariance system, 270, 271, 278

Eryngo, 254–256, 259–261, 263–265

Evidence-based conservation, 283

Extreme events, 12, 16, 40, 56, 69, 90, 96, 103

S. Sharma et al. (eds.), *Climate Change Adaptation, Risk Management and Sustainable Practices in the Himalaya*, https://doi.org/10.1007/978-3-031-24659-3

F
Forest management, 66, 69, 75, 77, 79–81, 83–85, 166, 187

G
Geographic information system (GIS), 6, 137, 154, 155, 339, 341, 351, 356, 358, 384
Geophysical disaster, 363, 364
Geophysical hazards, 90–104
Glacier inventory, 110–131
Glacier mapping, 110, 114, 117
Global warming, 2, 16, 22, 51, 53, 54, 56, 66, 94, 238–240, 366–368

H
Hargreaves and Samani's method, 408, 411–413, 415
Himachal Pradesh, 15, 52, 58, 59, 68, 71–75, 80–82, 92, 103, 111, 130, 158, 214, 215, 222, 231, 245, 290, 293, 311–315, 319, 321, 323, 339, 340, 358, 366, 371, 372, 375, 377, 382
Himalayan region, 22, 24, 25, 30, 39, 46, 47, 51, 52, 55, 57–59, 66, 69, 82, 92, 95, 96, 100, 103, 110, 111, 113, 118, 119, 127, 158, 159, 161, 179, 184, 188, 213, 214, 231, 239–242, 246, 254, 273, 282, 305, 338, 363, 365–371, 373–380, 387, 389–392
Himalayas, 12, 15, 22–26, 36, 40, 46, 47, 50, 51, 53, 54, 56, 58, 59, 67, 68, 90–99, 111, 112, 120, 124, 127, 137, 158, 162, 164–167, 169, 170, 175, 176, 183, 200, 203, 206, 239, 240, 242–244, 246, 247, 268, 269, 275, 279, 284, 304, 305, 309, 318–320, 322, 323, 325, 338–358, 362–366, 368, 370, 377, 378, 381, 384–388, 393, 409
Human-wildlife interactions (HWI), 282, 285

I
Indigenous, 55, 66, 69, 163, 214, 221, 227, 228, 239, 254, 256, 258, 259, 264, 391
Indo-Burma hotspot, 253–265
Invasive species, 76, 165, 243

K
Knowledge gaps, 57, 184, 306

L
Lahaul & Spiti, 213–233, 245
Landslide hazard zonation (LHZ), 137, 138, 147
Livelihoods, 46, 47, 51, 57, 67, 72, 73, 75, 78, 80–83, 96, 103, 158–188, 193, 195, 203, 206, 213–233, 239, 243, 246, 264, 265, 268, 293, 294, 296, 299, 304, 305, 323, 326, 387, 389

M
Major rivers, 46–60, 92, 200, 367
Management, 52, 54–59, 72, 77–81, 83–85, 91, 99, 104, 161, 166–168, 185, 186, 188, 201, 202, 209, 210, 214, 216, 228, 229, 231–233, 242, 246, 247, 282, 283, 294–304, 306, 307, 317, 318, 324, 325, 338, 339, 363, 384, 387, 388, 390, 409
Management strategies, 55–57, 209, 228, 323
Manipur, 24, 58, 158, 214, 254–260, 263, 264, 369, 371

N
Natives, 76, 84, 227, 231, 233, 254, 367
Natural disasters, 46, 50, 59, 83, 91, 97–99, 102, 103, 202–203, 205–209, 394
Net ecosystem exchange (NEE), 269, 271, 274–278

P
Paraglacial zone, 2–4, 10, 11, 16
Phenological changes, 159, 367
Precipitation, 4, 5, 8, 12, 15, 22–27, 29, 31, 33–40, 46–54, 56, 59, 90, 93, 94, 97, 103, 110–112, 121, 127, 164, 238, 240, 242, 243, 276, 340, 364–367, 370, 377, 384–387, 408, 409

Q
Quercus leucotrichophora, 268–270, 278

R
Ravi Basin Glacier Inventory (RBGI), 113, 119, 120, 124–129, 131
Reference crop evapotranspiration, 407–415
Remote sensing, 16, 30, 78, 117, 137, 139, 154, 339, 358, 366, 383

Risk assessment, 46, 48, 53–55, 210, 388, 389
Risk reduction, 53, 91, 93, 97–104, 202, 209, 210, 381, 388–392

S
Satellite images, 113, 116, 137, 138
Sikkim, 23, 24, 39, 55, 58, 81, 158–171, 174–176, 179–190, 193, 195, 214, 242, 366, 369, 370, 373, 377
Social vulnerability, 97, 200, 202, 203
Spice, 186, 222, 223, 225, 254–257, 259, 260, 264, 265
Sustainable development, 57, 70, 72, 74, 83, 188, 214, 216, 221, 325, 363, 390, 391
Sustainable management, 46, 49, 55–57, 71, 84, 167

T
Tehri reservoir, 136–155
Temperature anomaly, 29–31
Thornthwaite's method, 408, 410–415
Trend analysis, 8, 27, 29, 30, 33, 36
Turc's method, 408, 412, 413, 415

U
Upper Ganga catchment, 2–16

V
Vapour pressure deficit (VPD), 268, 269, 271–273, 275–278

W
Weights, 137, 147, 149–151, 153, 344, 345, 356–358
Western Himalaya, 24, 25, 53, 66, 68, 82, 83, 92–94, 114, 158, 215, 242, 268, 282–310, 313–316, 339, 365, 385
Wet days' frequency, 27, 36, 39

The manufacturer's authorised representative in the EU is Springer Nature Customer Service Centre GmbH, Europaplatz 3, 69115 Heidelberg, Germany. If you have any concerns regarding our products, please contact ProductSafety@springernature.com

Printed and bound by CPI Group (UK) Ltd, Croydon, CR0 4YY

30/09/2026

02225244-0015